KS3 MATHS NOW

Learn and Practice Book

Revised edition

William Collins' dream of knowledge for all began with the publication of his first book in 1819.

A self-educated mill worker, he not only enriched millions of lives, but also founded a flourishing publishing house. Today, staying true to this spirit, Collins books are packed with inspiration, innovation and practical expertise. They place you at the centre of a world of possibility and give you exactly what you need to explore it.

Collins. Freedom to teach.

Published by Collins
An imprint of HarperCollins*Publishers*
The News Building
1 London Bridge Street
London
SE1 9GF

HarperCollins*Publishers*
Macken House, 39/40 Mayor Street Upper,
Dublin 1, D01 C9W8, Ireland

Browse the complete Collins catalogue at
www.collins.co.uk

10 9

ISBN 978-0-00-836286-7 Revised edition

British Library Cataloguing-in-Publication Data
A catalogue record for this publication is available from the British Library.

Authors: Leisa Bovey, Belle Cottingham, Rob Ellis, Kath Hipkiss, Trevor Senior, Peter Sherran, Brian Speed and Colin Stobart
Contributing authors: Peter Derych, Kevin Evans, Keith Gordon, Michael Kent and Chris Pearce
Publisher: Katie Sergeant
Product Manager: Joanna Ramsay
Editorial production: Oriel Square Ltd
Cover designer: The Big Mountain Design Ltd
Illustrations: Ann Paganuzzi, Nigel Jordan, Tony Wilkins and Ken Vail Graphic Design Ltd
Typesetter: Ken Vail Graphic Design Ltd
Production controller: Katharine Willard
Printed and bound in India by Replika Press Pvt. Ltd.

With thanks to Charlie Jackson for his help on this project.

This book is produced from independently certified FSC™ paper
to ensure responsible forest management.

For more information visit: **www.harpercollins.co.uk/green**

CONTENTS

Contents

Contents

How to use this book

Learning objectives

Now I can statements outline the learning objectives and what you will practise throughout the topic. Check that you are confident with them in the *Now I can* review at the end of each chapter.

Structured progress

Work through each chapter to develop your maths fluency, practise using your knowledge to reason mathematically and apply that knowledge to solve problems.

Each topic is split into three sections:

Develop fluency

These sections will help you to consolidate your knowledge of each topic. The questions are structured to help you master the fundamental skills before you apply them to different contexts.

Key words

Important terms are highlighted the first time that they appear in the text, helping you to master the terminology that you need to express yourself fluently in maths. Definitions are provided in the glossary in the digital resources.

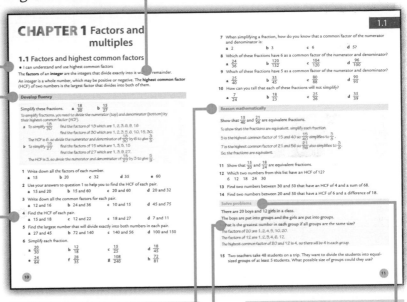

Reason mathematically

You will need to use the maths that you have practiced in the fluency section to follow a line of enquiry or find mathematical relationships. You may need to develop a justification, argument or proof using mathematical terminology.

Solve problems

In the final section of each topic you will need to apply your knowledge to a range of problems in different contexts or to multi-part questions.

Key questions

The starred 'key' questions are a guide to the depth and strength of your understanding of the topic. If you can answer a key question without having to look over notes or look back at previous exercise questions, you have a good grasp of the concepts underlying the topic and should feel confident that you can use and apply your new skills and understanding.

Worked examples

Each section begins with one or more worked examples in a yellow box. Read through the examples to see how you can approach the questions in that section.

Chapter reviews

At the end of each chapter, review the *Now I can* statements and see how confident you are with each skill. •

Answers

Check your work using the answers provided at the back of this book.

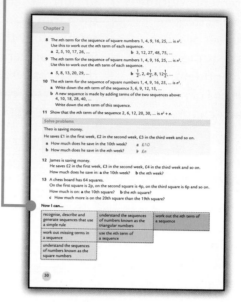

CHAPTER 1 Factors and multiples

1.1 Factors and highest common factors

● I can understand and use highest common factors

The **factors** of an **integer** are the integers that divide exactly into it with no remainder.

An integer is a whole number, which may be positive or negative. The **highest common factor** (HCF) of two numbers is the largest factor that divides into both of them.

Develop fluency

Simplify these fractions.   **a** $\dfrac{18}{30}$ **b** $\dfrac{15}{27}$

To simplify fractions, you need to divide the numerator (top) and denominator (bottom) by their highest common factor (HCF).

a To simplify $\dfrac{18}{30}$ find the factors of 18 which are 1, 2, 3, 6, 9, 18

   find the factors of 30 which are 1, 2, 3, 5, 6, 10, 15, 30.

   The HCF is 6, so divide the numerator and denominator of $\dfrac{18}{30}$ by 6 to give $\dfrac{3}{5}$.

b To simplify $\dfrac{15}{27}$ find the factors of 15 which are 1, 3, 5, 15

   find the factors of 27 which are 1, 3, 9, 27.

   The HCF is 3, so divide the numerator and denominator of $\dfrac{15}{27}$ by 3 to give $\dfrac{5}{9}$.

1 Write down all the factors of each number.

   a 15   **b** 20   **c** 32   **d** 35   **e** 60

2 Use your answers to question 1 to help you to find the HCF of each pair.

   a 15 and 20   **b** 15 and 60   **c** 20 and 60   **d** 20 and 32

3 Write down all the common factors for each pair.

   a 12 and 16   **b** 24 and 36   **c** 10 and 15   **d** 45 and 75

4 Find the HCF of each pair.

   a 15 and 18   **c** 12 and 22   **c** 18 and 27   **d** 7 and 11

5 Find the largest number that will divide exactly into both numbers in each pair.

   a 27 and 45   **b** 72 and 140   **c** 140 and 56   **d** 100 and 150

6 Simplify each fraction.

   a $\dfrac{20}{30}$ **b** $\dfrac{12}{18}$ **c** $\dfrac{15}{25}$ **d** $\dfrac{18}{45}$

   e $\dfrac{24}{64}$ **f** $\dfrac{28}{35}$ **g** $\dfrac{108}{240}$ **h** $\dfrac{72}{81}$

7 When simplifying a fraction, how do you know that a common factor of the numerator and denominator is:

   a 2   **b** 3   **c** 6   **d** 5?

8 Which of these fractions have 6 as a common factor of the numerator and denominator?

   a $\dfrac{24}{36}$ **b** $\dfrac{120}{132}$ **c** $\dfrac{104}{120}$ **d** $\dfrac{96}{100}$

9 Which of these fractions have 5 as a common factor of the numerator and denominator?

   a $\dfrac{25}{40}$ **b** $\dfrac{35}{45}$ **c** $\dfrac{80}{88}$ **d** $\dfrac{90}{95}$

10 How can you tell that each of these fractions will not simplify?

   a $\dfrac{7}{24}$ **b** $\dfrac{18}{23}$ **c** $\dfrac{25}{26}$ **d** $\dfrac{35}{39}$

Reason mathematically

Show that $\dfrac{15}{40}$ and $\dfrac{21}{56}$ are equivalent fractions.

To show that the fractions are equivalent, simplify each fraction.

5 is the highest common factor of 15 and 40 so $\dfrac{15}{40}$ simplifies to $\dfrac{3}{8}$.

7 is the highest common factor of 21 and 56 so $\dfrac{21}{56}$ also simplifies to $\dfrac{3}{8}$.

So, the fractions are equivalent.

11 Show that $\dfrac{15}{20}$ and $\dfrac{18}{24}$ are equivalent fractions.

12 Which two numbers from this list have an HCF of 12?

   6   12   18   24   30

13 Find two numbers between 30 and 50 that have an HCF of 4 and a sum of 68.

14 Find two numbers between 20 and 50 that have a HCF of 6 and a difference of 18.

Solve problems

There are 20 boys and 12 girls in a class.

The boys are put into groups and the girls are put into groups.

What is the greatest number in each group if all groups are the same size?

The factors of 20 are 1, 2, 4, 5, 10, 20.

The factors of 12 are 1, 2, 3, 4, 6, 12.

The highest common factor of 20 and 12 is 4, so there will be 4 in each group.

15 Two teachers take 48 students on a trip. They want to divide the students into equal-sized groups of at least 5 students. What possible size of groups could they use?

16 Mrs Roberts is taking 36 students on a visit to a museum. She doesn't want to divide them into groups of unequal size. What possible group sizes could she use?

17 Two teachers wanted to put their classes into groups of equal size. They didn't want to mix the classes. Class A had 36 students and class B had 27 students.
What is the largest group they can make and how many of them would there be?

18 There are 65 boys and 52 girls on a school trip.
They are put into equal-sized groups containing the same number of boys in each group.
a Work out the number of groups.
b Work out the number of boys and the number of girls in each group.

1.2 Multiples and common multiples

● I can understand and use lowest common multiples

A **multiple** of an integer is the result of multiplying that integer by another integer. You can find a **common multiple** for any pair of integers by multiplying one by the other. The **lowest common multiple** (LCM) of two numbers is the lowest number that is a multiple of both.

Develop fluency

Find the lowest common multiple (LCM) of each pair of numbers.

a 3 and 7   b 6 and 9

a Write out the first few multiples of each number.
3: 3, 6, 9, 12, 15, 18, 21, 24, 27, …
7: 7, 14, 21, 28, 35, …
You can see that the LCM of 3 and 7 is 21.

b Write out the multiples of each number.
6: 6, 12, 18, 24, …
9: 9, 18, 27, 36, …
You can see that the LCM of 6 and 9 is 18.

1 Write down the numbers from the list that are multiples of:
a 2   b 3   c 5   d 9
10   4   23   18   69   81   8   65   33   72   100

2 Write down the first ten multiples of each number.
a 4   b 5   c 8   d 15   e 20

3 Use your answers to question 2 to help you to find the LCM of each pair.
a 5 and 8   b 4 and 20   c 6 and 15   d 8 and 20

4   Find the LCM of the numbers in each pair.

   a 5 and 9   **b** 5 and 25   **c** 9 and 21   **d** 7 and 11

5   Find the LCM of the numbers in each set.

   a 2, 3 and 5   **b** 6, 9 and 15   **c** 5, 12 and 16   **d** 3, 7 and 11

6   Which of these sets of numbers has 24 as their LCM?

   a 6 and 12   **b** 3 and 8   **c** 3, 4 and 8   **d** 2, 4 and 6

7   Which of these sets of numbers has 42 as their LCM?

   a 6 and 14   **b** 2, 3 and 7   **c** 3, 4 and 7   **d** 2, 6 and 7

8   How can you tell that the LCM of 2, 5 and 7 must be even?

9   How can you tell that the LCM of two odd numbers must be odd?

10 How can you tell that the LCM of an odd number and an even number must be even?

Reason mathematically

A clock chimes every 15 minutes. A light flashes every 10 minutes.

At midnight the clock chimes and the light flashes together.

What time will they next chime and flash together?

The LCM of 15 and 10 is 30, so the next time the clock chimes and the light flashes at the same time is at 12.30 a.m.

11 A yellow light flashes every 10 seconds, a blue light flashes every 12 seconds.
    At the start they flash together.
    How many seconds is it before they flash together again?
    Show how you work out your answer.

12 Ali is singing the 9 times table. Sam is singing the 7 times table.
    a What is the first number that they both sing?
    b If they both start at the same time and sing at the same speed, who will sing the
        number first? Give a reason for your answer.

13 Chairs are put into rows of equal length in a school hall.
    There is room for up to 20 chairs in each row.
    Exactly 165 chairs are to be put out.
    Show that 15 is the most that can be put in each row.

14 Two numbers have an LCM of 20 and an HCF of 2.
    The numbers are not 2 and 20.
    Show that the sum of the numbers is 14.

Solve problems

A baker makes small buns, some of mass 15 g and some of mass 20 g.

He wants to sell them in bags that all have the same mass.

What is the smallest mass he could have in each of these bags?

The 15 g cakes could be put into batches of mass 15 g, 30 g, 45 g, 60 g, 75 g, … (all multiples of 15 g).

The 20 g cakes could be put into batches of mass 20 g, 40 g, 60 g, 80 g, … (all multiples of 20 g).

The smallest mass will be the lowest common multiple (LCM) of these numbers, which is 60 g.

15 In the first year-group of a large school, it is possible to divide the pupils exactly into groups of 24, 30 or 32. Find the smallest number of pupils in this first year-group.

16 Two model trains leave the station at the same time and travel around tracks of equal length. One completes a circuit in 14 seconds, the other in 16 seconds. How long will it be before they are together again at the station?

17 Three friends are walking along a straight pavement. Suzy has a step size of 14 cm, Kieran has a step size of 15 cm and Miguel has a step size of 18 cm.

They all set off, walking from the same point, next to each other. How far will they have gone before they are all in step with each other again?

18 Three men regularly visit their local gym in the evenings.

Joe goes every three days. John goes every seven days. James goes every four days.

How many days in a year are they likely all to be in the gym on the same evening?

1.3 Prime factors

- I can understand what prime numbers are
- I can find the prime factors of an integer

A **prime number** is an integer that has only two factors; itself and one.
A **prime factor** of an integer is a factor that is also a prime number.

Repeated prime factors can be written in **index form**: $18 = 2 \times 3^2$

A **factor tree** is a useful way to find the prime factors of any integer. Keep splitting the factors until there is a prime number at the end of each 'branch'. Whichever pair of factors you start with, you will always finish with the same set of prime factors.

An example for 18 is shown here.

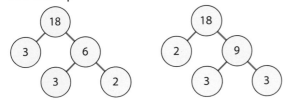

$18 = 2 \times 3 \times 3$

$\quad = 2 \times 3^2$

Develop fluency

Use the division method to find the prime factors of 24.

$2 \overline{)24}$
$2 \overline{)12}$
$2 \overline{)\,6}$
$3 \overline{)\,3}$
$\quad\;1$

So, $24 = 2 \times 2 \times 2 \times 3 = 2^3 \times 3$.

1 These numbers are written as products of their prime factors. What are the numbers?
   a $2 \times 2 \times 3$ b $2 \times 3 \times 3 \times 5$ c $2 \times 3^3 \times 5$

2 Use a factor tree to work out the prime factors of each number.
   a 8   b 28   c 35   d 52   e 180

3 Use the division method to work out the prime factors of each number.
   a 42   b 75   c 140   d 250   e 480

4 Find the prime factors of all the numbers from 2 to 20.

5 a Which numbers in question 4 only have one prime factor?
   b What special name is given to these numbers?

6 100 can be written as a product of its prime factors: $100 = 2 \times 2 \times 5 \times 5 = 2^2 \times 5^2$
   a Write down each number, as a product of its prime factors, in index form.
      i 200   ii 50   iii 1000
   b Write down the prime factors of one million, in index form.

7 a Write down the prime factors of 32, in index form.
   b Write down the prime factors of 64, in index form.
   c Write down the prime factors of 128, in index form.
   d Write down the prime factors of 1024, in index form.

8 Find the LCM of 84 and 140.

9 What do you know about the prime factors of an even number?

10 If 2 is not a prime factor of a number, what does this tell you?

Reason mathematically

$48 = 2^4 \times 3$

Use this fact to write 96 and 144 as a product of prime factors, in index form.
$96 = 2 \times 48$, so $96 = 2^5 \times 3$
$144 = 3 \times 48$, so $144 = 2^4 \times 3^2$

11 $36 = 2^2 \times 3^2$

Use this fact to write 18 and 72 as a product of prime factors, in index form.

12 $600 = 2^3 \times 3 \times 5^2$ and $216 = 2^3 \times 3^3$

Use these facts to show that $\frac{216}{600}$ simplifies to $\frac{9}{25}$.

13 The smallest number with exactly two different prime factors is $2 \times 3 = 6$.

    a What is the next smallest number with exactly two different prime factors?

    b What is the smallest number with exactly three different prime factors?

14 **a** What are the prime factors of: **i** 60   **ii** 100?

    b Use your answers to parts **i** and **ii** to find the LCM of 60 and 100.

    c Explain how you found your answer to part **b**.

Solve problems

$210 = 2 \times 3 \times 5 \times 7$ $210^2 = 44\,100$

    a Write 44 100 as a product of prime factors in index form.

    b Write 210^5 in index form.

    a $44\,100 = 210^2$ So, $44\,100 = (2 \times 3 \times 5 \times 7)^2 = 2^2 \times 3^2 \times 5^2 \times 7^2$ in index form.

    b $210^5 = 2^5 \times 3^5 \times 5^5 \times 7^5$ in index form.

15 In prime factor form, $105 = a \times b \times c$.

Work out the values of a, b and c.

16 In prime factor form, $539 = x \times y^2$.

Work out the values of x and y.

17 **a** Write 625 as a product of prime factors.

    b Use your answer to part **a** to write $\sqrt{625}$ as a product of prime factors.

18 **a** Write 324 as a product of prime factors.

    b Use your answer to part **a** to write $\sqrt{324}$ as a product of prime factors.

Now I can...

understand highest common factors	use highest common factors	find the prime factors of an integer
understand lowest common multiples	use lowest common multiples	
understand what prime numbers are		

CHAPTER 2 Sequences

2.1 Sequences and rules

● I can recognise, describe and generate sequences that use a simple rule

A **sequence** is a list of numbers that follow a pattern or rule. The numbers in the sequence are called **terms** and the starting number is called the **first term**. The rule may be called the **term-to-term rule**.

Sequences that increase or decrease by a fixed amount, from one term to the next, are called **linear sequences**. Sequences in which you find each term after the first term by multiplying or dividing by a fixed amount are called **geometric sequences**.

Develop fluency

a The rule for a linear sequence is 'add 3'. Write the first five terms of the sequence if the first term is: **i** 1 **ii** 2 **iii** 6.

b The rule for a geometric sequence is 'multiply by 2'. Write the first five terms of the sequence if the first term is: **i** 1 **ii** 3 **iii** 5.

a Rule: add 3
  i Starting at 1 gives the sequence 1, 4, 7, 10, 13, …
  ii Starting at 2 gives the sequence 2, 5, 8, 11, 14, …
  iii Starting at 6 gives the sequence 6, 9, 12, 15, 18, …

b Rule: multiply by 2
  i Starting at 1 gives the sequence 1, 2, 4, 8, 16, …
  ii Starting at 3 gives the sequence 3, 6, 12, 24, 48, …
  iii Starting at 5 gives the sequence 5, 10, 20, 40, 80, …

1 Use the term-to-term rule to work out the first five terms of each sequence. Start from a first term of 1.
  a add 3   **b** multiply by 3   **c** add 5   **d** multiply by 10
  e add 11   **f** multiply by 4   **g** add 8   **h** add 105

2 Use the term-to-term rule to work out the first five terms of each sequence. Start from a first term of 5.
  a add 3   **b** multiply by 3   **c** add 5   **d** multiply by 10
  e add 11   **f** multiply by 4   **g** add 8   **h** add 105

3 Work out the next two terms of each sequence. Describe the term-to-term rule you have used.
  a 2, 4, 6, …, …   **b** 1, 10, 100, …, …   **c** 2, 10, 50, …, …   **d** 0, 7, 14, …, …
  e 4, 9, 14, …, …   **f** 4, 8, 12, …, …   **g** 12, 24, 36, …, …   **h** 2, 6, 18, …, …

4 Work out the next two terms of each sequence.
Describe the term-to-term rule you have used.

   a 50, 45, 40, 35, 30, ..., ...
        b 35, 32, 29, 26, 23, ..., ...

   c 9, 5, 1, –3, –7, ..., ...
        d 6.5, 1.5, –3.5, –8.5, –13.5, ..., ...

5 Work out the first four terms of each sequence.

   a Start with a first term of 3. To work out the next term, multiply by 3 and then add 7.

   b Start with a first term of 32. To work out the next term, divide by 2 then subtract 4.

6 Work out the first four terms of each sequence.

   a Start with a first term of 3. To work out the next term, multiply by 3 then subtract 7.

   b Start with a first term of 5. To work out the next term, add 2 and then multiply by 4.

7 Work out the first four terms of each sequence.

   a Start with a first term of 5. To work out the next term, add 2 and then multiply by 3.

   b Start with a first term of 32. To work out the next term, divide by 4 and then add 4.

8 Work out the first four terms of each sequence.

   a Start with a first term of 4. To work out the next term, add 2 and then multiply by 3.

   b Start with a first term of 0. To work out the next term, divide by 4 and then add 4.

9 In this sequence of fractions, the numerators form a linear sequence and the denominators form a geometric sequence.
Work out the next two fractions in the sequence.

$$\frac{1}{2}, \frac{3}{4}, \frac{5}{8}, ..., ...$$

10 In this sequence of fractions, the numerators form a geometric sequence and the denominators form a linear sequence.
Work out the next two fractions in the sequence.

$$\frac{3}{5}, \frac{6}{15}, ..., ...$$

Reason mathematically

   a Work out the two terms between this pair of numbers to form a linear sequence.

      3, ..., ..., 21

   b Work out the missing term in this geometric sequence.

      5, ..., 45

   a *The difference between the first and fourth terms is 21 – 3 = 18.*

      So, the difference between one term and the next is 18 ÷ 3 = 6.

      The sequence is 3, 9, 15, 21 as the term-to- term rule is 'add 6'.

b First, work out how to get from 5 to 45: $45 \div 5 = 9$ so multiply by 9.

$9 = 3 \times 3$

So, the term-to-term rule is 'multiply by 3'.

The sequence is 5, 15, 45

11 Work out the two terms between each pair of numbers, to form a linear sequence. Describe the term-to-term rule you have used.

   a 1, ..., ..., 7   **b** 3, ..., ..., 12   **c** 5, ..., ..., 14   **d** 2, ..., ..., 38

12 Work out the missing terms between each pair of numbers, to form a linear sequence. Describe the term-to-term rule you have used.

   a 1, ..., ..., 13   **b** 4, ..., 5.5   **c** 80, ..., ..., ..., ..., 55 **d** 2, ..., ..., –1

13 Work out the missing term in each geometric sequence. Describe the term-to-term rule you have used.

   a 4, ..., 16   **b** 3, ..., 12   **c** 5, ..., 20   **d** 4, ..., 100

Solve problems

The term-to term rule for a sequence has two steps, 'multiply by a number' then 'add 2'.

3, 8, 18, 38, ...

Work out the term-to-term rule.

Subtracting 2 from the second term gives 6, so to get from 3 to 6 multiply by 2.

The term-to-term rule is 'multiply by 2 and then add 2'.

Checking this for the other terms gives:

$8 \times 2 + 2 = 16 + 2 = 18$ and $18 \times 2 + 2 = 36 + 2 = 38$

14 Complete the two-step term-to term rule for each sequence.

   a 7, 15, 31, 63, 127, ...   Multiply by 2 and then ....

   b 6, 4, 3, 2.5, 2.25, ... and then add 1

   c 3, 2, 0, –4, –12   Subtract 2 and then ....

15 For each pair of numbers, find at least two different sequences and write down the next two terms. Describe the term-to-term rule you have used.

   a 1, 4, ..., ...   **b** 3, 9, ..., ...   **c** 3, 6, ..., ...   **d** 5, 15, ..., ...

16 Each of these sequences uses a 'subtract', 'multiply' or 'divide' rule. Write the next two terms in each sequence. Describe the term-to-term rule you have used.

   a 1, 3, 9, 27, ...   **b** 25, 20, 15, 10, ...   **c** 1000, 100, 10, 1, ...

   d 36, 28, 20, 12, ...   **e** 5, 10, 20, 40, ...   **f** 20, 10, 5, 2.5, ...

2.2 Working out missing terms

● I can work out missing terms in a sequence

You need to know how to work out any term in a sequence.

Work out the fourth term in this sequence of patterns made with lolly sticks.

Term number

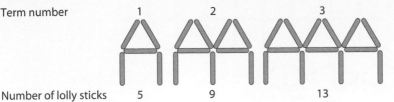

Number of lolly sticks   5   9   13

Look at the number sequence shown by this pattern.

You can see that four more lolly sticks are added each time.

You can draw the pattern for the fourth term and work out that it has 17 lolly sticks.

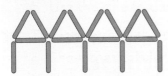

Alternatively, draw a table to find the number of lolly sticks for the patterns for all the terms.

Term	1	2	3	4
Lolly sticks	5	9	13	17

Work out the 5th term, the 25th term and the 50th term in the sequence:

7, 10, 13, 16, …,

You first need to know what the term-to-term rule is.

You can see that you add 3 from one term to the next.

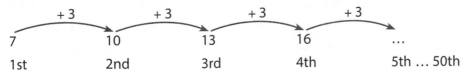

To get to the 5th term, you add 3 to the fourth term, which gives 19.

To get to the 25th term, you will have to add on 3 a total of 24 times (25 − 1) to the first term, 7.
This will give $7 + 3 \times 24 = 7 + 72 = 79$.

To get to the 50th term, you will have to add on 3 a total of 49 times (50 − 1) to the first term, 7.
This will give $7 + 3 \times 49 = 7 + 147 = 154$.

For questions 1–4:

a draw the next pattern of lolly sticks

b work out the number of lolly sticks in the 10th pattern.

1

2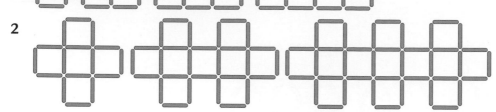

3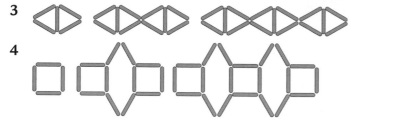

4

5 Work out the 5th and the 50th term in each sequence.

a 4, 6, 8, 10, …	**b** 1, 6, 11, 16, …	**c** 3, 10, 17, 24, …
d 5, 8, 11, 14, …	**e** 1, 5, 9, 13, …	**f** 2, 10, 18, 26, …
g 20, 30, 40, 50, …	**h** 10, 19, 28, 37, …	**i** 3, 9, 15, 21, …

6 Work out the 5th and the 50th term in each sequence.

a –2, –4, –6, –8, …	**b** –2, –7, –12, –17, …	**c** –4, –11, –18, –25, …
d 8, 5, 2, –1, …	**e** –1, –6, –11, –16, …	**f** –12, –20, –28, –36, …
g 9, 3, –3, –9, …	**h** –37, –28, –19, –10, …	**i** –21, –18, –15, –12, …

7 The first term of a linear sequence is 7. The term-to-term rule is +4.
Work out the 10th, 20th and 30th terms of the sequence.

8 The first term of a linear sequence is 10. The term-to-term rule is +3.
Work out the 10th, 20th and 50th terms of the sequence.

9 The first term of a linear sequence is 50. The term-to-term rule is –4.
Work out the 10th, 20th and 30th terms of the sequence.

10 The first term of a linear sequence is 6. The term-to-term rule is –4.
Work out the 10th, 20th and 50th terms of the sequence.

Reason mathematically

In a linear sequence, the 10th term is 34 and the 11th term is 38.

Work out the first term and the 100th term.

The term-to-term rule is 'add 4'.

The first term is: 10th term – 9 × 4 = 34 – 36 = –2

The 100th term is: first term + 99 × 4 = –2 + 396 = 394

11 For each sequence, work out the first and 100th term.
  a In a linear sequence, the 10th term is 40 and the 11th term is 42.
  b In a linear sequence, the 50th term is 254, the 51st is 259 and the 52nd is 264.
  c In a linear sequence, the 10th term is 40 and the 11th term is 37.
  d In a linear sequence, the 50th term is –20 and the 51st term is –21.

Solve problems

In this sequence, you have been given the fourth, fifth and sixth terms.

..., ..., ..., 20, 22, 24, ...

Work out the first term, then work out the 25th term.

The term-to-term rule is 'add 2'.

Working backwards, the sequence starts 14, 16, 18, 20, 22, 24, ...

So, the first term is 14.

The 25th term is 14 + 24 × 2 = 14 + 48 = 62

12 In each sequence, work out the first term, then work out the 25th term.
  In each case, you have been given the fourth, fifth and sixth terms.
  a ..., ..., ..., 13, 15, 17, ...   **b** ..., ..., ..., 18, 23, 28, ...
  c ..., ..., ..., 19, 23, 27, ...   **d** ..., ..., ..., 32, 41, 50, ...

13 This is a sequence of patterns made from mauve and white squares.
  The diagrams show the patterns for the third and fifth terms.

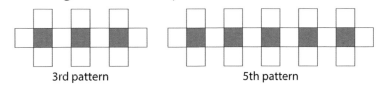

3rd pattern   5th pattern

  a How many mauve squares are there in the pattern for the fourth term?
  b How many white squares are there in the pattern for the fourth term?
  c Draw the pattern for the first term.

14 a The second and third terms of a linear sequence are 2 and 4.

..., 2, 4, ..., ...

Write down a rule for the way the sequence is building up and work out the first and fourth terms.

b The second and third terms of a geometric sequence are 2 and 4.

..., 2, 4, ..., ...

Write down a rule for the way this sequence is building up and work out the first and fourth terms.

15 Think about the 10th diagram in this pattern.

Diagram 1   Diagram 2   Diagram 3

Work out the number of:

a blue squares   b orange squares   c squares in total.

2.3 Other sequences

• I know and can understand the sequences of numbers known as the square numbers and the triangular numbers

Multiplying a number by itself produces a **square number**.
The first ten square numbers are 1, 4, 9, 16, 25, 36, 49, 64, 81 and 100.
You need to know the square numbers up to $15^2 = 225$.

Adding the integers, in sequence, produces the **triangular numbers**.

The first ten triangular numbers are 1, 3, 6, 10, 15, 21, 28, 36, 45 and 55.

Develop fluency

Write down the 11th row of each of these sequences.

a $1 \times 1 = 1$

$2 \times 2 = 4$

$3 \times 3 = 9$

b $1 = 1$

$1 + 2 = 3$

$1 + 2 + 3 = 6$

a $11 \times 11 = 121$

b $1 + 2 + 3 + 4 + 5 + 6 + 7 + 8 + 9 + 10 + 11 = 66$

1 Work out the square numbers up to 15×15.

2 Write each number as the sum of two square numbers.
The first two have been done for you.

a $5 = 1 + 4$ b $10 = 1 + 9$ c $13 = ... + ...$ d $17 = ... + ...$

e $20 = ... + ...$ f $25 = ... + ...$ g $26 = ... + ...$ h $29 = ... + ...$

i $34 = ... + ...$ j $45 = ... + ...$ k $50 = ... + ...$ l $52 = ... + ...$

3 Work out the triangular numbers up to the 15th triangular number.

4 Write each number as the sum of two triangular numbers.
The first two have been done for you.

a $4 = 1 + 3$ b $7 = 1 + 6$ c $9 = ... + ...$ d $11 = ... + ...$

e $13 = ... + ...$ f $16 = ... + ...$ g $24 = ... + ...$ h $25 = ... + ...$

i $27 = ... + ...$ j $29 = ... + ...$ k $31 = ... + ...$ l $34 = ... + ...$

5 Write down two numbers that are both square numbers and also triangular numbers.

6 Some sums of two square numbers are special because they give an answer that is also a square number. For example:
$3^2 + 4^2 = 9 + 16 = 25 = 5^2$
Which of these pairs of squares give a total that is also a square number?

a $5^2 + 12^2$ b $6^2 + 8^2$ c $7^2 + 9^2$ d $7^2 + 24^2$

7 a Add up the first 10 pairs of consecutive triangular numbers, starting with
$1 + 3, 3 + 6, 6 + 10, ...$

b What is special about the answers?

8 a Subtract the first 10 pairs of consecutive square numbers, starting with
$4 - 1, 9 - 4, 16 - 9, ...$

b What is special about the answers?

9 a Subtract the first 10 pairs of consecutive triangular numbers, starting with
$3 - 1, 6 - 3, 10 - 6, ...$

b What is special about the answers?

10 a Work out the values of this sequence:
$$\frac{1 \times 2}{2}, \frac{2 \times 3}{2}, \frac{3 \times 4}{2}, \frac{4 \times 5}{2}, \frac{5 \times 6}{2}$$

b What do you notice about your answers to part a?

Reason mathematically

There are five netball teams in a competition.

Each team plays every other team once.

Show that the number of games played is a triangular number.

Call the teams A, B, C, D and E.

Each team plays 4 games, so for team A they are: A v. B, A v. C, A v. D, A v. E.

But each game will be counted twice, for example, A v. B is the same as B v. A.

So, the number of games is $4 \times 5 \div 2 = 10$, which is a triangular number.

11 There are six football teams in a competition.
    Each team plays every other team once.
    Show that the number of games played is a triangular number.

12 **a** Add up the first five pairs of consecutive square numbers, starting with
        $1 + 4$, $4 + 9$, $9 + 16$, …

    b Is it possible to get an even number if you add any pair of consecutive square
        numbers? If not, explain why not.

13 Find at least three different sequences that begin 2, 3, 5, …
    Explain how each sequence works.

Solve problems

Look at this pattern of numbers.

$1^2 = 1^3$
$(1 + 2)^2 = 1^3 + 2^3$
$(1 + 2 + 3)^2 = 1^3 + 2^3 + 3^3$

a Write down the next line of this number pattern.

b Use the pattern to work out the value of $1^3 + 2^3 + 3^3 + 4^3 + 5^3$.

a $(1 + 2 + 3 + 4)^2 = 1^3 + 2^3 + 3^3 + 4^3$

b $1^3 + 2^3 + 3^3 + 4^3 + 5^3 = (1 + 2 + 3 + 4 + 5)^2$
$$= 15^2$$
$$= 225$$

14 Look at this pattern of numbers.

    1 $= 1 = 1^2$
    $1 + 3$ $= 4 = 2^2$
    $1 + 3 + 5$ $= 9 = 3^2$
    $1 + 3 + 5 + 7$ $= 16 = 4^2$
    $1 + 3 + 5 + 7 + 9$ $= 25 = 5^2$

    a Write down the next two lines of this number pattern.

    b What is special about the numbers on the left-hand side?

    c Without working them out, write down the answers to these calculations.

    i $1 + 3 + 5 + 7 + 9 + 11 + 13 + 15 + 17 + 19 = …$

    ii $1 + 3 + 5 + 7 + 9 + 11 + 13 + 15 + 17 + 19 + 21 + 23 + 25 + 27 + 29 = …$

15 Look at this pattern of numbers.

1	= 1
1 + 2	= 3
1 + 2 + 3	= 6
1 + 2 + 3 + 4	= 10
1 + 2 + 3 + 4 + 5	= 15

   a Write down the next two lines of this number pattern.

   b What is special about the numbers on the left-hand side?

   c What is special about the numbers on the right-hand side?

   d Without working them out, write down the answers to these calculations.

   i $1 + 2 + 3 + 4 + 5 + 6 + 7 + 8 + 9 + 10 = \ldots$

   ii $1 + 2 + 3 + 4 + 5 + 6 + 7 + 8 + 9 + 10 + 11 + 12 + 13 + 14 + 15 = \ldots$

16 The diagram shows some dominoes arranged in a pattern.

   a Draw the next column of dominoes in the pattern.

   b Write down the total number of spots in each of the four columns of dominoes.

   c Divide each of these totals by 3 and write down the new sequence of numbers.

   d Write down the next two numbers for the sequence you obtained in part **c**.

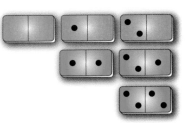

2.4 The *n*th term of a sequence

● I can use the *n*th term of a sequence

In a sequence, the first term is *a* and the difference between one term and the next is *d*.

The rule for finding the terms of a sequence is an **algebraic expression** called the **nth term**. In a sequence with *n*th term $3n + 2$, the number in front of *n* (3) is the **coefficient** of *n* and the number by itself in the expression (2) is the **constant term**.

The *n*th term of a sequence is given by $1 - 3n$.

   a Work out the first three terms of the sequence.

   b Work out the 60th term of the sequence.

   a When $n = 1$: $1 - 3 \times 1 = 1 - 3 = -2$

   When $n = 2$: $1 - 3 \times 2 = 1 - 6 = -5$

   When $n = 3$: $1 - 3 \times 3 = 1 - 9 = -8$

   So, the first three terms are −2, −5, −8.

   b When $n = 60$: $1 - 3 \times 60 = 1 - 180 = -179$

   So, the 60th term is −179.

1. Work out:
   i   the first three terms   ii   the 100th term
   for each sequence, when the nth term is given by:
   a $2n + 5$ b $5n - 3$ c $4n + 5$ d $10n + 1$
   e $7n - 1$ f $10 - n$ g $20 - 2n$ h $7 - 3n$.

2. For each sequence, with the given nth term, write down:
   i   the first four terms   ii   the first term, a iii   the difference, d.
   a $2n + 1$ b $2n + 2$ c $2n + 3$ d $2n + 4$

3. For each sequence, with the given nth term, write down:
   i   the first four terms   ii   the first term, a iii   the difference, d.
   a $5n - 1$ b $5n + 2$ c $5n - 4$ d $5n + 3$

4. For each sequence, with the given nth term, write down:
   i   the first four terms   ii   the difference, d.
   a $3n - 1$ b $4n + 2$ c $6n - 4$ d $10n + 3$

5. For each sequence, write down the first term, a, and the difference, d.
   a   4, 9, 14, 19, 24, 29, ...   b   1, 3, 5, 7, 9, 11, ...
   c   3, 9, 15, 21, 27, 33, ...   d   5, 3, 1, −1, −3, −5, ...

6. Here are the nth terms for different sequences.
   a $5n - 1$ b $8n + 2$ c $6n - 9$ d $10n + 3$
   For each sequence write down:
   i   the first four terms   ii   the difference, d.

7. For each sequence, write down the first term, a, and the difference, d.
   a   1, 8, 15, 22, 29, ...   b   2, 4, 6, 8, 10, ...
   c   3, 12, 21, 30, 39, ...   d   6, 3, 0, −3, −6, ...

8. Given the first term, a, and the difference, d, write down the first six terms of
   each sequence.
   a $a = 1, d = 8$ b $a = 5, d = 7$ c $a = 4, d = -2$
   d $a = 1.5, d = 0.5$ e $a = 10, d = -3$ f $a = 2, d = -0.5$

9. The nth term of a sequence is n^2.
   a   Write down the first four terms.   b   What is the special name of this sequence?

10. The nth term of a sequence is $\dfrac{n(n+1)}{2}$.
    a   Write down the first four terms.   b   What is the special name of this sequence?

Reason mathematically

The *n*th term of the sequence 7, 11, 15, 19, 23, ..., is given by the expression $4n + 3$.

a Show that this is true for the first three terms.

b Use the rule to decide whether 101 is in the sequence.

a When $n = 1$: $4 \times 1 + 3 = 4 + 3 = 7$ True ✓

     When $n = 2$: $4 \times 2 + 3 = 8 + 3 = 11$ True ✓

     When $n = 3$: $4 \times 3 + 3 = 12 + 3 = 15$ True ✓

b If $4n - 3 = 101$

     then $4n = 104$

           $n = 26$

     So, 101 is the 26th term in the sequence.

11 The *n*th term of the sequence 7, 10, 13, 16, ..., is given by the expression $3n + 4$.

     a Show that this is true for the first three terms.

     b Use the rule to decide whether 40 is in the sequence.

12 The *n*th term of the sequence 2, 5, 10, 17, ..., is given by the expression $n^2 + 1$.

     a Show that this is true for the first three terms.

     b Use the rule to decide whether 35 is in the sequence.

13 Look back at your answers to questions 2–4.

     What do you notice about the value of *d* and the coefficient of *n*?

14 Here are the *n*th terms for different sequences.

     a $4n - 1$ **b** $4n + 2$ **c** $4n - 4$ **d** $4n + 5$

     i For each sequence write down the first term, *a* and the difference, *d*.

     ii What do you notice about *d* and the coefficient of *n*?

15 Here are the *n*th terms for different sequences.

     a $5n - 1$ **b** $5n + 2$ **c** $5n - 3$ **d** $5n + 4$

     i For each sequence write down the first term, *a* and the difference, *d*.

     ii What do you notice about *d* and the coefficient of *n*?

Solve problems

Sequence A has *n*th term $5n - 2$.

Sequence B has *n*th term $4n + 3$.

What is the first number to appear in both sequences?

Listing some of the terms:

Sequence A: 3, 8, 13, 18, 23, ...

Sequence B: 7, 11, 15, 19, 23, ...

So, 23 is the first number to appear in both sequences.

16 Sequence A has nth term $2n + 7$.

Sequence B has nth term $3n - 1$.

What is the first number to appear in both sequences?

17 Sequence A has nth term $18 - 4n$.

Sequence B has nth term $15 - 3n$.

What is the first negative number to appear in both sequences?

18 Sequence A has nth term n^2.

Sequence B has nth term $\frac{n(n + 1)}{2}$.

What is the second number to appear in both sequences?

2.5 Finding the nth term

● I can work out the nth term of a sequence

In the sequence 5, 8, 11, 14, 17, 20, ..., $a = 5$ and $d = 3$ and the nth term is $3n + 2$.

The coefficient of n is 3, which is the same as the difference, d and the constant term, c, is 2.

Note that $c = a - d$.

Develop fluency

a Work out the nth term for this sequence.

7, 11, 15, 19, 23, 27, ...

b Work out the nth term for this sequence.

40, 38, 36, 34, 32, 30, ...

a Here $a = 7$ and $d = 4$.

So, the coefficient of n is 4 and $c = 7 - 4 = 3$.

The nth term of the sequence is $4n + 3$.

b Here $a = 40$ and $d = -2$.

So, the coefficient of n is -2 and $c = 40 - (-2) = 40 + 2 = 42$.

The nth term of the sequence is $-2n + 42$. (You would normally write this as $42 - 2n$.)

1 Work out the nth term for each sequence.

    a 4, 10, 16, 22, 28, ...   **b** 9, 12, 15, 18, 21, ...   **c** 9, 15, 21, 27, 33, ...

    d 2, 5, 8, 11, 14, ...   **e** 2, 9, 16, 23, 30, ...   **f** 8, 10, 12, 14, 16, ...

    g 10, 14, 18, 22, 26, ...   **h** 3, 11, 19, 27, 35, ...   **i** 9, 19, 29, 39, 49, ...

    j 4, 13, 22, 31, 40, ...

2 Work out the nth term for each sequence.

    a 90, 85, 80, 75, 70, ...   **b** 43, 36, 29, 22, 15, ...

    c 28, 25, 22, 19, 16, ...   **d** 44, 36, 28, 20, 12, ...

3 Work out the *n*th term for each sequence.
   a 4, 10, 16, 22, 28, ...   b 8, 11, 14, 17, 20, ...   c 9, 15, 21, 27, 33, ...
   d 4, 7, 10, 13, 16, ...   e 13, 20, 27, 34, 41, ...

4 Work out the *n*th term for each decimal sequence.
   a 2.5, 3, 3.5, 4, 4.5, ...   b 10.5, 13, 15.5, 18, 20.5, ...
   c 3.1, 3.2, 3.3, 3.4, 3.5, ...   d 8.0, 7.8, 7.6, 7.4, 7.2, ...

5 Work out the *n*th term for each pattern and use it to find the 40th term.

   a    b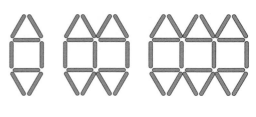

6 Without working out the terms of the sequences, match each sequence to its *n*th term expression.

a	1, 5, 9, 13, 17, ...	$6n$
b	10, 12, 14, 16, 18, ...	$4n - 3$
c	6, 12, 18, 24, 30, ...	$2n + 2$
d	4, 6, 8, 10, 12, ...	$2n + 8$

7 Without working out the terms of the sequences, match each sequence to its *n*th term expression.

a	4, 8, 12, 16, 20, ...	$5n - 1$
b	4, 9, 14, 19, 24, ...	$3n + 1$
c	4, 7, 10, 13, 16, ...	$n + 3$
d	4, 5, 6, 7, 8, ...	$4n$

Reason mathematically

The *n*th term for the sequence of square numbers 1, 4, 9, 16, 25, ... is n^2.

Use this to work out the *n*th term of each sequence.

   a 0, 3, 8, 15, 24, ...   b 2, 8, 18, 32, 50, ...

   a Each term is 1 less than a square number so the *n*th term is $n^2 - 1$.
   b Each term is double a square number so the *n*th term is $2n^2$.

8 The nth term for the sequence of square numbers 1, 4, 9, 16, 25, ... is n^2.
Use this to work out the nth term of each sequence.

   a 2, 5, 10, 17, 26, ...   **b** 3, 12, 27, 48, 75, ...

9 The nth term for the sequence of square numbers 1, 4, 9, 16, 25, ... is n^2.
Use this to work out the nth term of each sequence.

   a 5, 8, 13, 20, 29, ...   **b** $\frac{1}{2}$, 2, $4\frac{1}{2}$, 8, $12\frac{1}{2}$, ...

10 The nth term for the sequence of square numbers 1, 4, 9, 16, 25, ... is n^2.

   a Write down the nth term of the sequence 3, 6, 9, 12, 15, ...

   b A new sequence is made by adding terms of the two sequences above:
       4, 10, 18, 28, 40, ...

      Write down the nth term of this sequence.

11 Show that the nth term of the sequence 2, 6, 12, 20, 30, ... is $n^2 + n$.

Solve problems

Theo is saving money.

He saves £1 in the first week, £2 in the second week, £3 in the third week and so on.

   a How much does he save in the 10th week?   **a** £10

   b How much does he save in the nth week?   **b** £n

12 James is saving money.
He saves £2 in the first week, £3 in the second week, £4 in the third week and so on.
How much does he save in: **a** the 10th week?   **b** the nth week?

13 A chess board has 64 squares.
On the first square is 2p, on the second square is 4p, on the third square is 6p and so on.
How much is on: **a** the 10th square?   **b** the nth square?

   c How much more is on the 20th square than the 19th square?

Now I can...

recognise, describe and generate sequences that use a simple rule	understand the sequence of numbers known as the triangular numbers	work out the nth term of a sequence
work out missing terms in a sequence	use the nth term of a sequence	
understand the sequence of numbers known as the square numbers		

CHAPTER 3 Perimeter and area

3.1 Perimeter and area of a rectangle

- I can use a simple formula to calculate the perimeter of a rectangle
- I can use a simple formula to calculate the area of a rectangle

Develop fluency

A **rectangle** has two equal long sides (each equal to its **length**, l) and two equal short sides (each equal to its **width**, w).

The perimeter, P, equals $2 \times$ length $+ 2 \times$ width, and the **formula** is $P = 2l + 2w$.

The area, A, equals length $\times$ width, and the formula is $A = lw$.

a Work out the perimeter (P) and area (A) of a rectangle that is 6 cm long and 4 cm wide.

$P = 2 \times 6 + 2 \times 4$

$\quad = 12 + 8$

$\quad = 20\,cm$

$A = 6 \times 4$

$\quad = 24\,cm^2$

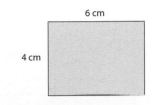

b Work out the perimeter (P) and area (A) of this square patio.

$P = 5 + 5 + 5 + 5$

$\quad = 4 \times 5$

$\quad = 20\,m$

$A = 5 \times 5 = 5^2$

$\quad = 25\,m^2$

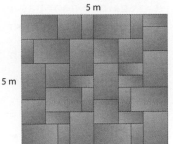

1 Work out the perimeter and area of each rectangle.

a 5 cm, 5 cm **b** 15 cm, 8 cm **c** 8 m, 7 m **d** 24 mm, 30 mm

2 A rectangular paving slab measures 0.8 m by 0.6 m.
Work out the perimeter of the slab.

3 Work out: **i** the area   and   **ii** the perimeter of each rectangle.

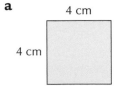
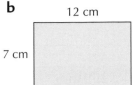
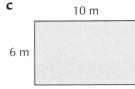
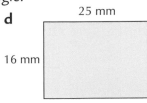

a 4 cm, 4 cm **b** 12 cm, 7 cm **c** 10 m, 6 m **d** 25 mm, 16 mm

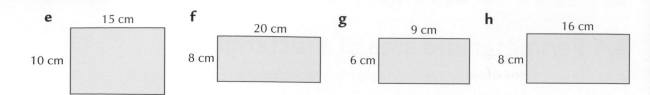

e 15 cm 10 cm

f 20 cm 8 cm

g 9 cm 6 cm

h 16 cm 8 cm

Reason mathematically

This fence has an area of $21\,m^2$.

Work out the height of the fence, shown as h on the diagram.

The fence is a rectangle and its height is the shorter side, so take this as its width.

Area = length × width

So, $21 = 7 \times w$

To find w, divide both sides of the equation by 7.

$\dfrac{21}{7} = w$

$3 = w$

So, the width (or height) of the fence is 3 metres.

You could simply work out $21 \div 7 = 3$.

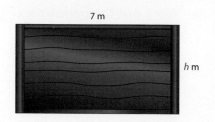

7 m

h m

4 A garage door has an area of $6\,m^2$. The length is 1 m longer than the width. Work out the length and width.

5 Work out the length of each rectangle.

a Area = 12 cm² 3 cm

b Area = 20 cm² 2 cm

c Area = 24 m² 4 m

d Area = 48 cm² 6 cm

6 Work out the perimeter of this square tile.

25 cm²

7 Copy and complete the table for rectangles *a* to *f*.

	Length	Width	Perimeter	Area
a	8 cm	6 cm		
b	20 cm	15 cm		
c	10 cm		30 cm	
d		5 m	22 m	
e	7 m			42 m²
f		10 mm		250 mm²

Solve problems

A farmer wants to build a sheep pen with the largest possible area.
He has 18 lengths of fencing, each of length 1 m.

What is the area of the largest pen he can build?

8 × 1 m pen, area of 8 m², perimeter 18 m

7 × 2 m pen, area of 14 m², perimeter 18 m

6 × 3 m pen, area of 18 m², perimeter 18 m

5 × 4 m pen, area of 20 m², perimeter 18 m. This is the largest area of the pen.

8 a Work out the perimeter of this room.

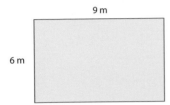

9 m

6 m

   b Skirting board is sold in 3-metre lengths.

      How many lengths are needed to go around the four walls of the room?

9 A swimming pool is 10 m wide and 40 m long. Draw a sketch of the pool.

   a As a warm-up for a swimming lesson, pupils are asked to swim around the perimeter of the pool. How far do they swim?

   b Emma wants to swim 500 m.

      i How many widths does she need to swim?

      ii How many lengths does she need to swim?

10 Can a square have the same numerical value for its perimeter and its area?

11 A farmer has 60 lengths of fencing, each of length 1 m, to make three sides of a rectangular sheep pen, with the fourth side being formed by an existing long wall. Work out the length and width of the pen that will make its area as large as possible.

3.2 Compound shapes

● I can work out the perimeter and area of a compound shape

Develop fluency

A **compound shape** is made from more than one shape. You can work out its perimeter and area by dividing it into the shapes that make it up.

Work out the perimeter and area of this compound shape.

First split the shape into two rectangles. This split depends on the information you are given. If you split this shape into rectangles A and B, as shown, you will be able to work out all the lengths you need.

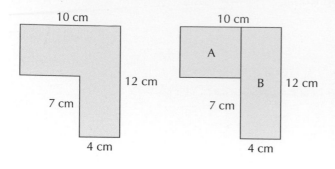

For rectangle A, the length is $(10 - 4) = 6\,cm$ and the width is $(12 - 7) = 5\,cm$.

For rectangle B, the length is $12\,cm$ and the width is $4\,cm$.

Perimeter $= 10 + 12 + 4 + 7 + 6 + 5$
$= 44\,cm$

Total area $=$ area of A $+$ area of B
$= 6 \times 5 + 12 \times 4$
$= 30 + 48$
$= 78\,cm^2$

1 Work out: **i** the perimeter **ii** the area of each compound shape.

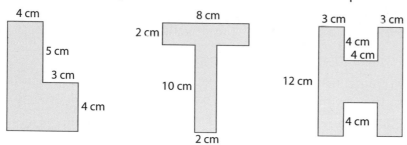

2 Work out: **i** the area **and** **ii** the perimeter of each compound shape.

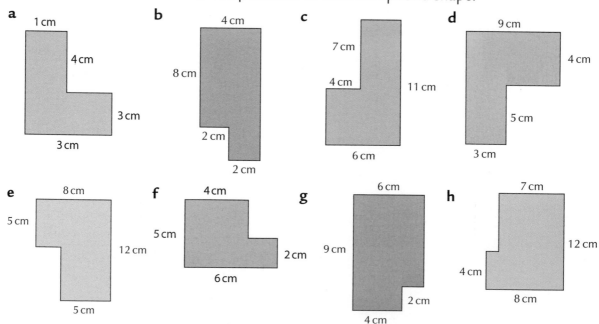

Reason mathematically

Work out area of the part of this shape that is shaded yellow.

Area of complete rectangle = 12 × 8

$\qquad$ = 96 cm²

Area of rectangle A, = 5 × 3

$\qquad$ = 15 cm²

So area of shaded region = 96 − 15

$\qquad$ = 81 cm²

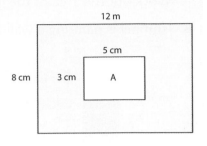

3 This is Zach's working for the area of this compound shape.

Area = 10 × 4 + 8 × 5

$\qquad$ = 40 + 40

$\qquad$ = 80 cm²

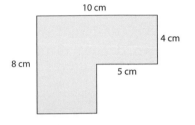

a Explain why he is wrong.

b Calculate the correct answer.

4 Nadia uses a rectangular piece of card to make a picture frame for a photograph of her favourite band.

a Work out the area of the photograph.

b Work out the area of the card she uses.

c Work out the area of the border.

5 This shape has a perimeter of 58 cm.

Find the length of the missing side.

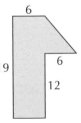

6 This shape has an area of 66 m².

Find the lengths of the unknown sides.

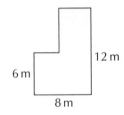

Solve problems

A mirror is surrounded by a wooden frame. Calculate the surface area of the mirror.

Length of mirror is 80 − 14 = 66 cm

Width of mirror is 50 − 14 = 36 cm

Area of mirror is 66 × 36 = 2376 cm²

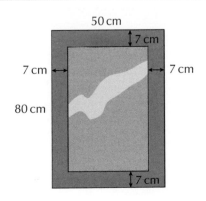

7 A garden is in the shape of a rectangle measuring 16 m by 12 m.
Work out the area of the grass in the garden.

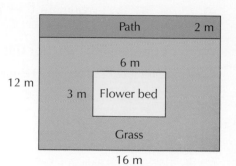

8 Ishmael is painting a wall in his house.

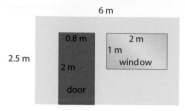

Work out the area of the wall that he needs to paint.

9 This compound shape is made from two identical rectangles.

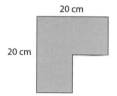

Work out the perimeter of the compound shape.

10 The four unmarked sides in this diagram are the same length. Work out the area of the compound shape.

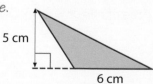

3.3 Area of a triangle

● I can work out the area of a triangle

Develop fluency

To work out the area of a triangle, you need to know the length of its base, b, and its perpendicular height, h.

The formula for the area of a triangle is $A = \frac{1}{2} \times b \times h = \frac{1}{2}bh$

a Work out the area of this triangle.

$A = \frac{1}{2} \times 8 \times 3$

$= 12\,cm^2$

b Work out the area of this obtuse-angled triangle.

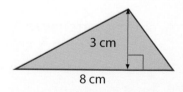

Notice that you must measure the perpendicular height outside the triangle.

$A = \frac{1}{2} \times 6 \times 5$

$= 3 \times 5$

$= 15\,cm^2$

For questions 1–2, work out the area of each triangle.

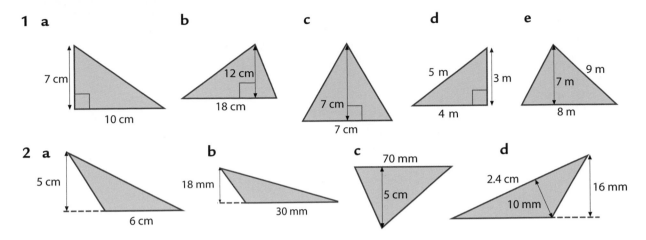

1 a 7 cm, 10 cm

b 12 cm, 18 cm

c 7 cm, 7 cm

d 5 m, 3 m, 4 m

e 9 m, 7 m, 8 m

2 a 5 cm, 6 cm

b 18 mm, 30 mm

c 70 mm, 5 cm

d 2.4 cm, 10 mm, 16 mm

3 Copy and complete the table for the triangles described in rows **a** to **c**.

4 For each of the below, draw a coordinate grid on centimetre-squared paper, with axes for x and y both numbered from 0 to 6. Then draw each triangle, with the given coordinates and work out the area of each triangle.

Triangle	Base	Height	Area
a	6 cm	5 cm	
b	8 cm	7 cm	
c	11 m	5 m	

a Triangle ABC with vertices A(2, 0), B(5, 0) and C(4, 4)

b Triangle DEF with vertices D(1, 1), E(6, 1) and F(3, 5)

c Triangle PQR with vertices P(2, 1), Q(2, 5) and R(5, 3)

d Triangle XYZ with vertices X(0, 5), Y(6, 5) and Z(4, 1)

5 Use squared paper to draw four different triangles, each with area 24 cm². Draw the base first, using a whole number of centimetres that is a factor of 48, for example, 8 cm. Then calculate the height of the triangle.

6 Calculate the area of each triangle.

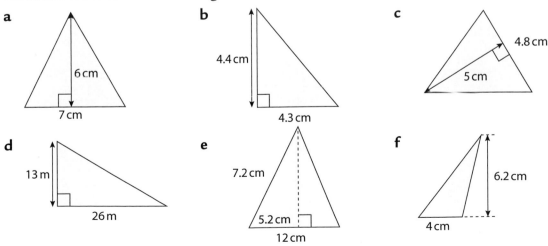

a 6 cm, 7 cm

b 4.4 cm, 4.3 cm

c 4.8 cm, 5 cm

d 13 m, 26 m

e 7.2 cm, 5.2 cm, 12 cm

f 6.2 cm, 4 cm

Reason mathematically

Work out the area of this compound shape.

To work out the area of a compound shape, divide it into simple shapes.

Divide the shape into a rectangle (A) and a triangle (B).

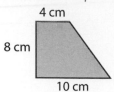

4 cm

8 cm

10 cm

4 cm

8 cm | A | B

10 cm

Area of A = $8 \times 4 = 32 \, cm^2$

Area of B = $\frac{1}{2} \times 6 \times 8 = 3 \times 8 = 24 \, cm^2$

So the area of the shape = 32 + 24 = 56 cm².

7 Work out the area of each compound shape.

a

4 m

2 m

2 m

b

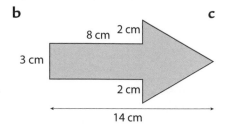

8 cm 2 cm

3 cm

2 cm

14 cm

c

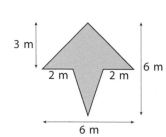

3 m

2 m   2 m

6 m

6 m

8 This right-angled triangle has an area of 36 cm².
Find other right-angled triangles, with different
measurements, that also have an area of 36 cm².

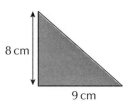

8 cm

9 cm

9 Here is Millie's working for the area of this triangle.

$A = \frac{1}{2}bh$

$A = \frac{1}{2} \times 5 \times 13$

$A = 32.5 \, cm^2$

Is Millie correct?

Explain your answer.

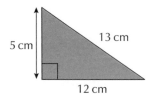

5 cm

13 cm

12 cm

10 Work out the area of each shape.

a

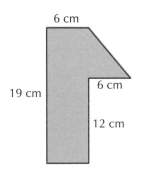

6 cm

19 cm

6 cm

12 cm

b

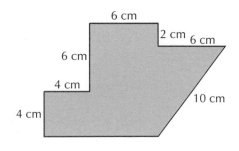

6 cm

2 cm 6 cm

6 cm

4 cm

10 cm

4 cm

Solve problems

The area of this triangle is 35 cm².
Calculate the height h.

$A = \frac{1}{2}bh$

$35 = \frac{1}{2} \times 7 \times h$

$70 = 7h$

$h = 10\,cm$

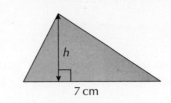

7 cm

11 The area of this triangle is 48 cm².
Calculate the length of base b.

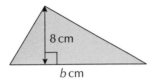

8 cm

b cm

12 The diagram shows the dimensions
of a symmetrical flower garden.
Work out the area of the garden.

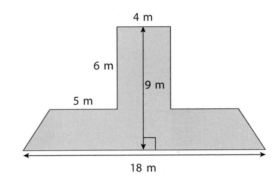

4 m

6 m

9 m

5 m

18 m

13 Copy and complete the table for the triangles described in rows a to c.

Triangle	Base	Height	Area
a	8 cm		32 cm²
b		9 cm	27 cm²
c	11 m		66 cm²

14 This composite shape is made up of a rectangle and a triangle.
The area of the rectangle is 30 cm².
Work out the area of the triangle.

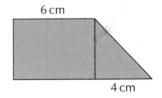

6 cm

4 cm

3.4 Area of a parallelogram

● I can work out the area of a parallelogram

Develop fluency

● To work out the area of a parallelogram, you need to know the length of its base, b, and its perpendicular height, h. The formula for the area of a parallelogram is $A = b \times h = bh$.

Work out the area of this parallelogram.

$A = 10 \times 6 = 60 \, \text{cm}^2$

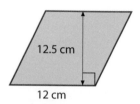

6 cm

10 cm

1 Calculate the area of each parallelogram.

a
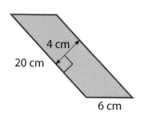
4 cm
9 cm

b
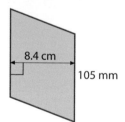
12.5 cm
12 cm

c
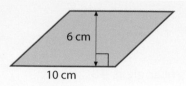
7 m
7 m
8 m

d
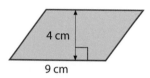
4 cm
20 cm
6 cm

e
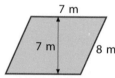
8.4 cm
105 mm

f
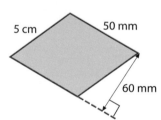
5 cm
50 mm
60 mm

2 Calculate the area and perimeter of each parallelogram.

a
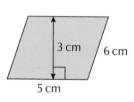
3 cm
6 cm
5 cm

b
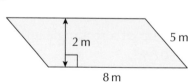
2 m
5 m
8 m

c
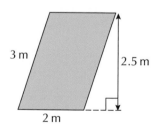
3 m
2.5 m
2 m

d
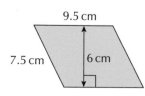
9.5 cm
7.5 cm
6 cm

e
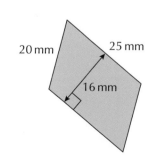
20 mm
25 mm
16 mm

f
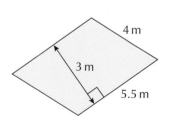
4 m
3 m
5.5 m

3　Copy and complete the table below for parallelograms **a** to **c**.

Parallelogram	Base	Height	Area
a	8 cm	4 cm	
b	17 cm	12 cm	
c	8 m	5 m	

4　For each of the below, draw a coordinate grid on centimetre-squared paper, with axes for x and y both numbered from 0 to 8. Then draw each parallelogram with the given coordinates and work out the area of each parallelogram.

　a　Parallelogram ABCD with vertices at A(2, 0), B(6, 0), C(8, 5) and D(4, 5)

　b　Parallelogram EFGH with vertices at E(1, 2), F(4, 2), G(7, 7) and H(4, 7)

　c　Parallelogram PQRS with vertices at P(1, 8), Q(7, 2), R(7, 6) and S(1, 4)

5　Use squared paper to draw four different parallelograms with area 48 cm^2.

Draw the base first, using a whole number of centimetres that is a factor of 48, for example, 8 cm. Then calculate the height of the parallelogram.

Reason mathematically

The area of this shape is 36 m^2. Work out the length of the base.

Area $A = bh$

$36 = b \times 4$

$36 \div 4 = b$

$b = 9$ m

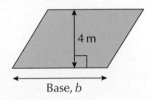

4 m

Base, b

6　The area of this parallelogram is 27 cm^2. Work out the perpendicular height, h, of the parallelogram.

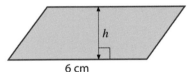

h

6 cm

7　The perpendicular height of a parallelogram is 8 cm and it has an area of 50 cm^2. Work out the length of the base of the parallelogram.

8　This parallelogram has an area of 45 m^2. Work out the perimeter of the parallelogram.

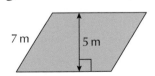

7 m　5 m

9　Work out the value of h in this diagram.

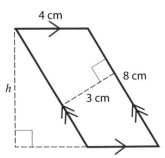

4 cm

8 cm

3 cm

h

Solve problems

This shape is made from three identical parallelograms.

The area of the shape is 168 cm². Work out the value of p.

Height of one parallelogram = 24 ÷ 3 = 8 cm

Area of one parallelogram = 168 ÷ 3 = 56 cm²

Area = bh

$56 = p \times 8$

$56 \div 8 = p$

$p = 7$ cm

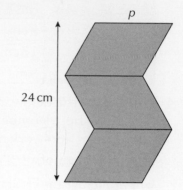

Note that you could have worked this out by calculating 168 ÷ 24 = 7 cm, because a single parallelogram could be constructed from the shape.

10 This shape is made from four identical parallelograms. Given that the area of the shape is 120 cm², work out the value of y.

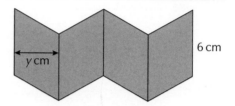

11 Each piece on this grid is a compound shape or a parallelogram. Each small square represents a square centimetre. Find the area of each puzzle piece. Show your calculations.

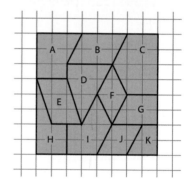

12 This logo is made from two parallelograms.

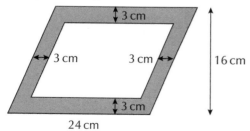

Work out the area of the yellow border.

13 This shape is made from four identical parallelograms.

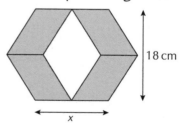

Given that the area of the area of the shape is 252 cm², work out length x.

3.5 Area of a trapezium

● I can work out the area of a trapezium

To work out the area of a trapezium, you need to know the length of its two parallel sides, *a* and *b*, and the perpendicular height, *h*, between the parallel sides.

The area is given by $\frac{1}{2}(a + b) \times h$

Work out the area of this trapezium.

$A = \frac{1}{2} \times (5 + 9) \times 4 = 28\,cm^2$

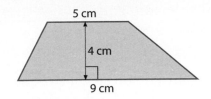

1 Work out the area of each trapezium.

a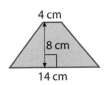

b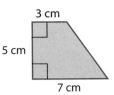

c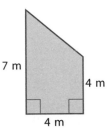

d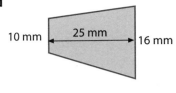

e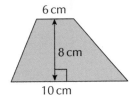

f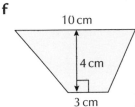

g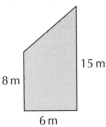

h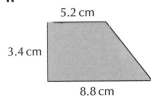

2 Find the area and perimeter of each trapezium.

a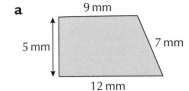

b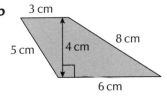

c

3   Copy and complete the table below for each trapezium **a** to **c**.

Trapezium	Length, *a*	Length, *b*	Height, *h*	Area, *A*
a	4 cm	6 cm	3 cm	
b	10 cm	12 cm	6 cm	
c	9 m	3 m	5 m	

4   The side of a swimming pool is a trapezium, as shown in the diagram.

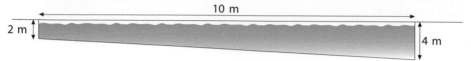

Work out its area.

5   Draw a coordinate grid on centimetre-squared paper, with axes for *x* and *y* both numbered from −6 to 6. Plot the points A(3, 5), B(−4, 4), C(−4, −2) and D(3, −5) and work out the area of the shape ABCD.

Reason mathematically

The area of this trapezium is 60 mm². Work out its height, *h*.

$A = \frac{1}{2}(a+b) \times h$

$60 = \frac{1}{2}(8 + 12) \times h$ Work out $8 + 12 = 20$

$60 = \frac{1}{2} \times 20 \times h$ Multiply $20 \times \frac{1}{2} = 10$

$60 = 10 \times h$ Divide both sides by 10

$6 = h$

$h = 6 \text{ mm}$

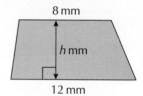

6   The area of the trapezium is 80 cm².
    Work out the height, *h*.

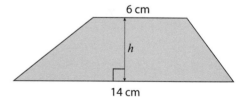

7   These three shapes have the same area.

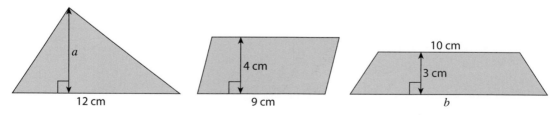

Work out the lengths marked *a* and *b*.

8 The end of a lean-to shed has a cross-sectional area of 4.5 m². Work out its height at its tallest point.

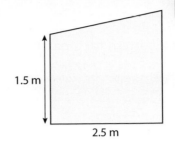

1.5 m

2.5 m

9 The area of this trapezium is 9 cm². Work out three different whole-number values of *a*, *b* and *h*, with *b > a*.

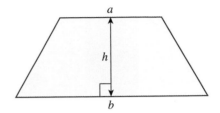

a

h

b

Solve problems

Work out the area of the blue region of this tile.

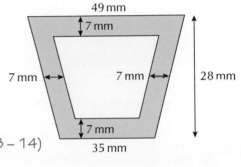

49 mm

7 mm

7 mm   7 mm

28 mm

7 mm

35 mm

Area of the whole shape $= \frac{1}{2}(a + b) \times h$

$= \frac{1}{2}(35 + 49) \times 28$

$= 1176 \text{ mm}^2$

Area of yellow trapezium $= \frac{1}{2}(a + b) \times h$

$= \frac{1}{2}((35 - 14) + (49 - 14)) \times (28 - 14)$

$= \frac{1}{2}(21 + 35) \times 14$

$= 392 \text{ mm}^2$

Area of blue region $= 1176 - 392$

$= 784 \text{ mm}^2$

10 The diagram shows a shaded symmetrical trapezium drawn inside a square.
What is the area of the trapezium?

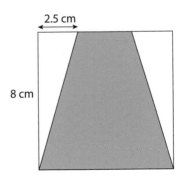

2.5 cm

8 cm

11 ABCD is a trapezium.
The height of the trapezium is half the length of AB.
CD is twice the length of AB.
Work out the area of the trapezium, when AB = 10 cm.

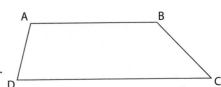

A   B

D   C

12 This is a pattern for a patchwork quilt.

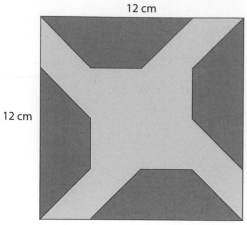

12 cm

12 cm

Each pattern is a square, measuring 12 cm by 12 cm.
These are the measurements for each trapezium.

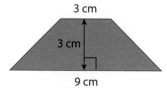

3 cm

3 cm

9 cm

Work out the area of the yellow part of the quilt.

13 The area of this trapezium is 6 cm² and $b > a$.
Work out different values of a, b and h.

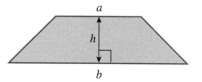

Now I can...

use a simple formula to calculate the perimeter of a rectangle	work out the area of a triangle	work out the area of a parallelogram
work out the perimeter of a compound shape	work out the area of a compound shape	work out the area of a trapezium
use a simple formula to calculate the area of a rectangle		

CHAPTER 4 Negative numbers

4.1 The number line

- I can use a number line to order positive and negative numbers, including decimals
- I can understand and use the symbols < (less than) and > (greater than)

Using a number line helps you to order negative and positive numbers.

Which number is greater, −7 or −3?

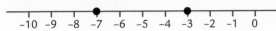

Because −3 is further to the right on the number line than −7 is, it is the larger number.

Dan wrote −6 < 4

His sister said she could write the same thing but with a different sign.
Explain how she could do this.

Swapping the numbers around and using the greater than sign gives 4 > −6, which is the same thing but with a different sign.

Write these temperatures in order, from lowest to highest.
8 °C, −2 °C, 10 °C, −7 °C, −3 °C, 4 °C

Draw a number line then mark the numbers on that line.

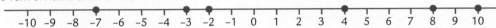

You can see that the order is: −7 °C, −3 °C, −2 °C, 4 °C, 8 °C, 10 °C

1 State whether each statement is true or false.

   a 5 < 10   **b** 5 > −10   **c** −3 > 6   **d** 3 < 6

2 Copy each statement and put < or > into the ☐ to make it true.

   a 6 ☐ 10   **b** 5 ☐ −2   **c** −1 ☐ 9   **d** −3 ☐ −1

3 Write down the lower temperature in each pair.

   a −1 °C, −8 °C   **b** 2 °C, −9 °C   **c** −1 °C, 9 °C   **d** −3 °C, −1 °C

4 Put these numbers into order, starting with the lowest.

   a −0.5, 0, −1   **b** −1, 2, −5   **c** −1.1, −1.6, −1.9

5 Write down the greater number in each pair.

   a −1, −3   **b** −6.5, −5.6   **c** −1.2, −2.1

6 Put these numbers into order, starting with the highest.

   a −3, −2, −5   **b** −1, 1, 0   **c** −3, −3.3, −1.3

7 Put these numbers into order, starting with the lowest.

   a −0.8, 0.5, −1.2   **b** −1.5, 2.3, −5.9   **c** −3.3, −2.7, −3.8

8 Write down the higher temperature in each pair.

   a −211 °C, −108 °C   **b** −58 °C, −29 °C   **c** −101 °C, −99 °C   **d** −73 °C, −61 °C

9 Write down the highest number in each set.

   a −5.5, −5.3, −5.6   **b** −7.5, −7.6, −7.9   **c** −4.2, −4.5, −4.1

Reason mathematically

Leo said, '−6 is larger than −4'.

His sister said he was wrong.

How could she explain to Leo why he was wrong?

She could draw a number line and place −6 and −4 onto the line, showing that −6 is further to the left than −4 and that the further left a number is on the line, the lower the number is.

10 Carla was told that the temperature outside would drop by 5 °C that night.
She said that it must be freezing outside then.
Explain why she might be incorrect.

11 A submarine is at 40 metres below sea level.
The captain is told to move to 20 metres below sea level.
Does the captain move the submarine down or up? Give a reason for your answer

12 Explain how you know that −5.1 is higher than −5.9.

13 James has a bank balance of £187. He pays his energy bill of £53.62, and his credit card bill of £228.51. Explain why his bank balance is now negative.

Solve problems

Work out the number that is halfway between −3 and 1.

Draw a number line showing both −3 and 1.

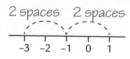

2 spaces   2 spaces

Count 4 spaces between −3 and 1, so halfway is 2 spaces from the −3, which is −1.

14 Bernie took the lift. He started on floor 2. First he went up two floors to marketing. Then he took the lift down seven floors to the canteen. Then he took the lift up two floors to the boardroom. Finally he took the lift up four floors to the IT department.
On which floor is the IT department?

15 Luke has £137 in the bank, he pays a bill of £155 and also pays £50 into his bank. Then pays another bill of £40. How much will he have in his bank account now?

4.2 Arithmetic with negative numbers

- I can carry out additions and subtractions involving negative numbers
- I can use a number line to calculate with negative numbers

It can be useful to use the number line to help you add and subtract.

Develop fluency

Use a number line to work out the answers.

a $5 - 13$ **b** $(-11) + 9$ **c** $6 - 12 - 3$

a

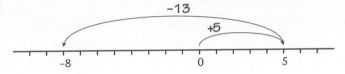

Starting at zero and 'jumping' along the number line to 5 and then back 13 gives an answer of −8.

b

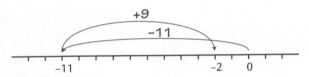

Similarly, $(-11) + 9 = -2$ Notice that **brackets** are sometimes used so that the negative sign is not confused with a subtraction sign.

c

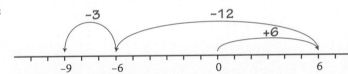

Using two steps this time, $6 - 12 - 3 = -9$

1 Use a number line to work out the answers.
  a $1 - 5$ **b** $-2 + -4$ **c** $-1 + 3$ **d** $-3 - 1$ **e** $-2 + -3$

2 Use a number line to work out the answers.
  a $4 - 6$ **b** $-5 + -2$ **c** $-4 + 3$ **d** $-3 - 3$ **e** $-1 + -4$

3 Without using a number line, write down the answers.
  a $1 - 3$ **b** $3 - 8$ **c** $7 - 9$ **d** $4 - 6$

4 Without using a number line, write down the answers.
  a $-9 + 3$ **b** $-8 + 4$ **c** $-2 - 5$ **d** $-3 - 2$

5 Use a number line to work out the answers.
  a $1 - 5 + 4$ **b** $2 - 7 + 3$ **c** $-3 + 1 - 5$ **d** $-6 + 8 - 4$

Reason mathematically

In a magic square, the numbers in any row, column or diagonal add up to give the same answer. Could this be a magic square? Give a reason for your answer.

		−6
	2	
−1	−3	

The diagonal adds up to −5.

So the missing number on the bottom row will be −1.

And the missing top number in the other diagonal will be −4.

From the middle vertical numbers, −1 and 2, the missing top middle number is −6.

From the two numbers in the top row, −6 and −4, the missing top number of the middle column should be 5.

Since −6 is not the same as 5, these rows and columns do not add up to the same number, so it is not a magic square.

6 In a magic square, the numbers in any row, column or diagonal add up to give the same answer. Could this be a magic square? Give a reason for your answer.

−3		−7
		4
		−5

7 Alf has £124 in the bank. He makes an online payment for £135.

How much has he got in the bank now? Explain your working.

Solve problems

In a quiz, two points are awarded for a correct answer, but three points are deducted for an incorrect answer.

Team A answers eight questions correctly, two incorrectly and doesn't answer the last two questions. Team B answers all the questions; nine correctly and three incorrectly.

Who wins the quiz?

Team A scores 8 × 2 points, giving 16, but lose 2 × 3 points, leaving them with 10 points.

Team B scores 9 × 2 points, giving 18, but lose 3 × 3 points, leaving them with 9 points.

So Team A wins the quiz.

8 In a maths competition, 5 points are awarded for each correct answer but 1 point is taken off for each incorrect answer. If a question is not answered then it scores zero points. There are ten questions in the competition.

   a What is the lowest score possible?
   b Given that Amy got four correct answers and only left out one question, how many points did she score?
   c Given that Barakat scored 34 points, how many questions did she get right?
   d Given that Conor scored 0 points but answered at least one question, how many questions did he not answer correctly?
   e Explain why it is impossible to score 49 points.

9 In this 4×4 magic square, all of the rows, columns and diagonals add to -18.
   Copy and complete the square.

-27			15
6			-12
		18	-3
	3	-15	

10 A hotel has 24 floors. Phil get in the lift at floor at floor 18. The lift takes him down 9 floors, then the lift goes up 12 floors before going down 22 floors where he gets out. What floor does Phil end up at?

4.3 Subtraction with negative numbers

● I can carry out subtractions involving negative numbers

Subtracting a negative number is the same as adding an appropriate positive number.

Develop fluency

   a $12 - (-15)$
   b $23 - (-17)$
   a $12 - (-15) = 12 + 15 = 27$
   b $23 - (-17) = 23 + 17 = 40$

1 Use a number line to work these out.
   a $2 - (-5)$ b $4 - (-1)$ c $7 - (-4)$ d $6 - (-8)$

2 Use a number line to work these out.
   a $-3 - (-4)$ b $-5 - (-3)$ c $-6 - (-5)$ d $-7 - (-3)$

3 Without using a number line, write down the answers.
   a $2 - (-5)$ b $3 - (-2)$ c $3 - (-8)$ d $6 - (-)7$

4 Without using a number line, write down the answers.
   a $-1 - (-6)$ b $-2 - (-3)$ c $-4 - (-7)$ d $-8 - (-9)$

5 Use a number line to work these out.

   a $8 - (-5) - 3$ **b** $6 - (-1) - 7$ **c** $8 - (-4) - 5$

6 Use a number line to work these out.

   a $-1 - (-3) + 5$ **b** $-2 - (-3) + 4$ **c** $-9 - (-5) + 3$

7 Use a number line to work these out.

   a $9 - (-2) - (-4)$ **b** $5 - (-3) - (-1)$ **c** $4 - (-6) - (-7)$

8 Use a number line to work these out.

   a $-2 - (-4) - (-3)$ **b** $-11 - (-4) - (-2)$ **c** $-8 - (-3) - (-4)$

9 Without using a number line, write down the answers.

   a $1 - (-5) - 4$ **b** $2 - (-7) - 3$ **c** $3 - (-9) - 2$

10 Without using a number line, write down the answers.

   a $-3 - (-1) - (-5)$ **b** $-6 - (-8) - (-4)$ **c** $-8 - (-2) - (-5)$

Reason mathematically

The deepest part of the English Channel is 174 metres below sea level.

The top of a mast of a yacht is 6 metres above sea level.

   a How much higher is the top of the mast than the lowest part of the English Channel?

   b If the yacht were sat on the bottom of the English Channel, how far would the top of the mast be below sea level?

   a $174 + 6 = 180$ metres

   b $174 - 6 = 168$ metres

11 A fish is 12 m below the surface of the water. A fish eagle is 17 m above the water. How many metres must the bird descend to get the fish? Explain your working.

12 Alice and Dipesh are playing a game with a dice. The dice has the numbers −4, −3, −2, −1, 1 and 2. They both start with a score of 20 and whatever number they roll is subtracted from their total. They take it in turns to roll the dice. The winner is the first player to score 40 or higher.

   a Alice rolls −2, −4 and 1. What is her score?

   b Dipesh rolls 2, 2 and −3 and Alice rolls 1, −1 and −2. Who is closer to 40 points?

   c What is the lowest number of rolls a player could roll to win the game?

Solve problems

This is Laz's solution to a problem

$(-15 + 3) - (3 - 5) = 12 - 2 = 10$

Laz has made some errors in his calculation.

Explain where he has made errors.

−15 + 3 should have been −12, 3 − 5 should have been −2 giving the answer as
−12 − −2 = −12 + 2 = −10

13 Choose a number from each list and subtract one from the other. Repeat for at least four pairs of numbers. What are the biggest and smallest answers you can find?

A	14	−17	−25	11	15
B	−23	9	−18	8	−14

14 This is Alia's solution to a problem: $(8 - 10) - (4 - 9) = 2 - 5 = 3$

She has the correct answer but made some errors. Explain where she has made errors.

4.4 Multiplication with negative numbers

● I can carry out multiplications involving negative numbers

Multiplying a positive number by another positive number gives a positive number.

Multiplying a negative number by a positive number gives a negative number.

Multiplying a negative number by a negative number results in a positive number.

Develop fluency

Work out the answers.

a -12×4 **b** -7×3.

a $-12 \times 4 = -48$ **b** $-7 \times 3 = -21$

1 Work out the answers.
a -2×1 **b** -3×4 **c** -6×7 **d** -1×7

2 Work out the answers.
a $2 \times (-3)$ **b** $3 \times (-5)$ **c** $6 \times (-5)$ **d** $4 \times (-7)$

3 Work out the answers.
a -1×3 **b** $3 \times (-7)$ **c** $-7 \times (-8)$ **d** $4 \times (-6)$

4 Work out the answers.
a -2.5×2 **b** -1.5×4 **c** -0.5×8 **d** -1.5×7

5 Work out the answers.

    a $2.5 \times (-3)$ **b** $3.1 \times (-4)$ **c** $6.2 \times (-3)$ **d** $4.3 \times (-4)$

6 Work out the answers.

    a -1.4×4 **b** $3.2 \times (-4)$ **c** $-7.1 \times (-9)$ **d** $4.3 \times (-5)$

7 Work out the answers.

    a -3×4.3 **b** $2 \times (-4.7)$ **c** $-5 \times (-9.1)$ **d** $3 \times (-5.4)$

8 Work out the answers.

    a $-2 \times (4 - 7)$ **b** $-3 \times (6 - 8)$ **c** $6 \times (2 - 7)$ **d** $-4 \times (5 - 9)$

Reason mathematically

Write down the next three numbers in each number sequence.

    a $-1, -2, -4, -8, ..., ..., ...,$ **b** $-1, 3, -9, 27, ..., ..., ...,$

    a *Multiplying by 2 each time to give missing numbers* $-16, -32, -64$

    b *Multiplying by −3 each time to give missing numbers* $-81, 243, -729$

9 Write down the next three numbers in each number sequence.

    a $1, -2, 4, -8, ..., ..., ...$ **b** $-1, -3, -9, -27, ..., ..., ...$

    c $-1, 5, -25, 125, ..., ..., ...$ **d** $1, -4, 16, -64, ..., ..., ...$

10 This is Tim's solution to a problem: $(7 - 11) \times (3 - 8) = 4 \times 5 = 20$

    He has the correct answer but made some errors. Explain where he has made errors.

Solve problems

Peter asked Kath to think of two integers smaller than eight and tell him their **product**.

Kath said, 'The product is −12'.

Peter said, 'There are four different possible sets of numbers that give that product.'

Write down the four possible pairs of numbers Kath could have been thinking of.

To have a negative product, one number must be positive while the other is negative.

To have a product of 12, you can have 3 × 4 or 2 × 6 only using integers less than eight.

So the four sets are −3 × 4, 3 × −4, −2 × 6, 2 × −6

11 Julie asked Chris to think of two integers less than ten with a product of −24.
    Write down the four different possible pairs of numbers Chris could have been
    thinking of.

12 a In each brick wall, work out the number to write in an empty brick by multiplying the numbers in the two bricks below it. Copy and complete each brick wall.

i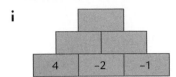

ii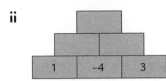

b Andy said: 'You will always have a positive number at the top of the brick wall if there is one negative number in the bottom layer.'

Is Andy correct? Explain your answer.

4.5 Division with negative numbers

● I can carry out divisions involving negative numbers

Dividing a positive number by another positive number results in a positive number.

Dividing a positive number by a negative number results in a negative number.

Dividing a negative number by a positive number results in a negative number.

Dividing a negative number by a negative number results in a positive number.

Develop fluency

Work out the answers.

a $-14 \div 2$ **a** $-14 \div 2 = -7$ **b** $-60 \div 5$ **b** $-60 \div 5 = -12$

1 For questions 1-10, work out the answers.

  a $-6 \div 2$ **b** $-8 \div 4$ **c** $-16 \div 4$ **d** $-21 \div 7$

2 a $12 \div (-3)$ **b** $35 \div (-5)$ **c** $60 \div (-5)$ **d** $28 \div (-7)$

3 a $-12 \div (-4)$ **b** $-18 \div (-6)$ **c** $-16 \div (-8)$ **d** $-45 \div (-9)$

4 a $-15 \div 3$ **b** $35 \div (-7)$ **c** $-27 \div (-9)$ **d** $42 \div (-6)$

5 a $-2.6 \div 2$ **b** $-4.4 \div 4$ **c** $-8.4 \div 2$ **d** $-1.5 \div 3$

6 a $4.5 \div (-3)$ **b** $12.8 \div (-4)$ **c** $6.9 \div (-3)$ **d** $4.8 \div (-4)$

7 a $-2.5 \div (-5)$ **b** $-12.4 \div (-4)$ **c** $-4.5 \div (-5)$ **d** $-2.1 \div (-7)$

8 a $-8.4 \div 4$ **b** $3.2 \div (-2)$ **c** $-8.1 \div (-9)$ **d** $6.5 \div (-5)$

9 a $(-4 \times -6) \div 8$ **b** $28 \div (-4 \times -1)$ **c** $-35 \div (-9 + 2))$ **d** $36 \div (-5 - 4))$

10 a $-12 \div (4 - 7)$ **b** $-45 \div (6 - 9)$ **c** $60 \div (2 - 7)$ **d** $-42 \div (5 - 7)$

Reason mathematically

This is Lee's solution to a problem:

$(5 - 11) \div (-6 + 3) = 6 \div 3 = 2$

Lee has the correct solution but has made some errors in his calculation.

Explain where he has made errors.

(5 − 11) should have been −6, and (−6 + 3) should have been −3 giving −6 ÷ −3 = 2

11 This is Toni's solution to a problem: $(8 - 52) \div (-7 + 3) = 46 \div 10 = 4.6$

Toni has made some errors in his calculation. Explain where he has made errors.

12 Ann said she has thought of two negative numbers that have a product of -8.

Is this possible? Explain your answer.

13 Dean said that if I divide a positive number by a negative number I will always get a negative answer.

Give four examples to illustrate that Dean is correct.

14 Abby was told that $111 \times 13 = 1443$

She said, 'So $1443 \div -13$ will be -111'.

  a Is Abby correct? Explain your answer.

  b Give another similar division that gives the answer -111.

Solve problems

What are the four different pairs of integers between -9 and 9 that will divide to give -4?

We have $-8 \div 2 = -4$, $8 \div -2 = -4$, $-4 \div 1 = -4$, $4 \div -1 = -4$

15 What are the six different pairs of integers less than 10 that will divide to give -3?

16 Write down five calculations involving division that give the answer -5.

17 Work out the answer to $(-8 \div 4) \times (-9 \div -3) \times (10 \div -2)$

Now I can...

use a number line to order positive and negative numbers, including decimals	carry out additions and subtractions involving negative numbers	carry out divisions involving negative numbers
understand and use the symbols < (less than) and > (greater than)	carry out multiplications involving negative numbers	solve problems with negative numbers
use a number line to calculate with negative numbers		

CHAPTER 5 Averages

5.1 Mode, median and range

- I can understand and calculate the mode, median and range of data

An **average** is a single or typical value that represents a whole set of values.

- The **mode** is the value that occurs most often. It is the only average that you can use for non-numerical data, such as favourite colours or football teams. Sometimes there may be no mode.

- The **median** is the middle value for a set of values when they are put in numerical order.

- The **range** is the difference between the largest and smallest values, so it is equal to the largest value minus the smallest.

A small range means that the values in the set of data are similar in size, whereas a large range means that the values differ a lot and therefore are more spread out.

Develop fluency

These are the ages of 11 players in a football squad.

23, 19, 24, 26, 27, 27, 24, 23, 20, 23, 26

Find: **a** the mode **b** the median **c** the range.

First, put the ages in order:

19, 20, 23, 23, 23, 24, 24, 26, 26, 27, 27

a The mode is the number that occurs most often.
So, the mode is 23.

b The median is the number in the middle of the set.
This will be the sixth of 11 values.
So, the median is 24.

c The largest value is 27, the smallest is 19.
As 27 − 19 = 8, the range is 8.

1 Find the mode of each set of data.
  a red, white, blue, red, white, blue, red, blue, white, red
  b rain, sun, cloud, fog, rain, sun, snow, cloud, snow, sun, rain, sun
  c E, A, I, U, E, O, I, E, A, E, A, O, I, U, E, I, E
  d ♠, ♣, ♥, ♦, ♣, ♠, ♥, ♣, ♦, ♥, ♣, ♥, ♦, ♥

2 Find the mode of each set of data.
  a 7, 6, 2, 3, 1, 9, 5, 4, 8, 4, 5, 5
  b 36, 34, 45, 28, 37, 40, 24, 27, 33, 31, 41, 34, 40, 34
  c 14, 12, 13, 6, 10, 20, 16, 8, 13, 14, 13
  d 99, 101, 107, 103, 109, 102, 105, 110, 100, 98, 101, 95, 104
  e 7, 6, 7, 5, 4, 5, 5, 4, 5, 8, 10, 5, 1, 5, 5, 5, 12, 13, 5, 4, 4, 5, 5, 5, 7, 6, 5, 5, 14, 3, 5, 4, 5, 3, 5, 3, 5

3 Find the range of each set of data.
   a 23, 37, 18, 23, 28, 19, 21, 25, 36   b 3, 1, 2, 3, 1, 0, 4, 2, 4, 2, 6, 5, 4, 5
   c 51, 54, 27, 28, 38, 45, 39, 50   d 95, 101, 104, 92, 106, 100, 97, 101, 99

4 Find the mode and range of each set of data.
   a £2.50, £1.80, £3.65, £3.80, £4.20, £3.25, £1.80
   b 23 kg, 18 kg, 22 kg, 31 kg, 29 kg, 32 kg
   c 132 cm, 145 cm, 151 cm, 132 cm, 140 cm, 142 cm
   d 32°, 36°, 32°, 30°, 31°, 31°, 34°, 33°, 32°, 35°

5 A group of nine Year 7 pupils had their lunch in the school cafeteria.
   These are the amounts that each pupil spent.
   £2.30, £2.20, £2.00, £2.50, £2.20, £2.90, £3.60, £2.20, £2.80
   a Find the mode for the data.   b Find the range for the data.

6 a Find the mode for the data.   b Find the range for the data.
   12, 13, 14, 15, 16, 18, 19, 20, 22, 25

7 The heights of 10 girls in Year 7, in cm, are
   155, 148, 134, 154, 149, 138, 148, 165, 132 and 163.
   What is the a range of their heights b mode of their heights c median height?

8 Mark asked some friends how many TVs they had in their homes.
   These are his results: 3, 5, 2, 0, 1, 3, 2, 1, 2, 3
   What was the modal number of TVs in a household?

9 Josh counted how many runs he had scored in his latest 7 cricket games.
   Here are his scores: 15, 32, 45, 34, 11, 23, 28
   Find the median and range of the numbers of runs he scored.

10 Luke recorded how long it took to get to school every day for a week:
    16 minutes, 11 minutes, 23 minutes, 15 minutes, 11 minutes
    Find the mode, median and range of his times, to the nearest minute.

Reason mathematically

There are four children at a party. The youngest is 11 years old. The range of their ages is 4 years. The modal age is 12 years. How old are the other three children?

Start with the youngest, 11, □, □, □.

Next fill in the modal ages, 11, 12, 12, □.

As the range is 4, add this on to the youngest age to leave 11, 12, 12, 15.

11 **a** There are two children in the Bishop family.
The range of their ages is exactly 5 years.
What could the ages of the two children be? Give an example.

b In the Patel family, two of the children are twins.
What is the range of their ages?

12 The median of this set of data is 7, but one value is missing. What is the missing value?
3, 10, 5, ☐, 8

13 There are three children in a family. The youngest is 3 years old. The range of their ages is 8 years. The median age is 7 years. How old are the other two children?

14 The range of this set of numbers is 6.
7, 9, 12, 7, 8, ☐
What are the two possible values for the missing number?

Solve problems

The median of a set of 5 numbers is 7, the mode is 2 and the range is 10.
What could the set of numbers be?

Start with what you know (the median) and write a list.

☐, ☐, 7, ☐, ☐

The mode is 2 so this is the most common value, so the first two numbers must be 2.

2, 2, 7, ☐, ☐

The range is the difference between the biggest and the smallest, so 2 + 10 = 12.

2, 2, 7, ☐, 12

Finally, the missing number must be bigger than 7 but smaller than 12. It cannot be either 7 or 12 as the mode has to be 2. It may be 8, 9, 10 or 11.

15 **a** Write down a list of seven numbers with a median of 10 and mode of 12.

b Write down a list of eight numbers with a median of 10 and mode of 12.

c Write down a list of seven numbers with a median of 10, mode of 12 and range of 8.

16 These are the names of the 12 people who work for a company.

| Abbas | Kathy | Yiiki | Suki | Brian | Kathy |
| Lucy | Tim | Kathy | James | Ryan | Tim |

a Which name is the mode?

b One person leaves the company. A different person joins the company.
Now the modal name is Tim.

i What is the name of the person who leaves?

ii What is the name of the person who joins?

17 Three numbers have a median of 7, a range of 7 and a total of 16.
Give the three numbers.

18 Alex is doing a survey of makes of car in a car park. He records them as:
Ford, Nissan, Toyota, Fiat, Vauxhall, Ford, Volkswagen, BMW, Honda, Renault, Audi,
Honda, Fiat, Ford, Skoda, Kia, Toyota.

  a Which make is the mode?

  b As he is recording the data, another car arrives. Alex realises there are now two modes.
     What makes could the other car be?

  c Before he has chance to finish his survey, two cars leave the car park, leaving Honda
     as the mode. What cars must have left? What does it tell you about your answer to
     part **b**?

  d Why is it not possible to find a median or range for this data?

  e Think of some other categories where you can only find a mode; no range, no median.

5.2 The mean

● I can understand and calculate the mean average of data

The **mean**, often called the **mean average** or just the **average**, is the most commonly used
average. It can be used only with numerical data.

$$\text{Mean} = \frac{\text{sum of all values}}{\text{number of values}}$$

The mean takes all of the values into account, but it can be distorted by an **outlier**, which is a
value that is much larger or much smaller than the rest. When there is an outlier, the median
is often used instead of the mean.

Develop fluency

Find the mean of 2, 7, 9, 10.

The mean is $\dfrac{2+7+9+10}{4} = \dfrac{28}{4} = 7$

1 Complete each calculation.

  a The mean of 3, 5, 10 is $\dfrac{3+5+10}{3} = \dfrac{\square}{\square} = \square$

  b The mean of 2, 5, 6, 7 is $\dfrac{2+5+6+7}{4} = \dfrac{\square}{\square} = \square$

2 Complete each calculation.

  a The mean of 5, 6, 10 is $\dfrac{5+6+10}{3} = \dfrac{\square}{\square} = \square$

  b The mean of 1, 3, 3, 5 is $\dfrac{1+3+3+5}{4} = \dfrac{\square}{\square} = \square$

3 Complete each calculation.

  a The mean of 1, 5, 6 is $\dfrac{1+5+6}{\square} = \dfrac{\square}{\square} = \square$

  b The mean of 1, 4, 7, 8 is $\dfrac{1+4+7+8}{\square} = \dfrac{\square}{\square} = \square$

4 Complete each calculation.

    a The mean of 2, 4, 11, 13, 15 is $\dfrac{\square+\square+\square+\square+\square}{\square}=\dfrac{\square}{\square}=\square$

    b The mean of 1, 3, 8, 8 is $\dfrac{\square+\square+\square+\square}{\square}=\dfrac{\square}{\square}=\square$

5 Find the mean of each set of data.

    a 8, 7, 6, 10, 4   **b** 23, 32, 40, 37, 29, 25

    c 11, 12, 9, 26, 14, 17, 16   **d** 2.4, 1.6, 3.2, 1.8, 4.2, 2.5, 4.5, 2.2

6 Calculate the mean of each set of data, giving your answers to 1 dp.

    a 6, 7, 6, 4, 2, 3   **b** 12, 15, 17, 11, 18, 16, 14

    c 78, 72, 82, 95, 47, 67, 77, 80   **d** 9.1, 7.8, 10.3, 8.5, 11.6, 8.9

7 Find the mean of these masses.
    6 kg, 8 kg, 7 kg, 6 kg, 8 kg, 13 kg

8 Find the mean of these times.
    13 s, 20 s, 27 s, 30 s, 25 s, 28 s, 30 s, 35 s

9 In three different shops the prices of a can of cola were 98p, 76p and 104p. What is the mean price for a can of cola?

10 Jenson times how fast his remote control car does laps around the course. His first attempt was 34 seconds, his second attempt was 31 seconds. His mean time for all three attempts was 32 seconds. How long did the third attempt take?

Reason mathematically

These are the distances, in metres, 7 students jumped in the long jump.

2.4, 1.8, 2.9, 1.9, 2.4, 3.1, 3.0

  a Calculate the mean distance jumped by the students.

  b What is the median distance jumped by the students?

  c Which average do you think is the best one to use? Explain your answer.

  a The mean is $\dfrac{2.4+1.8+2.9+1.9+2.4+3.1+3}{7}=2.5\,m$

  b The median is the number in the middle of the list: 1.8, 1.9, 2.4, 2.4, 2.9, 3.0, 3.1, which is 2.4 m

  c The best average to use would be the mean as it takes into account all the distances jumped.

11 These are the heights, in centimetres, of 10 children.
    132, 147, 143, 136, 135, 146, 153, 132, 137, 149

    a Calculate the mean height of the children.

    b What is the median height of the children?

    c What is the modal height of the children?

    d Which average do you think is the best one to use? Explain your answer.

12 These are the numbers of children in the families of Marie's class at school.
1, 1, 1, 1, 1, 1, 2, 2, 2, 2, 3, 3, 3, 3, 4
   a What is the mode?   **b** What is the median?   **c** What is the mean?
   d Which average do you think is the best one to use? Explain your answer.

13 These are the shoe sizes of all the girls in class 7JS.
3, 3, 3, 3, 3, 4, 4, 4, 5, 6, 6
   a What is the mode?   **b** What is the median?   **c** What is the mean?
   d Which average do you think is the best one to use? Explain your answer.

14 These are the lap times, in seconds, for eight model racing cars.
17, 20, 15, 16, 20, 15, 17, 20
   a Find the mean lap time.
   b Find the median lap time.
   c Find the modal lap time.
   d Which average do you think is the worst one to use? Explain your answer.
   e Why are so few lap times greater than the mean?

Solve problems

The ages of seven people are 40, 37, 34, 42, 45, 39, 35. Calculate their mean age.

The mean age is $\dfrac{40+37+34+42+45+39+35}{7} = \dfrac{272}{7} = 38.9\,(1dp)$

15 The table shows the numbers of eggs laid by Paddy's chickens during one week.

Monday	Tuesday	Wednesday	Thursday
7	6	8	5

   a Find the mean number of eggs laid from Monday to Thursday.
   b The chickens laid more eggs on Friday. The mean for all five days is 7.
      How many eggs did the chickens lay on Friday?

16 When the masses, in kg, of eight bags of grain were checked, these results were obtained.
2.53, 2.52, 2.49, 2.51, 2.53, 2.51, 2.53, 2.54
   a Find the mean mass of these eight bags of grain.
   b Each bag is supposed to weigh 2.12 kg.
      By how much, in kilograms, is the heaviest bag overweight?

17 There are 10 people doing a skills test. Their average score is 17.
Stan does the test and scores 25.
   a Does this increase or decrease the mean score?
   b What is the total score for all eleven people taking the test now?

18 These are the prices of chocolate bars on sale at a school fete.

50p, 70p, 45p, 60p, 90p, 65p, 55p, 60p, 55p, 60p

The mean price of the chocolate bars last year was 60p.

Is the mean price this year higher or lower? Explain your answer.

5.3 Statistical diagrams

● I can read and interpret different statistical diagrams

The most common ways to display data are:

● **pictograms**, in which small diagrams or **icons** represent the data
● **bar charts** with gaps between the bars to show data that has separate or distinct categories
● bar charts with no gaps between the bars to show data values that fall into ranges such as 1–5, 6–7
● **line graphs** to show trends and patterns in the data, generally over time
● **pie charts,** which are circular diagrams divided into sectors to show proportions of the data categories in the sample.

Develop fluency

Millie decides to find out her friends' favourite types of movie. Her results are shown in the table.

She decides to use a pie chart to display her data. Show how Millie uses the data to create a pie chart.

She divides each category by the total amount and then multiplies by 360, because that is how many degrees there are in a circle. Finally, she checks her degrees add up to 360.

Type of movie	Total
Comedy	13
Horror	5
Thriller	7
Other	5

Type of movie	Total	Degrees
Comedy	13	$\frac{13}{30} \times 360 = 156$
Horror	5	$\frac{5}{30} \times 360 = 60$
Thriller	7	$\frac{7}{30} \times 360 = 84$
Other	5	$\frac{5}{30} \times 360 = 60$
Total	30	360

1 The bar chart shows how the students in class 7PB travel to school.

   a How many students walk to school?

   b What is the mode for the way the students travel to school?

   c How many students are there in class 7PB?

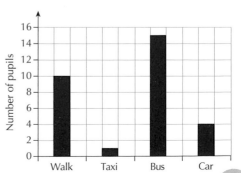

2 The pictogram shows the amounts of money collected for charity by different year groups in a school.

Key **£20** represents £20

a How much money was collected by Year 8?

b How much money was collected by Year 10?

c Which year group collected most money?

d How much money did the school collect altogether?

3 The pictogram shows how many songs five students have in their playlists.

a Who has the most songs in their playlist?

b How many songs does Jessica have in her playlist?

c How many songs does Ceri have in her playlist?

d How many more songs does Dipesh have than Tania?

e How many songs do the five students have altogether?

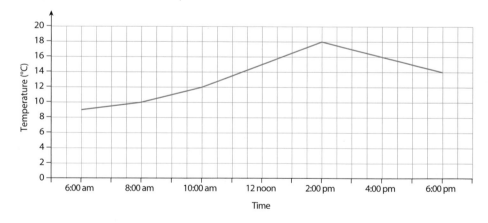

Key

 represents 4 songs

4 The line graph shows the temperature, in degrees Celsius (°C), in Bristol over a 12-hour period.

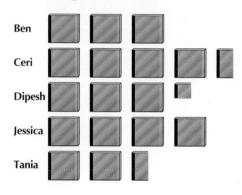

What was the temperature at **a** midday **b** 3:00 pm?

c Write down the range for the temperature over the 12-hour period.

d Explain why the line graph is a useful way of showing the data.

5 This compound bar chart shows the favourite colours for a sample of Year 7 students.

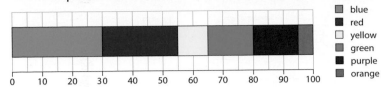

- blue
- red
- yellow
- green
- purple
- orange

    a Which colour was chosen by the greatest number of students?

    b What percentage of the students chose yellow?

    c Which two colours were equally liked by the students?

    d Explain why the compound bar chart is a useful way to illustrate the data.

6 In a survey, 60 people were asked which TV channels they watched most often.

The pie chart shows the results.

    a Which is the most popular channel?

    b Which is the least popular channel?

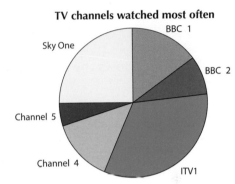

TV channels watched most often

7 The vertical line graph shows the shoe sizes of some KS3 students.

    a How many students wear size 9 shoes?

    b Which is the most common shoe size?

    c How many students were asked altogether?

    d Construct a pictogram to represent this data.

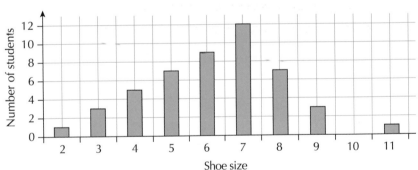

8 The table shows some students' preferred end-of-year activities.

    a Which was the most popular activity?

    b How many students are there in total?

    c How many boys are there in total?

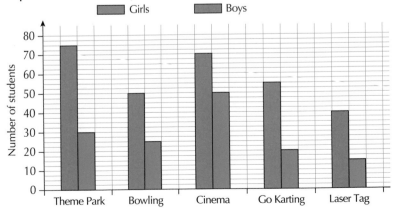

Reason mathematically

Looking at the dual bar chart, how many more girls exercised than boys?

Boys: 7 + 11 + 3 + 2 + 5 = 28

Girls: 5 + 15 + 3 + 5 + 6 = 34

6 more girls exercised compared to boys.

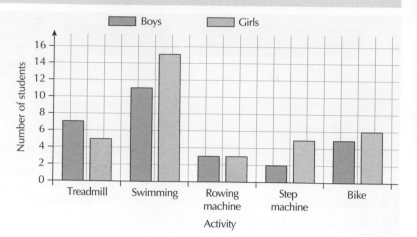

9 This table shows the numbers of accidents at a crossroads over a six-year period.

Year	2014	2015	2016	2017	2018	2019
Number of accidents	6	8	7	9	6	4

a Draw a pictogram for this data.

b Draw a bar chart for this data.

c Which diagram would you use if you were going to write to your local council to suggest that traffic lights should be installed at the crossroads? Explain why.

10 Lynda asked some people to name their favourite football team. She displayed her results in a pie chart.

a Albion had 40 supporters. How many supporters were there altogether?

b How many supported United?

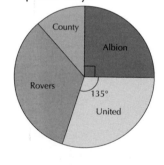

11 Boys and girls in Year 7 were asked to name their favourite takeaway food.

Draw a dual bar chart to represent this information.

Takeaway food	Girls	Boys
Pizza	4	9
Fish and chips	12	8
Burger	8	13
Other	5	6

Solve problems

This bar chart shows the numbers of students that exercised altogether.

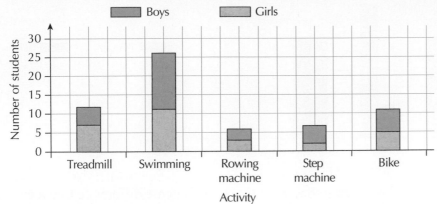

a How many students exercised altogether?

b What was the modal exercise for boys?

c How many more students went swimming than used the bike?

As this is a composite bar chart, you need to remember the height of each bar represents the total number of girls and boys doing the exercise.

a Add the frequencies within each bar together to get the total number of students.

12 + 26 + 6 + 7 + 11 = 62

b The modal exercise for the boys is the most common exercise for just the boys' parts of the bars: swimming in this case.

c 26 − 11 = 15 more went swimming than used the bike.

12 The bar chart shows the marks obtained in a mathematics test by the students in class 7RN.

a How many students are there in class 7RN?

b How many students got a mark over 60?

c Write down the smallest and greatest range of marks possible for the data.

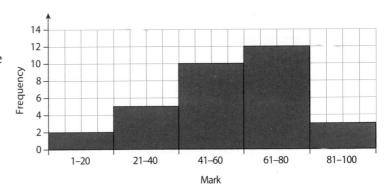

13 A teacher keeps records about students arriving late to her lesson.
The dual bar chart shows the number of late arrivals over the first four days, in one week.

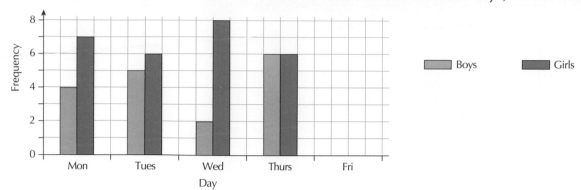

a On which day, from Monday to Thursday, was there the biggest difference between the numbers of boys and girls arriving late?

b How many late arrivals were there altogether by boys from Monday to Thursday?

c By the end of Friday 18 boys in total had arrived late.
Twice as many girls as boys had been late to lessons during the whole week.
Complete the chart for Friday.

14 The table shows the ages of people attending concerts of two pop groups.
Draw a composite bar chart to represent the data.

Age range	< 12	12–16	17–21	> 21
Number of people at concert 1	500	1900	1600	1100
Number of people at concert 2	2200	2900	2800	3900

15 The pictogram shows how far some people travelled in the Lincoln soapbox race.

Kath	🛒🛒🛒🛒🛒🛒🛒
Mark	🛒🛒🛒
Steve	🛒🛒

🛒 = 500 m

a How far did Mark travel?

b Kath completed the course. How far did she travel?

c How much further did Kath travel than Steve?

Now I can...

understand the mode, median and range of data	calculate the mode, median and range of data	interpret different statistical diagrams
understand the mean average of data	calculate the mean average of data	
read different statistical diagrams		

CHAPTER 6 Equivalent fractions

6.1 Equivalent fractions

- I can find equivalent fractions
- I can write fractions in their simplest form

In the fraction $\frac{3}{4}$, the number on the top, 3, is the **numerator** and the number on the bottom, 4, is the **denominator**.

You can find **equivalent fractions** by multiplying or dividing the numerator and the denominator of a fraction by the same number. To **simplify** a fraction to its **simplest form**, divide the numerator and the denominator by the highest **common factor**.

Develop fluency

Fill in the missing number in each of these equivalent fractions.

a $\frac{2}{3} = \frac{a}{15}$ **b** $\frac{40}{48} = \frac{b}{12}$

a $3 \times ? = 15$
$3 \times 5 = 15$ *Multiply the numerator by 5 to find the value of a.*
$\frac{2}{3} = \frac{10}{15}$ $a = 10$

b $48 \div ? = 12$ *In this case divide, because the denominator is smaller.*
$48 \div 4 = 12$ *Divide the numerator by 4 to find the value of b.*
$\frac{40}{48} = \frac{10}{12}$ $b = 10$

Write $\frac{18}{24}$ in its simplest form.

2 is a common factor of 18 and 24, so divide top and bottom by 2.
$\frac{18}{24} = \frac{9}{12}$ *This is not in the simplest form.*
3 is a common factor of 9 and 12, so divide top and bottom by 3.
$\frac{9}{12} = \frac{3}{4}$ *This is the simplest form.*
You could reach the answer in one step by noticing that 6 is the highest common factor of 18 and 24 and dividing by 6.

For questions 1–2, fill in the missing number in each pair of equivalent fractions.

1 a $\frac{2}{3} = \frac{a}{36}$ **b** $\frac{3}{8} = \frac{b}{24}$ **c** $\frac{6}{7} = \frac{c}{56}$ **d** $\frac{3}{7} = \frac{d}{56}$

2 a $\frac{42}{48} = \frac{a}{8}$ **b** $\frac{30}{36} = \frac{b}{6}$ **c** $\frac{30}{100} = \frac{c}{20}$ **d** $\frac{48}{132} = \frac{d}{11}$

3 Write each fraction in its simplest form.

a $\frac{8}{12}$ **b** $\frac{12}{18}$ **c** $\frac{12}{20}$ **d** $\frac{12}{60}$ **e** $\frac{240}{40}$ **f** $\frac{128}{240}$

For questions 4–5, write three fractions that are equivalent to each of these fractions.

4 **a** $\frac{1}{3}$ **b** $\frac{1}{4}$ **c** $\frac{3}{4}$ **d** $\frac{25}{35}$ **e** $\frac{15}{50}$ **f** $\frac{150}{200}$

5 Look at the fractions listed here.

$$\frac{5}{15} \quad \frac{8}{12} \quad \frac{45}{60} \quad \frac{14}{20} \quad \frac{18}{24} \quad \frac{30}{45} \quad \frac{14}{42} \quad \frac{49}{70} \quad \frac{19}{57} \quad \frac{75}{100} \quad \frac{36}{54} \quad \frac{42}{60}$$

Write down all the fractions, from the list, that are equivalent to:

a $\frac{1}{3}$ **b** $\frac{2}{3}$ **c** $\frac{3}{4}$ **d** $\frac{7}{10}$.

6 Find the missing number in each pair of equivalent fractions.

a $\frac{a}{24} = \frac{24}{32}$ **b** $\frac{b}{24} = \frac{12}{32}$ **c** $\frac{15}{21} = \frac{c}{70}$ **d** $\frac{6}{42} = \frac{d}{70}$

Reason mathematically

Write two fractions equivalent to $\frac{3}{2}$ and $\frac{5}{7}$ that have the same numerator.

A common multiple of 3 and 5 is 15.

$\frac{3}{2} = \frac{15}{\square}$ *Multiply the denominator by 5 to get* $\frac{3}{2} = \frac{15}{10}$

$\frac{5}{7} = \frac{15}{\square}$ *Multiply the denominator by 3 to get* $\frac{5}{7} = \frac{15}{21}$

7 **a** Write two fractions equivalent to $\frac{3}{4}$ and $\frac{2}{3}$ that have the same numerator.

    b Write two fractions equivalent to $\frac{3}{4}$ and $\frac{2}{3}$ that have the same denominator.

8 Here is part of a multiplication table.

7	14	21	28	35	42
8	16	24	32	40	48
9	18	27	36	45	54

    a Use the rows in the table to write five fractions that are equivalent to each of these fraction.

      i $\frac{7}{9}$ **ii** $\frac{8}{9}$ **iii** $\frac{7}{8}$ **iv** $\frac{8}{7}$

    b Write two fractions that are equivalent to $\frac{9}{14}$.

9 **a** One kilogram is 1000 grams. What fraction of a kilogram is 550 grams?

    b One hour is 60 minutes. What fraction of 1 hour is 33 minutes?

    c One kilometre is 1000 metres. What fraction of a kilometre is 75 metres?

     Give each of your answers to parts **a**, **b** and **c** in its simplest form.

Solve problems

Which of the fractions in the sequence $\frac{1}{14}, \frac{2}{14}, \frac{3}{14}, \dots \frac{13}{14}$ cannot be simplified? Explain why they cannot be simplified.

Any fraction in which the numerator and denominator do not have a common factor other than 1 cannot be simplified.

The numbers from 1 to 13 for which the only common factor with 14 is 1 are 1, 3, 5, 9, 11, and 13.

Therefore, none of $\frac{1}{14}, \frac{3}{14}, \frac{5}{14}, \frac{9}{14}, \frac{11}{14}, \frac{13}{14}$, can be simplified.

10   There are 23 fractions in the sequence $\frac{1}{24}, \frac{2}{24}, \frac{3}{24}, \frac{4}{24}, \dots \frac{23}{24}$

How many of them cannot be simplified?

11   This compass rose has eight divisions around its face.
What fraction of a turn takes you from:

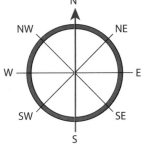

  a NW to SE clockwise   **b** E to SW anticlockwise
  c NE to SW clockwise   **d** SW to NE anticlockwise?
Write your fractions in their simplest form.

12   Work out these quantities, giving each answer in its simplest form.

  a Calculate the fraction of a metre that is equivalent to:
   i 85 cm   **ii** 250 mm   **iii** 40 cm.

  b Calculate the fraction of a kilogram that is equivalent to:
   i 450 g   **ii** 45 g   **iii** 975 g.

  c Calculate the fraction of a turn the minute hand of a clock goes through from:
   i 7:15 to 7:50   **ii** 8:25 to 9:10   **iii** 6:24 to 7:12.

13   What fraction of this shape is shaded?
Give your fraction in its simplest form.

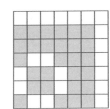

6.2 Adding and subtracting fractions

● I can add and subtract fractions with different denominators

To add or subtract fractions with the same denominator just add or subtract the numerators and leave the denominator the same. You may be able to simplify the answer.

If the denominators are different, find an equivalent fraction for one or both, so that you have two fractions with the same denominator.

Develop fluency

Work out the answers.

a $\dfrac{2}{3} - \dfrac{1}{6}$ **b** $\dfrac{2}{3} + \dfrac{1}{4}$

a $\dfrac{2}{3} - \dfrac{1}{6}$ A common multiple of 3 and 6 is 6 so change $\dfrac{2}{3}$ to sixths by multiplying the denominator and numerator by 2.

$$\dfrac{2}{3} - \dfrac{1}{6} = \dfrac{4}{6} - \dfrac{1}{6}$$
$$= \dfrac{3}{6}$$
$$= \dfrac{1}{2}$$

b $\dfrac{2}{3} + \dfrac{1}{4}$ A common multiple of 3 and 4 is 12.

Change both fractions to twelfths by multiplying the denominator and numerator of $\dfrac{2}{3}$ by 4 and the denominator and numerator of $\dfrac{1}{4}$ by 3.

$$\dfrac{2}{3} + \dfrac{1}{4} = \dfrac{8}{12} + \dfrac{3}{12}$$ $\dfrac{2}{3}$ is $\dfrac{8}{12}$ and $\dfrac{1}{4}$ is $\dfrac{3}{12}$
$$= \dfrac{11}{12}$$ This cannot be simplified.

1 Add these fractions. Simplify the answers as much as possible.

  a $\dfrac{3}{8} + \dfrac{1}{8}$ **b** $\dfrac{3}{10} + \dfrac{3}{10}$ **c** $\dfrac{5}{8} + \dfrac{3}{8}$ **d** $\dfrac{5}{12} + \dfrac{1}{12}$

2 Add these fractions. Write your answers as simply as possible.

  a $\dfrac{1}{2} + \dfrac{3}{10}$ **b** $\dfrac{1}{3} + \dfrac{5}{12}$ **c** $\dfrac{1}{12} + \dfrac{3}{4}$ **d** $\dfrac{5}{16} + \dfrac{3}{8}$

3 Subtract these fractions. Write your answers as simply as possible.

  a $\dfrac{1}{2} - \dfrac{1}{8}$ **b** $\dfrac{3}{4} - \dfrac{1}{4}$ **c** $\dfrac{5}{8} - \dfrac{1}{4}$ **d** $\dfrac{3}{4} - \dfrac{1}{8}$

4 Subtract these fractions. Write your answers as simply as possible.

  a $\dfrac{7}{8} - \dfrac{5}{12}$ **b** $\dfrac{4}{5} - \dfrac{2}{9}$ **c** $\dfrac{3}{7} - \dfrac{1}{8}$ **d** $\dfrac{1}{2} - \dfrac{2}{5}$

5 Add or subtract these fractions. Simplify the answers as much as possible.

  a $\dfrac{1}{2} + \dfrac{1}{3}$ **b** $\dfrac{1}{2} - \dfrac{1}{3}$ **c** $\dfrac{3}{4} - \dfrac{1}{6}$ **d** $\dfrac{2}{3} + \dfrac{1}{4}$

6 Add or subtract these fractions. Write your answers as simply as possible.

  a $\dfrac{2}{3} - \dfrac{1}{4}$ **b** $\dfrac{3}{4} + \dfrac{1}{8}$ **c** $\dfrac{3}{8} + \dfrac{4}{5}$ **d** $\dfrac{7}{8} - \dfrac{1}{12}$

Reason mathematically

Phoebe, Emily and Annabel share a pizza. Phoebe eats $\frac{1}{6}$ of the pizza and Emily eats $\frac{5}{12}$ of the pizza. How much pizza is left for Annabel?

Add together the amounts of pizza Phoebe and Emily eat.

$\frac{1}{6} + \frac{5}{12}$ *A common multiple of 6 and 12 is 12:* $\frac{1}{6} = \frac{2}{12}$

$\frac{2}{12} + \frac{5}{12} = \frac{7}{12}$

Then subtract the amount Phoebe and Emily ate, from 1, to find the remaining amount.

$1 - \frac{7}{12} = \frac{5}{12}$

$\frac{5}{12}$ *of the pizza is left for Annabel.*

7   In a book, $\frac{1}{12}$ of the space on the pages is taken up by photographs, $\frac{2}{3}$ is filled with text and the rest contains only diagrams.
What fraction of the book is taken up with diagrams?
Give your answer in its simplest form.

8   A survey of the students in one class showed that $\frac{3}{5}$ of them walked to school, $\frac{1}{4}$ came by bus and rest came by car.
What fraction of the class came by car?

9   Mohamed is buying ice cream for his class. $\frac{1}{3}$ of the class likes chocolate, $\frac{3}{5}$ class likes vanilla and the rest like strawberry. What fraction of the class like strawberry ice cream?

10 In a survey, $\frac{5}{9}$ of the class takes school dinners every day and $\frac{3}{8}$ of the class brings a packed lunch every day. The rest of the class sometimes has school dinners and sometimes has a packed lunch. What fraction of the class sometimes has school dinners?

Solve problems

Thomas is making a cake. The recipe calls for $\frac{1}{5}$ kg of flour. He has $\frac{3}{4}$ kg of flour. How much flour will he have left after he makes the cake? Give your answer as a fraction of a kg.

Subtract $\frac{3}{4} - \frac{1}{5}$ *A common multiple of 4 and 5 is 20:* $\frac{3}{4} = \frac{15}{20}$ *and* $\frac{1}{5} = \frac{4}{20}$.

So $\frac{3}{4} - \frac{1}{5} = \frac{15}{20} - \frac{4}{20} = \frac{11}{20}$

He has $\frac{11}{20}$ *kg of flour left.*

11 Desmond makes $\frac{3}{4}$ of a litre of squash in a jug. He pours $\frac{1}{8}$ of a litre into one glass and $\frac{1}{6}$ of a litre into another glass. What fraction of a litre is left in the jug?

12 John has £600. He gives $\frac{1}{4}$ to charity. He spends $\frac{2}{3}$ of the original amount. He puts half of the rest into a savings accounts.
What fraction does he save?

13 In a fridge there are two cartons of milk. One contains $\frac{2}{3}$ of a litre. The other contains $\frac{1}{4}$ of a litre.

  a How much milk is there altogether?

  b Milk is poured from one carton to the other so that they both contain the same amount.
How much milk was poured into the other carton? Give your answer as a fraction of a litre.

14 Kyra lives $\frac{3}{8}$ km from school. Sharon lives $\frac{9}{10}$ km from school. How much further from school does Sharon live than Kyra?

6.3 Mixed numbers and improper fractions

- I can convert mixed numbers to improper fractions
- I can convert improper fractions to mixed numbers

A number such as $2\frac{3}{4}$ that is made up of a whole number and a fraction is a **mixed number**.

There are 11 quarters in $2\frac{3}{4}$, written as $\frac{11}{4}$, which is an **improper fraction**.

You can **convert** (change) improper fractions to mixed numbers and mixed numbers to improper fractions. Addition of fractions can give an answer that is a mixed number.

Develop fluency

a Convert $3\frac{2}{5}$ into an improper fraction.   **b** Convert $\frac{22}{3}$ into a mixed number.

a $3 = \frac{15}{5}$ *3 × 5 gives 15 fifths.*

$3\frac{2}{5} = \frac{15}{5} + \frac{2}{5}$

$= \frac{17}{5}$

b $22 \div 3 = 7$ remainder 1 $\frac{22}{3}$ *will give 7 whole ones plus a remainder of $\frac{1}{3}$.*

$\frac{22}{3} = 7\frac{1}{3}$ *Write the remainder as a fraction.*

c Work out $\frac{3}{4} + \frac{2}{3}$.

c $\frac{3}{4} + \frac{2}{3} = \frac{9}{12} + \frac{8}{12}$ 4 and 3 are factors of 12 so change the fractions to twelfths.

$= \frac{17}{12}$ This is an improper fraction. It cannot be simplified.

$= 1\frac{5}{12}$ $17 \div 12 = 1$ remainder 5.

1 Convert each of these mixed numbers to an improper fraction.

a $2\frac{1}{2}$ **b** $3\frac{1}{2}$ **c** $5\frac{1}{2}$ **d** $11\frac{1}{2}$ **e** $4\frac{1}{9}$ **f** $4\frac{4}{9}$

2 Convert each of these mixed numbers to an improper fraction.

a $1\frac{11}{12}$ **b** $3\frac{5}{12}$ **c** $2\frac{4}{11}$ **d** $3\frac{4}{11}$ **e** $3\frac{7}{9}$ **f** $2\frac{3}{9}$

3 Convert each of these improper fractions to a mixed number.

a $\frac{29}{6}$ **b** $\frac{35}{6}$ **c** $\frac{45}{8}$ **d** $\frac{27}{8}$ **e** $\frac{11}{5}$ **f** $\frac{41}{5}$

4 Convert each of these improper fractions to a mixed number.

a $\frac{25}{4}$ **b** $\frac{27}{4}$ **c** $\frac{27}{3}$ **d** $\frac{25}{3}$ **e** $\frac{37}{5}$ **f** $\frac{37}{10}$

5 Add these fractions. Give your answers as improper fractions.

a $\frac{1}{3} + \frac{6}{7}$ **b** $\frac{3}{8} + \frac{6}{7}$ **c** $\frac{3}{8} + \frac{17}{24}$ **d** $\frac{3}{8} + \frac{8}{3}$ **e** $\frac{1}{2} + \frac{5}{8}$ **f** $\frac{1}{2} + \frac{7}{12}$

6 Subtract these fractions. Give your answers as improper fractions.

a $\frac{27}{5} - \frac{20}{8}$ **b** $\frac{30}{5} - \frac{20}{8}$ **c** $\frac{20}{8} - \frac{1}{4}$ **d** $\frac{17}{8} - \frac{1}{4}$ **e** $\frac{15}{8} - \frac{3}{4}$ **f** $\frac{15}{8} - \frac{2}{3}$

Reason mathematically

Put these numbers in order, from smallest to biggest.

$\frac{1}{3}, \frac{10}{4}, \frac{7}{3}, 2\frac{1}{8}, 5\frac{1}{6}, \frac{120}{11}, \frac{17}{5}, \frac{5}{8}$

Convert the improper fractions to mixed numbers.

$\frac{10}{4} = 2\frac{1}{2}$ $\frac{7}{3} = 2\frac{1}{3}$ $\frac{120}{11} = 10\frac{10}{11}$ $\frac{17}{5} = 3\frac{2}{5}$

Order the mixed numbers: look at the whole numbers first, then look at the fractions.

$\frac{1}{3}, \frac{5}{8}, 2\frac{1}{8}, 2\frac{1}{3}, 2\frac{1}{2}, 3\frac{2}{5}, 5\frac{1}{6}, 10\frac{10}{11}$

Rewrite the numbers in their original forms.

$\frac{1}{3}, \frac{5}{8}, 2\frac{1}{8}, \frac{7}{3}, \frac{10}{4}, \frac{17}{5}, 5\frac{1}{6}, \frac{120}{11}$

7 Put these numbers in order, from smallest to biggest.

$2\frac{1}{4}, \frac{3}{8}, \frac{8}{3}, \frac{7}{12}, 1\frac{1}{3}, \frac{27}{8}, 4$

8 Look at each sequence.

a $\dfrac{17}{10}, \dfrac{17}{11}, \dfrac{17}{12}, \dfrac{17}{13}, \dfrac{17}{14}$ **b** $\dfrac{17}{10}, \dfrac{17}{9}, \dfrac{17}{8}, \dfrac{17}{7}, \dfrac{17}{6}$

    i Work out if the fractions in the sequence are increasing or decreasing in value.

    ii Give a reason for your answer.

9 Arya says that $\dfrac{27}{10}$ is greater than $\dfrac{24}{5}$ because 27 is greater than 24.
Is she correct? Give a reason for your answer.

10 Ben says that $\dfrac{21}{15}$ is greater than $\dfrac{19}{15}$ because 21 is greater than 19.
Is he correct? Give a reason for your answer.

Solve problems

Vinny eats some chocolate every day after school. On Monday he eats $\dfrac{1}{4}$ of a bar of chocolate, on Tuesday he eats $\dfrac{2}{5}$ of a bar, on Wednesday he eats $\dfrac{3}{10}$ of a bar, on Thursday he eats $\dfrac{1}{20}$ of a bar and on Friday he eats $\dfrac{1}{4}$ of a bar. How many bars of chocolate did he eat throughout the week?
Give your answer as a mixed number in simplest form.

Add up the fractions he has eaten each day.

$\dfrac{1}{4} + \dfrac{2}{5} + \dfrac{3}{10} + \dfrac{1}{20} + \dfrac{1}{4}$ 20 is a common multiple of 4, 5, 10 and 20.

$\dfrac{5}{20} + \dfrac{8}{20} + \dfrac{6}{20} + \dfrac{1}{20} + \dfrac{5}{20} = \dfrac{25}{20}$

$\dfrac{25}{20} = 1\dfrac{5}{20} = 1\dfrac{1}{4}$

11 At a builders' merchant, sand is normally packed in 5 kg bags.
Three bags have fallen over and spilled their contents.

One bag is $\dfrac{3}{4}$ full. A second is $\dfrac{2}{3}$ full. A third is $\dfrac{5}{8}$ full.

Is there enough sand in these thee bags to fill two bags completely?
Give a reason for your answer.

12 Which of these fractions is nearer in value to 1: $\dfrac{7}{8}$ or $\dfrac{8}{7}$?
Show how you made your decision.

13 Write down a fraction with a value between $\dfrac{13}{6}$ and $\dfrac{13}{5}$.

14 Look at each sequence.

a $\dfrac{17}{10}, \dfrac{18}{11}, \dfrac{19}{12}, \dfrac{20}{13}, \dfrac{21}{14}$ **b** $\dfrac{17}{10}, \dfrac{16}{9}, \dfrac{15}{8}, \dfrac{14}{7}, \dfrac{13}{6},$

    i Work out if the fractions in the sequence are increasing or decreasing in value.

    ii Give a reason for your answer.

6.4 Adding and subtracting mixed numbers

- I can add two mixed numbers
- I can subtract one mixed number from another

When you add two mixed numbers you can add the whole-number parts and the fraction parts separately. Then you can combine the two parts to find your final answer.

You can subtract mixed numbers in a similar way.

Develop fluency

Add these mixed numbers.

a $3\frac{1}{2} + 1\frac{1}{4}$

a $3\frac{1}{2} + 1\frac{1}{4} = 3 + \frac{1}{2} + 1 + \frac{1}{4}$ Think of $3\frac{1}{2}$ as $3 + \frac{1}{2}$ and $1\frac{1}{4}$ as $1 + \frac{1}{4}$

$\qquad\qquad = 3 + 1 + \frac{1}{2} + \frac{1}{4}$ Add the whole numbers and the fractions separately.

$\qquad\qquad = 4\frac{3}{4}$

b $2\frac{3}{4} + 2\frac{5}{8}$

b $2\frac{3}{4} + 2\frac{5}{8} = 2 + \frac{3}{4} + 2 + \frac{5}{8}$

$\qquad\qquad = 4 + \frac{3}{4} + \frac{5}{8}$

Now $\frac{3}{4} + \frac{5}{8} = \frac{6}{8} + \frac{5}{8}$

$\qquad\qquad = \frac{11}{8}$

$\qquad\qquad = 1\frac{3}{8}$ This is what to do if the fractions add to more than 1.

So $2\frac{3}{4} + 2\frac{5}{8} = 4 + 1\frac{3}{8}$

$\qquad\qquad\qquad = 5\frac{3}{8}$

Subtract these mixed numbers.

a $3\frac{1}{2} - 1\frac{1}{4}$

a $3\frac{1}{2} - 1\frac{1}{4} = 3 + \frac{1}{2} - 1 - \frac{1}{4}$ Compare this with the example above.

$\qquad\qquad = 3 - 1 + \frac{1}{2} - \frac{1}{4}$ Subtract the whole numbers and the fractions separately.

$\qquad\qquad = 2\frac{1}{4}$ $3 - 1 = 2$ and $\frac{1}{2} - \frac{1}{4} = \frac{2}{4} - \frac{1}{4} = \frac{1}{4}$

b $4\frac{1}{4} - 2\frac{5}{8}$

b $4\frac{1}{4} - 2\frac{5}{8} = 4 + \frac{1}{4} - 2 - \frac{5}{8}$

$\qquad = 2 + \frac{1}{4} - \frac{5}{8}$ *Change $\frac{1}{4}$ to eighths.*

$\qquad = 2 + \frac{2}{8} - \frac{5}{8}$ *Watch out here, $\frac{2}{8} - \frac{5}{8} = -\frac{3}{8}$*

$\qquad = 2 - \frac{3}{8}$

$\qquad = 1\frac{5}{8}$

This means that $4\frac{1}{4} - 2\frac{5}{8} = 1\frac{5}{8}$

An alternative method, which avoids negative fractions, is to write both numbers as improper fractions.

$4\frac{1}{4} - 2\frac{5}{8} = \frac{17}{4} - \frac{21}{8}$ $4 \times 4 + 1 = 17$ and $2 \times 8 + 5 = 21$

$\qquad = \frac{34}{8} - \frac{21}{8}$ *Change the quarters to eighths.*

$\qquad = \frac{13}{8}$

$\qquad = 1\frac{5}{8}$

You can use either method.

Write your answers to questions 1–8 as mixed numbers in their simplest form.

1 Add these mixed numbers.

  a $1\frac{1}{3} + 2\frac{2}{3}$ **b** $2\frac{1}{6} + \frac{5}{6}$ **c** $4\frac{3}{10} + 2\frac{1}{10}$ **d** $3\frac{1}{9} + 5\frac{2}{9}$ **e** $\frac{1}{2} + 1\frac{1}{2}$

2 Add these mixed numbers.

  a $1\frac{2}{3} + 3\frac{1}{2}$ **b** $5\frac{1}{2} + \frac{2}{3}$ **c** $2\frac{1}{2} + 2\frac{7}{10}$ **d** $1\frac{7}{10} + 2\frac{3}{5}$ **e** $2\frac{3}{4} + \frac{1}{2}$

3 Subtract these mixed numbers.

  a $3\frac{5}{7} - 1\frac{2}{7}$ **b** $2\frac{5}{6} - \frac{1}{6}$ **c** $5\frac{9}{10} - 2\frac{3}{10}$ **d** $4\frac{8}{9} - 1\frac{2}{9}$

4 Subtract these mixed numbers.

  a $2\frac{3}{4} - 1\frac{3}{8}$ **b** $2\frac{1}{3} - 1\frac{1}{4}$ **c** $3\frac{3}{8} - 1\frac{1}{4}$ **d** $2\frac{7}{8} - 1\frac{3}{4}$ **e** $1\frac{5}{8} - \frac{1}{2}$

5 Add or subtract these mixed numbers.

  a $10\frac{1}{2} - 3\frac{3}{4}$ **b** $4\frac{3}{4} + 3\frac{5}{8}$ **c** $7\frac{1}{8} - 1\frac{5}{8}$ **d** $2\frac{2}{8} + 2\frac{2}{8}$ **e** $4\frac{1}{3} - 1\frac{1}{2}$

Reason mathematically

Amirah is putting together a large jigsaw puzzle. On Friday she works for $1\frac{1}{2}$ hours, on Saturday she works for $2\frac{1}{4}$ hours and on Sunday she works for $1\frac{3}{4}$ hours and completes it. How many hours has she spent working on the jigsaw in total?

Add up the mixed numbers.

$$1\frac{1}{2} + 2\frac{1}{4} + 1\frac{3}{4} = 1 + 2 + 1 + \frac{1}{2} + \frac{1}{4} + \frac{3}{4}$$

$$= 4 + \frac{1}{2} + \frac{1}{4} + \frac{3}{4}$$

Now $\frac{1}{2} + \frac{1}{4} + \frac{3}{4} = \frac{2}{4} + \frac{1}{4} + \frac{3}{4} = \frac{6}{4} = 1\frac{1}{2}$

So $1\frac{1}{2} + 2\frac{1}{4} + 1\frac{3}{4} = 4 + 1 + \frac{1}{2}$

$$= 5\frac{1}{2}$$

6 Sam watches two films. One lasts $1\frac{3}{4}$ hours and the other lasts $2\frac{1}{2}$ hours. How long do the two last altogether?

7 A flight is due to last $9\frac{1}{4}$ hours.

During the flight, Josie looks at her watch and sees that $3\frac{3}{4}$ hours have passed. How much longer is it before the flight ends?

8 Jason is carrying two bags. He has one of mass $3\frac{1}{2}$ kg in one hand and one of mass $2\frac{2}{5}$ kg in the other hand.
   a What is the total mass of the two bags altogether?
   b What is the difference in mass between the two bags?

Solve problems

Work out the perimeter of this trapezium.

Add the lengths of all the sides.

$$10\frac{3}{8} + 5\frac{2}{3} + 11 + 5\frac{2}{3} = 10 + 5 + 11 + 5 + \frac{3}{8} + \frac{2}{3} + \frac{2}{3}$$

$$= 31 + \frac{3}{8} + \frac{2}{3} + \frac{2}{3}$$

$$\frac{3}{8} + \frac{2}{3} + \frac{2}{3} = \frac{9}{24} + \frac{16}{24} + \frac{16}{24}$$

$$= \frac{41}{24} = 1\frac{17}{24}$$

So $10\frac{3}{8} + 5\frac{2}{3} + 11 + 5\frac{2}{3} = 31 + 1\frac{17}{24}$

$$= 32\frac{17}{24} \text{ cm}$$

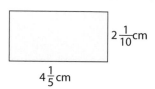

9   Work out the perimeter of this rectangle.

$4\frac{1}{5}$ cm $2\frac{1}{10}$ cm

10   Joe is tying two pieces of rope together. One piece is $1\frac{2}{5}$ m long and the other is $2\frac{3}{8}$ m long. When he ties them he loses $\frac{1}{2}$ m of length. How long is the combined length of the rope?

11   Katrina walks up a hill that is $500\frac{3}{4}$ m high then walks down $200\frac{3}{8}$ m. How high is she from the bottom of the hill?

12   Imran runs 4 laps of a running track. It takes him 14 minutes in total. His first lap takes him $2\frac{1}{2}$ minutes, the second lap takes $3\frac{1}{3}$ minutes, the third lap takes him $3\frac{5}{8}$ minutes. How long does it take him to run his fourth lap?

Now I can...

find equivalent fractions	convert mixed numbers to improper fractions	add and subtract fractions with different denominators
write fractions in their simplest form	subtract one mixed number from another	convert improper fractions to mixed numbers
add two mixed numbers		

CHAPTER 7 Algebraic expressions

7.1 Order of operations

- I can use the conventions of BIDMAS to carry out calculations

In mathematics, a calculation can be made up of various different **operations**, for example, adding, subtracting, multiplying or squaring. **BIDMAS** describes the **order of operations** – the order in which they must always be carried out.

Develop fluency

Circle the operation that you do first in each calculation. Then work out each one.

a $2 + 6 \div 2$ **b** $32 - 4 \times 5$ **c** $6 \div 3 - 1$ **d** $6 \div (3 - 1)$

a Division is done before addition, so you get $2 + \boxed{6 \div 2} = 2 + 3 = 5$.

b Multiplication is done before subtraction, so you get $32 - \boxed{4 \times 5} = 32 - 20 = 12$.

c Division is done before subtraction, so you get $\boxed{6 \div 3} - 1 = 2 - 1 = 1$.

d Brackets are done first, so you get $6 \div (\boxed{3 - 1}) = 6 \div 2 = 3$.

Calculate: **a** $8 - 3 - 2$ **b** $24 \div 6 \div 2$.

Work from left to right.

a $\boxed{8 - 3} - 2 = 5 - 2 = 3$ *b* $\boxed{24 \div 6} \div 2 = 4 \div 2 = 2$

1 Write down the operation that you do first in each calculation. Then complete the calculation.

    a $2 + 3 \times 6$ **b** $12 - 6 \div 3$ **c** $5 \times 5 + 2$ **d** $12 \div 4 - 2$

    e $(2 + 3) \times 6$ **f** $(12 - 3) \div 3$ **g** $5 \times (5 + 2)$ **h** $12 \div (4 - 2)$

2 Work these out, showing each step of the calculations.

    a $2 \times 3 + 4$ **b** $2 \times (3 + 4)$ **c** $2 + 3 \times 4$ **d** $(2 + 3) \times 4$

    e $4 \times 4 - 4$ **f** $5 + 3^2 + 6$ **g** $5 \times (3^2 + 6)$ **h** $3^2 - (5 - 2)$

For questions 3–8, work out the value of each expression.

3 a $(4 + 4) \div (4 + 4)$ **b** $(4 \times 4) \div (4 + 4)$ **c** $(4 + 4 + 4) \div 4$ **d** $4 \times (4 - 4) + 4$

    e $(4 \times 4 + 4) \div 4$ **f** $(4 + 4 + 4) \div 2$ **g** $4 + 4 - 4 \div 4$ **h** $(4 + 4) \times (4 \div 4)$

4 a $7 + 8 \times 3$ **b** $7 \times 3 - 2 \times 4$ **c** $3^2 + 4 \times 6$ **d** $54 \div 9 - 3$

5 a $3 \times 6 + 8 \times 2$ **b** $14 \div 2 - 10 \div 5$ **c** $(5 + 2) \times 4$ **d** $3 \times (5 + 6)$

6 a $3 \times (4 + 5)$ **b** $24 \div 4 + 4$ **c** $(6 + 2)^2$ **d** $6^2 + 2^2$

7 a $14 \div (9 - 2)$ **b** $(2 + 7) \div 3$ **c** $(11 - 6) \times 4$ **d** $6 - 2^2 \times 2$

8 a $22 - 2 \times (3^2 + 2)$ **b** $5 \times 6 - 3^2 \times 2$ **c** $(2^2 + 3) \times (2 + 3)$ **d** $(3 + 2)^2 + 3^2 + 2^2$

Reason mathematically

Work out each of these, showing each step of the calculation.

a $1 + 3^2 \times 4 - 2$ **b** $(1 + 3)^2 \times (4 - 2)$

a The order will be power, multiplication, addition, subtraction.
This gives:

$1 + \boxed{3^2} \times 4 - 2 = 1 + \boxed{9 \times 4} - 2 = \boxed{1 + 36} - 2 = 37 - 2 = 35$

b The order will be brackets (both of them), power, multiplication.
This gives:

$\boxed{(1 + 3)}^2 \times \boxed{(4 - 2)} = 4^2 \times 2 = 16 \times 2 = 32$

Put brackets into each equation to make the calculation true.

a $5 + 1 \times 4 = 24$ **b** $1 + 3^2 - 4 = 12$ **c** $24 \div 6 - 2 = 6$

Decide which operation has been done first.

a $(5 + 1) \times 4 = 24$ **b** $(1 + 3)^2 - 4 = 12$ **c** $24 \div (6 - 2) = 6$

9 Put brackets into each equation to make the calculation true.

a $2 \times 5 + 4 = 18$ **b** $2 + 6 \times 3 = 24$ **c** $2 + 3 \times 1 + 6 = 35$

d $5 + 2^2 \times 1 = 9$ **e** $3 + 2^2 = 25$ **f** $3 \times 4 + 3 + 7 = 28$

g $9 - 5 \times 2 = 8$ **h** $4 + 4 + 4 \div 2 = 6$ **i** $1 + 4^2 - 9 - 2 = 18$

10 One of the calculations $2 \times 3^2 = 36$ and $2 \times 3^2 = 18$ is wrong.
Which is it? How could you add brackets to make it true?

11 Beth orders six garden plants at £1.99 each. Delivery costs £2.50.
Write down the calculation you need to do to work out the total cost.
Work out the answer.

12 Jack wants to travel on three trams. He has €10.
The fares are €2.70, €4.20 and €2.70.

a Does he have enough money?

b Write down the calculation you need to do.

c Work out the answer.

Solve problems

Angharad buys 5 pens at £2 each and 2 reams of printer paper at £3 per ream.

Write down the calculation and work out the total cost of the items.

5×2 is the total cost of pens.

2×3 is the total cost of paper.

Total cost of all items is $5 \times 2 + 2 \times 3 = 10 + 6 = £16$.

13 Sarah is given three $10 notes and four $20 notes.
Write down the calculation you need to do to work out how much she is given altogether.
Then work out the answer.

14 Using only the numbers 2, 3, 4 and 5, write down a calculation to give an answer of 23.

15 Write the correct signs between the numbers to make the calculations correct.
   a $4 \square 2 \square 3 = 6$ **b** $2 \square 3 \square 4 = 20$

16 Using only four 4s and any operations, write an expression to give an answer of 20.

7.2 Expressions and substitution

- I can use algebra to write simple expressions
- I can substitute numbers into expressions to work out their value

In algebra, the letters or symbols used to represent numbers are **variables**, which can take different values. Variables and numbers can be combined to form **terms**, such as $2a$, b and 10. Terms can be combined to form **expressions** such as $2a + b + 10$. You can **substitute** different values for the variables, to work out the value of the expression.

Develop fluency

Write a term or expression to illustrate each sentence.
   a 6 more than w **b** 8 less than x **c** y multiplied by 4   **d** z multiplied by z

   a 6 more than w is written as $w + 6$.

   b 8 less than x is written as $x - 8$.

   c y multiplied by 4 is written as $4y$.

   b $z \times z$ is written as z^2.
   For example, $3^2 = 3 \times 3 = 9$ or $8^2 = 64$.

Work out the value of these expressions when $x = 8$ and $y = 4$.
   a $5x$ **b** $x + 3y$ **c** $x^2 - 2y$ **d** $\frac{x}{4}$ **e** $5(x + y)$

   a $5x$ means $5 \times x = 5 \times 8 = 40$.

   b $x + 3y$ means $8 + 3 \times 4 = 8 + 12 = 20$.

   c $x^2 - 2y$ means $8^2 - 2 \times 4 = 64 - 8 = 56$.

   d $\frac{x}{4}$ means $x \div 4 = 8 \div 4 = 2$.

   e Work out the value of the expression in brackets first.
   $5(x + y)$ means $5 \times (8 + 4) = 5 \times 12 = 60$.

1 Write terms, or expressions, to illustrate these sentences.
   a 4 more than m **b** t multiplied by 8   **c** y less that 9
   d m multiplied by itself   **e** n divided by 5   **f** t subtracted from seven
   g Multiply n by 3, then add 5.   **h** Multiply 6 by t.
   i Multiply m by 5, then subtract 5.   **j** Multiply x by x.

For questions 2–8, write down the values of each term for the three values of n.

2 **a** $4n$ where: **i** $n = 2$ **ii** $n = 5$ **iii** $n = 11$.

   b $\dfrac{n}{2}$ where: **i** $n = 6$ **ii** $n = 14$ **iii** $n = 8$.

   c n^2 where: **i** $n = 3$ **ii** $n = 6$ **iii** $n = 7$.

3 **a** $3n$ where: **i** $n = 7$ **ii** $n = 5$ **iii** $n = 9$.

   b $\dfrac{n}{5}$ where: **i** $n = 10$ **ii** $n = 5$ **iii** $n = 20$.

   c $3(n + 2)$ where: **i** $n = 2$ **ii** $n = 4$ **iii** $n = 8$.

4 **a** $n + 7$ where: **i** $n = 3$ **ii** $n = 4$ **iii** $n = -6$.

   b $n - 5$ where: **i** $n = 8$ **ii** $n = 14$ **iii** $n = 2$.

   c $10 - n$ where: **i** $n = 4$ **ii** $n = 7$ **iii** $n = -3$.

5 **a** $2n + 3$ where: **i** $n = 2$ **ii** $n = 5$ **iii** $n = 0$.

   b $5(n - 1)$ where: **i** $n = 3$ **ii** $n = 11$ **iii** $n = 1$.

   c $2(n + 8)$ where: **i** $n = 1$ **ii** $n = 5$ **iii** $n = 12$.

6 **a** $n^2 - 1$ where: **i** $n = 2$ **ii** $n = 3$ **iii** $n = 1$.

   b $n^2 + 1$ where: **i** $n = 5$ **ii** $n = 6$ **iii** $n = 10$.

   c $5 + n^2$ where: **i** $n = 8$ **ii** $n = 9$ **iii** $n = 0$.

7 **a** $\dfrac{n}{10}$ where: **i** $n = 20$ **ii** $n = 50$ **iii** $n = 100$.

   b $4n + 5$ where: **i** $n = 4$ **ii** $n = 3$ **iii** $n = 20$.

   c $n^3 - 1$ where: **i** $n = 2$ **ii** $n = 4$ **iii** $n = 1$.

8 **a** $3n - 3$ where: **i** $n = 2$ **ii** $n = 5$ **iii** $n = 0$.

   b $2(n + 3)$ where: **i** $n = 3$ **ii** $n = 11$ **iii** $n = 1$.

   c $5(n^2 - 4)$ where: **i** $n = 2$ **ii** $n = 3$ **iii** $n = 4$.

9 $v = u + at$

Use the values $u = 5$, $a = 2$ and $t = 3$ to work out the value of v.

10 $s = ut + 0.5at^2$

Use the values $u = 15$, $t = 2$ and $a = 10$ to work out the value of s.

Reason mathematically

The lengths of the sides of this quadrilateral are:
10 cm, a cm, a cm and b cm.

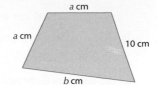

a Write down an expression for the perimeter of the quadrilateral.

b Work out the perimeter of the quadrilateral, given that
$a = 7.5$ and $b = 12$.

a The perimeter of the quadrilateral is the sum of the lengths of the sides.

Perimeter $= a + a + b + 10$

You can write $a + a$ as $2a$, so the perimeter of the quadrilateral is $2a + b + 10$ cm.

b Now substitute the values for the variables, a and b.

$2a + b + 10 = 2 \times 7.5 + 12 + 10 = 15 + 12 + 10 = 37$

The perimeter of the quadrilateral is 37 cm.

11 a Show that the perimeter of this shape is $a + 22$ cm.
b Work out the value of $a + 22$, given that $a = 8.5$.

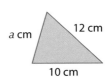

12 a Write down an expression for the perimeter of this triangle.
b Work out the value of your expression, if $p = 10.5$ and $q = 12.5$.

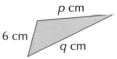

13 a Show that the perimeter of this parallelogram is $2(d + 14)$ cm.
b Work out the value of $2(d + 14)$, given that $d = 6$.

14 All the sides of this pentagon are s cm long.
a Write down an expression for the perimeter of the pentagon.
b Work out the value of your expression, given that $s = 25$.

Solve problems

In the rectangle ABCD, the side DC is 2 cm longer than the side AD.

a Write down an expression for the perimeter of the rectangle, using the variable x for the length of side AD.

b Write down an expression for the area of the rectangle, using the variable x for the length of side AD.

c Given that $x = 3$, calculate the perimeter and area of the rectangle.

a Perimeter $= 2x + 2(x + 2)$
$= 4x + 4$

b Area $= x(x + 2)$

c The perimeter is $4x + 4 = 4 \times 3 + 4 = 16$.
The area is $3(3 + 2) = 3 \times 5 = 15$.

15 The length of side AB of this triangle is x cm.

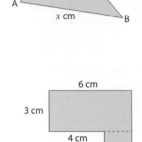

   a Side BC is 2 cm longer than side AB.
      Write down an expression for the length of BC.

   b Side AC is 1.5 cm shorter than side AB.
      Write down an expression for the length of AC.

   c Write an expression for the perimeter of the triangle.

   d Work out the value of your expression, given that $x = 20$.

16 The diagram shows an L-shape divided into two rectangles.

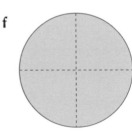

   a Show that the area of one of the rectangles is 18 cm².

   b Write down an expression for the area of the other rectangle.

   c Write down an expression for the area of the whole shape.

   d Show that the length of the missing side is $a + 3$ cm.

   e Work out an expression for the perimeter of the whole shape.

17 The areas of these two shapes are A cm² and B cm².
    Work out an expression for the area of each of these shapes.
    The first one has been done for you.

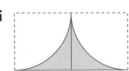

A cm²   B cm²

a

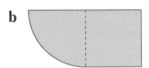

A + 2B cm²

b

c

d

e

f

g

h

i

18 Look back at the two shapes from question 17.
    Draw shapes that have these areas.

   a $3A$ cm²   **b** $3B$ cm²   **c** $3A + B$ cm²   **d** $2A + 4B$ cm²

7.3 Simplifying expressions

- I can simplify expressions

You can **simplify** the expression $2a + b + 3a + b$ by adding the terms in a and adding the terms in b separately.

$2a + 3a = 5a$ and $b + b = 2b$ Just add the coefficients of each variable.
Simplifying an expression in this way is called **collecting like terms**.

Develop fluency

Simplify each expression.

a $a + a + a$ **b** $3x + 7x$ **c** $9w - 4w$ **d** $2xy + 5xy - 3xy$

a $a + a + a$ simplifies to $3a$.

b $3x + 7x$ simplifies to $10x$.   Because $3 + 7 = 10$.

c $9w - 4w$ simplifies to $5w$.   Because $9 - 4 = 5$.

d $2xy + 5xy - 3xy$ simplifies to $4xy$.   Because $2 + 5 - 3 = 4$.

Simplify each expression.

a $3a + 5b + 6a - 2b$ **b** $x^2 + 4x + 7 - 7x + 3x^2 + 2$

a $3a + 5b + 6a - 2b = 3a + 6a + 5b - 2b$ Putting like terms together.
                $= 9a + 3b$ Simplifying.

    You cannot make this any simpler because $9a$ and $3b$ contain different variables.

b $x^2 + 4x + 7 - 7x + 3x^2 + 2$ simplifies to $4x^2 - 3x + 9$.

    Note that in the term $-3x$, the coefficient of x is -3.

Simplify each expression.

1 a $a + a$ **b** $b + 3b$ **c** $9c - c$ **d** $d + d + 3d$
    e $4c + 2c$ **f** $6d + 4d$ **g** $7p - 5p$ **h** $2x + 6x + 3x$

2 a $x + 2x - x$ **b** $8y - y$ **c** $z + z - z$ **d** $4q + q - q$

3 a $4t + 2t - t$ **b** $7m - 3m$ **c** $q + 5q - 2q$ **d** $5g - g - 2g$

4 a $2x + 2y + 3x + 6y$ **b** $4w + 6t - 2w - 2t$ **c** $4m + 7n + 3m - n$
    d $4x + 8y - 2y - 3x$ **e** $8 + 4x - 3 + 2x$ **f** $8p + 9 - 3p - 4$

5 a $2y + 4x - 3 + x - y$ **b** $5d + 8c - 4c + 7$ **c** $4f + 2 + 3d - 1 - 3f$

6 a $5x + 3y + 2x - 6y$ **b** $9x + 5y - 4x - 9y$ **c** $7a + 7b - 8a - b$
    d $5x - 8y - 7x - 5y$ **e** $10 + 5w - 13 + 7w$ **f** $11x + 4 - 12x - 8$

7 a $10x + 9 - 17x - 14$ **b** $12x - y + 4x - 8y$ **c** $7t - 8 - 6t - 6$

8 **a** $x^2 + x^2$ **b** $x^2 + 3x^2$ **c** $5x^2 - x^2$ **d** $x^2 + x^2 + x^2$
e $4x^2 + 2x^2$ **f** $4x^2 + 2x^2$ **g** $9x^2 - 5x^2$ **h** $x^2 + 3x^2 + 5x^2$

9 **a** $2x^2 + x^2 + 3$ **b** $4x^2 + 4 + 3x^2$ **c** $3x^2 - 4 - x^2$ **d** $5x^2 + 2x^2 - 7$
e $2x^2 + 3 + 2x^2 + 5$ **f** $3x^2 - 7 + 5 + 2x^2$
g $4x^2 + 2 - x^2 - 5$ **h** $6 + x^2 + 2 + 3x^2 + 8 + 5x^2$

10 **a** $3x^2 + x^2 + y$ **b** $x^2 + y + 5x^2$ **c** $2x^2 - 3y - x^2 + y$
d $6x^2 + 2x^2 + 5y^2 + 2y^2$ **e** $4x^2 + 2x + 3x + 7$ **f** $7x^2 - 6x + 3x + 5$
g $2y^2 + 2y^2 - 7y - 5y + 3$ **h** $4x^2 + 2x^2 + 5 + 3x + 6 + 4x$

Reason mathematically

Work out the missing numbers.

a $2x + y + \square x + 6y = 8x + \square y$ **b** $5x + 7y - 4x - 9y = \square x - \square y$

a $2x + y + 6x + 6y = 8x + 7y$
Check: coefficients of x give $2 + 6 = 8$, coefficients of y give $1 + 6 = 7$.
b $5x + 7y - 4x - 9y = x - 2y$
Check: coefficients of x give $5 - 4 = 1$, coefficients of y give $7 - 9 = -2$.

Calculate the perimeter of this regular pentagon.
There are five sides, each of length $2x - 3$.
$5 \times 2x + 5 \times -3 = 10x - 15$

$2x - 3$

11 Work out the missing coefficients.
a $2x + 8y + \square x + 6y = 6x + \square y$ **b** $5w - 5t - \square w - \square t = w - 9t$
c $8p + \square x - \square p + 2x = p + 7x$ **d** $8p + 9q - \square p - \square q = 5p - q$

12 Work out the missing coefficients.
a $3x + 5y + \square x + 2y = 7x + \square y$ **b** $8w - 4z - \square w - \square z = w - 9z$
c $12y + \square x - \square y + 2x = y + 5x$ **d** $9p + 5q - \square p - \square q = 3p - q$

13 Write an expression for the perimeter of each shape.

a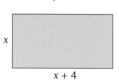
x
$x + 4$

b
$x + 3$

c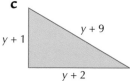
$y + 1$
$y + 9$
$y + 2$

14 Bethan and Arabella simplify this expression.
$2x^2 + 3x + 2y + 4y^2$
Arabella gets an answer of $5x^2 + 6y^2$.
Bethan says it can't be simplified.
Who is correct? Explain your answer.

Solve problems

In this number wall, the number in each brick is the sum of the numbers in the two bricks below it. Find the number missing from the top brick.

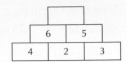

$4 + 2 = 6$ and $2 + 3 = 5$

The missing top number is 11.

15 Work out the top number or expression in each of these number walls. Write any expressions as simply as possible.

a

| 2 | 6 | 5 |

b

| a + 6 |
| a | 6 | 5 |

c

| 2 | 6 | c |

d

| 2 | d | 5 |

e

| x | 4 | x |

f

| x | x | 9 |

g

| y | 7 | y |

h

| n | n | n |

16 In each of these number walls, there is more than one variable. Work out the top expression. Write it as simply as possible.

a

| a | 8 | b |

b

| c | d | 9 |

c

| 12 | e | f |

d

| x | y | z |

17 A rectangle has a perimeter of $8x + 12$. Write down an expression for the length and width of the rectangle.

18 Work out the missing expressions in each of these number walls.

a

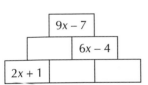

| 9x − 7 |
| | 6x − 4 |
| 2x + 1 | | |

b

| 15x + 12y |
| 10x + 5y | |
| | 4x | |

7.4 Using formulae

● I can use formulae

A **formula** (plural **formulae**) is a mathematical rule connecting variables. It may be written in words or using letters.

Develop fluency

Use the formula $c = 200 + 6n$ to calculate the cost of hiring a hall for a wedding attended by 70 people, where c is the cost and n is the number of people.

$c = 200 + 6 \times 70$
$\quad = 200 + 420$
$\quad = £620$
The cost is £620.

The formula could also be written as $c = 6n + 200$.
Then the calculation is:
$c = 6 \times 70 + 200$
$\quad = 420 + 200$
$\quad = 620$
The result is the same as before.

1 A cleaner uses the formula:
$c = 5 + 8h$
where c is the charge (in £) and h is the number of hours worked.
Calculate what the cleaner charges to work for:
   a 5 hours   **b** 3 hours   **c** 8 hours.

2 A mechanic uses the formula:
$c = 25t + 20$
where c is the charge (in £) and t is the time, in hours, to complete the work.
Calculate what the mechanic charges to complete the work in:
   a 1 hour   **b** 3 hours   **c** 7 hours.

3 A singer uses the formula:
$c = 8 + 5s$
where c is the charge (in £) and s is the number of songs sung.
Calculate what the singer charges to sing:
   a 2 songs   **b** 4 songs   **c** 8 songs.

4 The formula for the average speed of a car is:
$S = \dfrac{D}{T}$ where S is the average speed (in miles per hour)
D is the distance travelled (in miles) and
T is the time taken (in hours).
Use the formula to calculate the average speed for each journey below.
   a 300 miles in 6 hours   **b** 200 miles in 5 hours   **c** 120 miles in 3 hours

5 The formula for the cost of a newspaper advert is:
$C = 5W + 10A$
where C is the charge (in £)
W is the number of words used and
A is the area of the advert (in cm^2).
Use the formula to calculate the charge for the following adverts.
   a 10 words with an area of 20 cm^2 **b** 8 words with an area of 6 cm^2
   c 12 words with an area of 15 cm^2 **d** 17 words with an area of 15 cm^2

6 The formula for the perimeter, P cm, of a rectangle is:

$P = 2(l + w)$

where l is the length (in centimetres) and

w is the width (in centimetres).

   a Work out the perimeter of a rectangle measuring 5 cm by 9 cm.

   b Work out the perimeter of a rectangle measuring 14 cm by 12 cm.

7 The formula for calculating resistance, R ohms, in an electrical circuit is:

$R = \dfrac{V}{I}$

where V = voltage in volts and I = current in amps.

   a Calculate the resistance when $V = 12$ and $I = 4$.

   b Calculate the resistance when $V = 15$ and $I = 1.5$.

8 The formula to calculate the cost, £C, of a household water bill is:

$C = 0.3v + 0.1d$

where v = volume of water used and d = charge per day.

   a Calculate the cost of using 50 m³ of water over a 20-day period.

   b Calculate the cost of using 100 m³ of water over a 50-day period.

9 The formula for converting temperatures from Celsius degrees (°C) into Fahrenheit degrees (°F) is:

$F = \dfrac{9}{5} C + 32$

   a Convert 20 °C into Fahrenheit.   b Convert –10 °C into Fahrenheit.

10 The formula $v = u + at$ is used to calculate the velocity, v, of an object after t seconds, when it has a starting velocity u and acceleration a.

Find the value of v when: a $u = 10$, $a = 2$ and $t = 3$ b $u = 18$, $a = 1.5$ and $t = 10$.

Reason mathematically

The formula to calculate the sum, S, of the interior angles in any polygon is:

$S = 180(n - 2)$ where n is the number of sides.

Calculate S for an octagon.

An octagon has 8 sides.

$S = 180(8 - 2)$

$S = 180 \times 6$

$S = 1080°$

11 The diagram shows a regular five-sided polygon and a
    regular six-sided polygon.
    All the angles of a regular polygon are the same size.
    For a regular polygon with n sides, the size of the angle
    marked $a°$ is given by the formula:

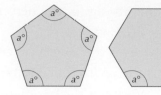

$A = 180 - \dfrac{360}{n}$ In this formula, $\dfrac{360}{n}$ means $360 \div n$.

a Show that the angle of a regular polygon with 5 sides is 108°.

b Work out the angle of a regular polygon with 6 sides.

c Work out the angle of a regular polygon with 8 sides.

d Work out the angle of a regular polygon with 10 sides.

12 The formula to calculate the number of faces, F, on a 3D shape is:
    $F = E - V + 2$
    where E is the number of edges and V is the number of vertices (corners).
    Use the formula to show that:

    a a cube has 6 faces   b a pyramid has 5 faces.

13 A formula to calculate the total resistance, R, in an electrical circuit is:
    $$R = \dfrac{R_1 R_2}{R_1 + R_2}$$
    where R_1 and R_2 are the values of two resistors in the circuit in ohms Ω.
    Calculate the resistance in a circuit when:

    a $R_1 = 100\,\Omega$ and $R_2 = 60\,\Omega$ b $R_1 = 250\,\Omega$ and $R_2 = 50\,\Omega$.

14 Another version of the formula to calculate the total resistance, R, in an electrical
    circuit is:
    $$\dfrac{1}{R} = \dfrac{1}{R_1} + \dfrac{1}{R_2}$$
    where R_1 and R_2 are the values of two resistors in the circuit in ohms Ω.
    Calculate the resistance in a circuit when:

    a $R1 = 12\,\Omega$ and $R2 = 4\,\Omega$ b $R1 = 3\,\Omega$ and $R2 = 6\,\Omega$.

Solve problems

The rectangle below has a perimeter of 28 cm.

a Write down a formula for the perimeter of the shape.

b Work out the length and width of the rectangle.

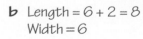

a $2x + 2(x + 2) = 28$ b Length $= 6 + 2 = 8$
   which simplifies to $4x + 4 = 28$ Width $= 6$

   Solving: $4x = 24$
             $x = 6$

15 A rectangle has an area of $30\,\text{cm}^2$ and a perimeter of $22\,\text{cm}$.

    a Write down a formula for area, A, and perimeter, P, in terms of length, L, and width, W.

    b Work out the length and width of the rectangle.

16 The perimeter of this isosceles triangle is P.

    a Write a formula to calculate side x in terms of P and y.

    b Given that $x = 7$ and $P = 25$, use your formula to calculate the value of y.

17 Ivor Flex, the electrician, charges £A for a call-out charge and then £B per hour when he does a job.

He uses the formula: $C = A + B \times (\text{number of hours})$

He does a job for Mrs Smith that takes 3 hours and charges her £118.

He does a job for Mr Khan that takes 4 hours and charges him £150.

    a How much extra did the fourth hour cost Mr Khan?

    b Work out the values of A and B.

18 **a** Write a formula for the perimeter, P, of this shape in terms of x.

       A square has the same perimeter as the shape shown.

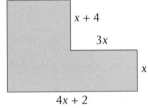

    b Write down an expression for one side of the square in terms of x.

    c Write down a formula for the area, A, of the square in terms of x.

    d Given that $x = 2$, calculate the area of the square.

7.5 Writing formulae

● I can write formulae

Sometimes you will need to create your own formula to represent a statement or to solve a problem.

Write down a formula for the sum, S, of three consecutive whole numbers, starting from n.

Consecutive whole numbers go up in ones, so they can be represented by n, $n + 1$ and $n + 2$.

So, the formula is $S = n + n + 1 + n + 2$

$S = 3n + 3$

I think of a number, n, multiply it by 4 and subtract 3. Write down a formula for the answer, A.

Multiplying n by 4 gives: $4 \times n = 4n$

Subtract 3 from $4n$ to give: $A = 4n - 3$

1 Write down a formula for the product, P, of two consecutive numbers.

2 Write down a formula for the product, P, of two numbers a and b.

3 Write down a formula for the sum, S, of two consecutive numbers.

4 Write down a formula for the number of weeks, W, in Y years.

5 Write down a formula for the difference, d, between the ages of Bert, who is b years old, and his little brother Ernie, who is e years old.

6 Write down a formula for the total mark, M, scored in an examination with two papers, when you score X marks on paper 1 and Y marks on paper 2.

7 I think of a number, n, double it and add 3. Write down a formula for the answer, A.

8 I think of a number, n, triple it and subtract 5. Write down a formula for the answer, A.

9 Write down a formula for the total cost, £C, of P pens priced at 25p each.

Reason mathematically

Look at this number wall.

a Work out a formula for the top number, t, in terms of r and s.

b Work out the value of t, when r is 15 and s is 25.

c Work out the value of t, when r and s add up to 30.

a First complete the number wall.

$t = r + 12 + s + 12$

$\quad = r + s + 24$

b When $r = 15$ and $s = 25$, t is $15 + 25 + 24 = 64$.

c When $r + s = 30$, t is $30 + 24 = 54$.

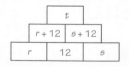

10 Look at this number wall

a Work out a formula for the top number, r, in terms of p and q.

b Work out the value of r, when p is 5 and q is 15.

c Work out the value of r, when p is –2 and q is 4.

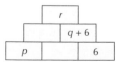

11 A boy is *b* years old. His mother is *m* years old.

   a Write down a formula for the difference, *d*, between their ages.

   b Write down a formula for the sum, *s*, of their ages.

   c Work out the value of *s* when $b = 9$ and $m = 41$.

   d Write down an expression for the boy's age in six years' time.

   e Write down an expression for his mother's age in six years' time.

   f Work out a formula for the sum, *t*, of their ages in six years' time.

12 Look at this number wall.

   a Work out a formula for the top number, *t*, in terms of *c* and *d*.

   b Work out the value of *t*, when *c* is 10 and *d* is 15.

   c Work out the value of *t*, when *c* is 15 and *d* is 10.

13 This number wall has four layers.

   a Work out a formula for the top number, *t*, in terms of *a* and *b*.

   b Work out the value of *t*, when *a* is 8 and *b* is 2.

   c Work out the value of *t*, when *a* is 2 and *b* is 8.

   d Work out the value of *t*, when *a* and *b* are both 5.

Solve problems

A mechanic charges £15 per hour plus a call-out fee of £30.

A job lasts *h* hours and the cost is £*c*.

Write down a formula for *c* in terms of *h*.

The cost is:
number of hours × 15 + call-out fee of £30.
The formula is: $c = 15h + 30$

14 For each case below, there are two variables. Write a formula to connect them. The first one has been done for you.

   a It costs £6 per hour to hire a rowing boat.
Pete rents a boat for *h* hours.
The cost is £*c*.
A formula for *c* in terms of *h* is $c = 6h$.

   b Petrol costs £1.40 per litre.
Sally buys *b* litres and the cost is £*a*.
Write a formula for *a* in terms of *b*.

   c There are seven days in a week.
Building a house takes *w* weeks or *d* days.
Write a formula for *w* in terms of *d*.

   d It takes 5 minutes to answer each question in an examination.
Jack answers *q* questions.
It takes him *m* minutes.
Write a formula for *m* in terms of *q*.

   e Apples cost 20p each. Oranges cost 30p each.
Alice buys *x* apples and *y* oranges.
She pays *t* pence altogether.
Write a formula for *t* in terms of *x* and *y*.

   f The time to cook a chicken is 30 minutes plus 20 minutes per kilogram.
The mass of a chicken is *x* kg.
It takes *m* minutes to cook.
Write a formula for *m* in terms of *x*.

15 a Work out the lengths of the sides of this rectangle, when $n = 3$.

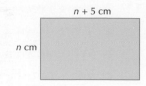

  b Work out the perimeter of the rectangle, when $n = 13$.

  c Work out a formula for the perimeter, p cm, of the rectangle.

  d Work out the value of p, when $n = 30$.

16 This is a quadrilateral.

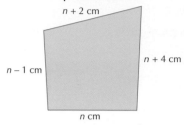

  a Work out the lengths of the four sides of the quadrilateral, when $n = 13.5$.

  b Work out a formula for the perimeter, p cm, of the quadrilateral. Write your formula as simply as possible.

  c Work out the value of p, when $n = 50$.

17 I start with the number 4. I multiply it by x and then add y. The result is r.

  a Write down a formula for r.

  b Work out the value of r, given that $x = 8$ and $y = 5$.

  c Work out the value of r, given that $x = 6$ and $y = 13$.

  d Work out the value of r, given that $x = 13$ and $y = 6$.

Now I can...

use the conventions of BIDMAS to carry out calculations	use formulae	substitute numbers into expressions to work out their value
use algebra to write simple expressions	write formulae	
simplify expressions		

CHAPTER 8 Angles

8.1 Calculating angles

- I can calculate **angles at a point**
- I can calculate **angles on a straight line**
- I can calculate opposite angles

You can **calculate** unknown angles, usually denoted by letters such as a, b, c, in a diagram from the information given. Use the facts that:

- **angles at a point** add up to 360°
- **angles on a straight line** add up to 180°
- **opposite angles** (or **vertically opposite angles**) are equal
- a **right angle** is 90° and is marked by a square symbol.

Remember that diagrams are not usually drawn to scale.

Develop fluency

Calculate the size of the angle marked x.

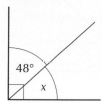

$x = 90 - 48$

$\quad = 42°$

Note that the square symbol indicates that the angle is 90°.

Calculate the size of the angle marked a.

The three angles add up to 360° so:

$a = 360 - 150 - 130$

$\quad = 80°$

Calculate the size of the angle marked b.

The two angles on a straight line add up to 180° so:

$b = 180 - 155$

$\quad = 25°$

1 Work out the size of each unknown angle.

a **b** **c** **d**

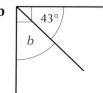

2 Calculate the size of each unknown angle.

a **b** **c** **d**

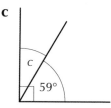

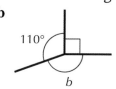

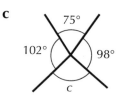

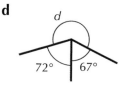

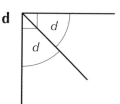

3   Calculate the size of each angle marked with a letter.

a

130° a

b

50° b

c
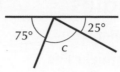
75° 25° c

d

d 49°

e
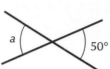
a 50°

f

25° c

g

e d 37°

h

39° a

i

b 51°

j
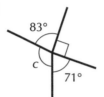
83° c 71°

k
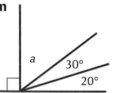
50° a a

l

c c

m
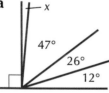
a 30° 20°

n
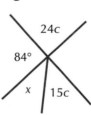
b 25°

o
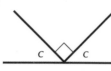
3c c

4   Calculate the size of the angle labelled x in each diagram.

a
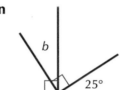
x 47° 26° 12°

b
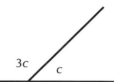
19° x 16°

c
24c 84° x 15c

d
d + 15° x 115°

5   Calculate the values of x and y in each diagram.

a
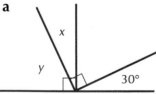
x y 30°

b
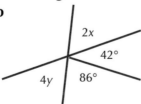
2x 42° 4y 86°

6 Calculate the value of *x* in each of these diagrams.

a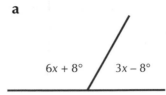

$6x + 8°$ $3x - 8°$

b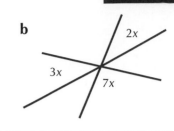

$2x$ $3x$ $7x$

Reason mathematically

Calculate the sizes of the angles marked *d* and *e*.

Give reasons for your answers.

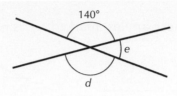

140°   *e*   *d*

$d = 140°$ *Opposite angles are equal*

Angles on a straight line sum to 180°, so:

$e = 180 - 140$

$ = 40°$

7 Is the line segment *XY* a straight line? Give reasons for your answer.

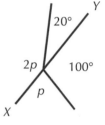

Y   20° $2p$ 100°   *p*   X

8 Choose three of these angles that would fit together to make a straight line.

 35°   45°   55°   60°   85°

9 In the diagram, AXB is a straight line.

  a Find the value of *m* when *n* is 165°.

  b Find the sizes of both angles when *n* is five times *m*.

  c Given that *m* is one quarter of the whole angle on the line, find the sizes of both angles.

  d Given that angle *n* is twice angle *m*, what are both angles?

  e Look at the sequence of answers for *m*. Which times table does it represent?

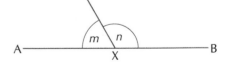

Y   *m*   *n*   A ———————— B   X

10 Calculate the size of each unknown angle. Explain your answers.

a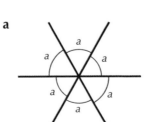

a   *a*   *a*   *a*   *a*   *a*

b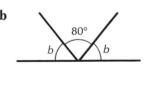

80°   *b*   *b*

Solve problems

XY is a straight line. Calculate the value of angle b.

XY is a straight line so $3a + 78 = 180$

$3a = 180 - 78 = 102$

$a = 102 \div 3 = 34$

$b = 180 - 34 = 146°$

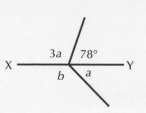

11 Which of these angles, marked c–f, is not a multiple of 10?

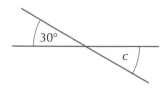 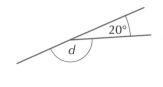 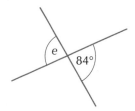 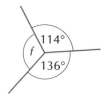

12 In the diagram, x and y lie on a straight line and the value
of y is 10° more than the value of x.
Work out the values of a pair of angles to make this statement true.

13 In the diagram: $b = 2a$, $c = 2b$ and $d = 2c$.
What is the value of a.

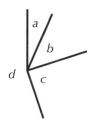

14 Alice is wondering whether there is a set of angles that are six
consecutive multiples of an integer that will fit on a straight
line. She draws this diagram as a starting point.
In the diagram, $180 = 6x + 5x + 4x + 3x + 2x + x$
$180 = 21x$

$x = \dfrac{60}{7}$ which is not an integer.

Alice now tries $180 = 7x + 6x + 5x + 4x + 3x + 2x$

$180 = 27x$

$x = \dfrac{20}{3}$ which is also not an integer.

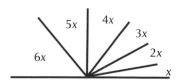

Alice decides that there must not be any six consecutive multiples of x where x is an
integer. She is wrong. Why?

8.2 Angles in a triangle

● I can use the fact that the sum of the angles in a triangle is 180°

In any **triangle**, the sum of the angles is 180°.
In an **isosceles** triangle, the angles opposite the equal sides are equal.

Develop fluency

Calculate the size of the angle marked c.

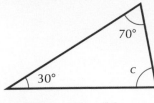

$c = 180 - 70 - 30$

$c = 80°$

Calculate the sizes of the unknown angles in this triangle.

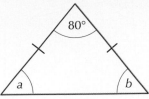

$180 - 80 = 100$

The angles labelled a and b are the same size because the triangle is isosceles.

$100 \div 2 = 50$

So $a = b = 50°$

1 Calculate the size of the unknown angle in each triangle.

a 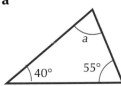 **b** 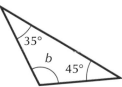 **c**   **d**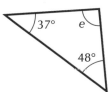

2 Calculate the size of the unknown angle in each triangle.

a 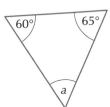 **b** 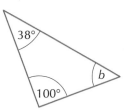 **c**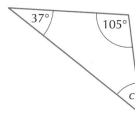

d 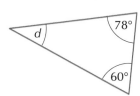 **e** 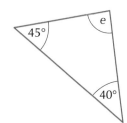 **f**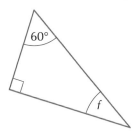

3 Calculate the sizes of the unknown angles in each isosceles triangle.

a

b
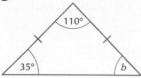

c
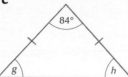

4 Calculate the sizes of the unknown angles in each diagram.

a
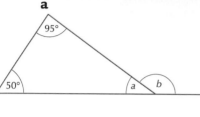

b
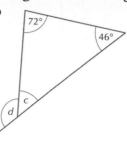

c
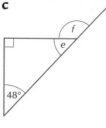

5 Calculate the size of the unknown angles in each diagram.

a
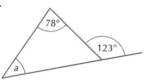

b
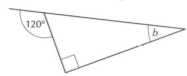

c

6 Calculate the size of the angle marked with a letter in each of these diagrams.

a

b
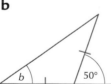

c

d

7 Calculate the values of x and y in each of these diagrams.

a

b

8 Calculate the angle marked with a letter in each of these diagrams.

a 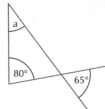   **b** 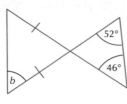   **c**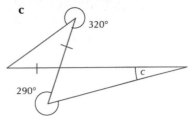

9 Calculate the size of angle *x* in these diagrams.

a 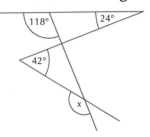   **b**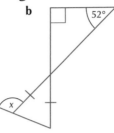

10 Calculate the values of *x* and *y* in each of these diagrams.

a 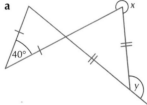   **b**

Reason mathematically

Calculate the sizes of the angles marked *a* and *b*.

Give reasons for your answers.

$a = 180 - 75 - 40$

So, $a = 65°$ Sum of angles in a triangle = 180°

$b = 180 - 65$

So, $b = 115°$ Sum of angles on a line = 180°

11 Calculate the size of the angles, *a* and *b*, in this diagram.

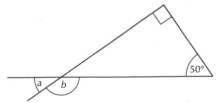

12 One angle in an isosceles triangle is 42°.
Calculate the possible sizes of the other two angles.
Use diagrams to explain your answer.

13 In this triangle, the value of *b* is 10° more than the value of *a*.

The value of *c* is 10° more than the value of *b*.
What are the sizes of the three angles?

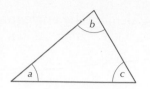

14 A bridge is to be constructed according to the plan in the diagram. Use your knowledge of angles and symmetry to work out the value of the lettered angles.

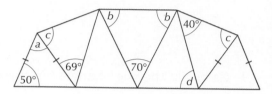

Solve problems

Calculate the size of the unknown angles in this diagram.

The angles in a triangle add up 180°, so:

$180 = 2x + 3x + 40$

$180 - 40 = 2x + 3x$

$140 = 5x$

$x = 140 \div 5$

$x = 28$

So, the unknown angles in the triangle are $2 \times 28 = 56°$ and $3 \times 28 = 84°$.

The angle on a straight line is 180°, therefore

$180 = y + 84$

$y = 180 - 84 = 96°$

15 Calculate the size of the unknown angle in each diagram.
Give reasons for your answers.

a

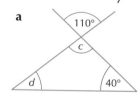

b

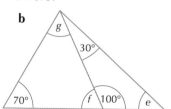

16 Calculate the size of each unknown angle.

a

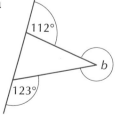

b

c

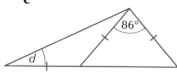

17 Calculate the size of the unknown angles in each of these diagrams.

a

b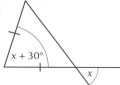

18 a Calculate the size of the unknown angles in this triangle.

 b A triangle has angles of $(3x - 22)°$, $(2x + 2)°$ and $(4x - 16)°$.
 Is this triangle isosceles? Give reasons for your answer.

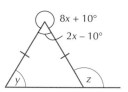

8.3 Angles in a quadrilateral

● I can use the fact that the sum of the angles in a quadrilateral is 360°

The sum of the angles in any **quadrilateral** is 360°.

Develop fluency

Calculate the sizes of the angles marked *a* and *b* on the diagram.

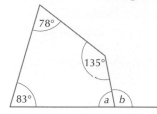

The angles in a quadrilateral add up to 360°.

So $a = 360 - 135 - 78 - 83 = 64°$

The angles *a* and *b* on the straight line add up to 180°.

So, $b = 180 - 64 = 116°$

Calculate the size of the angles marked *m* on the diagram.

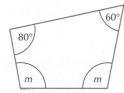

The angles in a quadrilateral add up to 360°.

So $360 = m + m + 80 + 60$

$360 = 2m + 140$

$360 - 140 = 2m$

$220 = 2m$

$220 ÷ 2 = m$

$m = 110°$

1 Calculate the size of each angle marked by a letter in each quadrilateral.

a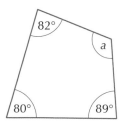

b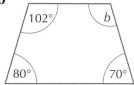

c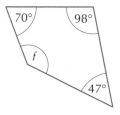

2 Calculate the size of the angle marked by a letter in each quadrilateral.

a 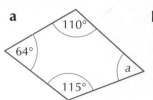 **b** 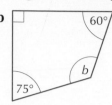 **c** 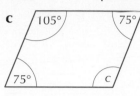 **d**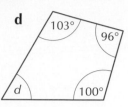

3 Calculate the size of the angle marked by a letter in each quadrilateral.

a  **b**  **c**

4 Calculate the size of the angle marked by a letter in each quadrilateral.

a  **b**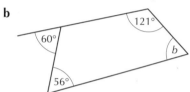

5 Which of these sets of four angles could be the interior angles of a quadrilateral?
a 97°, 43°, 114°, 106° **b** 68°, 126°, 97°, 79° **c** 72°, 113°, 91°, 84°

6 Calculate the size of the unknown angle marked by a letter in each quadrilateral.

a 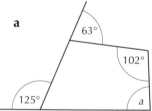 **b** 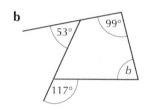 **c**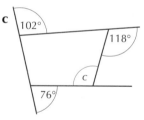

7 Calculate the size of the unknown angle marked by a letter in each quadrilateral.

a 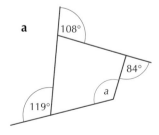 **b** 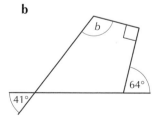 **c**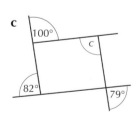

8 Calculate the size of the unknown angles marked by a letter in each quadrilateral.

a 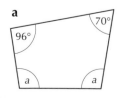 **b** 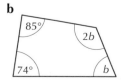 **c**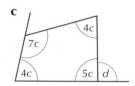

9 Calculate the size of each unknown angle.

a 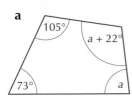   **b** 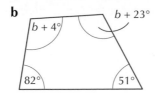   **c**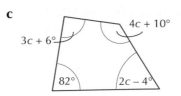

10 Sven says that these two quadrilaterals are similar because the angles in them match.
Is he correct? Give a reason for your answer.

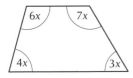   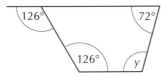

Reason mathematically

Ivan says that this quadrilateral is a rectangle. Is he correct?
Explain your answer.

Triangles BEC, CED, DEA and AEB are all isosceles.

$\angle EBC = \angle ECB = (180 - 50) \div 2 = 65°$

$\angle EAD = \angle EDA = (180 - 50) \div 2 = 65°$

The angle on a straight line is 180°, so $\angle BEA = 180 - 50 = 130°$

$\angle EAB = \angle EBA = (180 - 130) \div 2 = 25°$

$\angle EDC = \angle ECD = (180 - 130) \div 2 = 25°$

So, $\angle DAB$, $\angle ABC$, $\angle BCD$, $\angle CDA$, $= 25 + 65 = 90°$, meaning all the angles of the shape ABCD are 90°.

So, $4 \times 90 = 360°$, meaning the shape is a quadrilateral and only two types of quadrilateral have four angles of 90° squares and rectangles.

It is not a square because the diagonals are not perpendicular to each other, so the shape must be a rectangle.

11 Paula looks at the rectangle in the worked example above.
She says, 'The shape has four triangles drawn in it. There are 180° in each triangle, so the total of its interior angles must be $4 \times 180 = 720°$.'
What mistake is Paula making? Explain your answer.

12 Calculate the size of the angle marked by a letter in each quadrilateral.
Give a reason for each of your answers.

a 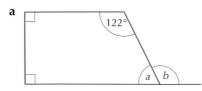   **b**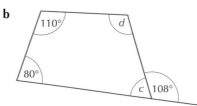

13 The diagram shows a kite.

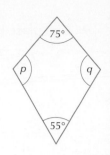

   a Make a sketch of the kite and draw on it a line that divides it into two equal parts (its line of symmetry).

   b What does this tell you about the angles marked p and q?

   c Use this information to work out the size of the angles marked p and q.

14 Can a quadrilateral have exactly three right angles?
Give a reason for your answer.

Solve problems

This is the cross-section of a length of wood for a boat.

$\angle ADC = \angle DCB = 110°$

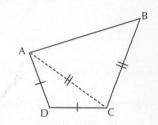

   a What is the size of $\angle ABC$?

The designer decides that $\angle ABC$ must be equal to half $\angle ADC$.
$\angle ACB$ will need to be changed.

   b Should $\angle ACB$ be made bigger or smaller?

   c What size will $\angle ACB$ need to be?

   a Angles in a triangle add up to 180° and triangle ADC is isosceles.

     So, $\angle ACD = (180 - 110) \div 2 = 35°$

     $\angle BCA = 110 - 35 = 75°$

     Triangle ABC is isosceles, so $\angle ABC = (180 - 75) \div 2 = 52.5°$

   b $\angle ABC$ should be 55° so $\angle ACB$ should be made smaller so that $\angle ABC$ will become bigger.

   c $\angle ABC$ and $\angle BAC$ should be 55°, so $\angle ACB$ should be $180 - 55 - 55 = 70°$

15 A quadrilateral can always be made from two triangles. What special quadrilaterals are created when you join these pairs of identical triangles?

   a two equilateral triangles   **b** two right-angled isosceles triangles

   c two right-angled scalene triangles   **d** two obtuse-angled scalene triangles

16 The diagram shows a yard surrounded by buildings with a window between the two dashes.

   a How many acute, obtuse and reflex angles are there?

   b Shade in the part of the yard that cannot be seen from the window.

   c Given that the two smallest angles in the yard are equal, work out the size of the angles in the shaded area.

17 Medhi is given some measurements for a quadrilateral shaped window frame for a car. The angles are: 95°, 100°, 97° and 86°.

  a Medhi tells the designer that something is wrong. Why?

The designer tells Medhi that one of the angles has had its digits reversed when writing it down.

  b Which angle has been written incorrectly. What should it be?

18 A jeweller has a number of quadrilateral shaped diamonds as shown.

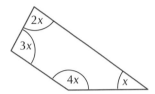

  a What are the angles of this shape?

The jeweller is going to put some of her diamonds together to form a circular design, as in this example, where they meet at the centre with the same angle.

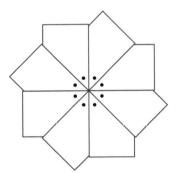

  b What is the smallest number of her diamonds she can use? Explain your answer.

8.4 Angles within parallel lines

- I can calculate angles in parallel lines

When a **transversal** intersects a set of parallel lines several different sets of equal angles may be formed.

The two angles marked on this diagram add up to 180° and are **allied angles**.

The two angles marked on this diagram are equal and are **corresponding angles**.

The two angles marked on this diagram are equal and are **alternate angles**.

Develop fluency

Look at this diagram.

a Name pairs of angles that are alternate angles.

b Name pairs of angles that are corresponding angles.

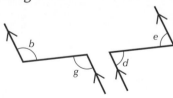

a There are two pairs of alternate angles.

$b = g$ $d = e$

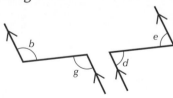

b There are four pairs of corresponding angles.

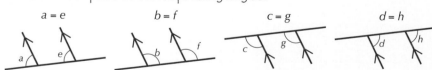

$a = e$ $b = f$ $c = g$ $d = h$

1   Quadrilateral ABCD is a parallelogram.
    What is special about:
    a ∠ABC and ∠BCD **b** ∠EAB and ∠FBG
    c ∠ABC and ∠DCI **d** ∠LAD and ∠ADC
    e ∠JDC and ∠DCI **f** ∠HCB and ∠CBA?

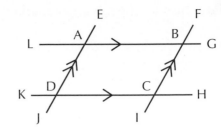

2   Copy and complete each sentence.
    a *a* and ... are corresponding angles.
    b *b* and ... are corresponding angles.
    c *c* and ... are corresponding angles.
    d *d* and ... are corresponding angles.
    e *e* and ... are alternate angles.
    f *f* and ... are alternate angles.

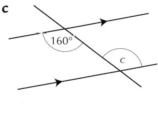

3   Work out the size of each angle marked with a letter.

a
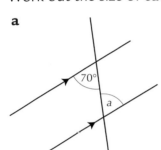

b

c

d
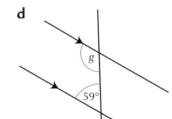

e
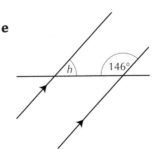

4   Calculate the size of each angle marked with a letter.

a
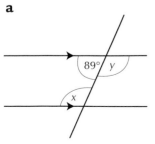

b
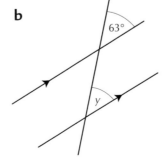

c
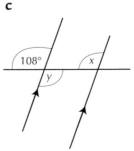

5 Calculate the size of each angle marked with a letter.

a

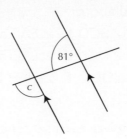

72°

a

b

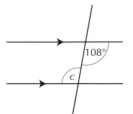

116°

b

c

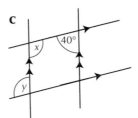

81°

c

6 Calculate the size of each angle marked with a letter.
   State whether it is an alternate angle or a corresponding angle.

a

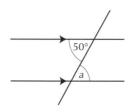

50°

a

b

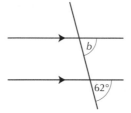

b

62°

c

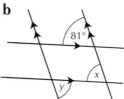

108°

c

7 Calculate the size of each angle marked with a letter.

a

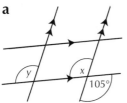

y   x

105°

b

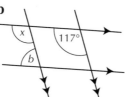

81°

x

y

c

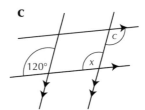

x   40°

y

8 In each of these diagrams, finding the value of x is the first step to finding values for
   a, b and c.
   State the relationship of x to the given angle and then the connection between the value
   of x and the final answer.

a

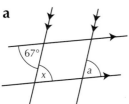

67°

x   a

b

x   117°

b

c

120°   x

c

9 Calculate the size of each angle marked with a letter.

a

a

2a

b

3b

2b

Reason mathematically

Calculate the sizes of the angles labelled *p*, *q* and *r*.

Give reasons for your answers.

a 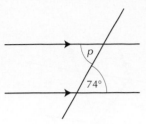 **b** 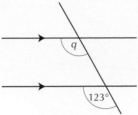 **c**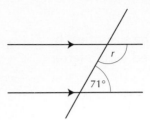

a *p* = 74°   Alternate angles are equal.

b *q* = 123°   Corresponding angles are equal.

c The angle adjacent to *r* is 71°.   Alternate angles are equal.
So *r* = 109°.   Angles on a straight line add up to 180°.

10 Calculate the size of each unknown angle marked on the diagrams.
Explain how you worked out your answers.

a  **b**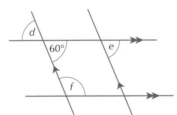

11 Which diagram is the odd one out? Give a reason for your answer.

a  **b**  **c**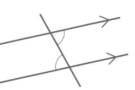

d 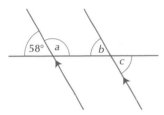 **e**  **f**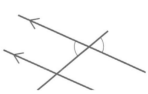

g 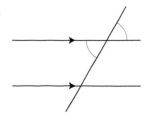 **h** 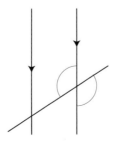 **i**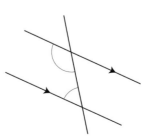

12 Calculate the value of *f* on the diagram.
Draw a diagram to explain your answer.

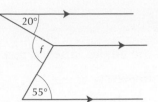

13 Calculate the size of each
angle marked with a letter.

a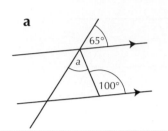

b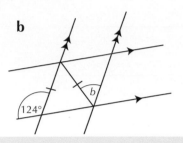

Solve problems

Calculate the size of the two angles.

These are allied angles, so they add up to 180°

$3x + 5 + 7x + 15 = 180$

$10x + 20 = 180$

$10x = 160$

$x = 160 \div 10$

$x = 16°$

So the two angles are $(3 \times 16) + 5 = 53°$ and $(7 \times 16) + 15 = 127°$

14 Calculate the size of the angles
in these diagrams.

a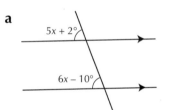

b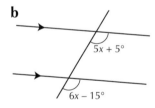

15 Calculate the size of the
angles marked in these diagrams.

a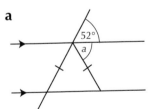

b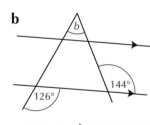

16 One of the expressions in this parallelogram is incorrect.
Which one? Work out the correct size of each angle.

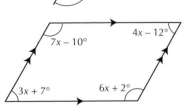

17 Using the diagram in question 1 above, write a proof to show that the opposite angles of
a parallelogram are equal.

8.5 Constructions

- I can construct the mid-point and the perpendicular bisector of a line
- I can construct an angle bisector
- I can construct a perpendicular to a line from or at a given point
- I can construct a right-angled triangle

Geometric **constructions** are useful because they give exact measurements. Always use a sharp pencil, a ruler, compasses and a protractor, and leave all your construction lines on the diagrams.

You need to know how to find the **mid-point** of one line and construct a second line at right angles to it at that point. This line is the **perpendicular bisector**. You also need to know how to **bisect** an angle accurately and draw the **angle bisector**.

Develop fluency

Construct the mid-point and the perpendicular bisector of the line AB.

- Draw a line segment AB of any length.

  A ———————————— B

- Set your compasses to a radius greater than half the length of AB and draw two arcs with the centre at A, one above and one below AB.

- With your compasses still set at the same radius, draw two arcs with the centre at B, to intersect the first two arcs at C and D.

- Join C and D to intersect AB at X. X is the mid-point of the line AB.

  The line CD is the perpendicular bisector of the line AB.

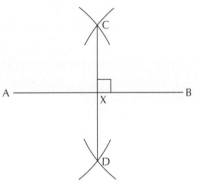

Construct the bisector of the angle ABC.

- Draw an angle ABC of any size.

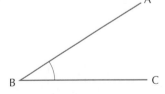

- Set your compasses to any radius, and with the centre at B, draw an arc to intersect BC at X and AB at Y.

- With compasses set to any radius, draw two arcs with the centres at X and Y, to intersect at Z.

- Join BZ. BZ is the bisector of the angle ABC.

- Then ∠ABZ = ∠CBZ.

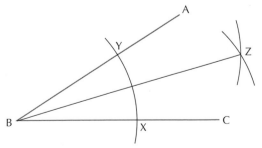

Construct the perpendicular from a point P to a line segment AB.

- Set your compasses to any suitable radius and draw arcs from P to intersect AB at X and Y.

- With your compasses set at the same radius, draw arcs with the centres at X and Y to intersect at Z below AB.
- Join PZ
- PZ is perpendicular to AB.

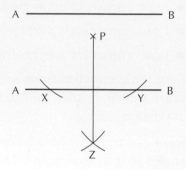

1   Draw a line AB that is 8 cm long. Using compasses, construct the perpendicular bisector of the line.

2   Draw a line CD of any length. Using compasses, construct the perpendicular bisector of the line.

3   Use a protractor to draw an angle of 70°. Now, using compasses, construct the angle bisector of this angle.
    Measure the two angles formed to check that they are both 35°.

4   Copy the line AB and the point X.
    Construct the perpendicular from the point X to the line AB.

A ——————————————— B

5   Draw a circle of radius 6 cm. Label the centre O.
    Draw a line AB of any length across the circle, as in the diagram.
    AB is called a **chord**.
    Construct the perpendicular from O to the line AB. Extend the perpendicular to make a diameter of the circle.

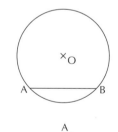

6   Copy the line XY and the point A. Construct the perpendicular from the point A on the line segment XY.

7   Construct each of these right-angled triangles. Remember to label all the sides.

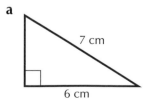   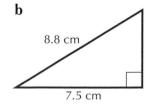   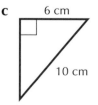

8 a Draw a triangle ABC with sides of any length.

  b Construct the angle bisectors for each of the three angles.
The three angle bisectors will meet at a point O in the centre of the triangle.

  c Using O as the centre, draw a circle to touch the three sides of the triangle.
This is called an **inscribed circle**.

9 Draw a line *xy* that is 8 cm in length. Mark a point *z* on the line *xy* that is 3 cm from *x*.
Construct a perpendicular to *xy* at point *z*.

Reason mathematically

Draw a line AB. Construct an angle of 45° at point A.

Draw a line AB that is 4 cm long. Construct an angle of 45° at point A.

Draw the line AB and, using a thin construction line, extend the line out, away from A.

Construct an angle of 90° (a perpendicular) at A.

Bisect this angle to produce an angle of 45°.

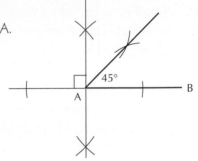

10 Construct an equilateral triangle of side length 5 cm.

11 Construct an angle of 30°

12 Draw a line XY that is 7 cm long.

  a Construct the perpendicular bisector of XY.

  b By measuring the length of the perpendicular bisector, draw a rhombus with diagonals of length 7 cm and 5 cm.

13 A ladder of length 10 m leans against a wall, with its foot on the ground 3 m from the wall.

  a Use a scale of 1 cm to 1 m to construct an accurate scale drawing.

  b Work out how far up the wall the ladder actually reaches.

  c Measure the angle the ladder makes with the ground.

14 Draw a circle with a radius of 3 cm, and centre O.
Mark a point, A ,8 cm away from O and draw two tangents from A to meet the circumference of the circle at opposite sides, *x* and *y*.
Bisect the ∠xAy and extend the bisecting line through the circle.
What do you notice about the bisector?

Solve problems

Two trees are marked on a map as A and B. A footpath is going to be made that runs between the trees. One condition of making the footpath is that anyone on the path must always be equidistant from both trees. How can you decide on a route for the footpath? Draw your explanation.

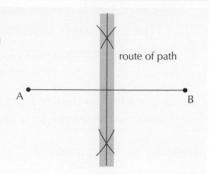

route of path

Points that are always equidistant from the two trees will form the perpendicular bisector of line AB so connect A to B and construct the perpendicular bisector of it.

15 A ball is to be rolled from point C so that it will always be equidistant from points A and B.
Where will the ball travel? Draw your explanation.

A •

• C

• B

16 ABCD is a rectangle. X and Y are points on AB and AD respectively.
Use a suitable construction to work out the shortest distance from C to XY.

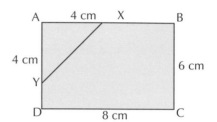

17 Ahmed creates a circle by drawing around the end of a cylindrical tin can.
He wants to find the middle of the circle. Explain how he can do this if he starts by making three markers on the circumference.

18 A zip wire ride is to be built from the top of a giant tree to an activity centre as shown.
The tree is 200 m away from the centre and an angle of 15° is required between the wire and the ground.

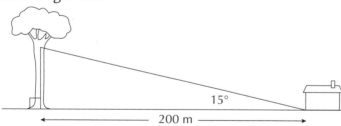

15°

200 m

Using compasses and a scale of 1 cm = 20 m, draw a diagram to work out how high up the tree the wire will reach and how long the wire will need to be.

Now I can...

calculate angles at a point and on a straight line	use the fact that the sum of the angles in a quadrilateral is 360°	construct an angle bisector
calculate opposite angles	calculate angles in parallel lines	construct a perpendicular to a line from or at a given point
use the fact that the sum of the angles in a triangle is 180°	construct the mid-point and the perpendicular bisector of a line	construct a right-angled triangle

CHAPTER 9 Decimals

9.1 Rounding numbers

- I can round numbers to a given degree of accuracy

To **round** numbers – to a given number of **decimal places** or to the nearest 10, 100 or 1000 – look at the digit to the right of the digit being rounded. If the digit is 4, 3, 2, 1 or 0, **round down**. If the digit is 5, 6, 7, 8 or 9, **round up**.

When you are **rounding**, you can use the special symbol, ≈, which means 'approximately equal to'. For example, if you round 34 to the nearest 10, you use the **units digit** and you could write '34 ≈ 30', which means '34 is approximately equal to 30'.

Develop fluency

Round each number to the nearest: **i** 10   **ii** 100.

a 521 **b** 369 **c** 605

i Look at the units digit.

If its value is less than 5, round the number down. If its value is 5 or more, round up.

a 521 ≈ 520 **b** 369 ≈ 370 **c** 605 ≈ 610

ii Look at the tens digit. Round down or up, as before.

a 521 ≈ 500 **b** 369 ≈ 400 **c** 605 ≈ 600

Round each number to the nearest whole number.

a 2.356 **b** 4.75 **c** 6.5 **d** 8.49

Look at the number after the decimal point. Round down or up as before.

a 2.356 ≈ 2 **b** 4.75 ≈ 5 **c** 6.5 ≈ 7 **d** 8.49 ≈ 8

Round each number to one decimal place.

a 9.615 **b** 4.532 **c** 18.857 **d** 12.99

Look at the digit in the second decimal place (hundredths). Round down or up as before.

a 9.615 ≈ 9.6 **b** 4.532 ≈ 4.5 **c** 18.857 ≈ 18.9 **d** 12.99 ≈ 13.0

1 Round each number to the nearest 10.
   a 16 **b** 22 **c** 43 **d** 78 **e** 157 **f** 269 **g** 145

2 Round each number to the nearest 100.
   a 642 **b** 354 **c** 529 **d** 718 **e** 851 **f** 962 **g** 54

3 Round each number to the nearest 1000.
   a 6322 **b** 3674 **c** 5519 **d** 7098 **e** 501

4 Round each number to the nearest million.
   a 3 786 120 **b** 1 570 235 **c** 2 500 000 **d** 12 055 555 **e** 1 495 999

5 Round each number to the nearest whole number

    a 15.41   **b** 12.64   **c** 43.75   **d** 72.86   **e** 7.17   **f** 167.91

6 Round each number to one decimal place.

    a 1.531   **b** 14.78   **c** 45.26   **d** 73.65   **e** 2.445   **f** 43.967

7 The English Academy in Kuwait has 1057 students.

    How many students is this, to the nearest   **a** 100 **b** 10?

8 The price of a freezer is £285.

    What is the price to the nearest   **a** £100 **b** £10?

9 A car is advertised for sale at £8965.

    What is the price to the nearest   **a** £100 **b** £1000 **c** £10?

10 These are the attendances at the Champions League quarter-final matches in 2013.

    Copy and complete the table.

Match	Attendance	Attendance (to nearest 100)	Attendance (to nearest 1000)
Paris Saint-Germain *v.* FC Barcelona	45 336	45 300	
FC Barcelona *v.* Paris Saint-Germain	96 022		
Bayern Munich *v.* Juventus	68 047	68 000	68 000
Juventus *v.* Bayern Munich	40 823		
Malaga CF *v.* Borussia Dortmund	28 548		
Borussia Dortmund *v.* Malaga FC	65 829		
Real Madrid *v.* Galatasaray	76 462	76 500	
Galatasaray *v.* Real Madrid	49 975		

Reason mathematically

Gary measured a length as 175.32 cm.

He was asked to round this off.

  a What is the largest value he could round this to?

  b What is the smallest value he could round this to?

  a *If he rounded to the nearest 100, this would round to 200 cm.*

  b *Rounding to the nearest 10 gives 180 cm, rounding to nearest integer gives 175 cm,*

    rounding to one decimal place gives 175.3 cm.

    The smallest value he could round to is 175 cm.

11 Kirsty said she ran a race in 145.8 seconds.

She needed to round this off.

  a What is the longest time she could round this to?

  b What is the shortest time she could round this to?

12 a The table shows the diameters of the planets, in kilometres. Round each diameter to the nearest 1000 km. Then place the planets in order of size, starting with the smallest.

Planet	Earth	Jupiter	Mars	Mercury	Neptune	Pluto	Saturn	Uranus	Venus
Diameter (km)	12 800	142 800	6780	5120	49 500	2284	120 660	51 100	12 100

  b What would happen if you rounded the diameters to the nearest 10 000 km?

13 Mae thinks of a number and says it rounds to 100.

Amber says: 'Is your number 45?'

Explain why Amber cannot be correct.

14 Greenland, the world's largest island country, is estimated to be 840 000 square miles in area. What is the lowest and highest area it could be (in whole numbers) if this has been rounded to:

  i the nearest 10 000   **ii** the nearest 1000?

Solve problems

Theo and Laz are thinking of whole numbers.

Theo says: 'When I round my number to the nearest 10 it is 350.'

Laz says: 'When I round my number to the nearest 100 it is 400.'

How many possible numbers are there, given that Theo's number is larger than Laz's number?

Theo's number is between 345 and 354.

Laz's number is between 350 and 449.

Theo's number is larger than Laz's, so the numbers are between 350 and 354,

that is 350, 351, 352, 353 and 354 so there are five possible numbers.

15 Starla and Morgan are thinking of whole numbers.

Starla says: 'When I round my number to the nearest 10 it is 460.'

Morgan says: 'When I round my number to the nearest 100 it is 500.'

How many possible answers are there, given that Morgan's number is smaller than Starla's number?

16 Max sells potatoes in bags with a mass that rounds to 5 kg.
He is asked to supply a restaurant with 100 kg.
How many of these 5 kg bags must he supply to be sure it's at least 100 kg?

17 Tara says it takes her 30 seconds, to the nearest 10 seconds, to prepare one envelope.
She has two hours to spend preparing envelopes.
a What is the maximum number of envelopes she can prepare?
b What is the least number of envelopes she can prepare?

9.2 Multiplying and dividing by powers of 10

● I can multiply and divide decimal numbers by 10, 100 and 1000

You use the same method to multiply or divide decimals as for multiplying and dividing whole numbers by 10, 100 and 1000.

Remember, the number of zeros in the power of ten you are multiplying by is the number of places the digits move left.

For example, $4.15 \times 10 = 41.5$, $4.15 \times 100 = 415$, $4.15 \times 1000 = 4150$

Remember also, the number of zeros in the power of ten you are dividing by is the number of places the digits move right.

For example, $621.8 \div 10 = 62.18$, $621.8 \div 100 = 6.218$, $621.8 \div 1000 = 0.6218$

Develop fluency

Work out:

a 3.5×100 **b** 4.7×10.

a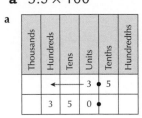

b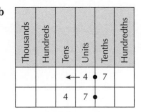

Work out:

a $23 \div 1000$ **b** $13.6 \div 10$.

a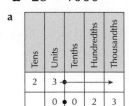

b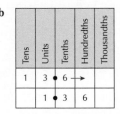

The digits move two places to the left when you multiply by 100.

$3.5 \times 100 = 350$

The digits move one place to the left when you multiply by 10.

$4.7 \times 10 = 47$

The digits move three places to the right when you divide by 1000.

$23 \div 1000 = 0.023$

The digits move one place to the right when you divide by 10.

$13.6 \div 10 = 1.36$

Work out the answers to questions 1–4 without using a calculator.

1 a 27×10 **b** 63×100 **c** 9.7×1000 **d** 4.2×100 **e** 3.8×10

2 a $37 \div 10$ **b** $53 \div 100$ **c** $88 \div 1000$ **d** $5.2 \div 10$ **e** $3.8 \div 1000$

3 a 3.42×10 **b** 1.97×1000 **c** 13.4×1000 **d** 37.4×100

4 a $7 \div 10$ **b** $7.5 \div 100$ **c** $5.83 \div 1000$ **d** $37.4 \div 1000$ **e** $48.5 \div 100$

5 Work out the missing number in each case.

    a $3 \times 10 = \square$ **b** $3 \times \square = 300$ **c** $3 \div 10 = \square$ **d** $3 \div \square = 0.03$

6 Work these out without using a calculator.

    a 4.51×10 **b** 0.62×10 **c** 0.032×100 **d** 5.18×1000 **e** 0.037×10

7 Work these out without using a calculator.

    a $8.5 \div 10$ **b** $0.3 \div 10$ **c** $1.08 \div 10$ **d** $7.9 \div 1000$ **e** $0.04 \div 100$

8 Work out the missing number in each case.

    a $0.3 \times 10 = \square$ **b** $0.3 \times \square = 300$ **c** $\square \times 10 = 300$

    d $\square \times 1000 = 30\,000$ **e** $\square \times 10 = 0.3$ **f** $\square \times 10 = 0.03$

9 Work out the missing number in each case.

    a $0.3 \div 10 = \square$ **b** $0.3 \div \square = 0.003$ **c** $\square \div 10 = 30$

    d $0.3 \div \square = 0.03$ **e** $\square \div 100 = 0.3$ **f** $\square \div 10 = 0.003$

10 Fill in the missing operation in each case.

    a $0.37 \square 37$ **b** $567 \square 5.67$ **c** $0.07 \square 70$

    d $650 \square 65$ **e** $0.6 \square 0.006$ **f** $345 \square 0.345$

Reason mathematically

Tim told his brother Pete that when he multiplied a number by 100 he should move the digits three places to the right.

Is Tim correct? Explain your answer.

No Tim is incorrect, you should move the digits two places to the left.

11 Rykie told her sister that when she divided a number by 1000 she should move the digits four places to the left.

    Is Rykie correct? Explain your answer.

12 How would you explain to someone how to multiply and divide a number by one million?

13 A trillion is $1\,000\,000\,000\,000$. How would you multiply and divide by a trillion?

14 You multiply by a quintillion by moving the digits eighteen places to the left. Write down a quintillion in digits.

Solve problems

To change centilitres, cl, to litres, you divide by 100.

For example, $350\,\text{cl} = 350 \div 100 = 3.5$ litres.

Change these centilitres to litres.

  a 450 cl   **b** 175 cl   **c** 51 cl   **d** 2 cl

  a $450 \div 100 = 4.5$ litres   **b** $175 \div 100 = 1.75$ litres

  c $51 \div 100 = 0.51$ litres   **d** $2 \div 100 = 0.02$ litres

15 To change kilowatts, kW, to watts, you multiply by 1000.
   For example, $7.2\,\text{kW} = 7.2 \times 1000 = 7200\,\text{W}$.
   Change these kilowatts to watts.
   a 1.3 kW   **b** 15.5 kW   **c** 0.44 kW   **d** 0.07 kW

16 Work out the total shopping bill.
   1000 sweets at £0.03 each
   100 packets of mints at £0.45 each
   10 cans of cola at £0.99 each

17 To change metres to centimetres, you multiply by 100.
   For example, $4.7\,\text{m} = 4.7 \times 100\,\text{cm} = 470\,\text{cm}$.
   Change these lengths to centimetres.
   a 3.9 m   **b** 1.75 m   **c** 23.5 m   **d** 0.7 m   **e** 0.25 m   **f** 2.08 m

18 To change grams to kilograms, you divide by 1000.
   For example, $250\,\text{g} = 250 \div 1000\,\text{kg} = 0.25\,\text{kg}$.
   Change these masses to kilograms.
   a 375 g   **b** 75 g   **c** 4550 g   **d** 5250 g   **e** 615 g   **f** 2008 g

9.3 Putting decimals in order

● I can order decimal numbers according to size

The table shows heights of six people. Suppose you want to put them in **order** of height.

Name	Leroy	Myrtle	Shehab	Baby Jane	Alf	Doris
Height	170 cm	1.58 m	189 cm	0.55 m	150 cm	1.80 m

First, rewrite each set of measures in the same units. There are 100 centimetres in a metre so rewrite heights given in centimetres as heights in metres by dividing by 100.

Name	Leroy	Myrtle	Shehab	Baby Jane	Alf	Doris
Height	1.70 m	1.58 m	1.89 m	0.55 m	1.50 m	1.80 m

Now you can compare the sizes of the numbers by considering the **place value** of each digit.

Develop fluency

Put the numbers 2.33, 2.03 and 2.304 in order, from smallest to largest.

It helps to put the numbers in a table like this one.

Thousands	Hundreds	Tens	Units	Tenths	Hundredths	Thousandths
			2	3	3	0
			2	0	3	0
			2	3	0	4

Use zeros to fill the missing decimal places.

Working across the table from the left, you can see that all of the numbers have the same units digit. Two of them have the same tenths digit, and two have the same hundredths digit. But only one has a digit in the thousandths.

The smallest is 2.03 because it has no tenths, which both of the other numbers have.

Next is 2.304 because it has fewer hundredths than 2.33, even though it has the same number of tenths.

So, 2.33 is the largest.

The order is 2.03, 2.304, 2.33.

1 a Copy the table in the example above (but leave out the numbers).
Write these numbers in the table, placing each digit in the appropriate column.
4.57, 45, 4.057, 4.5, 0.045, 0.5, 4.05

   b Use your answer to part **a** to write the numbers in order, from smallest to largest.

2 Write each set of numbers in order, from smallest to largest.
   a 0.73, 0.073, 0.8, 0.709, 0.7   b 1.203, 1.03, 1.405, 1.404, 1.4
   c 34, 3.4, 0.34, 2.34, 0.034

3 Put these amounts of money in order, from smallest to largest.
   a 56p, £1.25, £0.60, 130p, £0.07   b $0.04, $1.04, $10, $0.35, $1

4 Put these times in order.
   1 hour 10 minutes, 25 minutes, 1.5 hours, half an hour

5 Write each set of numbers in order, from smallest to largest.
   a 0.82, –0.82, 0.8, –0.708, –0.7   b –5.14, 5.07, –5.44, 5.11, –5.12
   c –1.7, 1.82, –1.65, 1.8, –1.73

6 One centimetre (cm) is 10 millimetres (mm).
   Change these lengths to a single common unit and then order them, from smallest to largest.
   3 cm, 134 mm, 16 mm, 14 cm, 1700 mm

7 One metre is 100 centimetres.
   Change these lengths to a single common unit and then order them, from smallest to largest.
   6 m, 269 cm, 32 cm, 27 m, 3400 cm

8 One kilogram is 1000 grams.

Change these masses to a single common unit and then order them, from smallest to largest.

467 g, 1 kg, 56 g, 5 kg, 5500 g

9 One litre is 100 centilitres.

Change these capacities to a single common unit and then order them, from smallest to largest.

8 litres, 876 cl, 98 cl, 17 litres, 8300 cl

10 One tonne (t) is 1000 kilograms (kg).

Change these masses a single common unit and then order them, from smallest to largest.

678 kg, 6 t, 67 kg, 6600 kg, 6.09 t, 650.9 kg

Reason mathematically

Put the correct sign, > or <, between the numbers in each pair.

a 6.05 and 6.046   **b** 0.06 and 0.065

a *Rewrite both numbers with the same number of decimal places.*

6.050 ☐ 6.046

Both numbers have the same units digits and tenths digits, but the hundredths digit is bigger in the first number.

So, the answer is 6.05 > 6.046.

b *Rewrite both numbers with the same number of decimal places.*

0.060 ☐ 0.065

Both numbers have the same units, tenths and hundredths digits, but the second number has the bigger thousandths digit, as the first number has a zero in the thousandths.

So, the answer is 0.06 < 0.065.

11 Put the correct sign, > or <, between the numbers in each pair.

    a 0.315 ☐ 0.325   **b** 0.42 ☐ 0.402   **c** 6.78 ☐ 6.709

    d 5.25 ☐ 5.225   **e** 0.345 ☐ 0.4   **f** 0.05 ☐ 0.7

12 Put the correct sign, > or <, between the amounts of money in each pair.

    a £0.51 ☐ 52p   **b** £0.62 ☐ 60p   **c** £0.05 ☐ 7p

    d 42p ☐ £0.40   **e** 14p ☐ £0.04   **f** 8p ☐ £0.07

13 Write each statement in words.

    a 3.1 < 3.14 < 3.142   **b** £0.07 < 32p < £0.56

14 Shehab said that 0.508 was smaller than 0.51.

Is Shehab correct? Explain your answer.

Solve problems

The table shows the heights of eight boys.

Boy	Sam	Alec	Bri	Joe	Dave	Ali	Jed	Kal
Height	1.3 m	1.04 m	98 cm	85 cm	112 cm	108 cm	1.1 m	89 cm

The four tallest went sailing, the four smallest went canoeing.

a Who went sailing?

b Who went canoeing?

Put the boys in order of size.

Boy	Joe	Kal	Bri	Alec	Ali	Jed	Dave	Sam
Height	85 cm	89 cm	98 cm	1.04 m	108 cm	1.1 m	112 cm	1.3 m

a *Sam, Dave, Jed and Ali go sailing.*

b *Joe, Kal, Bri and Alec go canoeing.*

15 The table shows the times of eight girls in a race.

Girl	Sam	Mae	Gill	Kay	Ann	Preya	Lucy	Bran
Time (seconds)	9.06	9.12	9.1	9.09	9.3	9.21	9.17	9.09

The four girls with the shortest times are in the final.

Which girls are in the final?

16 The map shows the positions and heights of six mountains in the UK.
Write the names of the six mountains in order of height, starting with the highest.

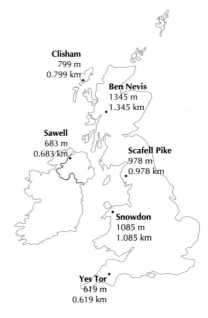

Clisham
799 m
0.799 km

Ben Nevis
1345 m
1.345 km

Sawell
683 m
0.683 km

Scafell Pike
978 m
0.978 km

Snowdon
1085 m
1.085 km

Yes Tor
619 m
0.619 km

17 Tony was asked to use all the prime numbers less than ten, a zero and a decimal point to write the smallest possible number he could.
What is the smallest number Tony could write?

18 In a marrow growing competition, each entrant had to state the mass of their marrow. This table shows the entrants and the masses.

Entrant	Ken	Bill	Kate	Beth	Pete	Pam	Roy	Les
Marrow mass	5.13 kg	5078 g	5209 g	5.08 kg	5.1 kg	5125 g	5.15 kg	5027 g

Who entered the **a** heaviest marrow **b** second heaviest marrow **c** lightest marrow?

9.4 Estimates

● I can estimate calculations in order to recognise possible errors

Suppose you were telling a friend about the game between Town and City. Which numbers would you **round up** to a sensible **approximation**? Which would you **round down**? Which ones must you give exactly?

TOWN v CITY

Crowd	41 923
Score	2 – 1
Time of first goal	42 min 13 sec
Price of a pie	£2.95
Children 33% off normal ticket prices	

·   **Round** numbers to **estimate** whether the answer to a calculation that you have completed is about right.

·   For a multiplication, check that the final digit is correct.

·   Round numbers and do a mental calculation to see if an answer is about the right size.

·   You can use the **inverse operation**, such as taking a **square root**.

Develop fluency

Estimate the answers to these calculations.

a $\dfrac{21.3 + 48.7}{6.4}$ **b** 31.2×48.5 **c** $359 \div 42$

a *Round the numbers on the top, 20 + 50 = 70. Round 6.4 to 7. Then 70 ÷ 7 = 10.*

b *Round to 30 × 50, which is 3 × 5 × 100 = 1500.*

c *Round to 360 ÷ 40, which is 36 ÷ 4 = 9.*

1 Estimate the answer to each problem.
   a 2768 + 392   **b** 2317 + 1808   **c** 4701 + 3087

2 Estimate the answer to each problem.
   a 3767 – 491   **b** 4124 – 1028   **c** 2571 – 783

3 Estimate the answer to each problem.
   a 68 × 39   **b** 231 × 18   **c** 792 × 37   **d** 863 × 41
   e 423 × 423   **f** 1172 × 48   **g** 4086 × 19   **h** 342 × 29

4   Estimate the answer to each problem.
   a $98 \div 21$ **b** $431 \div 19$ **c** $991 \div 46$ **d** $968 \div 51$
   e $921 \div 33$ **f** $2577 \div 53$ **g** $2080 \div 18$ **h** $3052 \div 31$

5   Estimate the answer to each problem.
   a $\dfrac{39 + 42}{23}$ **b** $\dfrac{128 - 88}{8}$ **c** $\dfrac{283 + 27}{18}$ **d** $\dfrac{98 - 14}{22}$

6   Estimate the answer to each problem.
   a $38.9 - 27.06$ **b** $3.456 + 19.21$ **c** $175.9 - 13.98$ **d** $26.78 + 4.979$

7   Estimate the answer to each problem.
   a 7.9×1.8 **b** 3.42×27.2 **c** $89.7 \div 8.7$ **d** $3.68 \div 0.87$
   e 51.7×51.7 **f** $126.83 \div 11.5$ **g** $301.7 \div 19.3$ **h** 5.81×1.08

8   Estimate the answer to each problem.
   a $\dfrac{28.7 + 31.8}{12.7}$ **b** $\dfrac{17.6 - 8.98}{9.34}$ **c** $\dfrac{17.9 + 76.8}{19.8}$ **d** $\dfrac{89.3 - 53.6}{21.8}$

9   Estimate the answer to each problem.
   a $\dfrac{338.5 + 116.8}{387 - 189.5}$ **b** $\dfrac{762.8 - 112.3}{17.3 \times 3.36}$ **c** $\dfrac{135.75 - 68.2}{15.8 - 8.9}$ **d** $\dfrac{38.9 \times 61.2}{39.6 - 18.4}$

10   Estimate the number each arrow is pointing to.
   a 18   20   **b** −2   8   **c** −1.2   0.8

Reason mathematically

Use the inverse operation to explain how each calculation can be checked.
   a $450 \div 6 = 75$ **b** $310 - 59 = 249$

   a   *By the inverse operation, $450 = 6 \times 75$.*
      Check mentally.
      $6 \times 70 = 420, 6 \times 5 = 30, 420 + 30 = 450$, so is correct.

   b   *By the inverse operation, $310 = 249 + 59$.*
      This must end in 8 as $9 + 9 = 18$, so the calculation cannot be correct.

Explain why these calculations must be wrong.
   a $23 \times 45 = 1053$ **b** $19 \times 59 = 121$

   a   *The last digit should be 5, because the product of the last digits (3 and 5) is 15.*
      That is, $23 \times 45 = \ldots 5$, so the calculation is wrong.

   b   *The actual answer is roughly $20 \times 60 = 1200$, so the calculation is wrong.*

11 Explain why each calculation must be wrong.

    a $24 \times 42 = 1080$ **b** $51 \times 73 = 723$ **c** $\dfrac{34.5 + 63.2}{9.7} = 20.07$

    d $360 \div 8 = 35$ **e** $354 - 37 = 323$

12 Amy bought 6 cans of lemonade at 86p per can. The shopkeeper asked her for £6.16. Without working out the correct answer, explain how Amy can tell that this is wrong.

13 A cake costs 47p. I need 8 cakes.
Will £4 be enough to pay for them? Explain your answer clearly.

14 In a shop I bought a chocolate bar for 53p and a model car for £1.47. The total on the till read £54.47. Why?

Solve problems

A farmer bought 518 kg of fertiliser at £1.89 per kilogram.

What is the approximate total cost of this fertiliser?

Round 518 to 500.

Round £1.89 to £2.

Then the approximate total cost will be £2 × 500 = £1000.

15 A merchant bought 293 kg of grain at $3.74 per kilogram.
What is the approximate total cost of this grain?

16 Leo wanted to find out approximately how many bricks there were in a large, circular chimney that was being knocked down. He counted 218 bricks on one row, all the way around the chimney. He thought he counted 147 rows of bricks up the chimney. Approximately how many bricks would Leo expect there to be?

17 Delroy had £20. In his shopping basket he had a magazine costing £3.65, some batteries costing £5.92 and a DVD costing £7.99.

    a Without adding up the numbers, how could Delroy be sure he had enough to buy the goods in the basket?

    b Explain a quick way for Delroy to find out if he could also afford a 45p bar of chocolate.

18 These are the amounts of money a family of four brought home, as wages.

Thomas: $280 per week   Dechia: $6250 per month

Joseph: $490 per week   Sheena: $880 per month

To qualify for an educational bursary, they need to be bringing home a total of less than $128 000 per year. Make an estimate to decide whether or not you think they will qualify. Show how you estimated.

9.5 Adding and subtracting decimals

● I can add and subtract decimal numbers

The method for adding and subtracting decimals is the same as for whole numbers. As with whole numbers, it is important to align the **decimal point** and place values.

Develop fluency

Work out: **a** $4 + 0.86 + 0.07$ **b** $6 - 1.45$.

a Whole numbers have no decimal places, but it can be helpful to write a decimal point after the units digit and show the decimal place values with zeros. Then you can line up the decimal points and place values of all the numbers, like this:

$$
\begin{array}{r}
4.00 \\
0.86 \\
+\,0.07 \\
\hline
4.93 \\
{\scriptstyle 1}
\end{array}
$$

b As in part **a**, put a decimal point and zeros after the whole number to show the place values, and line up the decimal points.

$$
\begin{array}{r}
{\scriptstyle 5\ 9\ 1} \\
\cancel{6}.\cancel{0}0 \\
-\,1.45 \\
\hline
4.55
\end{array}
$$

1 Work these out without using a calculator.
   a $3.5 + 4.7$ **b** $6.1 + 2.8$ **c** $9.3 + 6.1$ **d** $27.65 + 16.47$

2 Work these out without using a calculator.
   a $4.5 - 3.7$ **b** $13.41 - 7.69$ **c** $8.3 - 5.1$ **d** $47.56 - 17.74$

3 Work these out without using a calculator.
   a $4 - 2.38$ **b** $5 - 1.29$ **c** $7 - 1.08$ **d** $15 - 6.09$

4 Work these out without using a calculator.
   a $7 - (1.4 + 2.48)$ **b** $11 - (3.8 + 4.26)$ **c** $11 - (5.4 + 3.76)$

5 Work these out without using a calculator.
   a $9 - (3.4 - 1.48)$ **b** $8 - (5.06 - 3.7)$ **c** $13 - (7.4 - 4.36)$

6 Work these out without using a calculator.
   a $(9.2 - 1.08) + (9.4 - 3.47)$ **b** $(5.43 - 0.89) + (7.9 - 4.36)$
   c $(8.04 - 3.21) + (7.07 - 4.8)$ **d** $(12.16 - 2.76) + (8.3 - 4.17)$

7 The three legs of a relay race are 3 km, 4.8 km and 1800 m.
   How long is the race altogether?

8 Three packages have mass 4 kg, 750 grams and 0.08 kg.
   How much is their total mass altogether?

9 Estimate which of these has the greatest value, then check if you were right.
   a $14.82 + 7.56$ **b** $31 - 8.84$ **c** $106.55 + 135.45$

10 Three sticks have length 24 cm, 1.07 m and 0.27 m.
    What is their total length altogether?

Reason mathematically

Nazia has completed 4300 m of a 20 km bike ride.

  a How can she find out how far she still has to go?

  b Calculate this distance.

  a She will need to make the distances the same units and then subtract.

  b Make the units for the distances both kilometres, then subtract.

    Nazia still has to go 15.7 km.

$$\begin{array}{r} {\scriptstyle 1\ 9\ 1} \\ 2\!\!\!/0.0 \\ -\ \ 4.3 \\ \hline 15.7 \end{array}$$

11 The mass of a Christmas cake is 2 kg. Arthur takes a slice of mass 235 grams.
    How much is left?

12 Jack pours 23 cl of water from a jug containing 3 litres. (1 litre = 100 cl)
    How much is left?

13 Jasmine cuts 1560 millimetres of ribbon from a piece that is 5 metres long.
    (1 metre = 1000 mm)
    How much ribbon is left?

14 Ben does the calculation 8.208 – 3.17 and gets the answer 5.38.
    The answer is incorrect. Explain what he has done wrong and what the correct answer
    should be.

Solve problems

Kilroy finds his suitcase has a mass of 25 kg, but he needs it to only be 22 kg.
So far, he has removed 650 grams.

How much more does he need to take out?

*He needs to remove 3 kg. The units need to be made the same. So, change 650 grams
into 0.65 kg. This gives:*

Kilroy still has to remove 2.35 kg.

$$\begin{array}{r} {\scriptstyle 2\ 9\ 1} \\ 3\!\!\!/.00 \\ -\ 0.65 \\ \hline 2.35 \end{array}$$

15 Toby has 8 litres of fruit juice. He needs 9.75 litres.
    He finds a bottle of fruit juice that contains 85 cl.
    How many more litres of fruit juice does he need?

16 The diagram shows the lengths of the paths in a park.

a How long are the paths altogether?

b Sean wants to visit all the points A, B, C, D, E and F in this order. He wants to start and finish at A and go along each and every path.

Explain why he cannot do this in the distance you worked out for part **a**.

c Work out the shortest distance Sean could walk if he wanted to visit each point, starting and finishing at A.

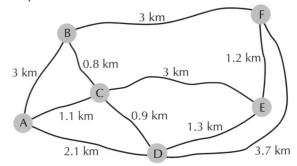

17 Talika checks her restaurant bill. She thinks it is wrong.

a Use estimation to check if the bill is about right.

b Calculate if the bill is actually correct or not.

2 drinks	£11.60
Soup of the day	£5.95
Madras Curry	£13.95
Cumberland Sausage	£11.85
Rib eye steak	£12.55
Grilled mushrooms	£3.85
Potato wedges	£3.85
Battered onion rings	£3.85
3 desserts	£18.15
Total	£90.45

18 Laz wants to make statues using plaster of Paris.
The amounts of plaster needed for the statues are:
a horse: 3500 g,   a zebra: 2.4 kg,   a wolf: 1.25 kg,   a rabbit: 700 g,   a cow: 2800 g

a How many kilograms of plaster of Paris does Laz need to make all the statues?

b Laz has an 8 kg bag of plaster of Paris. He wants to have as little plaster left over as possible. Which statues should he make?

c How much plaster would he use?

d How much plaster will he have left over?

9.6 Multiplying and dividing decimals

● I can multiply and divide decimal numbers by any whole number

The method of multiplying and dividing decimals by whole numbers is similar to that for any other type of multiplication and division but you need to take care where you put the decimal point.

As a general rule, there will be the same number of decimal places in the answer as there were in the original problem.

Develop fluency

Work these out.

  a 5×3.7 **b** 6×3.5 **c** $22.8 \div 6$ **d** $26.2 \div 5$

Each of these can be set out in columns or as short divisions.

$$
\begin{array}{llll}
\textbf{a} \quad 3.7 & \textbf{b} \quad 3.5 & \textbf{c} \quad 6\overline{)22^4.8}^{\,3\ .8} & \textbf{d} \quad 5\overline{)26^1.2^20}^{\,5\ .2\ 4} \\
\quad\ \underline{\times 5} & \quad\ \underline{\times 6} & & \\
\quad 18.5 & \quad 21.0 & & \\
\quad\ \ _3 & \quad\ \ _3 & &
\end{array}
$$

1 Work these out without using a calculator.

  a 3.14×5 **b** 1.73×8 **c** 6×3.35 **d** 9×5.67

2 State if each calculation below is true or false.

  a $4.25 \times 3 = `12.75$ **b** $2.84 \times 4 = 11.46$ **c** $2.57 \times 6 = 15.42$

  d $3.09 \times 8 = 24.72$ **e** $7 \times 4.47 = 32.19$ **f** $8 \times 4.56 = 36.48$

3 Work these out without using a calculator.

  a $17.04 \div 8$ **b** $30.6 \div 5$ **c** $25.88 \div 4$ **d** $4.44 \div 3$

4 State if each calculation below is true or false.

  a $34.16 \div 8 = 4.27$ **b** $79.4 \div 7 = 11.32$ **c** $54.4 \div 4 = 13.6$

  d $61.2 \div 6 = 11.2$ **e** $65.8 \div 5 = 13.16$ **f** $8.88 \div 3 = 2.69$

5 Which of these is the odd one out?

  a 15.2×4 **b** $182.4 \div 3$ **c** 7.8×8

6 Which of these has the largest answer?

  a 27.3×3 **b** 16.4×5 **c** 13.7×6

7 Which of these has the smallest answer?

  a $108.48 \div 8$ **b** $68.05 \div 5$ **c** $95.06 \div 7$

8 Work these out without using a calculator.

  a $\dfrac{18.5 \times 8}{5}$ **b** $\dfrac{23.7 \times 4}{3}$ **c** $\dfrac{45.2 \times 7}{4}$ **d** $\dfrac{19.5 \times 8}{5}$

9 Work these out without using a calculator.

  a $(4.87 + 21.6) \times 9$ **b** $(31.8 - 17.9) \times 8$ **c** $(87.4 - 18.8) \div 4$

Reason mathematically

A water bottle holding 1.44 litres is poured into six equal glasses.

How many litres are there in each glass?

Divide 1.44 by 6 to find an answer of 0.24.

Each glass has 0.24 litres.

10 A piece of wood, 2.8 metres long, is cut into five equal pieces.
    How long is each piece?

11 Five bars of metal each have mass 2.35 kg.
    How much is their total mass altogether?

12 Eight bottles of cola cost £6.24.
    What is the price of one bottle?

Solve problems

A cup of coffee and a bun cost £2.20 together. A cup of coffee and two buns cost £3.05 together.

How much change would I get from £10 if I bought four cups of coffee and five buns?

First find the cost of one bun by subtracting the cost of a coffee and one bun from a coffee and two buns. This is £3.05 – £2.20 = £0.85.

Now find the cost of a coffee by subtracting the price of a bun from the price of a coffee and a bun. This is £2.20 – £0.85 = £1.35.

So, the cost of four coffees and five buns will be (4 × £1.35) + (5 × 0.85) = £9.65.

This will give us £10 – £9.65 = £0.35 change.

13 A can of orange and a chocolate bar cost £1.30 together. Two cans of orange and a chocolate bar cost £2.15 together.
    How much would three cans of orange and four chocolate bars cost?

14 A shop was selling a range of DVDs at £7.49 each, or £19.98 for three.
    How much would you save if you buy three DVDs together, rather than buying them separately?

15 Eight packs of cards cost £12.80. How much would five packs cost?

Now I can...

round numbers to a given degree of accuracy	estimate calculations in order to recognise possible errors	multiply and divide decimal numbers by any whole number
multiply and divide decimal numbers by 10, 100 and 1000	add and subtract decimal numbers	
order decimal numbers according to size		

CHAPTER 10 Linear graphs

10.1 Coordinates

● I can understand and use coordinates to locate points in all four quadrants

You can use coordinates to locate a point on a grid consisting of two **axes**, called the **x-axis** and the **y-axis**, which are perpendicular (at right angles) to each other. The two axes meet at a point called the **origin**, which is numbered 0 on both axes. The axes divide the grid into four **quadrants**.

Coordinates are written in pairs. The first number is the **x-coordinate**, which shows how far along the x-axis the point is, and second is the **y-coordinate**, which shows how far along the y-axis it is. The coordinates of the origin are (0, 0). In general terms you can write the coordinates of any point as (x, y).

Develop fluency

Write down the coordinates of the points
A, B, C and D on this grid.

On this grid:

* point A has the coordinates (4, 2)
* point B has the coordinates (−2, 3)
* point C has the coordinates (−3, −1)
* point D has the coordinates (1, −4).

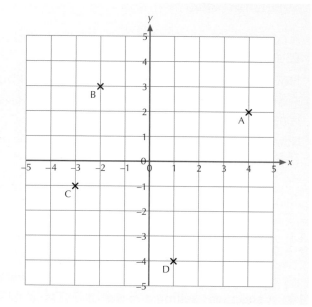

1   Look at this grid.
    Write down the coordinates of the points
    A, B, C, D, E, F, G and H.
    For questions 2–10, copy the grid from
    question 1 but omit the points A–H.

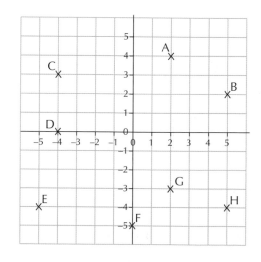

2 a Plot the points A(–3, –2), B(–3, 3), C(1, –2), and D(1, 3).

  b Join the points, in the order given. What letter have you drawn?

3 a Plot the points A(–4, –1), B(–4, 4), C(–2, 2), D(0, 4) and E(0, –1).

  b Join the points, in the order given. What letter have you drawn?

For questions 4–9, plot and join the given points and then name the shape.

4 A(–4, 1), B(1, 1), C(1, 3), D(1, –4)   **5** E(5, 2), F(3, 4), G(1, 2), H(3, –3)

6 J(–4, –4), K(–3, –2), L(1, –2), M(0, –4)   **7** N(1, 1), P(–1, 1), Q(–1, –1), R(1, –1)

8 S(–4, –1), T(–1, –1), U(–2, 0), V(–3, 0)   **9** W(4, –1), X(3, –3), Y(4, –5), Z(5, –3)

Reason mathematically

a Plot the points A(–4, –3), B(0, –5) and C(4, 3) on a grid.

b The points form three vertices of a rectangle ABCD. Plot the point D and draw the rectangle.

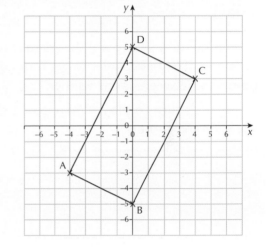

10 a Copy the grid in question 1, omitting the points A–H. Plot the points W(2, –2), X(2, 3) and Y(–3, 3).

  b The points form three vertices of a square WXYZ. Plot the point Z and draw the square.

  c What are the coordinates of the point Z?

  d Draw in the diagonals of the square. What are the coordinates of the point of intersection of the diagonals?

11 a Draw a grid, numbering the x-axis from –10 to 10 and the y-axis from –5 to 5.

  b Using letters with straight sides and drawing them so that their vertices fall on coordinate points on your grid, write a short message.

  c Write a list of coordinates to represent each letter, when its coordinates are joined up. In each word, separate the letters with a forward slash (/).

  d Swap coordinate messages with a partner and decode each other's messages.

12 The diagram shows the plan of a garden.

   a Write down the coordinates of the marked points in:

      i the pond   **ii** the vegetable plot

      iii the hedge   **iv** the flower bed

      v the patio   **vi** the lawn.

   b Where are these points?

      i (4, 4)   **ii** (3, 8)   **iii** (8, 3)

      iv (6, 8)   **v** the origin **vi** (0, 7)

      vii (3, 0)

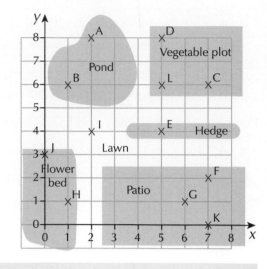

Solve problems

In a game of '4 in a row', players take turns to place counters on a grid. The first player to get four counters in a row, with no points in between, is the winner.

It is black's turn to play. Where could she go to make sure that she wins the game?

She could place her counter at (3, −2), (0, 0), (0, −2) or (3, −3).

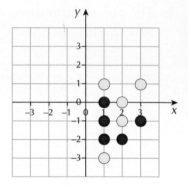

13 In a game of '4 square', players take turns to place counters on a grid. The first player to get four counters in a square, with no points in between, is the winner.

Tasha and Matt are playing the game. Tasha has white counters and Matt has black counters. It is Tasha's turn to play.

   a Where must Tasha go to stop Matt winning?

   b Matt can make certain he wins by placing his next counter at (2, 1). Explain why and write down the two coordinates that he could use to win.

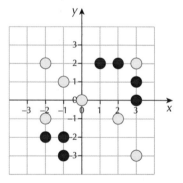

   c Tasha says they ought to allow squares of any size, as she would already have won after her turn in part a. Can you find the coordinates of the four counters that give her a larger winning square?

14 (−3, −2), (−2, −1) and (−1, −2) are three vertices of a square.
Work out the fourth vertex of the square.

15 (3, 5), (1, 4) and (3, −3) are three vertices of a kite.
Work out the fourth vertex of the kite.

10.2 Graphs from formulae

● I can draw a graph for a simple relationship

When you draw graphs from **relationships** expressed as equations, you need to work out the (x, y) coordinates by **substituting** values for x into the equation, to find the corresponding values of y. It is helpful to set these out in a table of values. It is possible to draw a line graph after plotting two points, but you should always plot at least three, to avoid errors.

Develop fluency

Draw the graph of the relationship $y = 2x + 1$ for $-2 \leq x \leq 4$.

Remember that $\leq$ means 'less than or equal to'.

You are asked to draw the graph for values of x from -2 to 4.

Calculate the value of y for each of the x-values of -2, -1, 0, 3 and 4.

Set up a table of values.

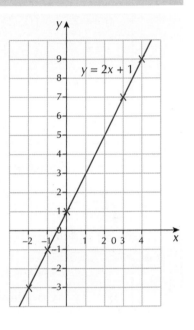

x	$y = 2x + 1$
-2	-3
-1	-1
0	1
3	7
4	9

Write the pairs of values as coordinates.

$(-2, -3), (-1, -1), (0, 1),$ $(3, 7), (4, 9)$

Plot these coordinates and join them in a straight line.

Always label the line.

For questions 1 to 6, copy and complete this table for the given equation.

1 $y = x$

2 $y = x + 3$

3 $y = x + 5$

4 $y = 4x$

5 $y = 6x$

6 $y = \frac{1}{2}x$

x	y	Coordinates
-2		(,)
0		(,)
2		(,)

7 For each relationship:

   i complete the table of values   **ii** complete the list of coordinates

   iii plot the coordinates and draw the graphs, numbering the *x*-axis from –3 to 7.

   Draw all the graphs on the same set of axes.

a

$y = x + 2$		
x	y	Coordinates
–1 → 1		(–1, 1)
0 → 2		(0, 2)
1 →		(1,)
2 →		(2,)
3 →		(3,)
4 →		(4,)

b

$y = x + 4$		
x	y	Coordinates
–1 → 3		(–1, 3)
0 → 4		(0, 4)
1 →		(1,)
2 →		(2,)
3 →		(3,)
4 →		(4,)

c

$y = x - 3$		
x	y	Coordinates
–1 → –4		(–1, –4)
0 → –3		(0, –3)
1 →		(1,)
2 →		(2,)
3 →		(3,)
4 →		(4,)

8 For each relationship:

   i complete the table of values   **ii** complete the list of coordinates

   iii plot the coordinates and draw the graphs, numbering the *x*-axis from –3 to 7.

   Draw all the graphs on the same set of axes.

a

$y = 2x$		
x	y	Coordinates
–1 → –2		(–1, –2)
0 → 0		(0, 0)
1 →		(1,)
2 →		(2,)
3 →		(3,)
4 →		(4,)

b

$y = 3x$		
x	y	Coordinates
–1 → –3		(–1, –3)
0 → 0		(0, 0)
1 →		(1,)
2 →		(2,)
3 →		(3,)
4 →		(4,)

c

$y = 5x$		
x	y	Coordinates
–1 → –5		(–1, –5)
0 → 0		(0, 0)
1 →		(1,)
2 →		(2,)
3 →		(3,)
4 →		(4,)

   d On the same axes, draw the graph of $y = x$.

9 **i** Draw a grid, numbering the *x*-axis from –3 to 4 and the *y*-axis from –10 to 15.

   ii Use *x*-values of –2, 0 and 3 to work out the corresponding *y*-values in each relationship.

   iii Write each pair of values as a set of coordinates.

   iv Plot the coordinates and draw the graphs.

     a $y = 3x - 2$ **b** $y = 2x + 2$ **c** $y = 4x + 1$

10 a Copy and complete this table for the relationship $y = -2x + 3$.

x	$y = -2x + 3$	Coordinates
-2	7	(-2, 7)
0		(0,)
2		(2,)

b Draw a grid, numbering the x-axis from -2 to 2 and the y-axis from -2 to 8. Plot the coordinates and draw the graph of $y = -2x + 3$.

c Find some coordinates to help you draw the graphs of these relationships.

i $y = -x - 1$ **ii** $y = -3x + 1$

Reason mathematically

Look at the graphs.

a What do you notice about lines $y = x$ and $y = x - 4$?

b What do you notice about lines $y = 2x$ and $y = -2x$?

a They are parallel.

b They are reflections of each other in the y-axis.

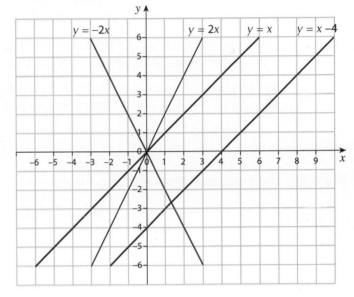

11 Look again at your graphs in question 7. What do you notice about the lines?

12 Look again at your graphs in question 8. What do you notice about the lines?

13 Choose some of your own starting points and create a graph from each relationship.

a $y = 2x + 3$ **b** $y = 3x + 2$ **c** $y = 2x - 3$ **d** $y = 4x - 3$

14 Choose some of your own starting points and create a graph for each relationship.

a $y = -2x + 1$ **b** $y = -3x + 2$ **c** $y = -2x - 1$ **d** $y = -4x$

Solve problems

A straight line passes through the points (0, 0), (1, 2), and (2, 4).

What is its equation?

As each y-coordinate is double the corresponding x-coordinate, the equation is $y = 2x$.

15 A straight line passes through the points (0, 0), (1, 1), and (2, 2).
What is its equation?

16 A straight line passes through the points (0, 1), (1, 2), and (2, 3).
What is its equation?

17 A straight line passes through the points (1, 3), (2, 5), and (3, 7).
What is its equation?

10.3 Graphs of $x = a$, $y = b$, $y = x$ and $y = -x$

● I can recognise and draw line graphs with fixed values of x and y

● I can recognise and draw graphs of $y = x$ and $y = -x$

The graph of the equation $x = a$ is a straight line, parallel to the y-axis, passing through the point $(a, 0)$ on the x-axis. Every point on the line has x-coordinate a.

The graph of the equation $y = b$ is a straight line, parallel to the x-axis, passing through the point $(0, b)$ on the y-axis. Every point on the line has y-coordinate b.

The graph of the equation $y = x$ is a straight line passing through all the points where the values of x and y are the same, such as $(-5, -5)$, $(0, 0)$ and $(4, 4)$.

The graph of the equation $y = -x$ is a straight line passing through all the points where the values of x and y add to 0, such as $(-5, 5)$, $(0, 0)$ and $(4, -4)$.

Develop fluency

Write down the equation of each of these lines.

a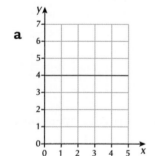

a $y = 4$

b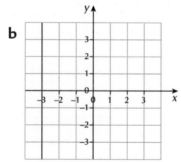

b $x = -3$

c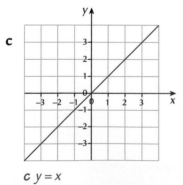

c $y = x$

d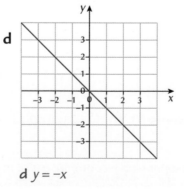

d $y = -x$

1   Write down the equation of the straight line that
    goes through each pair of points on this diagram.

    a   A and B   b   C and A
    c   E and H   d   G and C
    e   L and J   f   K and H
    g   D and F   h   D and E
    i   G and I   j   A and K

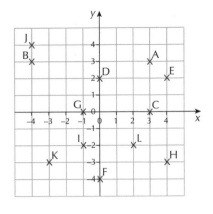

2   Draw a grid, numbering both axes from –5 to 5.
    Draw each of these graphs on the same grid. Remember to label them.

    a $y = -1$ b $y = -4$ c $y = x$ d $x = -1$ e $x = 3$ f $x = -2$ g $y = 4$ h $y = -x$

3   Write down the letters that are on each line.

    a $x = -2$ b $y = 1$
    c $y = -3$ d $y = x$
    e $x = 3$ f $x = -4$
    g $y = -1$ h $x = -1$
    i $y = -2$ j $y = -x$

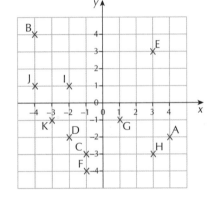

For questions 4–7, there is no need to draw a graph.

4   Write down the equation of the straight line passing through (2, 0), (2, 1) and (2, 22).

5   Write down the equation of the straight line passing through (4, 0) and (4, 12).

6   Write down the equation of the straight line passing through (–1, –1) and (3, 3).

7   Write down the equation of the straight line passing through (2, –2) and (–4, 4).

Reason mathematically

Write down the equation of a line that goes along the x-axis.

The y-coordinate of every point on the x-axis is 0, for example, (0, 0), (1, 0) and so on.

Therefore the equation is y = 0.

8   Write down the equation of a line that goes along the y-axis.

9 a Draw each pair of lines on the same grid, using a suitable set of coordinate axes. Write down the coordinates of the point where they cross.

    i $x = -1$ and $y = 3$ **ii** $x = 4$ and $y = -1$ **iii** $y = -5$ and $x = -6$

  b Write down the coordinates of the point where each pair of lines cross. Do not draw them.

    i $x = -9$ and $y = 5$ **ii** $x = 28$ and $y = -15$ **iii** $y = -23$ and $x = -48$

10 a Draw the graphs of $y = x + 1$, $y = 3x + 1$ and $y = 5x + 1$.

  b What do the three lines have in common?

  c Write down another equation that will do the same.

11 a Explain why the lines $y = 2x$, $y = 2x + 3$ and $y = 2x - 1$ will never intersect.

  b Write the equation of another line that will not intersect with the lines above.

Solve problems

You are given three points: P(1, –4), Q(4, –4) and R(1, –1).

Write the equation for the straight line that passes through both points in each pair. Then show the points and lines on a graph, as a check.

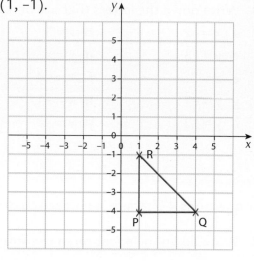

  a P and Q   **b** P and R   **c** Q and R

  a $y = -4$ **b** $x = 1$ **c** $y = -x$

12 These are the coordinates of 12 points.

  A(–2, –3)   B(–3, –5)   C(7, –3)   D(–2, –5)   E(–2, –2)   F(6, –6)

  G(3, 7)   H(7, –4)   I(–3, –4)   J(7, 7)   K(–5, 5)   L(–3, –6)

  Write down the equation for the straight line that passes through both points in each pair. Then draw a suitable set of axes and plot the points on the graphs, to check your answers.

  a A and C **b** B and D   **c** C and H   **d** A and D   **e** E and J

  f F and L   **g** B and I   **h** C and J   **i** F and K   **j** H and I

13 A rectangle is enclosed by the lines $x = 2$, $x = 7$, $y = -1$ and $y = 3$. Work out the area of the rectangle.

10.4 Graphs of the form $x + y = a$

● I can recognise and draw graphs of the form $x + y = a$

The graph of an equation such as $x + y = a$, where a is a constant, passes through the points $(a, 0)$ and $(0, a)$.

Develop fluency

Complete the table of values for $x + y = 4$ and draw its graph.

x	y	Total	Coordinates
1		4	(1,)
2		4	(2,)
–1		4	(–1,)
6		4	(6,)

First fill in the y-coordinates:

x	y	Total	Coordinates
1	3	4	(1, 3)
2	2	4	(2, 2)
–1	5	4	(–1, 5)
6	–2	4	(6, –2)

Now plot the coordinates to show the graph passing through (4, 0) and (0, 4).

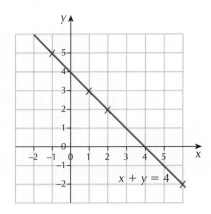

For questions 1–4, copy and complete the table of values for each equation.

1 $x + y = 5$

x	y	Total	Coordinates
1		5	(1,)
2		5	(2,)
–1		5	(–1,)

2 $x + y = 7$

x	y	Total	Coordinates
1			(1,)
2			(2,)
–1			(–1,)

3 $x + y = -1$

x	y	Total	Coordinates
1			(1,)
2			(2,)
–1			(–1,)

4 $x + y = 0$

x	y	Total	Coordinates
1			(1,)
2			(2,)
–1			(–1,)

For questions 5–7, copy and complete each statement.

5 The line $x + y = 8$ passes through the points (..., 0) and (0, ...).

6 The line $x + y = 2$ passes through the points (..., 0) and (0, ...).

7 Draw up a table of values to show ways in which x and y can add up to 9.
Remember to include some negative values.
Draw a grid, numbering both axes from –3 to 10.
Plot the coordinates on a grid and join up the points.
Label the line $x + y = 9$.

8 Draw a grid, numbering both axes from –6 to 10.
Draw up a table of values and use it to draw the line for each relationship.
   a $x + y = 3$ b $x + y = -2$

9 Use the points where each line intercepts the x-axis and y-axis to sketch the graph of
   a $x + y = 9$ b $x + y = -3$.

Reason mathematically

These are the graphs of $x + y = 2$ and $x + y = 5$.

   a What do you notice about the lines?

   b What can you say about all lines of the form $x + y = a$?

   a *They are parallel.*

   b *All the lines are parallel.*

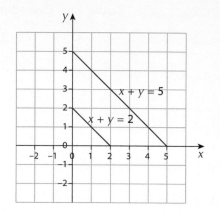

10 Explain why the lines $x + y = 2$ and $x + y = 7$ will never meet.

11 Which equation is the odd one out?
   $x + y = 6$ $x = y + 6$ $y = -x + 6$
   Give a reason for your answer.

12 There are 16 points marked on this grid.
   Write down the relationship for the line that passes
   through both points in each pair.
   a Q and H   b B and D
   c A and J   d L and N
   e C and P   f E and F
   g K and M   h G and I

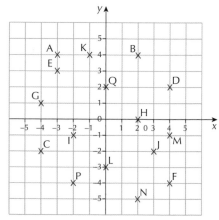

13 Which of these lines is the same as $y = x$?
There may be more than one answer.

   a $x = y$ **b** $y + x = 0$ **c** $x = -y$ **d** $y - x = 0$ **e** $y = x + 0$

Solve problems

A triangle is enclosed by the x-axis, the y-axis and the line $x + y = 4$.
Work out the area of the triangle.

Draw the graph.

Area of a triangle $= \dfrac{1}{2} \times base \times height$

$\qquad\qquad\qquad = \dfrac{1}{2} \times 4 \times 4$

$\qquad\qquad\qquad = 8$ square units

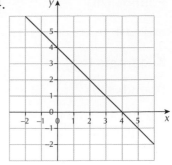

14 A triangle is enclosed by the x-axis, the y-axis and the line $x + y = 6$.
Work out the area of the triangle.

15 Without drawing a graph, work out the coordinates of the point where each pair of lines intersect.

   a $x + y = 4$ and $y = x$ **b** $x + y = 11$ and $y = x$ **c** $x + y = -8$ and $y = x$

10.5 Conversion graphs

- I can understand how graphs are used to represent real-life situations
- I can draw and use real-life graphs

Develop fluency

A **conversion graph** shows a direct relationship between two quantities, such as pounds and kilograms, or different currencies. Conversion graphs are usually straight-line graphs.

This conversion graph shows the relationship between pounds (lb) and kilograms (kg).

Use it to complete this table.

Kilograms (kg)	1	2			5
Pounds (lb)			6.6	8.8	

Reading from the graph, the completed table looks like this.

Kilograms (kg)	1	2	3	4	5
Pounds (lb)	2.2	4.4	6.6	8.8	11

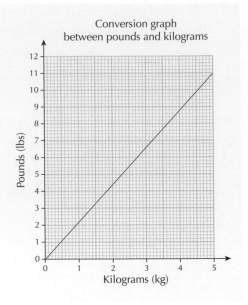

Conversion graph between pounds and kilograms

1 a Use the graph to convert these distances to kilometres.

    i 3 miles   **ii** 4.5 miles   **iii** 1 mile

  b Use the graph to convert these distances to miles.

    i 2 km   **ii** 4 km   **iii** 6 km

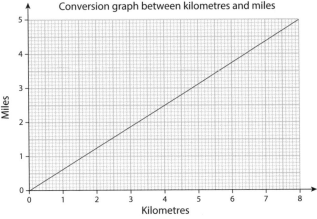

Conversion graph between kilometres and miles

2 The graph shows the distance travelled by a cyclist over 4 hours.

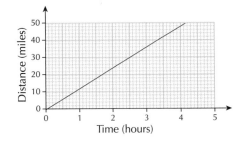

  a How far has he travelled after 3 hours?

  b How long did he take to travel 30 miles?

3 a How many pounds would you get for:

    i $10   **ii** $25?

  b How many dollars would you get for:

    i £12   **ii** £18?

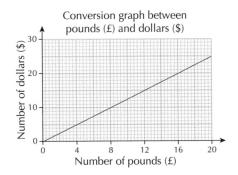

Conversion graph between pounds (£) and dollars ($)

4 The graph shows the distance travelled by a car during an interval of 5 minutes.

   a Work out the distance travelled in 5 minutes.

   b Work out the time taken to travel 1 km.

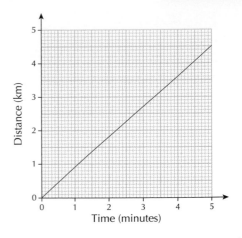

5 Draw a conversion graph to convert pounds (£) to euros (€), using £100 = €120.

6 Draw a conversion graph to convert kilometres per hour to metres per second using 36 km/h = 10 m/s.

7 Draw a conversion graph to convert gallons to litres, using 10 gallons = 45 litres.

Reason mathematically

Use this graph to convert 30 kilograms (kg) to pounds (lb).

3 kg = 6.6 lb

So, multiplying by 10 gives 30 kg = 66 lb.

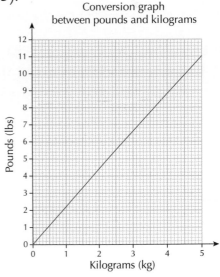

Conversion graph between pounds and kilograms

8 Look again at the kilometre–mile conversion graph in question 1.
Use the graph to convert 40 km to miles.

9 Look again at the 'dollars–pounds' graph in question 3.
Use the graph to work out how many pounds you could exchange for £70 dollars.

10 Look again at the car graph in question 4.
Show that the distance travelled during the third minute of the journey is the same as the distance travelled in the first minute.

Solve problems

Last month Mr Shock was charged £60 for 400 units of electricity.

This month Mr Shock used 280 units.

Use the conversion graph to work out how much he was charged this month.

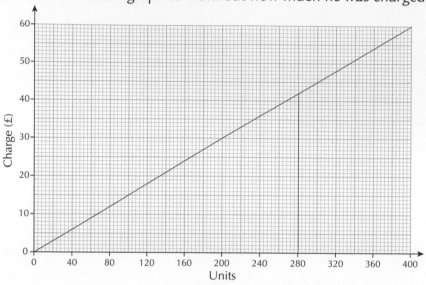

Reading from the graph, he is charged £42.

11 Last month Mrs Fume was charged £50 for 125 units of gas.
This month Mrs Fume used 150 units.
Use a conversion graph to work out how much she was charged this month.

12 Mr Leak was charged £2.40 for 20 bottles of water.
Use a conversion graph to work out the cost of 50 bottles of water.

13 One month, the exchange rate between the pound and the euro was €1 to £0.90.
Use a conversion graph to work out the number of euros (€) that are equivalent to £360

Now I can...

understand and use coordinates to locate points in all four quadrants	recognise and draw graphs of $y = x$ and $y = -x$	recognise and draw graphs of the form $x + y = a$
draw a graph for a simple relationship	understand how graphs are used to represent real-life situations	draw and use real-life graphs
recognise and draw line graphs with fixed values of x and y		

CHAPTER 11 Percentages

11.1 Fractions, decimals and percentages

● I can understand the equivalence between a fraction, a decimal and a percentage

Fractions, **decimals** and **percentages** can all be used to compare quantities or measurements.

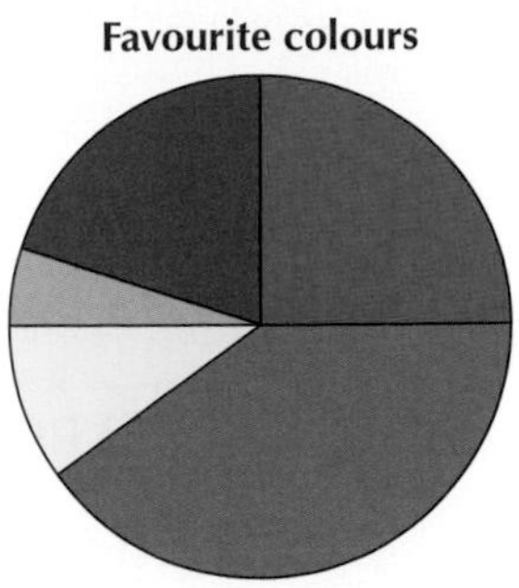

This pie chart shows the favourite colours of a group of children.

Favourite colours

The red sector is a quarter of the whole circle, which you can write as $\frac{1}{4}$ or 0.25 or 25%.

To convert between percentages, fractions and decimals, remember that per cent means 'for every 100', so 25% is 25 parts out of 100, or $\frac{25}{100}$, which can be simplified to $\frac{1}{4}$ and as a decimal it can be expressed as 0.25.

To convert a fraction that is not written as hundredths to a decimal, sometimes you can use equivalent fractions to write the fraction as hundredths. If it is not easy to convert the fraction to hundredths, divide the numerator by the denominator to get a decimal then multiply by 100.

Remember that percentages can also include decimals or fractions.

Develop fluency

In the pie chart shown above, the blue sector is $\frac{2}{5}$ of the whole.

Write $\frac{2}{5}$ as a decimal and as a percentage.

$\frac{2}{5} = \frac{40}{100}$ *Use equivalent fractions to write $\frac{2}{5}$ as hundredths.*

$\frac{40}{100} = 0.4$

$\frac{40}{100} = 40\%$

1 Change these fractions to percentages.

   a $\frac{1}{2}$ **b** $\frac{1}{4}$ **c** $\frac{3}{4}$ **d** $\frac{3}{10}$ **e** $\frac{7}{20}$

2 Write these decimals as percentages.

   a 0.5   b 0.05   **c** 0.8   **d** 0.08

3 Write these percentages as fractions.
   Give your answers as simply as possible.

   a 20%   **b** 30%   **c** 90%   **d** 95%

4 Copy and complete this table. Simplify fractions where possible.
   The first row has been done for you.

Percentage	10%	70%			
Fraction				$\frac{15}{25}$	$\frac{9}{25}$
Decimal	0.1		0.35		

5 $\frac{1}{3}$ is $33\frac{1}{3}$%.

   Write $\frac{2}{3}$ as a percentage.

6 Match each fraction with a percentage.
   One has been done for you.

$\frac{3}{5}$ $\frac{4}{5}$ $\frac{7}{10}$ $\frac{7}{20}$ $\frac{37}{50}$

74%   60%   35%   80%   70%

7 Percentages can be larger than 100%.
   1 = 100%   1.5 = 150%   2 = 200%
   Write these decimal numbers as percentages.
   a 1.2   **b** 1.9   **c** 1.53   **d** 1.74   **e** 2.74

8 Write these percentages as decimals.
   a 130%   **b** 140%   **c** 185%   **d** 285%   **e** 216%

9 Write these percentages as mixed numbers.
   a 150%   **b** 125%   **c** 130%   **d** 160%   **e** 275%

10 20% + 75% = 95%
    a Rewrite this as an addition of decimals.   **b** Rewrite this as an addition of fractions.

Reason mathematically

A survey of Year 7 students at a school showed that 47% of them take the bus to
school, 16% are driven to school and 22% ride their bikes to school. The rest of the
students walk to school. What fraction of students walk to school? Write your answer
as simply as possible.

The total percentage adds to 100, so 100% – 47% – 16% – 22% = 15% of students walk to school.

As a fraction, 15% = $\frac{15}{100}$ = $\frac{3}{20}$.

11 This pie chart shows the most popular colours chosen by a group of children to paint a pattern.
Three of the sectors of this pie chart represent 10%, 20% and 30%.

Colours to paint a pattern

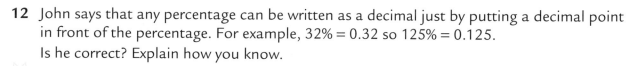

   a What percentage does the fourth sector represent?

   b Write down which colour represents the following percentages and explain how you know.

   i 10%   ii 20%   iii 30%

   c What fraction of the whole is each of the four sectors?

   d What percentage of the pie chart is not red?

   e What colour is 0.3 of the pie chart?

12 John says that any percentage can be written as a decimal just by putting a decimal point in front of the percentage. For example, 32% = 0.32 so 125% = 0.125.
Is he correct? Explain how you know.

13 An ice-cream stall offers hundreds and thousands, chocolate sauce or squirty cream as toppings. Their records show that of customers who order one topping, 18% choose hundreds and thousands, $\frac{24}{50}$ choose chocolate sauce and the rest choose squirty cream.
Write the fraction of customers who choose squirty cream as:

   a a fraction   b a decimal   c a percentage.

Solve problems

Write 18.7% as a fraction in its simplest form.

$$18.7\% = \frac{18.7}{100} = \frac{187}{1000}$$

A fraction can't have a decimal in it, so be careful not to leave the answer as $\frac{18.7}{100}$.

14 Write these percentages as decimals and fractions in their simplest form.

   a 18.5%   b 20.5%   c 20.8%   d 21.8%

15 $\frac{1}{4}$ = 25%

   $\frac{1}{8}$ is half of $\frac{1}{4}$

   a Write $\frac{1}{8}$ as a percentage.   b Write $\frac{3}{8}$, $\frac{5}{8}$ and $\frac{7}{8}$ as percentages.

16 Here is a sequence of percentages.
10%, 20%, 30%, 40%, ..., ..., ..., ..., 90%

   a Copy the sequence and fill in the missing numbers.

   b Write each percentage as a fraction, as simply as possible.

   c Which percentages give the fractions that have the smallest denominator?

11.2 Fractions of a quantity

● I can find a fraction of a quantity

You can work out a fraction of a number or a **quantity** by using multiplication and division. You can also find fractions of quantities when the fractions are larger than 1.

Develop fluency

There are 360 animals on a farm.

$\frac{1}{5}$ of the animals are cows and $\frac{3}{8}$ are sheep.

How many sheep are there?

$360 \div 8 = 45$ To work out $\frac{3}{8}$ first work out $\frac{1}{8}$ by dividing by 8.

$45 \times 3 = 135$ Then $\frac{3}{8}$ of $360 = \frac{1}{8}$ of 360×3.

There are 135 sheep.

Don't forget to include the units or name of the object in your answer.

1 Work out these quantities.

   a $\frac{1}{10}$ of £600   **b** $\frac{3}{10}$ of £600   **c** $\frac{7}{10}$ of £600   **d** $\frac{9}{10}$ of £600

2 Work out these quantities.

   a $\frac{1}{4}$ of £96   **b** $\frac{1}{2}$ of £96   **c** $\frac{1}{3}$ of £96   **d** $\frac{2}{3}$ of £6

3 Work out these quantities.

   a $\frac{1}{5}$ of 120 cm   **b** $\frac{3}{5}$ of 120 cm   **c** $\frac{3}{10}$ of 120 cm   **d** $\frac{3}{12}$ of 120 cm

4 Work out these quantities. Give your answers in pence.

   a $\frac{1}{2}$ of £1.20   **b** $\frac{1}{4}$ of £1.20   **c** $\frac{1}{8}$ of £1.20   **d** $\frac{5}{8}$ of £1.20

5 There are 100 centimetres in a metre. Work out these quantities and write your answers in metres.

   a $\frac{1}{2}$ of 800 cm   **b** $\frac{1}{4}$ of 800 cm   **c** $\frac{1}{8}$ of 800 cm   **d** $\frac{3}{8}$ of 800 cm

Reason mathematically

Write $\frac{1}{3}$ of 4 hours in hours and minutes.

4 hours is $4 \times 60 = 240$ minutes.

$\frac{1}{3}$ of $240 = 80$ minutes   Divide $240 \div 3 = 80$.

80 minutes = 1 hour 20 minutes

6 Work out each quantity.

a $\frac{1}{12}$ of 2 hours   Give your answer in minutes.

b $\frac{5}{6}$ of 2 days   Give your answer in hours.

c $\frac{2}{3}$ of 2 minutes   Give your answer in seconds.

7 Tom has £600.

He spends $\frac{3}{5}$ of it on a bed.

He spends $\frac{3}{4}$ of the remainder on curtains.

Show that he has £60 left.

8 The complete angle at the centre of a circle is 360°.

a One-third of this circle is coloured blue.
What is the angle at the centre of the blue sector?

b $\frac{5}{12}$ of this circle is coloured yellow.

What is the angle at the centre of the yellow sector?

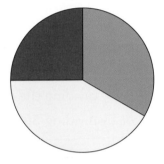

9 Jasmine saves up £2000. She buys a car with $\frac{3}{4}$ of the money and pays for her insurance

upfront with $\frac{7}{10}$ of the remaining money. She also wants new speakers costing £300

for the car sound system. Does she have enough money left for the speakers?
Show your working.

Solve problems

Find $\frac{8}{5}$ of 120.

$\frac{1}{5}$ of 120 = 24   120 ÷ 5 = 24

So, $\frac{8}{5}$ of 120 is 8 × 24 = 192.

10 a Work out each quantity.

i $\frac{2}{5}$ of 80   **ii** $\frac{4}{5}$ of 80   **iii** $\frac{6}{5}$ of 80   **iv** $\frac{8}{5}$ of 80

b What do you notice about the answers?

11 A marathon is approximately 26 miles.

a How far is half a marathon?

b Janice says: 'When I have run 20 miles I shall have completed more than three-quarters of the marathon.'
Is this correct? Give a reason for your answer.

12 Hazel earns £2600 per month. She spends $\frac{2}{5}$ of her income on her mortgage and $\frac{1}{3}$ of what is left on food.

   a How much more does Hazel spend on her mortgage than on food?

   b How much money does Hazel have left after paying for her mortgage and food?

11.3 Percentages of quantities

● I can find a percentage of a quantity

You can find a percentage of a quantity in a similar way to how you find a fraction of a quantity.

To find a percentage of a quantity without a calculator, remember that you can find:

- 1% by dividing the quantity by 100
- 10% by dividing the quantity by 10
- 5% by dividing 10% by 2
- 50% by dividing the quantity by 2
- 25% by dividing 50% by 2.

Develop fluency

Work out 60% of 200.

There are two methods.

Method 1: 60% is $\frac{60}{100}$ so find $\frac{3}{5}$ of 200.

$\frac{1}{5}$ of $200 = 40$

So, $\frac{3}{5}$ of 200 is $3 \times 40 = 120$.

Method 2: Find 10% by dividing by 10.

10% of $200 = 20$

So, 60% of 200 is $20 \times 6 = 120$.

1 Work out 10% of each amount.

   a £100   **b** £150   **c** £1500   **d** £300

2 Work out 5% of each amount.

   a £200   **b** £280   **c** £2000   **d** £2500

3 Work out 1% of each amount.

   a £500   **b** £600   **c** £800   **d** £1200

4 Work out these percentages of 320 kg.

   a 50%   **b** 25%   **c** 20%   **d** 10%

5 Find these amounts of 6400 cm.

   a 5%   **b** 1%   **c** 2%   **d** 14%

6 There are 1000 g in a kg. Find these amounts and write your answer in kg and g.

    a 1% of 180 000 g   **b** 5% of 180 000 g   **c** 10% of 10.8 kg   **d** 50% of 10.8 kg

7 Work out 63% of each quantity.

    a £1   **b** £2   **c** £10   **d** £20

8 Work out these percentages of £300.

    a 1%   **b** 9%   **c** 19%   **d** 99%

9 You can also find a percentage of an amount when the percentage is over 100.
To find 125% of £600, find 25% of 600 = 150 and add it to 600: 600 + 150 = 750.
Work out these percentages of £120.

    a 150%   **b** 120%   **c** 101%   **d** 202%

Reason mathematically

Work out 10% and 1% of £150.

Use your answer to find 32% of £150.

10% of £150 = £15   Remember you can find 10% by dividing by 10.

1% of 150 = £1.50   Remember you find 1% by dividing by 100.

32% = 3 × 10% + 2 × 1%

So, 32% of £150 is 3 × £15 + 2 × £1.50 = £48.

10 a Work out 10% of 32 kg.

    b Use your answer to part a to find these amounts.

       i 20% of 32 kg   **ii** 30% of 32 kg   **iii** 5% of 32 kg   **iv** 90% of 32 kg

11 a Find: **i** 25% of £28   **ii** 10% of £28.

    b Use your answers to part a to work these out.

       i 35% of £28   **ii** 15% of £28   **iii** 30% of £28

       iv 5% of £28   **v** 12.5% of £28

12 Copy these calculations.

    a ...% of 40 = 18   **b** ...% of 300 = 96   **c** ... % of 250 = 100

    The missing percentages are 32%, 40% and 45%.

    Match the calculations with the correct percentages.

13 42% of £375 is £157.50.

    What is: **a** 84% of £375   **b** 4.2% of £375   **c** 58% of £375?

Solve problems

Matilda buys a new mobile phone that cost £450. She pays a 20% deposit on the phone. Work out how much she still needs to pay after paying the deposit.

There are two methods.

Method 1: Work out the deposit and subtract it from £450.

20% of £450 = £90

Then £450 – £90 = £360.

Method 2: Work out the remaining percentage.

100% – 20% = 80%

So, find 80% of £450: 80% of £450 = £360

14 The cost of a holiday for a couple is £1320.
They must make a first payment of 30%.
   a How much is the first payment?   **b** How much is left to pay?

15 38% of £49.00 is £18.62.
Use this fact to work out 62% of £49.00.

16 At a conference, 25% of the people attending are men.
There are 30 men.
How many people are there at the conference altogether?

17 A sign in a shop states that the prices of all items are reduced by 15%.
   a How much would be saved on a toaster that used to cost £70?
   b What is the new price of a dishwasher that used to cost £450?

11.4 Percentages with a calculator

● I can use a calculator to find a percentage of a quantity

● I know when it is appropriate to use a calculator

Sometimes you can work out a percentage easily, without using a calculator.

If the calculation is complicated it may be more efficient to use a calculator.

To use a calculator, convert the percentage to a decimal and multiply by the quantity.

Develop fluency

In an election, 850 people vote. 28% vote for Ms White.

How many people vote for Ms White?

Write 28% as a decimal and multiply by 850. Use a calculator to do this.

28% of 850 = 0.28 × 850 = 238

238 people vote for Ms White.

1 Work out these percentages of 780 cm.

   a 42%   **b** 4%   **c** 29%   **d** 94%

2 Work out these percentages of 187 kg.

   a 79%   **b** 3%   **c** 36%   **d** 27%

3 Work out these percentages of £1502.25.

   a 18%   **b** 27%   **c** 93.2%   **d** 30.4%

4 Work out these percentages of 183.8 cm.

   a 21%   **b** 7%   **c** 71.3%   **d** 27.8%

5 1000 m = 1 km. Work out these percentages of 15 km. Write your answers in km and m.

   a 26%   **b** 86%   **c** 64%   **d** 45%

6 In an election, 17 600 people voted.

   The Red party gained 23% of the votes.

   The Blue party gained 36%.

   The Yellow party gained 19%.

   How many votes did each party get?

7 A politician is talking to a meeting of 350 people.

   He says: 'I know that 95% of the people in this room agree with me.'

   How many people is that?

Reason mathematically

According to the Office for National Statistics, the population of England and Wales increased by 7.1 per cent between 2001 and 2011. In 2001 the population of England and Wales was 52.4 million.

Use a single calculation to work out the population of England and Wales in 2011. Give your answer in millions to one decimal place.

An increase of 7.1% means the new population is 107.1% of the old population.

To find 107.1% on a calculator, convert it to a decimal, 1.071.

52.4 million × 1.071 = 56.1 million (1 dp)

8 The information on a cereal box says 'contains 25% more cereal than before'. The boxes used to contain 428 g. How many grams are there in a new box?

9 a Work out:

     i 17% of 4300 people   **ii** 54% of 4300 people   **iii** 29% of 4300 people.

  b Show that the three answers in part a add up to 4300. Explain why.

10 a Use a calculator to work out 32% of 76 g.

    b Find a simple way of working these out, without using your calculator.

     i 16% of 76 g   **ii** 32% of 38 m   **iii** 16% of £38

11 For a law to be passed, 55% of votes need to be in favour. There are 626 voters.
Work out which of these laws were passed.

   a 'Free chocolate for all' received 319 votes

   b 'All hair must be more than five inches long' received 328 votes

   c 'Everyone must eat broccoli' received 341 votes

   d 'An extra day off school every week' received 349 votes.

Solve problems

An airline adds on an extra 1.7% of the ticket price if you pay using a company credit card.

A ticket from London to Barcelona cost £250.32.

Work out the extra charge for paying with a company credit card.

1.7% is 0.017 as a decimal.

250.32 × 0.017 = £4.26

The airline charges an extra £4.26 for paying with a company credit card.

12 a Write 1.5% as a decimal.

   b A concert hall charges an extra 1.5% if you pay for tickets with a credit card.
   Jason buys a ticket for £68.
   How much will he be charged for paying with a credit card?

13 a Work out:

   i 64% of 380 kg   **ii** 32% of 380 kg   **iii** 16% of 380 kg   **iv** 8% of 380 kg.

   b The questions and answers in part a form a sequence.
   What is the next term in the sequence?

14 In the general election in the UK, in 2010, 29.6 million people voted.
This table shows the percentage of voters that voted for
the three main parties.

   a Work out how many people voted for each party. Give
   your answers correct to one decimal place (1 dp).

   b How many of the people who voted did not vote for
   one of the main parties?
   Give your answer in millions, correct to one decimal place.

Conservative	36%
Labour	29%
Liberal Democrat	23%

Now I can...

understand the equivalence between a fraction, a decimal and a percentage	find a percentage of a quantity	know when it is appropriate to use a calculator
find a fraction of a quantity	use a calculator to find a percentage of a quantity	

CHAPTER 12 3D shapes

12.1 Naming and drawing 3D shapes

- I am familiar with the names of 3D shapes and their properties
- I can use isometric paper to draw 3D shapes made from cubes

You need to be able to recognise and name these 3D shapes.

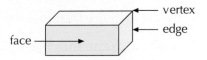

Cube Cuboid Square-based pyramid Tetrahedron Cone Cylinder Sphere Hemisphere Triangular prism Pentagonal prism Hexagonal prism

You can use **isometric paper** to draw 3D shapes.

This paper has dots that form a 60° grid of small triangles.

Develop fluency

How many faces, edges and vertices does a cuboid have?

A cuboid has 6 faces, 12 edges and 8 vertices.

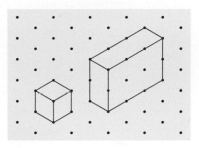

Draw a cube and a cuboid on isometric paper.

Start by drawing a vertical line for the nearest edge and then build the diagram from there.

Draw what you would see.

All of your lines will be either vertical or at 60° to the vertical.

1 Copy and complete the table. Use the pictures above to help you.

	Number of faces	Number of edges	Number of vertices
Cube			
Cuboid			
Square-based pyramid			
Tetrahedron			
Triangular prism			
Pentagonal prism			
Hexagonal prism			

2 Copy and complete the table.

		Number of faces	Number of edges	Number of vertices
Square-based pyramid				5
Pentagonal-based pyramid		6		
Hexagonal-based pyramid			12	

3 Which two shapes have 6 faces, 8 vertices and 12 edges?

4 Name a shape that has the same number of faces and vertices.

5 Name a shape that has only one vertex.

6 Draw a cube with edge length 2 cm on an isometric grid.

7 Draw each cuboid accurately on an isometric grid.

a
5 cm
2 cm
1 cm

b
2 cm
2 cm
5 cm

c
6 cm
5 cm
4 cm

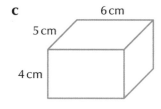

Reason mathematically

Decide whether the following statement is always true, sometimes true or never true.

A cube has 12 identical edges.

Always true. All faces are squares of equal size.

8 Decide whether each statement is always true, sometimes true or never true.
  a A cuboid has 8 vertices.
  b A prism has 9 edges.
  c A cube has 8 identical faces.
  d A cuboid is a prism.

9 This cuboid is drawn on an isometric grid.

It is made from three cubes.

On a copy of the grid:

a add another cube to make an L-shape

b add another two cubes to make a T-shape

c add another two cubes to make a +-shape.

10 Explain why cones and cylinders do not have just 'faces'.

Solve problems

The diagram shows an octahedron.

How many of each of the following does the shape have?

a faces   **b** vertices   **c** edges

a 8 faces   **b** 6 vertices   **c** 12 edges

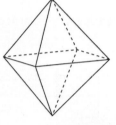

11 This 3D shape is made by putting together a cuboid and a pyramid.

How many of each of the following does the shape have?

a faces   **b** vertices   **c** edges

12 This 3D shape is made by putting together a cuboid and two pyramids.

How many of each of the following does the shape have?

a faces   **b** vertices   **c** edges

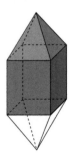

13 How many cubes are required to make this 3D shape?

Use an isometric grid to draw other similar 3D shapes of your own.

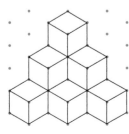

12.2 Using nets to construct 3D shapes

- I can draw nets of 3D shapes
- I can construct 3D shapes from nets

A **net** is a 2D shape that can be folded to **construct** a 3D object.

Develop fluency

State the 3D shape that can be made from the net.

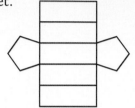

The shape is a pentagonal-based prism.

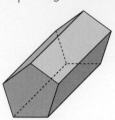

For questions 1–8, state the 3D shape that can be made from the net.

1

2

3
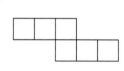

4

5

6

7

8
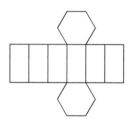

Reason mathematically

Give a reason why this is not the net of a cuboid.

For example, these edges should be the same length.

9 Give a reason why this is not the net of a square-based pyramid.

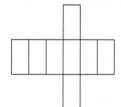

10 Which of these is the odd one out? Give a reason for your answer.

a

b

c

11 Theo says this is the net of a triangular prism.
Is he correct?
Give a reason for your answer.

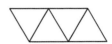

Solve problems

Here is a net of a
square-based pyramid.

Which edge will join to
edge A?

Label the edge B.

12 The diagram shows a net for a square-based pyramid.

   a Which edge is Tab 1 glued to? On a copy of the diagram,
label this A.

   b Which edge is Tab 4 glued to? On a copy of the diagram,
label this B.

   c The vertex marked with a red dot meets two other vertices.
Label these with dots.

13 Here is a net of a square-based pyramid.
The vertex marked with a red dot meets another vertex.
Label this with a dot.

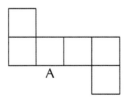

14 This is a net of a cube.
Which edge meets the edge labelled A?
Label the edge B.

12.3 Volume of a cuboid

- I can use a simple formula to work out the volume of a cuboid
- I can work out the capacity of a cuboid

Volume is the amount of space occupied by a three-dimensional (3D) shape.

The volume of a cuboid = length × width × height

You can write this as a formula, as $V = l \times w \times h = lwh$

Capacity is the amount of space inside a hollow 3D shape.

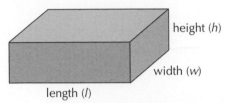

Develop fluency

Calculate the volume of this cuboid.

The formula for the volume of a cuboid is:

$V = lwh$

$= 5 \times 4 \times 3$

$= 60 \, cm^3$

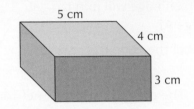

1 Work out the volume of each cuboid.

a
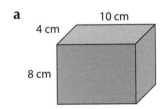

b
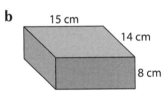

c
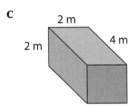

2 Work out the volume of a hall that is 30 m long, 20 m wide and 10 m high.

3 The diagram shows the dimensions of a swimming pool.
Work out the volume of the pool, giving the answer in cubic metres.

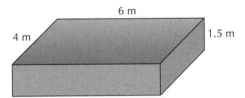

4 Work out the volume of this block of wood, giving your answer in cubic centimetres.

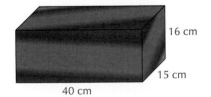

5 The volume of a cuboid is $120 \, cm^3$.
Its length is 3 cm and its width is 4 cm.
Work out its height.

6 Copy and complete the table of cuboids **a** to **e**.

	Length	Width	Height	Volume
a	6 cm	4 cm	1 cm	
b	3.2 m	2.4 m	0.5 m	
c	8 cm	5 cm		120 cm³
d	20 mm	16 mm		960 mm³
e	40 m	5 m		400 m³

7 Work out the volume of cubes with these edge lengths.

a 2 cm   b 5 cm   c 12 cm

8 A box in the shape of a cuboid measures 2 cm × 4 cm × 6 cm.
How many cubes with edge 2 cm will fit in the box?

Reason mathematically

Calculate the volume of the shape shown.

The shape is made up of two cuboids measuring
7 m by 3 m by 2 m and 2 m by 3 m by 6 m.

So, the volume of the shape is given by:

$V = (7 \times 3 \times 2) + (2 \times 3 \times 6)$

$= 42 + 36$

$= 78 \, m^3$

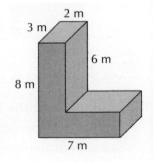

9 Work out the volume of each compound 3D shape.

a

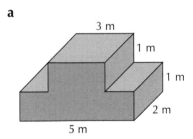

b

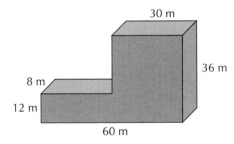

10 A tank measures 30 cm by 40 cm by 50 cm.
It is filled with water to half its height.
1 litre = 1000 cm³
Show that there are 30 litres of water in the tank.

11 Which is bigger, a cube with edge 5 cm or a cuboid of length 4 cm, width 5 cm and height 6 cm?
Show your working.

Solve problems

How many small cuboids measuring 2 cm by 3 cm by 4 cm can fit in a large cuboid measuring 6 cm by 12 cm by 20 cm?

$6 \div 2 = 3$, so you can fit 3 along the length.

$12 \div 3 = 4$ so you can fit 4 along the width.

and $20 \div 4 = 5$, so you can fit 5 in the height.

This gives $3 \times 4 \times 5 = 60$ small cuboids.

12 How many packets of sweets that each measure 8 cm by 5 cm by 2 cm can be packed into a cardboard box that measures 32 cm by 20 cm by 12 cm?

13 The diagram shows three different packaging boxes.

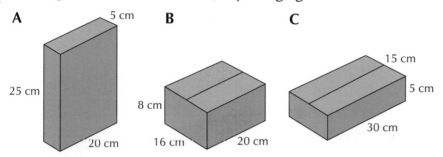

Which box has the greatest volume?

14 Look again at the boxes in question 13.
Will box B fit inside box A?
Show your working.

15 Look again at the boxes in question 13.
Will box A fit inside box B?
Show your working.

12.4 Surface area of a cuboid

● I can find the surface areas of cuboids

A **cuboid** is a 3D shape that has six rectangular faces. Their length, width and height can all be different. You can find the **surface area** of a cuboid by working out the total area of its six faces.

Area of top and bottom faces = 2 × length × width = $2lw$

Area of front and back faces = 2 × length × height = $2lh$

Area of the two sides = 2 × width × height = $2wh$

So the surface area of a cuboid = $S = 2lw + 2lh + 2wh$.

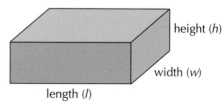

Develop fluency

Work out the surface area of this cuboid.

The formula for the surface area of a cuboid is:

$S = 2lw + 2lh + 2wh$

$= (2 \times 5 \times 4) + (2 \times 5 \times 3) + (2 \times 4 \times 3)$

$= 40 + 30 + 24$

$= 94 \, cm^2$

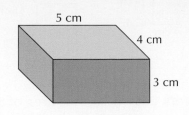

Work out the surface area of this cube.

There are six square faces and each one has an area of $3 \times 3 = 9 \, cm^2$.

So the surface area of the cube is $6 \times 9 = 54 \, cm^2$.

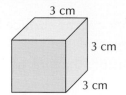

1   Work out the surface area of each cuboid.

a

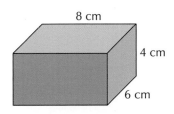

b

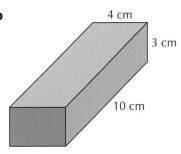

c

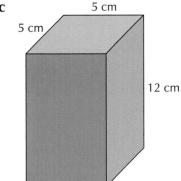

d

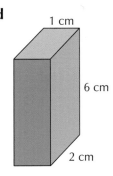

2   A cuboid measures 5 cm by 6 cm by 7 cm. Work out its surface area.

3   A cuboid measures 9 cm by 6 cm by 7 cm. Work out its surface area.

4   A cuboid measures 9 cm by 6 cm by 12 cm. Work out its surface area.

5   Work out the surface area of the cereal packet.

6 Work out the surface area of each cube.

 a 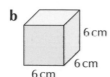 4 cm 4 cm 4 cm   b 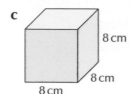 6 cm 6 cm 6 cm   c  8 cm 8 cm 8 cm   d 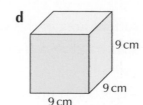 9 cm 9 cm 9 cm

7 Work out the surface area of a cube with an edge length of:

   a 1 cm   b 5 cm   c 19 cm   d 12 cm.

8 Work out the surface area of the outside of this open water tank.

Consider the tank as a cuboid without a top.

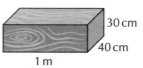
3 m, 1 m, 8 m

9 A cuboid measures 6 cm by 6 cm by 10 cm.

Work out the area of one of the larger faces.

10 A cuboid measures 4 cm by 6 cm by 8 cm. Work out the area of one of the smaller faces.

Reason mathematically

The volume of a cube is 64 cm³.

Show that the surface area is 96 cm².

$4^3 = 64$ so each edge is 4 cm in length.

The area of one face is $4 \times 4 = 16$ cm².

So the total surface area is $16 \times 6 = 96$ cm².

11 The volume of a cube is 27 cm³.
Show that the surface area is 54 cm².

12 Show that the surface area of this block of wood is 16 400 cm².

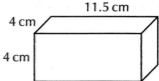
30 cm, 40 cm, 1 m

13 Which has the greater surface area; the cuboid or the cube, and by how much?

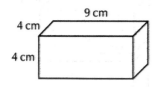
9 cm, 4 cm, 4 cm   5 cm

14 A cube has the same surface area as this cuboid.
Show that one edge of the cube is 6 cm long.

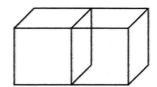
11.5 cm, 4 cm, 4 cm

15 Two identical cubes are stuck together, as shown, to make a cuboid.

Give a reason why the surface area of the cuboid is not double the surface area of one cube.

Solve problems

Work out the surface area of this 3D shape.

For the bottom cuboid

Area of base is $10 \times 5 = 50\,\text{cm}^2$

$\qquad$ = area of front face

$\qquad$ = area of rear face

Area of left face is $5 \times 5 = 25\,\text{cm}^2$

$\qquad$ = area of right face

Area of top face showing is $5 \times 4 = 20\,\text{cm}^2$

For the top cuboid

Area of top face is $6 \times 5 = 30\,\text{cm}^2$

Area of left face is $6 \times 5 = 30\,\text{cm}^2$

$\qquad$ = area of right face

Area of front face is $6 \times 6 = 36\,\text{cm}^2$

$\qquad$ = area of rear face

Total surface area is $(50 + 50 + 50 + 25 + 25 + 20) + (30 + 30 + 30 + 36 + 36) = 382\,\text{cm}^2$

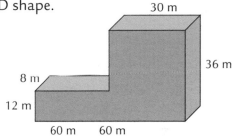

6 cm

6 cm

5 cm

5 cm

10 cm

16 Work out the surface area of this 3D shape.

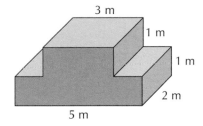

3 m

1 m

1 m

2 m

5 m

17 Work out the surface area of this 3D shape.

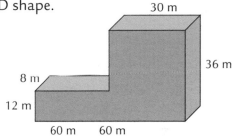

30 m

36 m

8 m

12 m

60 m $\quad$ 60 m

18 The surface area of a cube is $150\,\text{cm}^2$.
Two of these cubes are stuck together as shown.
Work out the surface area of the cuboid.

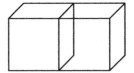

Now I can...

be familiar with the names of 3D shapes and their properties	construct 3D shapes from nets	use a simple formula to work out the volume of a cuboid
draw nets of 3D shapes	find the surface areas of cuboids	work out the capacity of a cuboid
use isometric paper to draw 3D shapes made from cubes		

CHAPTER 13 Introduction to probability

13.1 Probability words

- I know and can use the correct words about probability

Rolling a dice is an **event** that has several possible results, called **outcomes**. **Probability** helps you to decide how likely it is that each outcome will happen. An outcome may be impossible, very unlikely, unlikely, an even chance, likely, very likely or certain. These words can be shown on a probability scale.

Impossible Very unlikely Unlikely Evens Likely Very likely Certain

Other words you can use when describing probability, include 50–50 chance, probable, uncertain, good chance, poor chance.

If you flip a coin, or throw a dice, the outcome is **random**, which means you cannot predict the outcome. Taking a coloured ball from a bag without looking at it is 'taking the ball **at random**'.

Develop fluency

Match each of these outcomes to a position on a probability scale.

a It will snow in the winter in London.

b You will come to school in a helicopter.

c The next person to walk through the door will be male.

d When a normal dice is thrown the score will be 8.

a It usually snows in the winter, but not always, so this event is very likely.

b Unless you are very rich this is very unlikely to happen.

c As the next person to come through the door will be either male or female, this is an evens chance.

d An ordinary dice can only score from 1 to 6 so this is impossible.

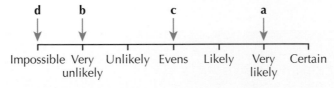

d b c a

Impossible Very unlikely Unlikely Evens Likely Very likely Certain

1 Match the events to the probability outcomes.

Events

A You are younger today than you were last year.
B You will score an odd number on an ordinary dice.
C Most people in your class went to sleep last night.
D Someone in your class has a pet dog.
E Someone in your class has a pet snake.
F Someone in your class has a birthday this month.
G Someone in your class has three sisters.

Probability outcome

a certain
b impossible
c fifty–fifty chance
d very unlikely
e likely
f very likely
g unlikely

Grid 1

Grid 2

2 A shape is picked at random from one of these two grids.
Copy and complete these sentences.
a Picking a … from grid … is impossible.
b Picking a … from grid … is likely.
c Picking a … from grid … is unlikely.
d Picking a … from grid … is very unlikely.
e Picking a … from grid … is fifty-fifty.

3 You pick one of these cards at random.
Put these outcomes in order of how likely they are,
with the least likely first.
A The number on the card will be a factor of 24.
B The number on the card will be a multiple of 2.
C The number on the card will be a multiple of 5.
D The number on the card will be a square number.
E The number on the card will be a multiple of 3.

4 These shapes are cut from card and put into a bag.

Triangle   Rectangle   Trapezium   Square   Kite   Parallelogram

Ben takes one shape out at random. Choose the most appropriate word from the list to complete each sentence.

a Taking a shape with four sides is … .

b Taking a shape with right angles is … .

c Taking a shape with three sides is … .

d Taking a shape with at least one acute angle is … .

impossible, very unlikely, unlikely, evens, likely, very likely, certain

5 Mae rolled a normal dice. State whether each statement below is true or false.

a There is a good chance she rolls a number greater than 4.

b There's an even chance that she rolls an odd number.

c It's certain she'll roll a number less than eight.

d It's unlikely that she'll roll a six.

6 Ollie is in a class of thirty students, all aged twelve.

There are 18 boys and 12 girls.

21 of the students have dark hair, the rest have fair hair.

4 of the students wear glasses.

Ollie sits next to another student at random.

Complete these sentences by referring to the information given about the class.

a Ollie sitting next to … is very unlikely.   b Ollie sitting next to … is likely.

c Ollie sitting next to … is certain.   d Ollie sitting next to … is unlikely.

7 Sophia travels to school in the centre of Manchester.

Pick a word from the list to complete each sentence.

a The chance that Sophia arrives by helicopter is … .

b The chance that Sophia arrives by bus is … .

c The chance that Sophia arrives by boat is … .

impossible, very unlikely, unlikely, evens, likely, very likely, certain

8 Theo is booked for a school trip to a country park in Yorkshire in the middle of July.

State whether each statement is true or false.

a There is a good chance the weather will be fine.   b It is certain to rain.

c It is very unlikely that it will snow.   d It is likely to be cloudy day.

9 Amber has a bag of mixed sweets that contains 2 toffees, 10 jellies, 20 mints and 28 chocolates.

While in the cinema, Amber chooses a sweet at random from the bag.

Pick a word from the list to complete each sentence.

a The chance that Amber chooses a toffee is … .

b The chance that Amber chooses a chocolate is almost … .

c The chance that Amber chooses a mint is … .

d The chance that Amber doesn't chooses a jelly is … .

impossible, very unlikely, unlikely, evens, likely, very likely, certain

10 Andrew is asked to choose a positive integer less than 10.

Put these outcomes in order of how likely they are, with the least likely first.

A The number chosen will be even.

B The number chosen will be prime.

C The number chosen will be a multiple of 3.

D The number chosen will be smaller than 8.

E The number chosen will be factor of 7.

Reason mathematically

Jess shuffles a normal pack of cards and says: 'The chance that I will draw a heart is $\frac{1}{4}$.'

Is Jess correct? Explain your answer.

There are 52 cards in a normal deck and 13 hearts. So, the chance of drawing a heart is $\frac{13}{52}$. This cancels down to $\frac{1}{4}$.

11 Bag A contains 10 red marbles, five blue marbles and five green marbles.

Bag B contains eight red marbles, two blue marbles and no green marbles.

A girl wants to pick a marble at random from a bag.

Which bag should she choose to have the better chance of picking:

a a red marble   b a blue marble   c a green marble?

Explain your answers.

12 A coin has been flipped five times and has landed 'heads' each time. If it is flipped again, is the chance of scoring another head likely, unlikely or fifty–fifty? Explain your answer.

13 Three pupils roll a dice. Eve rolls first, then Rob rolls and finally Helen rolls the dice.

Eve says she has the best chance of rolling a six as she is the first to roll.

Is she right? Explain your answer.

14 Matlock Town play Sheffield Wednesday in a cup game.

The Matlock manager says: 'We have an evens chance of winning as we either win or lose.'

Is he correct? Explain your answer.

Solve problems

It is claimed that if you ask people to pick a single-digit number (1–9) they are most likely to pick 7.

Laz carried out a survey. These are his results.

3, 6, 7, 1, 8, 3, 9, 2, 7, 5, 6, 3, 8, 1, 3, 7, 5, 2, 9, 7, 3, 4, 5, 8, 1, 3, 9, 8, 6, 7, 5, 3, 1, 4

Is the claim supported by these results?

Put the results in a table.

Number	1	2	3	4	5	6	7	8	9
Tally	4	2	7	2	4	3	5	4	3

These results show that 3 is the most likely number to be chosen, so this survey does not support the statement.

15 You have five rods with lengths, in metres (m), as shown.

A —2 m— B —3 m— C —4 m— D —5 m— E —6 m—

Three rods are picked at random and, if it is possible, are made into a triangle.

For example if A, C and D are chosen they make a triangle. If A, B and E are picked they do not make a triangle.

There are 10 different combinations of the three rods that can be picked from the five rods, for example, ABC, ABE, DCE.

Work out the 10 different combinations.

Which is more likely; that the three rods chosen will make a triangle or the three rods chosen will not make a triangle?

16 In a game Divock rolls two dice.

He wins the game if he rolls a double.

He loses the game if he rolls a total greater than eight, unless he rolls a double.

Is he more likely to win or lose the game?

17 Jess saw a February weather chart for 1950, the year her Grandad was born.

Monday	Tuesday	Wednesday	Thursday	Friday	Saturday	Sunday
		1st snow	2nd snow	3rd snow	4th rain	5th sunny
6th sunny	7th sunny	8th rain	9th rain	10th snow	11th snow	12th snow
13th snow	14th cloudy	15th rain	16th rain	17th cloudy	18th snow	19th snow
20th cloudy	21st sunny	22nd sunny	23rd rain	24th rain	25th cloudy	26th snow
27th sunny	28th sunny					

Her Grandad said it was snowing when he was born. What day of the week was he most likely to have been born?

18 Joe chooses a positive integer less than 20. Which is he more likely to have chosen; a prime number or an even number?

13.2 Probability scales

- I know about and can use probability scales from 0 to 1
- I can work out probabilities based on equally likely outcomes

Probability can be shown on a **probability scale** from 0 to 1.

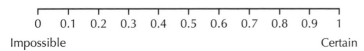

Impossible   Certain

The probability of an outcome is written as

$$P(\text{outcome}) = \frac{\text{number of ways the outcome can occur}}{\text{total number of possible outcomes}}$$

where P stands for 'the probability of' and the outcome is written inside the brackets.

For a fair coin, each outcome is equally likely to happen so $P(H) = \frac{1}{2}$ and $P(T) = \frac{1}{2}$.

This is the **probability fraction** for the event.
The probability can also be given as $P(H) = 0.5$ or $P(H) = 50\%$.

Probability fractions can be simplified, but sometimes it can be better to leave them unsimplified, especially if you are comparing them.

Faith picks a card at random from a normal pack of 52 cards.

Find:

a P(king)   **b** P(spade)   **c** P(picture card)

d P(queen of hearts)   **e** P(diamond picture)   **f** P(red card).

a Each of the four suits has a king, so $P(\text{king}) = \frac{4}{52}$.

b Each suit has 13 cards, so $P(\text{spade}) = \frac{13}{52}$.

c Each suit has three picture cards, jack, queen and king, so $P(\text{picture card}) = \frac{12}{52}$.

d There is only one queen of hearts in the pack, so $P(\text{queen of hearts}) = \frac{1}{52}$.

e There are three picture cards in each suit, so $P(\text{diamond picture}) = \frac{3}{52}$.

f The hearts and diamonds are both red suits, 13 in each suit, so $P(\text{red card}) = \frac{26}{52}$.

1 **a** A shape is chosen at random from grid 1.
       What is the probability it will be:

   i a triangle   **ii** a circle

   iii a square   **iv** a rectangle

   v a hexagon?

   b A shape is chosen at random from grid 2.
       What is the probability it will be:

   i a triangle   **ii** a circle

   iii a square   **iv** a rectangle

   v red?

   Give your answers as fractions in their
   simplest form.

Grid 1

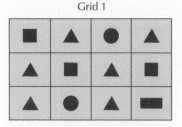

Grid 2

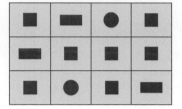

2 A bag contains 10 red marbles, five blue marbles and five green marbles.
   If a marble is chosen at random from the bag, what is the probability it will be:

   a a red marble   **b** a blue marble   **c** a green marble?

   Give your answers as fractions in their simplest form.

3 Cards numbered 1 to 10 are placed in a box.
   A card is drawn at random from the box.
   Find the probability that the card drawn will be:

   a a 5   **b** an even number   **c** a number in the 3 times table

   d a 4 or an 8   **e** a number less than 12   **f** an odd number.

   Give your answers as decimals.

4 A bag contains five red discs, three blue discs and two green discs.
   Linda takes out a disc at random.
   Find the probability that she takes out:

   a a red disc   **b** a blue disc   **c** a green disc

   d a yellow disc   **e** a red or blue disc.

   Give your answers as decimals.

5 Syed is using a fair, eight-sided spinner in a game.
   What is the probability that he scores:

   a 0   **b** 1   **c** 2   **d** 3?

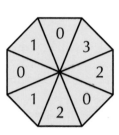

6 50 rings are placed into a bag. Ten are gold, four are silver, eight are copper and the rest
   are plastic. A ring is chosen at random from the bag.
   What is the probability that it is:

   a gold   **b** silver   **c** copper   **d** plastic?

   Give your answers as decimals.

7 Hugh has 12 dominoes.

Hugh choses one domino at random. What is the probability that it:

a is a double

b has a six

c has a total of six

d has a one

e has no dots

f has a total less than 10?

8 Joy kept a June weather chart.

Monday	Tuesday	Wednesday	Thursday	Friday	Saturday	Sunday
1st sunny	2nd sunny	3rd sunny	4th cloudy	5th sunny	6th sunny	7th sunny
8th rain	9th rain	10th sunny	11th sunny	12th sunny	13th sunny	14th cloudy
15th rain	16th rain	17th cloudy	18th sunny	19th sunny	20th cloudy	21st sunny
22nd sunny	23rd rain	24th rain	25th cloudy	26th sunny	27th sunny	28th sunny
29th rain	30th rain					

What was the probability of a day in June being:

a sunny

b cloudy

c rainy?

9 Mr Charlton has a draw full of ties. He has twelve blue ties, three red ties, eight green ties and two yellow ties. One morning he takes a tie out at random.

What is the probability that this tie is:

a blue

b red

c green

d yellow?

10 Padmini plants 40 tulip bulbs, eight yellow, twelve red, the rest are purple.

What is the probability of the first one to flower is:

a red

b yellow

c white

d purple?

Give your answers as fractions in their simplest form.

Reason mathematically

A bag contains eight red beads, four blue beads, three green beads and one white bead.

Evie takes out a bead at random.

Explain how she knows the chance of choosing a red bead is exactly the same as the chance of choosing a bead that is blue, green or white.

She sees that the number of blue, green and white beads is 8, the same as for red, so both have a chance of $\frac{8}{16}$.

11 Adam picks a card at random from a normal pack of 52 playing cards.

    a Find:

      i P(jack)   **ii** P(heart)   **iii** P(queen or king)

      iv P(ace of spades)   **v** P(9 or 10)   **vi** P(black card).

    Give your answers as fractions with a denominator of 52.

    b How can you tell easily that the probability of choosing a jack is higher that the probability of choosing the ace of spades?

12 Sandhu rolls a fair dice.

    a Calculate:

      i P(3)   **ii** P(odd number)   **iii** P(5 or 6)

      iv P(even number)   **v** P(6)   **vi** P(1 or 6).

    b Explain how you can tell that P(prime number) is the same as P(odd number).

13 Emma and Nazir are playing a game with spinners.

    a Who has more chance of spinning a three?

    b Suggest a way in which they could change the spinners, so that they both have the same chance of scoring a three.

 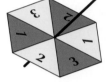

Emma's spinner   Nazir's spinner

14 Alia rolls a fair dice that has more than six faces.

    She rolls it fifty times and it lands on a 6, six times.

    Briony says: 'There's a good chance that was a dice with eight faces.'

    Is Briony correct? Explain your answer.

Solve problems

The letters of the word MATHS are each written on a card and put into a bag.

Sam takes out three letters at random.

What is the probability that he takes out the three letters in his name?

Find the possible combinations of three different letters.

MAT, MAH, MAS, MTH, MTS, MHS, ATH, ATS, AHS, THS

There are ten different combinations, but only one of these gives the letters SAM.

The probability that Sam takes out letters to make SAM is $\frac{1}{10}$.

15 Mr Evans has a box of 25 calculators, but five of them do not work very well.

    What is the probability that the first calculator taken out of the box at random does work very well?

    Write your fraction as simply as possible.

16 Tim chooses a ticket at the start of a tombola. There are 300 tickets inside the drum with 60 winning tickets available.

Halfway through the tombola, Kath chooses a ticket. There are 150 tickets inside the drum with 35 winning tickets.

Who has the better chance of choosing a winning ticket?

17 A weather forecaster estimates the chance of rain to be 15%, with a 0.9 chance of black ice, and a 1 in 8 chance of snow.

What is the probability of: **a** snow   **b** no rain   **c** no black ice?

Give each answer as a fraction.

18 Naz and Mal have a bag containing one £1, two 50p, three 10p and four 20p coins. Naz takes a coin out.

a What is the probability that he takes a 10p?

Naz puts the coin back and Mal takes two coins,

b What is the probability he takes out 30p?

13.3 Experimental probability

- I can understand experimental probability
- I can understand the difference between theoretical probability and experimental probability

Develop fluency

A probability calculated from equally likely outcomes is a **theoretical probability**. If you carry out a series of experiments, recording the results in a frequency table, you can find the **experimental probability**. You must repeat the experiment a number of times and each separate experiment is a **trial**.

$$\text{Experimental probability of an outcome} = \frac{\text{number of times the outcome occurs}}{\text{total number of trials}}$$

When you repeat an experiment, the experimental probability will be slightly different each time. The experimental probability is an estimate for the theoretical probability. As the number of trials increases, the value of the experimental probability gets closer to the theoretical probability.

An electrician wants to estimate the probability that a new light bulb lasts for less than 1 month.

He fits 20 new bulbs and 3 of them fail within 1 month.

What is his estimate of the probability that a new light bulb fails?

3 out of 20 bulbs fail within 1 month, so his experimental probability $= \frac{3}{20}$.

1   Paul flips a coin 50 times and recorded his results in this table.

	Tally	Frequency
Head	ⵌ ⵌ ⵌ ⵌ ⵌ I	
Tail	ⵌ ⵌ ⵌ ⵌ IIII	

What is Paul's experimental probability of scoring:

a a head   **b** a tail?

2   Jade threw a dice 100 times. Her results are shown in the table.

Number on dice	1	2	3	4	5	6
frequency	16	17	19	17	16	15

What is the experimental probability of Jade rolling:

a a six   **b** a three   **c** a five   **d** a two?

3   Morwena dropped a drawing pin 50 times to see how many times it landed point up.
    She recorded her results in this table.

	Tally	Frequency
Point up	ⵌ ⵌ ⵌ ⵌ ⵌ ⵌ III	
Point down	ⵌ ⵌ ⵌ II	

What is the experimental probability of the drawing pin landing:

a point up   **b** point down?

4   Kieron shuffled a pack of cards, chose a card at random and recorded its suit in a tally
    chart. He then reshuffled the pack and repeated this until he had done it 100 times.
    His results are shown in the table.

Suit	Clubs	Diamonds	Hearts	Spades
Frequency	26	24	27	23

What is the experimental probability of Kieron choosing:

a a club   **b** a spade   **c** a diamond   **d** a heart?

5   Peda kept a tally of how many days it rained in May. His results are shown in the table.

	Tally	Frequency
Rain	ⵌ ⵌ I	
No rain	ⵌ ⵌ ⵌ ⵌ	

What is the experimental probability of there being:

a a day with some rain in May   **b** a day with no rain in May?

6 Tara did an experiment of dropping a box, which can land in three different ways.
Her results are shown in the table.

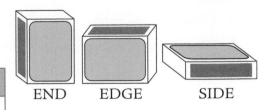

END   EDGE   SIDE

	Tally	Frequency
Side	ⅢⅢ ⅢⅢ ⅢⅢ ⅢⅢ IIII	
Edge	II	
End	I	

What is the experimental probability of the box landing:

a on its end   **b** on its edge   **c** on its side?

7 David often travelled on the train for work and made his tally of when the train was on time, up to ten minutes late and over ten minutes late.
His results are shown below.

	Tally	Frequency
On time	ⅢⅢ ⅢⅢ ⅢⅢ III	
Up to ten mins late	ⅢⅢ IIII	
Over ten mins late	III	

What is the experimental probability of the train being:

a more that 10 ten minutes late   **b** up to 10 minutes late   **c** on time?
Give your answers as decimals.

8 Brian complained that he was blamed for everything. He kept a tally of whenever he or one of his brothers was blamed.
Over one week he counted the number of times somebody was blamed.
Brian was blamed 18 times, David 11 times, Malcolm 4 times and Kevin only once!
Something went wrong, what is the experimental probability for blaming:

a Kevin   **b** Malcolm   **c** David   **d** Brian?

Reason mathematically

A company manufactures items for computers.
The numbers of faulty items are recorded in this table.

Number of items produced	Number of faulty items	Experimental probability
100	8	0.08
200	20	
1000	82	

a   Complete the table.

b   Which is the best estimate of the probability of an item being faulty? Explain your answer.

a

Number of items produced	Number of faulty items	Experimental probability
100	8	0.08
200	20	0.1
1000	82	0.082

b   The last result (0.082) is the best estimate, as it is based on more results.

9   A girl wishes to test whether a dice is biased. She rolls the dice 60 times.
    The results are shown in the table.

Score	1	2	3	4	5	6
Frequency	6	12	10	9	15	8

a   Do you think the dice is biased?
Give a reason for your answer.

b   How could she improve the experiment?

c   From the results, estimate the probability of rolling a 2.

d   From the results, estimate the probability of rolling a 1 or a 4.

10   Alec spun a coin on the table 30 times and recorded how it landed.
    It landed on a head ten times.   Alec said this coin is biased.
    Is Alec correct? Explain your answer.

11   Chris recorded the colour of the traffic lights at a junction each time he went through it for a week.
    They were green 12 times, amber twice and red 21 times.
    Chris said: 'There's always a good chance that I will get a red light at the junction.'
    Is Chris correct? Explain your answer.

12   Kirsty caught a bus to school each day. She kept a record of how late it was over one month.
    She found it was on time eight times, up to five minutes late 12 times and over five minutes late twice.
    Kirsty told her dad: 'The bus is generally on time.'
    Is Kirsty correct? Explain your answer.

Solve problems

A dentist keeps a record of the fillings he gives to 100 patients over a week.

36 patients had one filling, 7 patients had more than one filling. The rest had no fillings.

What is the probability of a patient not needing a filling?

Calculate the number that don't need a filling as 100 − (36 + 7) = 57

Experimental probability will be $\frac{57}{100}$

13 Eva rolled two dice. What is the probability that she will roll:

    a a double six   **b** a total greater than nine   **c** any double?

14 The number of days it rained in Wales was recorded over different periods.
The results are below.

Period of days	20	40	60	100	200
Number of days it rained	8	22	28	51	96

    a What is the best estimate of the probability of it raining in Wales?

    b Estimate the probability of it not raining.

    c Is there a greater chance of it not raining or raining in Wales?

15 Jake rolls a fair six-sided dice.
Is he most likely to roll a prime number, a square number or a factor of six?

16 Joy had a bag of sweets containing 8 toffees and 12 mints.
Her brother takes some toffees out of the bag.
The probability of Joy now taking a mint at random is 0.8.
How many toffees did Joy's brother take?

Now I can...

use the correct words about probability	use probability scales from 0 to 1	understand the difference between theoretical probability and experimental probability
know about probability scales from 0 to 1	work out probabilities based on equally likely outcomes	
understand experimental probability		

CHAPTER 14 Ratio, proportion and rates of change

14.1 Introduction to ratio

- I can use ratio notation
- I can use ratios to compare quantities

A **ratio** is a mathematical way to compare **quantities**.

Develop fluency

The mass of a lion is 150 kg.	The lion is heavier than the cat.
The mass of a domestic cat is 5 kg.	150 ÷ 5 = 30 so the lion is 30 times as heavy as the cat.
What is the ratio of the mass of the lion to the mass of the cat?	The ratio of the mass of the lion to the mass of the cat is 30 : 1.

1   Theo is making a chain with beads.
    He uses 20 white beads and 4 black beads.
    Complete the sentence.
    The ratio of white beads to black beads is ... : 1

2   Ade has saved £25 and Bea has saved £100.
    a   What is the missing number in this sentence?
        Bea has saved ... times as much as Ade.
    b   Work out the ratio of Ade's savings to Bea's savings.
    c   Work out the ratio of Bea's savings to Ade's savings.

3   Gary buys 500 g of rice, 250 g of pasta and 125 g of coffee.
    Work out the ratio of:
    a   the mass of rice to the mass of pasta   b   the mass of pasta to the mass of rice
    c   the mass of rice to the mass of coffee   d   the mass of coffee to the mass of pasta.

4   These are the ingredients to make 8 cheese scones.
    Work out the ratio of:
    a   the mass of flour to the mass of cheese
    b   the mass of cheese to the mass of butter
    c   the mass of butter to the mass of flour.

Flour	200 g
Butter	25 g
Cheese	100 g
Eggs	1
Milk	2 tablespoons

5   To make concrete, a website gives these ingredients.
    a   Write down these ratios.
        i   gravel : cement   ii   gravel : water
        iii   sand : water   iv   water : cement
    b   Kaspar uses three buckets of cement. How many
        buckets of water does he need?

6 parts of gravel
5 parts of sand
2 part of cement
1 part of water

6 The number of pages in a maths book is 25% of the number of pages in a science book. Work out the ratio of the numbers of pages in the two books.

7 a Cara's age is half of Dani's age. What is the ratio of their ages?　**b** Elfine's age is two-thirds of Frank's age. What is the ratio of their ages?

8 This pie chart shows an election result. There are four parties. Work out the ratio of:

  a Union votes to Liberal votes

  b Liberal votes to Democrat votes

  c Democrat votes to Social votes.

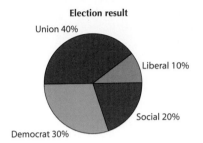

Election result

Union 40% · Liberal 10% · Social 20% · Democrat 30%

9 25% of the spectators at a football match are female. The rest are male. Work out the ratio of males to females.

10 There are 50 men and 200 women at a show. Write the ratio of men to women as simply as possible.

Reason mathematically

Bottled water comes in two sizes, 500 millilitres (ml) and 750 ml.

Show that the ratio of the two sizes is 2 : 3.

The larger bottle is $1\frac{1}{2}$ times bigger than the smaller one.

$1\frac{1}{2} = \frac{3}{2}$

Imagine each bottle, divided into 250 ml sections.

There are 2 sections in the smaller bottle and 3 sections in the larger bottle.

So the ratio of the smaller size to the larger size is 2 : 3.

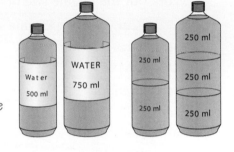

11 The number of pages in a Michael Morpurgo book is 75% of the number of pages in an Anthony Horowitz book.

Show that the ratio of the number of pages in these two books is 3 : 4.

12 There are 20 times as many people in India as in the UK.

The population of the UK is 60 million.

Sanjay says that this means the population of India is 1200 million.

Is he correct?

Give a reason for your answer.

13 The ratio of males to females in the UK is 97 : 100.
Which statement is correct?
A: There are more men than women in the UK.
B: There are the same number of men and women in the UK.
C: There are more women than men in the UK.
Give a reason for your answer.

Solve problems

An electric car has a choice of battery sizes: 30 kWh, 40 kWh or 60 kWh.

Which battery sizes are in the ratio 2 : 3?

Show your working.

$30:40 = 3:4$

$30:60 = 1:2$

$40:60 = 2:3$

so it is the 40 kWh and the 60 kWh

14 The size of the engine in a car is given in litres.
The table shows the sizes of some car engines.
a How many times larger than the engine of the Seat Ibiza is the engine of the AC Cobra?
b Work out the ratios of these engine sizes.
i Seat Ibiza : AC Cobra   ii Jaguar XJ8 : Ford Granada   iii Ford Granada : Seat Ibiza
c The ratio of the engine size of a Lexus LFA to the engine size of a Seat Ibiza is 3 : 1.
Work out the engine size of a Lexus LFA.

Car	Engine size (litres)
Seat Ibiza	1.4
Mazda MX5	1.6
Ford Granada	2.1
Jaguar XJ8	4.2
AC Cobra	7.0

15 The table shows the populations of some countries.

Country	Population (millions)
Pakistan	180
Nigeria	160
Germany	80
UK	60
Iraq	30
Sri Lanka	20

a Work out the ratios of the populations of:
i the UK to Sri Lanka
ii the UK to Pakistan
iii Nigeria to Germany
iv Nigeria to Sri Lanka.
b Suppose the population of the Sri Lanka is P.
Write the population of each of the other countries, in terms of P.

16 On a draughts board, there are two black pieces for every one white piece.
There are 18 pieces on the board.

    a How many black pieces are there?   **b** What proportion of the pieces is white?

14.2 Simplifying ratios

● I can write a ratio as simply as possible

Ratios such as $5:1$ or $3:2$ are in simplest form as all the terms are integers and are as simplified as much as possible.

A ratio such as $8:4$ is not in simplest form because it will simplify to $2:1$.

A ratio such as $2\frac{1}{2}:1$ or $2.5:1$ is not in simplest form because it will simplify to $5:2$.

Develop fluency

The heights of two office blocks are 12 metres and 32 metres.
What is the ratio of the heights of the office blocks?

You can write the ratio as $12:32$ as the units are the same.

You can simplify a ratio in the same way as a fraction: divide both numbers by a common factor.

4 is a common factor of 12 and 32.

$12:32 = 3:8$

This is the ratio in its simplest form.

1 Simplify these fractions as much as possible.

    a $\dfrac{10}{15}$ **b** $\dfrac{16}{80}$ **c** $\dfrac{27}{18}$ **d** $\dfrac{150}{250}$ **e** $\dfrac{7}{210}$

2 Simplify these ratios as much as possible.

    a $5:20$ **b** $12:18$ **c** $100:50$ **d** $32:4$ **e** $40:60$

3 In a wood there are 40 birch trees and 100 ash trees.

    a Work out the ratio of birch trees to ash trees.

    b Work out the ratio of ash trees to birch trees.
       Give your answers in their simplest form.

4 A drink is made from 50 ml of squash and 175 ml of water.
Work out the ratio of squash to water.

5 A council gardener plants these bulbs.

    250 daffodils   100 crocuses   150 snowdrops

    Work out the ratio of the number of:

    a daffodil bulbs to crocus bulbs   **b** crocus bulbs to snowdrop bulbs

    c snowdrop bulbs to daffodil bulbs.

    Give your answers in their simplest form.

6 Two suitcases weigh 16 kg and 24 kg.

Work out the ratio of the masses of the suitcases.

Give your answer in its simplest form.

7 The Gherkin building in London is 180 metres high. The London Eye is 135 metres high.

Work out the ratio of their heights.

Give your answer in its simplest form.

8 An Embraer 190 aircraft has 98 seats. An Embraer 170 aircraft has 76 seats.

Work out the ratio of the numbers of seats as simply as possible.

9 A display has 300 coloured lights.

120 are white, 60 are blue, 45 are green and the rest are yellow.

Work out the ratio of:

a white to blue   **b** blue to green   **c** green to yellow.

10 Write each ratio in its simplest form.

a $1\frac{1}{2}:1$ **b** $1\frac{1}{2}:2$ **c** $2\frac{1}{2}:1$ **d** $2\frac{1}{2}:3$ **e** $5:1\frac{1}{4}$

Reason mathematically

The body of an adult female is about 60% water.

Show that the ratio of water to other substances in the body of an adult female is $3:2$.

60% is water so 100% − 60% = 40% is other substances.

So the ratio is 60 : 40 which simplifies to 6 : 4 and then to 3 : 2.

11 The body of an adult male is about 65% water.

Show that the ratio of water to other substances in the body of an adult male is $13:7$.

12 Show that the ratio of yellow squares to white squares in this pattern is $4:3$.

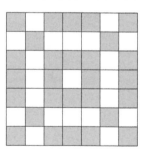

Solve problems

A pizza is shared in the ratio $2:3$.

What fraction of the pizza is the smaller share?

There are 5 parts altogether, so the smaller share is $\frac{2}{5}$ of the pizza.

13 Jack plays bowls.

He wins $\frac{1}{4}$ of his matches.

There are no draws.

Work out the ratio of wins to losses.

14 These are the ingredients to make a 900 ml mixture for Yorkshire puddings.

> 300 grams of plain flour
>
> 6 eggs (this is about 300 ml)
>
> 300 ml milk

  a If a millilitre and a gram is of equivalent proportion, what fraction of the mixture is milk?

  b What is the ratio of milk to other ingredients?

15 You can compare the masses of different elements by using their relative atomic masses. The table shows the values for some common elements.

Element	Relative atomic mass
hydrogen	1
carbon	12
oxygen	16
sulphur	32
calcium	40

  a Work out the ratio of the relative atomic masses of:

    i calcium to hydrogen

    ii calcium to oxygen

    iii calcium to sulphur.

  b Find two elements with relative atomic masses in the ratio $2:1$.

14.3 Ratios and sharing

● I can use ratios to find totals or missing quantities

As well using ratios to compare two different quantities, you can apply them when sharing or dividing a given quantity into different amounts.

Develop fluency

Divide £20 000 in the ratio 1 : 4.

The total number of parts is 1 + 4 = 5.

Imagine that £20 000 is divided into five equal parts.

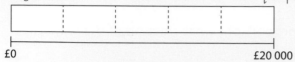

£0   £20 000

1 part is $\frac{1}{5}$ of the money and 4 parts is $\frac{4}{5}$ of the money.

$\frac{1}{5}$ of 20 000 = 4000

$\frac{4}{5}$ of 20 000 = 4 × 4000 = 16 000

So, £20 000 divided in the ratio 1 : 4 gives £4000 and £16 000.

In the example above you knew the total amount. If you know one of the shares, you can use this to work out the total amount.

Alesha makes a drink for her children by mixing juice and lemonade in the ratio 3 : 5.

She uses 250 ml of lemonade.

How much of the drink will there be altogether?

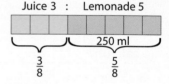

The number of parts is 3 + 5 = 8.

Juice is $\frac{3}{8}$ of the drink and lemonade is $\frac{5}{8}$ of the drink.

$\frac{5}{8}$ of the drink = 250 ml

So, $\frac{1}{8}$ of the drink = 250 ÷ 5 = 50 ml

And $\frac{8}{8}$ of the drink = 50 × 8 = 400 ml

There are 400 ml of the drink altogether.

1   Divide each amount in the given ratio.

   a 60 kg in the ratio 1 : 2   **b** £40 in the ratio 3 : 1

   c 180 litres in the ratio 5 : 1   **d** 90 grams in the ratio 5 : 4

   e 30 cm in the ratio 2 : 3   **f** 24 hours in the ratio 3 : 4 : 1

2   In each question part, the amount given in brackets represents the smaller share. Work out the amount of the larger share.

   a 1 : 3 (20 cm)   **b** 5 : 1 (12 litres)   **c** 2 : 3 (£8)

3   In each part, the amount given in brackets represents the larger share. Work out the amount of the smaller share.

   a 2 : 1 (10 kg)   **b** 7 : 2 (28 miles)   **c** 3 : 2 (36 g)

4 In a class of children, the ratio of swimmers to non-swimmers is 5 : 1.

  a What fraction of the class are swimmers?

  b There are 30 children in the class. How many are swimmers?

5 A farmer has grown oranges and lemons.
The ratio of oranges to lemons is 4 : 1.

  a What fraction of the fruit are oranges?

  b There are 2000 oranges. How many lemons are there?

6 Two waitresses shared their tips in the ratio 3 : 2.

  a What fraction is the smaller share?

  b The total of all the tips was £45.00. How much is the smaller share?

7 In a small library, the ratio of fiction to non-fiction books is 1 : 8.

  a What fraction are fiction books?

  b There are 120 fiction books. How many books are there all together?

8 100 people see a film at a cinema.
The numbers of children and adults are in the ratio 1 : 4.
How many children see the film?

9 In a fishing contest the numbers of trout and carp caught were in the ratio 1 : 2.
The total number of trout and carp caught was 72.
How many carp were caught?

Reason mathematically

There are 120 trollies at a supermarket.

The ratio of small trollies to big trollies is 5 : 3.

How many more small trollies than big trollies are there?

There are 5 + 3 = 8 parts altogether.

So, 8 parts is 120 trollies.

1 part is 120 ÷ 8 = 15 trollies.

5 parts – 3 parts is 2 parts so there are 15 × 2 = 30 more small trollies than big trollies.

Note: 5 parts is 5 × 15 = 75 and 3 parts is 3 × 15 = 45. So, 75 small and 45 big trollies.

10 At a concert the number of men to women is in the ratio 3 : 2.
There are 150 people altogether.
How many more men than women are at the concert?

11 Jai has £36 more than Kay.
The money they have is in the ratio 5 : 1.
How much do they have altogether?

12 Harriet and Richard go shopping separately.
They buy 66 items altogether.
Harriet buys twice as many items as Richard.
How many items does Harriet buy?

Solve problems

The scale on a map is 1 cm to 5 km.

  a Write this as a ratio in its simplest form.

  b Work out the actual distance that is represented by 3.5 cm on the map.

  a 1 km = 1000 m = 100 000 cm

    So, 5 km = 500 000 cm.

    So, ratio is 1 : 500 000.

  b 3.5 cm represents 5 km × 3.5 = 17.5 km.

13 The scale on a map is 1 cm to 4 km.

  a Write this as a ratio in its simplest form.

  b Work out the actual distance that is represented by 6 cm on the map.

14 The scale on a map is 1 cm to 2 km.

  a Write this scale as a ratio in its simplest form.

  b Work out the distance on the map that represents 7 km.

15 Dominic has 20 coloured sweets left in a packet. These are in the ratio red : green = 1 : 4.

  a How many of each colour are there?

  b If he eats one red sweet and four green sweets, what is the new ratio of red : green?
Comment on your answer.

  c If he only ate one of each colour, what would the ratio change to?

14.4 Ratios in everyday life

● I can understand the connections between fractions and ratios

● I can understand how ratios can be useful in everyday life

Ratios are often useful in practical situations.

Develop fluency

150 students are going on a school trip.

The teacher-to-student ratio must be 1 : 8 or better.

What is the smallest possible number of teachers that can go on the trip?

If the ratio of teachers to students is 1 : 8 then the students make up $\frac{8}{9}$ of the total number and the teachers make up $\frac{1}{9}$ of the total number.

Teachers 1 :   Students 8

150

$\frac{1}{9}$ $\frac{8}{9}$

If $\frac{8}{9}$ of the total is 150, then $\frac{1}{9}$ of the total is $150 \div 8 = 18.75$.

The smallest possible number of teachers is 19. (You cannot take $\frac{3}{4}$ of a teacher!)

1   There are 350 students in a primary school.
    The ratio of boys to girls is 3 : 2.
    How many students are boys and how many are girls?

2   Freda has downloaded some music tracks.
    75% of them are dance tracks.
    What is the ratio of dance tracks to other music?

3   Two-thirds of the passengers on a bus one morning are children on the way to school.
    What is the ratio of schoolchildren to other passengers?

4   Sand and cement are mixed in the ratio 4 : 1.
    A builder has five 25 kg bags of cement.
    How much sand does he need if he uses all the cement?

5   In a band, the ratio of brass to wind players is 2 : 3.
    There must be at least 6 brass players in the band.
    What is the smallest number of players in the band altogether?

6   A drink is made from orange juice and water in the ratio 1 : 3.
    How many litres of the drink can be made from a 500 ml bottle of orange juice?

7   10% of students in a school are left-handed.
    1080 are right-handed.
    Work out the number of left-handed students.

8   The ratio of women to men at a sewing club is 5 : 1.
    There are 7 men.
    How many women are there?

9 A piece of rope is 10 metres long.

It is cut into two pieces so that the ratio of the lengths is 3 : 1.

How long is the bigger piece?

Reason mathematically

Three-quarters of a class are boys.

a What is the ratio of boys to girls?

b There are fewer than 30 pupils in the class altogether.

Show that there are fewer than 8 girls in the class.

a One quarter are girls so the ratio of boys to girls is 3 : 1.

b 8 girls would mean 3 × 8 = 24 boys which gives 8 + 24 = 32 in the class.

So, there must be fewer than 8 girls.

Note: 7 girls means 3 × 7 = 21 boys which gives 7 + 21 = 28 in the class.

10 Repairs to a house roof cost £8000.

Of this, the materials cost £3200. The rest is a labour charge.

Show that the ratio of the cost of materials to labour is 2 : 3.

11 a Marco's age is $\frac{2}{3}$ of Pierre's age.

What is the ratio of their ages?

b The total of their ages is 60 years.

Show that Pierre is 24 years old.

12 This pie chart shows the populations of each country in the United Kingdom.

a Use a ratio to compare the populations of England and Scotland.

b Use a ratio to compare the populations of Wales and Northern Ireland.

c What fraction of the population of the UK lives in England?

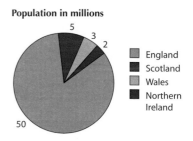

Population in millions

England
Scotland
Wales
Northern Ireland

13 Arthur has saved £720 and Guinevere has saved £1080.

a What is the ratio of their savings?

b What fraction of their total savings does Arthur have?

They both spend £360.

c What is the ratio of their savings now?

d What fraction of their total savings does Arthur have now?

Solve problems

The ratio of £1 to £2 coins in a jar is 5 : 3.

There are 96 of these coins in the jar.

How much are these coins worth?

5 + 3 is 8 parts altogether.

So, 8 parts is 96 coins and 1 part is 96 ÷ 8 = 12 coins.

Number of £1 coins is 5 × 12 = 60.

Number of £2 coins is 3 × 12 = 36.

The coins are worth £60 + 36 × £2 = £60 + £72 = £132.

14 James is saving 5p and 10p coins. He has 75 coins.
   The ratio of 5p to 10p coins is 7 : 8.
   How much are his coins worth?

15 There are 15 green and brown bottles on a wall.
   The ratio of green bottles to brown bottles is 1 : 4.
   One of the green bottles accidentally falls.
   What is the ratio of green bottles to brown bottles now?

16 On a train the ratio of first class seats to standard class seats is 1 : 6.
   The total number of seats is 140.
   110 people board the train.
   16 of these have first class seats.
   How many spare seats are there in
   a first class   **b** standard class?

Now I can...

use ratio notation	understand the connections between fractions and ratios	understand how ratios can be useful in everyday life
write a ratio as simply as possible	use ratios to find totals or missing quantities	
use ratios to compare quantities		

CHAPTER 15 Symmetry

15.1 Reflection symmetry

- I can recognise shapes with reflective symmetry
- I can draw lines of symmetry on a shape

If you can fold a 2D shape along a line, so that one half of the shape fits exactly over the other half, the fold line is a **line of symmetry** or **mirror line**. The shapes on each side of the line **reflect** each other. This property is called **reflection symmetry**. Some shapes have no lines of symmetry.

Develop fluency

Describe the symmetry of this shape.

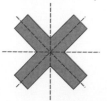

Look to see where you can fold the shape exactly over itself.

You will see four possible ways of folding the shape, which gives you the four lines of symmetry.

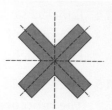

In questions 1–3, copy each shape and draw any lines of symmetry.

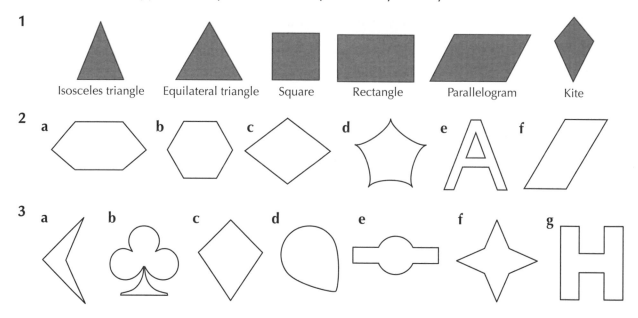

1
Isosceles triangle   Equilateral triangle   Square   Rectangle   Parallelogram   Kite

2 a   b   c   d   e   f

3 a   b   c   d   e   f   g

4 Copy the capital letters and draw in any lines of symmetry.

a **M** b **A** c **T** d **H** e **S** f **V** g **I** h **E** i **W**

Reason mathematically

Describe the symmetry of this shape.

This L-shape has no lines of symmetry.
You can use tracing paper or a mirror to check.

5 Write down the number of lines of symmetry for each shape.

   a   **b**   **c**   **d**

   e   **f**   **g**   **h**

6 Write down the number of lines of symmetry for each of these road signs.

   a   **b**   **c**   **d**

7 Here is a sequence of symmetrical shapes.

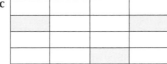

   a There is one line of symmetry that works for all these shapes.
Copy the shapes and draw this line of symmetry in for each one, looking carefully at the shapes formed.

   b Try to draw the next two shapes in the sequence.

   c Explain how the sequence is formed.

8 Copy each shape and shade parts of each one so they have exactly two lines of symmetry.

   a 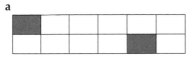   **b**   **c**

Solve problems

Make a three letter word out of the letters O and X that has two lines of symmetry.

OXO has a line of symmetry vertically through X and a horizontal line of symmetry through the middle of the word.

9 Here are three identical rectangles.
Put the three rectangles together to make a
shape that has:

   a no lines of symmetry   b exactly one line of symmetry

   c exactly two lines of symmetry.

10 The points A(2, 5), B(4, 7) and C(6, 5) are shown on the grid.
Copy the grid onto squared paper and plot the points A, B
and C.

   a Plot a point D so that the four points have no lines
      of symmetry.

   b Plot a point E so that the four points have exactly one line
      of symmetry.

   c Plot a point F so that the four points have exactly four
      lines of symmetry.

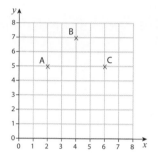

11 Some words have line symmetry horizontally across their middle.
Find at least four more words that have line symmetry similar
to this.

12 Some words have line symmetry vertically
through their middle.
Find at least four more words that have line
symmetry similar to this.

15.2 Rotation symmetry

- I can recognise shapes that have rotational symmetry
- I can find the order of rotational symmetry of a shape

A 2D shape has **rotation symmetry** if it can be rotated so that it looks exactly the same in a
new position. The **order of rotation symmetry** is the number of different positions in which
the shape looks the same, as it is rotated through one complete turn (360°).

A shape has no rotational symmetry if it has to be rotated through one complete turn to look
exactly the same. Such a shape has rotational symmetry of order 1.

Develop fluency

Describe the symmetry of this shape.

This shape has rotational
symmetry of order 3.

Describe the symmetry of this shape.

This shape has no
rotational symmetry.

Therefore, it has rotational
symmetry of order 1.

1 Copy each of these capital letters and write down its order of rotational symmetry.

a 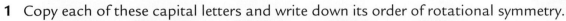 **H** b **M** c **N** d **S** e **X**

2 Write down the order of rotational symmetry for each shape.

a 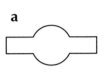 b 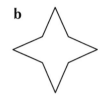 c d  e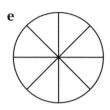

3 Write down the order of rotational symmetry for each shape.

a  b  c d  e

4 State the order of rotational symmetry of each capital letter.

a **H** b **C** c **N** d **S** e **T** f **X**

5 Copy and complete the table for each of these regular polygons.

a b c d e

Shape	Number of lines of symmetry	Order of rotational symmetry
a Equilateral triangle		
b Square		
c Regular pentagon		
d Regular hexagon		
e Regular octagon		

What do you notice?

6 Describe the rotational symmetry of each playing card shown below.

Reason mathematically

Joe said the word OXO has rotational symmetry.

Is he correct and, if so, what is the order of rotational symmetry?

He is correct, the order of rotational symmetry is 2.

7 These patterns are from Islamic designs.
Write down the order of rotational symmetry for each pattern.

a    b 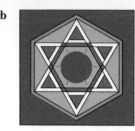   c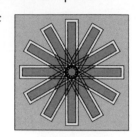

8 Copy these shapes and shade parts in each one so they have rotation symmetry of order 2.

a 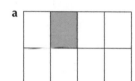   b   c

9 Copy each shape onto squared paper. Shade extra squares to give the shape rotation symmetry, order 4. Mark the centre of rotation.

a   b 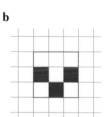   c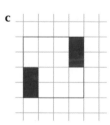

10 Eve said her name had rotational symmetry.
Her friend Ziz said Eve's name didn't have rotational symmetry, but his name did.
Who is correct? Explain your answer.

Solve problems

Shade in some small triangles on this shape to give it:

a rotational symmetry of order 1

b rotational symmetry of order 3.

a Shade any small triangle to give order 1.   *b Shade in three small triangles to give order 3.*

Note that there are many other different ways to correctly answer these two questions.

11 Here are four identical right-angled triangles.

2 cm

4 cm

Copy the four triangles and put them together to make a shape that has rotational symmetry of:

a order 1   **b** order 2   **c** order 4.

12 Copy this grid.

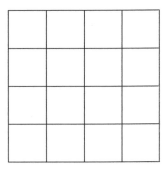

   a Shade in one square to give the shape rotational symmetry order 1.

   b Shade in another square to give the shape rotational symmetry order 2.

   c Shade in more squares to give the shape rotational symmetry order 4.

13 Which capital letters have rotational symmetry higher than 1?

14 Write down four numbers that have rotational symmetry higher than 1.

15.3 Properties of triangles and quadrilaterals

● I can understand the properties of parallel, intersecting and perpendicular lines

● I can understand and use the properties of triangles

● I can understand and use the properties of quadrilaterals

Any two lines either are **parallel** or **intersect** in a point. Lines that intersect at right angles are **perpendicular**.

Triangles have three sides and three angles or **vertices**.

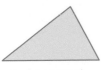  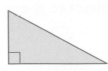

| Scalene triangle (no equal sides) | Obtuse-angled triangle | Right-angled triangle | Isosceles triangle (two equal sides) | Equilateral triangle (three equal sides) |

Quadrilaterals have four sides and four vertices.

 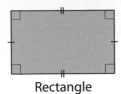 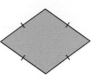

Square   Rectangle   Parallelogram   Rhombus

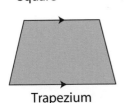  

Trapezium   Kite   Arrowhead

Develop fluency

Describe the geometrical properties of the triangle ABC.

You can see that AB = AC

And that ∠ABC = ∠ACB.

The triangle is isosceles.

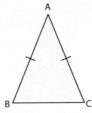

1 Match these triangles with the correct names.

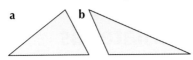 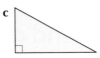

| equilateral   right-angled   obtuse-angled   scalene   isosceles |

2 Describe the geometrical properties of the triangle ABC.

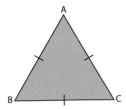

3 Describe the geometrical properties of an obtuse-angled triangle.

4 Describe the geometrical properties of a scalene triangle.

5 Describe the geometrical properties of a right-angled triangle.

6 Which quadrilaterals have the following properties?

   a four equal sides
   b two different pairs of equal sides
   c two pairs of parallel sides
   d only one pair of parallel sides

7 Describe the geometrical properties of the parallelogram ABCD.

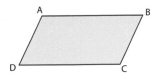

8 Describe the geometrical properties of:

   a a square   b a rhombus.

9 Describe the geometrical properties of:

   a a rectangle   b a trapezium

10 Describe the geometrical properties of a kite.

Reason mathematically

Mandy said that a scalene triangle could also be isosceles.

Is she correct? Explain your answer.

No, Mandy is incorrect as a scalene triangle has three different sides and an isosceles triangle has two sides equal.

11 Explain the difference between:

   a a square and a rhombus   b a rhombus and a parallelogram
   c a trapezium and a parallelogram.

12 A line that joins two vertices of a shape is called a diagonal.
   This quadrilateral ABCD has two diagonals, AC and BD.

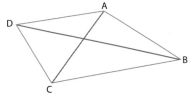

Name the quadrilaterals in the diagram below in which the diagonals:

   i are equal in length   ii bisect each other
   iii are perpendicular to each other.

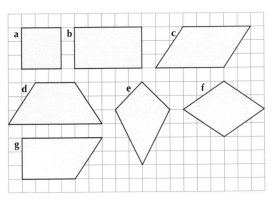

13 Neil looked at this 3 by 3 pin-board and said: 'I cannot use the dots to draw an equilateral triangle, only an isosceles triangle.'

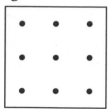

Is Neil correct? Explain your answer.

14 Gena looked at this 3 by 3 pin-board and said: 'I can use the dots to draw a parallelogram but not a kite.'

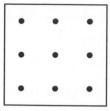

Is Gena correct? Explain your answer.

Solve problems

Draw:

a an obtuse-angled scalene triangle

b an obtuse-angled isosceles triangle.

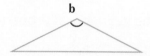

15 How many distinct quadrilaterals can be constructed on this 3 by 3 pin-board?
Draw each one and write down what type of quadrilateral it is.

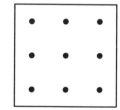

16 Draw:

a a scalene right-angled triangle

b an isosceles right-angled triangle.

17 Draw a kite and then show how you can draw a line to create:

a two isosceles triangles

b two scalene triangles.

18 Show how you can cut an equilateral triangle to create four equilateral triangles.

Now I can...

recognise shapes with reflective symmetry	find the order of rotational symmetry of a shape	understand and use the properties of triangles
recognise shapes that have rotational symmetry	understand the properties of parallel, intersecting and perpendicular lines	understand and use the properties of quadrilaterals
draw lines of symmetry on a shape		

CHAPTER 16 Solving equations

16.1 Finding unknown numbers

- I can find missing numbers in simple calculations

You can use **algebra** to write expressions and equations in which letters represent **unknown numbers**. You can use these equations to find the value of the unknowns.

Develop fluency

a Work out the value of the letter a if $a + 4 = 20$.

b Work out the value of the letter b if $2b + 4 = 20$.

a $a + 4 = 20$ *What number do you add to 4 to make 20? The answer is 16.*

  So, $a = 16$.

b $2b + 4 = 20$ *In this case the answer is 8 because $2 \times 8 + 4 = 20$.*

  So, $b = 8$.

For questions 1–9, work out the number that each letter represents.

1 a $5 + x = 7$ **b** $6 + y = 12$ **c** $5 + d = 19$ **d** $9 + g = 31$

2 a $2v = 12$ **b** $2u = 28$ **c** $3t = 36$ **d** $4r = 80$

3 a $a - 2 = 6$ **b** $b - 7 = 9$ **c** $c - 3 = 11$ **d** $d - 1 = 9$

4 a $a + 6.1 = 9$ **b** $r + 8.5 = 9$ **c** $w + 1.8 = 8$ **d** $t + 1.1 = 9$

5 a $4x = -12$ **b** $4y = -20$ **c** $7z = -7$ **d** $8w = -56$

6 a $a - 2 = -1$ **b** $b - 8 = -4$ **c** $c - 2 = -2$ **d** $d - 3 = -1$

7 a $9 + a = 45$ **b** $7 + r = 31$ **c** $9 + w = 23$ **d** $12 + t = 75$

8 a $3 - e = -2$ **b** $3 - f = -3$ **c** $4 - g = -2$ **d** $4 - h = -3$

9 a $\dfrac{x}{6} = 2$ **b** $\dfrac{y}{6} = 10$ **c** $\dfrac{12}{t} = 3$ **d** $\dfrac{18}{w} = 6$

Reason mathematically

In this number wall, the number in each brick is the sum of the numbers in the two bricks below it. Work out the values of a, b and c.

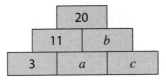

The top number is always the sum of the two numbers below it.

$11 + b = 20$, so b must be 9.

In the same way, $11 = 3 + a$ so $a = 8$.

Finally, $a + c = b$, so $8 + c = 9$.

So, $c = 1$.

10 In this number wall, the number in each brick is the sum of the numbers in the two bricks below it.
Work out the values of a, b and c.

		25		
	12		b	
5		a		c

11 Work out the values of d, e and f in this number wall.

		25		
	16		e	
10		d		f

12 This number wall has four rows and three missing numbers.
Work out the values of a, b and c.

		24				
		a		11		
	b		7		4	
c		4		3		1

Solve problems

Molly thinks of a number. She subtracts 5 and then multiplies by 3.
The answer is 24.

a Call Molly's number m. Write down an expression to show what Molly did.

b Find the value of m that makes the expression equal to 18.

a If you subtract 5 from m you get $m - 5$.
If you multiply this by 3 you get $3(m - 5)$.
The brackets show that you do the subtraction first.

b Now you want to find the value of m that makes $3(m - 5) = 18$.
You can see that $m - 5 = 6$, so m must be 11.

13 Jason thinks of a number. He multiplies it by 5. Then he adds 3.
  a Call Jason's number j and write down an expression to show what Jason did.
  Find the value of j, if the answer is: **b** 13 **c** 23 **d** 38.

14 Nina thinks of a number.
  She adds 6 then multiplies by 2.
  a Call Nina's number k and write down an expression to show what Nina did.
  Find the value of k, if the answer is: **b** 18 **c** 24 **d** 40.

15 I think of a number. I subtract 4 from my number.
  a Call my number a and write down an expression to show what I did.
  Find the value of a, if the answer is: **b** 9 **c** 17 **d** 35.

16 Kahlen thinks of a number. She doubles the number and adds 3.
  a Call Kahlen's number k and write down an expression to show what Kahlen did.
  Find the value of k, if the answer is: **b** 15 **c** 11 **d** 53.

16.2 Solving equations

- I can understand what an equation is
- I can solve equations involving one operation

Develop fluency

When you find the value of the unknown in an **equation**, you are **solving** the equation.

Solve these equations.

a $x + 17 = 53$

b $5y = 45$

a $x + 17 = 53$ This means 'a number + 17 = 53'.

You could try to guess the answer, but it is easier do it by subtraction.

$x = 53 - 17$ The number is $53 - 17$.

   $= 36$

b $5y = 45$ This means '5 × a number = 45'.

$y = 45 \div 5$ The number is $45 \div 5$.

   $= 9$

Solve these equations.

1   **a** $x + 3 = 11$ **b** $x + 5 = 13$ **c** $x - 7 = 12$ **d** $x - 9 = 0$

2   **a** $3x = 12$ **b** $2m = 18$ **c** $4x = 16$ **d** $5t = 25$

3   **a** $\dfrac{a}{2} = 6$ **b** $\dfrac{b}{5} = 8$ **c** $\dfrac{x}{12} = 2$ **d** $\dfrac{y}{4} = 20$

4   **a** $a + 16 = 41$ **b** $r + 18 = 32$ **c** $w + 13 = 54$ **d** $t + 11 = 60$

5   **a** $4v = 92$ **b** $8u = 96$ **c** $5t = 95$ **d** $3r = 84$

6   **a** $a - 12 = 8$ **b** $b - 18 = 4$ **c** $c - 12 = 12$ **d** $d - 13 = 12$

7   **a** $9 + a = 5$ **b** $7 + r = 3$ **c** $9 + w = 2$ **d** $12 + t = 7$

8   **a** $9 - a = -3$ **b** $1 - b = -2$ **c** $1 - c = -3$ **d** $1 - d = -6$

9   **a** $\dfrac{x}{6} = 42$ **b** $\dfrac{y}{3} = 17$ **c** $\dfrac{42}{t} = 3$ **d** $\dfrac{28}{w} = 4$

Reason mathematically

a Write down an expression for the perimeter of this triangle, in centimetres.

b The perimeter of the triangle is 100 cm. Write an equation to express this fact.

c Solve the equation.

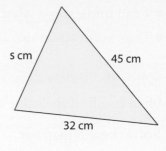

a The perimeter is $s + 45 + 32 = (s + 77)$ cm.

b The perimeter is 100 cm so the equation is $s + 77 = 100$.

c Solving the equation means finding the value of s.

$s + 77 = 100$

Then $s = 100 - 77$

$ = 23$

10 a Write down an expression for the perimeter of this triangle.

b Suppose the perimeter of the triangle is 37 cm.

  i Write down an equation involving f.

  ii Solve the equation to find the value of f.

c Suppose the perimeter of the triangle is 45 cm.

  i Write down an equation involving f.

  ii Solve the equation.

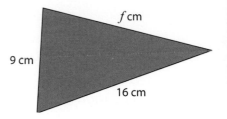

11 a Write down an expression for the area of this rectangle.

b Suppose the area of the rectangle is 88 cm².

  i Write down an equation involving r.

  ii Solve the equation.

c Suppose the area of the rectangle is 120 cm².

  i Write down an equation involving r.

  ii Solve the equation.

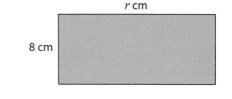

12 a Show that an expression for the number in the top brick of this number wall is $x + 48$.

b Suppose the number in the top brick is 54.

  i Write down an equation involving x.

  ii Solve the equation to find the value of x.

c Suppose the number in the top brick is 65.

  i Write down an equation involving x.

  ii Solve the equation to find the value of x.

d Suppose the number in the top brick is 90.

  i Write down an equation involving x.

  ii Solve the equation to find the value of x.

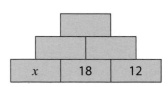

13 a Work out an expression for the number in the top brick of this number wall. Write it as simply as possible.

b Suppose the number in the top brick is 50.
   i Write down an equation involving y.
   ii Solve the equation.

c Suppose the number in the top brick is 100.
   i Write down an equation involving y.
   ii Solve the equation.

d Suppose the number in the top brick is 86. Work out the value of y.

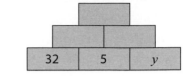

Solve problems

The area of this rectangle is 91 cm². What is the length y?

The area is $7 \times y = 7y$ cm².

The area is 91 cm² so we have the equation $7y = 91$.

Then $y = 91 \div 7$
$\qquad = 13$

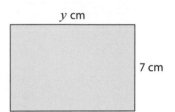

y cm

7 cm

14 a Write down an expression for the sum of the angles in this quadrilateral.

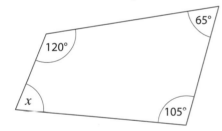

b The angles in a quadrilateral add up to 360°.
Find the value of x.

15 Work out the value of x in this number wall.

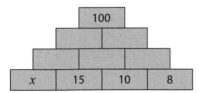

Try to find an equation.

16 A rectangle has sides of 4 cm and h cm.
The area of the rectangle is 28 cm².
Write an equation involving h and solve it.

17 A triangle has angles of 70°, 55° and $z°$.
Write down an equation involving z and solve it.

16.3 Solving more complex equations

● I can solve equations involving two operations

Some equations, such as $2n + 9 = 17$, have two operations. This example includes the two operations multiplication and addition. To solve equations like this, you need to use inverse operations.

Develop fluency

Solve these equations.

a $4a - 9 = 35$ **b** $3a + 8 = 41$ **c** $3(k + 2) = 27$ **d** $\dfrac{d}{4} - 6 = 14$

a $4a - 9 = 35$

$4a = 35 + 9$ First add 9 to both sides.

$4a = 44$

$a = 44 \div 4$ Now divide both sides by 4.

$a = 11$

You can see that to 'undo' a subtraction you do an addition.

To 'undo' a multiplication you do a division.

These are inverse operations.

b $3a + 8 = 41$

$3a = 33$ First subtract 8 from both sides.

$a = 33 \div 3$ Now divide both sides by 3.

$a = 11$

Here, to 'undo' an addition, you do a subtraction.

c $3(k + 2) = 27$

$k + 2 = 9$ First divide both sides by 3: $27 \div 3 = 9$

$k = 7$ Then subtract 2 from both sides: $9 - 2 = 7$

In this case, because of the brackets, you do the division first.

The inverse of 'multiply by 3' is 'divide by 3'.

The inverse of 'add 2' is 'subtract 2'.

d $\dfrac{d}{4} - 6 = 14$

$\dfrac{d}{4} = 20$ First add 6 to both sides: $14 + 6 = 20$

$d = 80$ Then multiply both sides by 4: $20 \times 4 = 80$

The inverse of 'divide by 4' is 'multiply by 4'.

Solve each equation.

1 a $2x + 5 = 13$ **b** $4m + 3 = 19$ **c** $2m + 17 = 33$ **d** $7x + 5 = 19$

2 a $2x - 5 = 13$ **b** $4m - 5 = 15$ **c** $2m - 17 = 13$ **d** $7x - 2 = 19$

3 a $2(x + 1) = 10$ **b** $3(y + 4) = 21$ **c** $4(z + 4) = 40$ **d** $2(w + 3) = 30$

4 a $3(x - 1) = 12$ **b** $4(y - 4) = 20$ **c** $3(z - 4) = 27$ **d** $5(w - 3) = 55$

5 a $4(8 - x) = 16$ **b** $5(3 + y) = 30$ **c** $4(15 - z) = 28$ **d** $6(4 + w) = 54$

6 a $\dfrac{x}{3} + 4 = 10$ **b** $\dfrac{x}{5} - 1 = 2$ **c** $\dfrac{x}{4} - 3 = 2$ **d** $\dfrac{m}{3} - 2 = 1$

7 a $3x + 4 = 19$ **b** $3(x + 4) = 30$ **c** $\dfrac{m}{4} + 3 = 12$ **d** $5n - 7 = 23$

8 a $3(x+1) = -6$ **b** $4(y+4) = -12$ **c** $5(z+4) = -40$ **d** $3(w+3) = -9$

9 a $5(x-1) = -10$ **b** $6(y-4) = -18$ **c** $5(z-4) = -25$ **d** $7(w-3) = -56$

10 a $6(8-x) = 60$ **b** $7(3+y) = -35$ **c** $2(15-z) = 36$ **d** $8(4+w) = -48$

Reason mathematically

In this pentagon, four of the sides are the same length.

a Write an expression for the perimeter of the pentagon, in terms of a.

b The perimeter of the rectangle is 98 cm.
 Write down an equation involving a.

c Solve the equation to find the value of a.

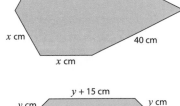

a The perimeter is $a + a + a + a + 30 = 4a + 30$ cm.

b If the perimeter is 98 cm, then $4a + 30 = 98$.

c $4a + 30 = 98$
 $4a = 68$ Subtract 30 from both sides: $98 - 30 = 68$
 $a = 17$ Divide both sides by 4: $68 \div 4 = 17$

11 a Write down an expression for the perimeter of this hexagon.
 The perimeter of the hexagon is 152 cm.

b Aaran said: 'The value of x is 18.'
 Is Aaran correct? Explain your answer

12 In this octagon, six sides are the same length.

a Write down an expression for the perimeter of the octagon.

b The perimeter of the octagon is 214 cm.
 Write down an equation for this.

c Show that the longest side is 38 cm.

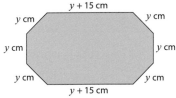

13 Joy thinks of a number, she doubles it and adds 15 to get 99.
 Theo said the number was 84.
 Is Theo correct? Explain your answer.

14 Ben was asked his age.
 He said: 'If I add 12 to my age and multiply by 5, then I get 100.'
 How old is Ben?

Solve problems

a Work out an expression involving x for the number in the top brick of this number wall.

b The top number is 21. Write an equation and solve it to find the value of x.

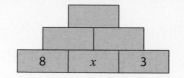

a The expressions on the second row are $x + 8$ and $x + 3$.

The expression in the top brick is $x + 8 + x + 3$.

This simplifies to $2x + 11$.

b The equation is $2x + 11 = 21$.

$2x = 21 - 11$ First subtract 11 from both sides. 'Subtract 11' is the inverse of 'add 11'.

$2x = 10$

$x = 10 \div 2$ Then divide by 2. 'Divide by 2' is the inverse of 'multiply by 2'.

$x = 5$

15 For each of these number walls:

  i use the number in the top brick to write an equation involving x

  ii solve the equation to find the value of x.

  Part **a** has been done for you.

a 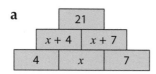   **b** 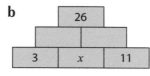   **c**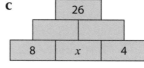

d 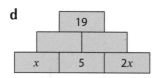   **e** 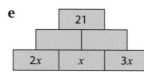   **f**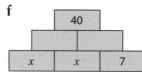

16 For each of these number walls:

  i use the number in the top brick to write an equation involving x

  ii solve the equation to find the value of x.

a 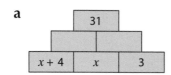   **b**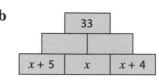

17 This is a page from Emma's exercise book.

1. $3x + 7 = 22$
Subtract 7 $3x = 18$
Divide by 3 $x = 6$
Check $3 \times 6 + 7 = 15 + 7 = 22$ ✓

2. $\frac{x}{5} - 8 = 2$
Add 8 $\frac{x}{5} = 10$
Multiply by 5 $x = 2$
Check $2 \times 5 - 8 = 10 - 8 = 2$ ✓

3. $\frac{x}{2} - 5 = 6$
Multiply by 2 $x - 5 = 12$
Add 5 $x = 17$
Check $17 - 5 = 12, 12 \div 2 = 6$ ✓

4. $3(x + 6) = 18$
Subtract 6 $3x = 12$
Divide by 3 $x = 4$

She has made a mistake when solving each equation.
Write down a correct solution to each problem.

18 The length of a rectangle is x cm longer than its width of 5 cm.

  a Write an expression for the length of the rectangle in terms of x.
     The area of the rectangle is 35 cm².

  b Write an equation involving w.

  c Solve this equation.

  d What is the perimeter of the rectangle?

16.4 Setting up and solving equations

● I can use algebra to set up and solve equations

Many real-life problems can be solved by first setting up an equation.

Develop fluency

Mark thinks of a number, doubles it and subtracts 8.

  a Use m to represent Mark's number.

     Write down an expression for the answer.

  b Mark's answer is 42.

     Write down an equation and solve it to find Mark's initial number.

  a Mark's number is $2m - 8$.

  b The equation is $2m - 8 = 42$.

     This solves to $2m = 50$.

     Giving Mark's number as 25.

1 Dan thinks of a number, doubles it and adds 5.
   a Use d to represent Dan's number.
      Write down an expression for the answer.
   b Dan's answer is 59.
      Write down an equation and solve it to find Dan's initial number.

2 Mike has m coloured pencils.
   Jon has 12 fewer pencils than Mike.
   a Write down an expression, in terms of m, for the total number of pencils.
   b They have 48 pencils altogether.
      Write an equation and solve it to find the value of m.
   c How many pencils does each boy have?

3 Josie is 24 years older than her son Tom.
   a If her son is t years old, write down an expression for Josie's age.
   b Their total age is 122 years.
      Write an equation, in terms of t, to show this.
   c Solve the equation to find the value of t.
   d How old is Josie?

4 Marie thinks of a number. She adds 14 then multiplies by 6.
   a Call Marie's number m.
      Write down an expression for her answer.
   b Marie's answer is 120.
      Write down an equation in terms of m.
   c Work out Marie's original number.

5 There are y girls in a school.
   The number of girls is 28 fewer than the number of boys.
   There are 1176 students altogether.
   a Write down an equation, in terms of y, based on this information.
   b Work out the number of girls and the number of boys in the school.

6 Ollie thinks of a number. He multiplies the number by 6 and subtracts 11.
   a Use x to represent Ollie's number.
      Write down an expression for the result.
   b Ollie's result is 37. Write down an equation and solve it to find his initial number.

7 Andrew has nine more marbles than Amber. Between them they have 39 marbles.
   a Using m to represent the number of marbles that Andrew has, write down how many marbles Amber has.
   b Write down an expression for how many marbles they have between them.
   c Work out how many marbles Andrew has.

8 Joy is fifteen years older than James.

   a Using a to represent Joy's age, write down an expression, in terms of a, for James's age.

   b Their combined age is 73 years.
     Write down an equation and use it to work out how old each is.

9 The teaching staff of a school is made up of 71 men and women.
  There are 19 more women than men.

   a Where w is the number of female teachers, write down an expression, in terms of w, to represent the number of male teachers.

   b Set up and solve an equation to work out the number of male teachers in the school.

10 Mae thinks of a number. She multiplies it by 8 then adds 27.

   a Where m is Mae's number, write down an expression for her result.

   b Mae's result is 83.
     Work out Mae's original number.

Reason mathematically

The cost of renting a hostel is £25 per night plus £30 per person staying.

  a Write an expression for the cost, in pounds, for x people.

  b The bill for a group of people for one night is £415.

    Show how you can use this information to find out there were 13 people in the group.

  a *The cost in pounds is $30x + 25$.*

  b *Set up the equation $30x + 25 = 415$.*

    $30x = 415 - 25 = 390$

    $x = 390 \div 30 = 13$

    There were 13 people in the group.

11 This is the plan of the floor of a room. The lengths are in metres.

   a Work out an expression for the area of this shape.

   b The area is 29 m². Write an equation to show this.

   c Solve the equation to find the value of a.
     Find the area of each rectangle.

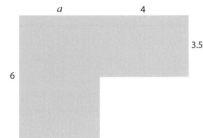

12 On Monday k cm of snow fell.
  The snowfall on Tuesday was 5 cm less than on Monday.
  On Wednesday there was twice as much snow as on Tuesday.

   a Write an expression, in terms of k, for the number of centimetres of snow on Wednesday.

   b In fact there were 30 cm of snow on Wednesday. Write an equation for this.

   c Solve the equation to find the value of k.

   d How many centimetres of snow fell altogether on the three days?

Solve problems

Kate thinks of a number. First she subtracts 13, then she multiplies by 4.

a Call Kate's number k. Write an expression for her answer.

b Kate's answer is 72. Write an equation to show this.

c Solve the equation to find Kate's initial number.

a After subtracting 13 from her number, Kate has $k - 13$.

She then multiplies this by 4 to get $4(k - 13)$.

The expression for Kate's answer is $4(k - 13)$.

b The equation for Kate's answer is $4(k - 13) = 72$.

c $4(k - 13) = 72$

$k - 13 = 18$ First divide both sides by 4: $72 \div 4 = 18$

$k = 31$ Then add 13 to both sides: $18 + 13 = 31$

Kate's initial number is 31.

---

13 Mrs George said, 'Think of a number, subtract 3 and then multiply the answer by 4. Tell me your answer.'

Alfie's answer was 52 and Harry's answer was 96.

    a Set up an equation and work out the number Alfie thought of.

    b Set up an equation and work out the number Harry thought of.

    Choose a letter for Alfie's number.

14 Buns cost £b each. Coffees cost £1.79 each.

    a Ravi buys four buns and six coffees. Write down an expression for the total cost, in pounds.

    b The cost for Ravi is £19.34. Write down an equation to show this.

    c Work out the cost of one bun.

15 Gary said, 'Robert Percy scored twelve more goals than Dwayne van Mooney last season!'

Mike answered, 'Yes but between them they scored 56 goals.'

Set up and solve an equation to work out how many goals Robert Percy scored last season.

16 Louis has £8. He buys five ice-creams that cost c pounds each. He gets 75p change.

Set up and solve an equation to work out the cost of one ice-cream.

Now I can...

find missing numbers in simple calculations	solve equations involving one operation	use algebra to set up and solve equations
understand what an equation is	solve equations involving two operations	

CHAPTER 17 Using data

17.1 Interpreting pie charts

● I can work out the size of sectors in pie charts by their angles at the centre

Pie charts show proportions in a set of data. They are a very good way to clearly show the distribution of the data.

Develop fluency

The **pie chart** shows the types of housing on a new estate.

All together there were 540 new houses built.

How many were: **a** detached   **b** semi-detached   **c** bungalows   **d** terraced?

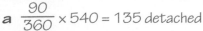

a $\frac{90}{360} \times 540 = 135$ detached

b $\frac{120}{360} \times 540 = 180$ semi-detached

c $\frac{40}{360} \times 540 = 60$ bungalows

d $\frac{110}{360} \times 540 = 165$ terraced

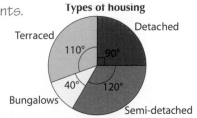

Types of housing

1   All 900 pupils in a school were asked to vote for their favourite subject.
    The pie chart illustrates their responses.
    How many voted for:

    a PE

    b Science

    c Maths?

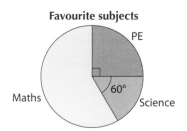

Favourite subjects

2   One weekend in Edale, a café sold 300 drinks.
    The pie chart illustrates the proportions of different drinks that were sold.
    Of the drinks sold that weekend, how many were:

    a soft drinks   **b** tea

    c hot chocolate   **d** coffee?

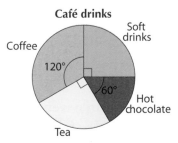

Café drinks

3   Priya did a survey about fruit and nut chocolate.
    She asked 30 of her friends.
    The pie chart illustrates her results.
    How many of Priya's friends:

    a never ate fruit and nut chocolate

    b sometimes ate fruit and nut chocolate

    c said that fruit and nut was their favourite chocolate?

Fruit & Nut

4 In one week, a supermarket sold 540 kilograms of butter.
The pie chart shows the proportions that were sold on
different days.
How many kilograms of butter were sold on:

  a Monday   **b** Tuesday   **c** Wednesday   **d** Thursday

  e Friday   **f** Saturday   **g** Sunday?

Butter sales

5 The pie chart shows the daily activities of Joe one Wednesday.
It covers a 24-hour time period.
How long did Joe spend:

  a at school   **b** at leisure   **c** travelling   **d** asleep?

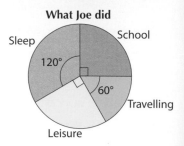

What Joe did

6 On a train one morning, there were 240 passengers.
The guard inspected all the tickets and reported how
many of each sort there were.
The pie chart illustrates his results.
How many of the tickets that he saw that morning were:

  a open returns   **b** season tickets   **c** day returns

  d travel passes   **e** super savers?

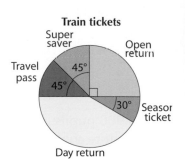

Train tickets

7 The pie chart shows the results of a survey of 216 children,
when they were asked about their favourite foods.
How many chose:

  a chips   **b** pizza   **c** pasta   **d** curry?

Favourite food

8 150 children went on one of four school summer holidays.
How many children chose:

  a camping   **b** France   **c** pony trek   **d** Disney World?

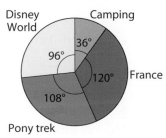

Reason mathematically

The pie chart shows the results of a survey of 180 pupils, when they were asked about their summer holiday.

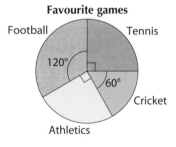

Holidays

a What type of holiday was chosen by 45 people?

b Tara said: 'A third of them stayed in the UK for their holiday.'

Is Tara correct? Explain your answer

a 45 is a quarter of 180, and 90° is a quarter of the pie chart, so a cruise was chosen by 45 pupils.

b The UK selection is shown by a sector of 120°, this represents $\frac{120}{360} = \frac{1}{3}$, so Tara is correct.

9 The pie chart shows the results of a survey of 144 children, when they were asked about their favourite TV programmes. Which statements are correct? Explain your answers.

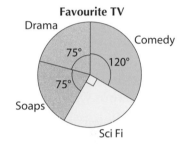

Favourite TV

a Jess said one third of them chose comedy.

b Just as many voted for Drama as soaps.

c 90 children voted for Sci Fi.

10 The pie chart shows the results of a survey of 120 children, when they were asked about their favourite sports.

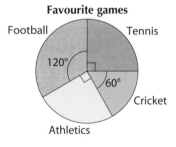

Favourite games

a Which sport was chosen by 20 people?

b Dara said: 'Tennis is more popular than athletics.'
Is Dara correct? Explain your answer

11 A group of children were asked what their favourite colour was. The pie chart shows the results.

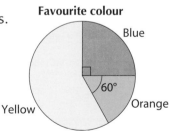

Favourite colour

a What colour was chosen by one quarter of the children?

b Jenny said that three times as many people chose yellow than orange.
Is Jenny correct? Explain your answer.

12 These pie charts show how two companies first contacted new customers.

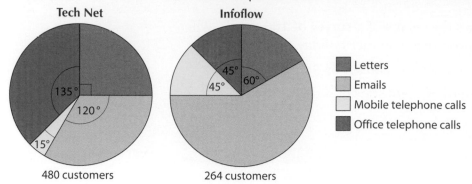

Tech Net

Infoflow

135°

120°

15°

45°

45°

60°

☐ Letters
☐ Emails
☐ Mobile telephone calls
☐ Office telephone calls

480 customers

264 customers

a Work out how many letters each company sent.

b Which company sent the most emails? Explain your answer.

c Make two more comparisons between the companies.

Solve problems

The pie chart shows how one country dealt with 3000 kg of dangerous waste in 2013.

How much waste did the country dispose of by:

a putting it into landfill

b burning it

c dumping it at sea

d chemical treatment?

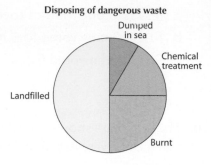

Disposing of dangerous waste

Dumped in sea

Chemical treatment

Landfilled

Burnt

a $\frac{1}{2}$ of the waste was put into landfill.

$\frac{1}{2}$ of 3000 kg = 1500 kg

So, 1500 kg of the waste was put into landfill.

c The angle of this sector is 30°.

Then the fraction of the circle is:

$\frac{30}{360} = \frac{1}{12}$

Therefore $\frac{1}{12}$ of the waste was dumped at sea.

3000 kg ÷ 12 = 250 kg

So, 250 kg of the waste was dumped at sea.

b $\frac{1}{4}$ of the waste was burned.

$\frac{1}{4}$ of 3000 kg = 3000 kg ÷ 4 = 750

So, 750 kg of the waste was burned.

d The angle of this sector is 60°.

Then the fraction of the circle is:

$\frac{60}{360} = \frac{1}{6}$

Therefore $\frac{1}{6}$ of the waste was treated by chemicals.

3000 kg ÷ 6 = 500 kg

So, 500 kg of the waste was treated by chemicals.

13. To motivate the pupils, a headteacher placed this pie chart on the school noticeboard.

    Emma decided to use the chart to find out how much money each year group had raised.

    Help Emma by estimating how much each year group raised.

Christmas charity collection
total so far … £1690

14. Brendan saw this pie chart in a magazine.

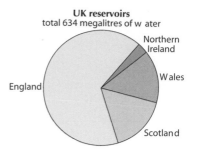

UK reservoirs
total 634 megalitres of water

Northern Ireland	11°
Wales	53°
Scotland	58°
England	238°

He measured and recorded the angles of the pie chart in the table shown.

Calculate how many megalitres of water there were in the reservoirs of each country.

Give your answers correct to two significant figures.

15. A dog rescue centre provides refuge for four types of dog as shown in the pie chart.

    They currently have six collies.

    Estimate the angles for each sector and work out how many of each breed there are and how many dogs they have in total.

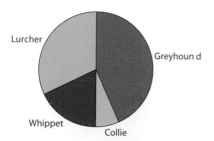

16. An international event was attended by people from five European countries as shown in the diagram.

    There were 240 Germans. Estimate how many people attended all together.

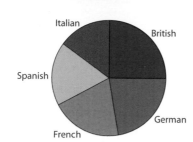

17.2 Drawing pie charts

● I can use a scaling method to draw pie charts

Sometimes you will have to draw a pie chart to display data that is given in a **frequency table**.

The size of the sector will depend on the **frequency** for that sector. You can use the **scaling method** to find the angle for each sector.

Develop fluency

Draw a pie chart to represent the data showing how a group of people travel to work.

Set the data out in a frequency table and write the calculations in it.

Now draw the pie chart.

When drawing a pie chart, draw the smallest angle first and try to make the largest angle the last one you draw, then any cumulative error in drawing will not be so noticeable.

Sector (type of travel)	Frequency	Calculation	Angle
Walk	24	$\frac{24}{240} \times 360° = 36°$	36°
Car	84	$\frac{84}{240} \times 360° = 126°$	126°
Bus	52	$\frac{52}{240} \times 360° = 78°$	78°
Train	48	$\frac{48}{240} \times 360° = 72°$	72°
Cycle	32	$\frac{32}{240} \times 360° = 48°$	48°
Total	240		360°

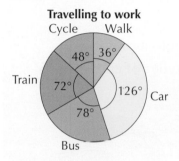

Travelling to work

1 The results of a transport survey show the various ways students go school. copy the table and fill in the gaps. Then draw a pie chart to show the results.

Transport	Frequency	As 120 pupils are represented by 360°, each pupil will be represented by 360 ÷ 120 = °
Car	23	$23 \times 3 = 69°$
Bus	17	$17 \times 3 =$
Train	25	$25 \times \quad =$
Bicycle		$\times 3 =$
Walk	40	$\times \quad =$
Total	120	Check the total $= 360°$

2 Draw a pie chart to represent the numbers of birds spotted on a field trip.

Bird	Crow	Thrush	Starling	Magpie	Other
Frequency	19	12	8	2	19

3 Draw a pie chart to represent the favourite subjects of 36 pupils.

Subject	Maths	English	Science	Languages	Other
Frequency	12	7	8	4	5

4 Draw a pie chart to represent the types of food that 40 people usually eat for breakfast.

Food	Cereal	Toast	Fruit	Cooked	Other	None
Frequency	11	8	6	9	2	4

5 Draw a pie chart to represent the numbers of goals scored by an ice hockey team in 24 matches.

Goals	0	1	2	3	4	5 or more
Frequency	3	4	7	5	4	1

6 Draw a pie chart to represent the favourite colours of 60 Year 8 pupils.

Colour	Red	Green	Blue	Yellow	Other
Frequency	17	8	21	3	11

7 **a** Draw a pie chart to represent the sizes of dresses sold in a shop during one week.

Size	8	10	12	14	16	18
Frequency	3	7	10	12	6	2

  b What size are a quarter of the dresses?

  c What fraction of dresses are size 16 or above?

  d What percentage are size 14?

Reason mathematically

Andrew was given data about the numbers of bottles of water sold at a test match.

Day	Thursday	Friday	Saturday	Sunday	Monday
Number of bottles	190	350	440	410	30

He said: 'I can't draw a clear pie chart as the Monday angle will be too small.'

Comment on Andrew's statement.

He is wrong as the angle needed for Monday on the pie chart will be just under 8°, which although small will still show the comparison.

8 Trains arriving at Blackpool station were monitored to see how well they ran on time. The results are shown in the table.

  a Draw a pie chart to display the results.

  b How many trains were late?

  c What proportion were early?
   The railway promises that no more than 10% of trains will be more than 5 minutes late, and no more than 5% will be over 10 minutes late.

  d Write a brief report to explain whether or not they achieved this.

Early	4
On time	18
Up to 5 minutes late	14
5 to 10 minutes late	3
Over 10 minutes late	1

9 Paul has to draw a pie chart to represent the eye colours in his school from the following data.

Colour	Blue	Brown	Green	Hazel	Grey	Amber
Frequency	192	130	20	8	9	1

Paul says: 'The angle for Amber will be too small to draw, shall I combine some colours to make a category of "other colours" to make a clear pie chart?'

comment on Paul's statement.

10 Helen is asked to draw a pie chart to represent the time she spent on activities in a day.

Activity	Sleeping	Lessons	Travelling	Eating	Playing	Reading
Time (hours)	9	5	1	2	5	2

Helen said: 'The angle for travelling will be 15°, so the angle for sleeping will be 130°.'
Is Helen correct? Explain your answer.

11 There are 120 pupils in a year group.
12 of these pupils wear glasses.

Who wears glasses

Wear glasses

Do not wear glasses

a The pie chart to show this is not drawn accurately.
What should the angles be? Show your working.
Exactly half of the 120 pupils in the school are boys.

b From this information, is the percentage of boys in this school that wear glasses 5%, 6%, 10%, 20%, 50% or is it not possible to tell?

Solve problems

This is data about the number of siblings each pupil had in a school.

a Why would it be extremely difficult to draw a pie chart to illustrate this information?

b Combine some groups so that you could draw a pie chart to represent this information.

Number of siblings	Number of applicants
None	323
One	672
Two	40
Three	3
Four	1

a The data for three and four siblings would require a very small angle to be drawn.

b

Number of siblings	Number of applicants
None	323
One	672
More than one	44

12 A player entered an 18-hole golf tournament. The table shows how well she did on the first day (par is the target, birdie is one less, eagle is two less, bogey is one over par and double bogey is two over par).

Eagle	1
Birdie	4
Par	11
Bogey	2
Double bogey	0

   a Draw a pie chart to illustrate her results.

   b Was her overall score better than par or worse? By how much?

   c Par for one round on this course is 72. Major tournaments play four rounds of 18 holes. If she repeats these results each day, what would her total score be?

13 This is an estate agent's waiting list for rental flats on one day in a large city.

   a Why would it be extremely difficult to draw a pie chart to illustrate this information?

   b Combine the groups so that you could draw a pie chart to represent this information.

   c Draw the pie chart.

Type of flat	Number of applicants
1 bedroom	646
2 bedrooms	1344
3 bedrooms	80
4 bedrooms	6
5 bedrooms	2

14 This diagram was shown in a business magazine. It illustrates the funds received by a charity one year. Redraw it as an accurate pie chart.

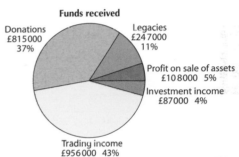

Funds received

Donations £815 000 37%
Legacies £247 000 11%
Profit on sale of assets £108 000 5%
Investment income £87 000 4%
Trading income £956 000 43%

15 Match the pie charts with their equivalent bar charts.

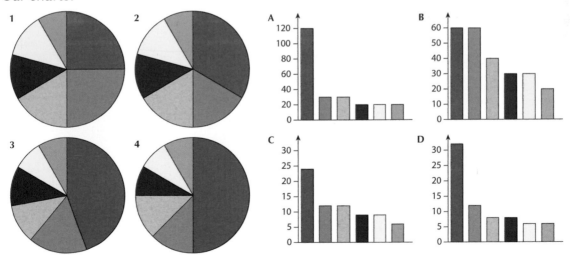

17.3 Grouped frequencies

● I can understand and use grouped frequencies

Sometimes you are given too many different values to make a sensible bar chart, so you need to organise them into a **grouped frequency table**. You sort the data into groups called classes. Where possible, you should always keep the classes the same size. You can then draw a bar chart.

You cannot find a **mode** for grouped data so, instead, you must use the **modal class**. This is the class with the **highest frequency**.

Develop fluency

A group of pupils are asked the number of times they have walked to school this term. Here are their replies.

6   3   5   20   15   11   13   28   30   5   2   6   8   18   23
22   17   13   4   2   30   17   19   25   8   3   9   12   15   8

  a   Organise the data into a grouped frequency table.

  b   Draw a bar chart.

  c   What is the modal class?

  a

Times walked to school	1–5	6–10	11–15	16–20	21–25	26–30
Frequency	7	6	6	5	3	3

  b

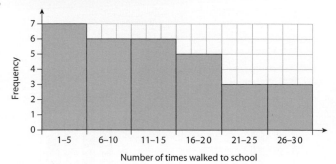

  c   The modal class is 1–5 times.

1 Two classes carried out a survey to find out how many text messages each pupil had sent the day before. These are the results of this survey.

4	7	2	18	1	16	19	15	13	0	9	17
4	6	10	12	15	8	3	14	2	14	15	18
5	16	3	6	5	18	12	5	9	19	5	17
17	16	5	10	19	7	10	17	16	10	7	19
3	16	16	18	6	5	8	9	3			

a Copy and complete this grouped frequency table with a class size of 5.

Number of texts	Tally	Frequency
0–4		
5–9		
10–14		
15–19		
Total:		

b Draw a bar chart of the data.

c What is the modal class?

2 A doctor kept a record of the waiting times, in minutes, of patients. These are the results of the survey.

5	8	3	19	2	17	15	16	14	1	10	18
5	7	11	13	16	9	4	15	3	15	16	19
18	17	6	11	19	8	11	18	17	11	8	15

a Copy and complete this grouped frequency table with a class size of 5.

Number of minutes	Tally	Frequency
0–4		
5–9		
10–14		
15–19		
Total:		

b Draw a bar chart of the data.

c What is the modal class?

3 A leader of a youth club asked her members: 'How many times this week have you played video games?'
These were their responses.

3	6	9	2	23	18	6	8	29	27	2	1
0	5	19	23	13	21	7	4	23	8	7	1
0	25	24	8	13	18	15	16	3	7	11	5
27	23	6	9	18	17	6	6	0	6	21	26
25	12	4	24	11	11	5	25				

a Create a grouped frequency table with a class size of 5.

Number of times	Tally	Frequency
0–4		
5–9		
10–14		
15–19		
20–24		
25–29		
Total:		

b Draw a bar chart of the data.

c What is the modal class?

4 Penny kept a record of how many minutes she had to wait for her bus every day for one month.
These are the results.

2	5	8	1	12	17	5	7	28	26	1	0
0	4	18	22	12	10	6	3	22	7	6	0
1	24	4	7	3	17	14	15	2	6	10	4

a Create a grouped frequency table with a class size of 5.

b Use the data above to complete your table.

c Draw a bar chart of the data.

d What is the modal class?

5 In a class sponsorship, the pupils raised these amounts of money, in pounds (£).

12.25	06.50	09.75	23.00	01.86	05.34	16.75	11.32
06.45	02.50	05.00	18.65	05.90	04.34	02.17	08.89
07.86	19.70	21.55	13.87	23.12	14.67	11.98	13.60
04.75	19.00	16.41	01.90	06.89	08.33		

Create a grouped frequency table:

a with a class size of £4, i.e. £0–£4, £4.01–£8, £8.01–£12,...

b with a class size of £6, i.e. £0–£6, £6.01–£12, £12.01–£18,... .

6 Theo recorded how much in pounds (£) he spent at a shop each day one month.

11.15	05.40	08.65	14.00	00.76	04.24	15.65	10.22
05.35	01.40	04.00	17.55	04.80	03.24	01.07	07.79
06.76	18.60	20.00	12.77	18.02	13.57	10.88	12.50
03.65	18.00	17.31	00.80	05.79	07.23	05.75	

  a Create a grouped frequency table:

    i with a class size of £4, i.e. £0–£4, £4.01–£8, £8.01–£12,...

    ii with a class size of £6, i.e. £0–£6, £6.01–£12, £12.01–£18,... .

  b What is the modal class for each group?

Reason mathematically

Look back at Question 6.

  a Select the better class size to illustrate the data.

  b Explain why you chose that class size.

  a ii

  b *The class shows a greater difference between frequencies.*

7 Look back at Question 5.

  a What is the modal class for each table that you created in your answer?

  b The teacher was asked to make a display illustrating how well the pupils had done. Select the better class size to illustrate the data.

  c Explain why you chose that class size.

8 The average ages of families at a small holiday resort are shown below.

31	17	31	21	32	29	27	39	25	25	40
27	36	28	46	19	38	32	23	28	35	19
41	30	24	34	55	26	20	36	43	51	

  a Which of the options shown is the most sensible format for this data?
Explain your answer.

  b Copy and complete the table you chose in part **a**.

  c Draw a bar chart to illustrate the data.

Average age	Tally	Frequency
10–14		
15–19 etc.		

Average age	Tally	Frequency
0–19		
20–39 etc.		

Average age	Tally	Frequency
10–19		
20–29 etc.		

9 In a club sponsorship, the members raised these amounts of money (£).

24.50	13.00	19.50	45.00	03.72	10.68	03.50	22.64
12.90	05.00	10.00	37.10	11.80	08.68	04.34	17.78
15.72	39.40	43.10	27.74	44.24	29.34	23.96	27.20
09.50	38.00	32.82	03.80	13.78	16.66	15.85	08.15

a Create a grouped frequency table:

   i with a class size of £5.00, i.e. £0.00–£5.00, £5.01–£10.00, £10.01–£15.00,...

   ii with a class size of £10.00, i.e. £0.00–£10.00, £10.01–£20.00, £20.01–£30.00,... .

b What is the modal class for each group?

c Select the best class size to illustrate the data. Explain your decision.

Solve problems

This grouped bar chart shows the number of people attending a yoga class over the year.

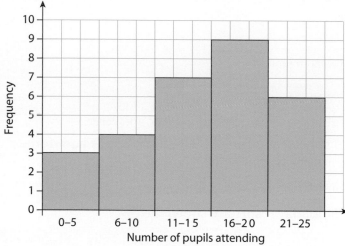

a What is the modal class?

b How many times did fewer than 11 people attend?

c How many times did over 15 people attend?

d How many classes were there over the year?

a 16–20 is the modal class.

b Add the two classes less than 11 giving 3 + 4 = 7.

c Add the two classes over 15 giving 9 + 6 = 15.

d Add the frequency in each class to give 3 + 4 + 7 + 9 + 6 = 29 classes.

10 This grouped bar chart shows the lifetimes of some batteries in a test.

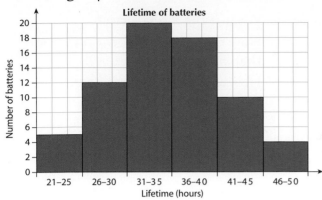

a Make a frequency table for the data.
b What is the modal class?
c What is the total number of batteries tested?
d How many batteries had a lifetime of 40 hours or less?
e What is the maximum and minimum possible range?

11 A weather station records the minutes of direct sunlight each day for a month.

Class interval	Tally	Frequency
0–39	/	1
40–79	////	5
80–119	//// //// //	12
120–159	//// ///	8
160–199	////	4
200–239	/	1
	Sum =	31

a What could the longest amount of direct sunshine have been, in hours and minutes?
b At what time of year do you think this recording took place?
c What was the chance of having no direct sunlight at all: 1 in 31, 1 in 39 or 1 in 40?
d Estimate how many days you think will have had 3 hours or more sunshine.

12 Mr Roberts gave his class a test and made the grouped bar chart below from the data.

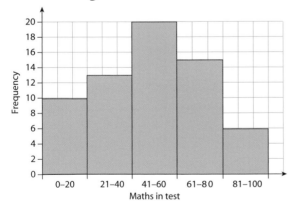

a What is the modal class?
b How many pupils scored fewer than 41 marks?
c How many pupils scored more than 60 marks?

13 Tom's morning train was always late. He made this grouped bar chart.

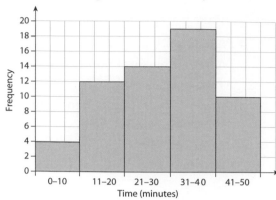

a What is the modal class?

b How many times was the train more than ten minutes late?

c How many times was the train over thirty minutes late?

d From this data what's the highest possible number of times the train was on time?

17.4 Continuous data

● I can understand and work with continuous data

Data such as masses or heights of all the pupils in a class often has a wide range of possible values. This type of data is **continuous data** and you have to group it together, to find any pattern. In a **grouped frequency** table, you arrange information into classes or groups of data. You can create frequency diagrams from grouped frequency tables, to illustrate the data.

Develop fluency

These are the journey times, in minutes, for a group of 16 railway travellers.

25 47 12 32 28 17 20 43 15 34 45 22 19 36 44 17

Construct a frequency table to represent the data.

Looking at the data, 10 minutes is a sensible size for the class interval.

There are six times in the group $10 < T \le 20$:	12, 17, 20, 15, 19 and 17.
There are three times in the group $20 < T \le 30$:	25, 28 and 22.
There are three times in the group $30 < T \le 40$:	32, 34 and 36.
There are four times in the group $40 < T \le 50$:	47, 43, 45 and 44.

Note how you write the class interval in the form $10 < T \le 20$.

$10 < T \le 20$ is a short way of writing the time interval of 10 minutes to 20 minutes.

The possible values for T include 20 minutes but not 10 minutes.

Put all the information in a table.

Time, T (minutes)	Frequency
$10 < T \le 20$	6
$20 < T \le 30$	3
$30 < T \le 40$	3
$40 < T \le 50$	4

1   This table shows the lengths of time 25 customers spent in a shop.

a   One of the customers was in the shop for exactly 20 minutes. In which class was this customer recorded?

b   Which is the modal class?

Time, T (minutes)	Frequency
$0 < T \le 10$	12
$10 < T \le 20$	7
$20 < T \le 30$	8

2   These are the heights (in metres) of 20 people.

1.65	1.53	1.71	1.72	1.48	1.74	1.56	1.55	1.80	1.85
1.58	1.61	1.82	1.67	1.47	1.76	1.79	1.66	1.68	1.73

a   Copy and complete the frequency table.

b   What is the modal class?

Height, h (metres)	Frequency
$1.40 < h \le 1.50$	
$1.50 < h \le 1.60$	
$1.60 < h \le 1.70$	
$1.70 < h \le 1.80$	
$1.80 < h \le 1.90$	

3   The table shows the waiting times for the bus of each pupil in a class.

a   One pupil had to wait exactly 10 minutes. In which class was this pupil recorded?

b   Which is the modal class?

Time, T (minutes)	Frequency
$0 < T \le 10$	11
$10 < T \le 20$	8
$20 < T \le 30$	5

4   The table shows the masses of the marrows in a competition.

a   Tim's marrow weighs exactly 5 kilograms. In which class was this marrow recorded?

b   Which is the modal class?

Mass, M (kilograms)	Frequency
$0 < M \le 5$	3
$5 < M \le 10$	9
$10 < M \le 15$	4

5   These are the lengths, in metres, jumped in a competition.

2.75	2.64	2.81	2.62	2.48	2.85	2.65	2.66	2.90	2.53
2.47	2.72	2.51	2.78	2.56	2.87	2.89	2.78	2.76	2.64

a   Copy and complete the frequency table

b   What is the modal class?

Length, L (metres)	Frequency
$2.40 < L \le 2.50$	
$2.50 < L \le 2.60$	
$2.60 < L \le 2.70$	
$2.70 < L \le 2.80$	
$2.80 < L \le 2.90$	

6 These are the noon temperatures, in degrees Celsius, °C, recorded in Bude during August.

16 18 19 19 18 17 19 20 22 24 22 20 25 24 23 21
19 18 17 17 16 16 18 19 20 21 21 20 19 18 22

  a Copy and complete the frequency table.

  b What is the modal class?

  c Joy was in Bude when the noon temperature was exactly 21°C. In which class was this temperature recorded?

Temperature, T (°C)	Frequency
$15 < T \le 17$	
$17 < T \le 19$	
$19 < T \le 21$	
$21 < T \le 23$	
$23 < T \le 25$	

7 The heights, in cm, of flowers in a collection were measured.

35.2   28.3   29.1   19.5   18.9   27.1   19.7   20.8   22.3   24.5
22.2   20.9   25   24.4   27.3   23.6   19.8   18.9   27.3   27.1
18.2   26.3   28.7   29.6   20.8   21.4   21.2   30.3

  a Copy and complete the frequency table

  b What is the modal class?

  c How many flowers were taller than 30 cm?

Height, H (cm)	Frequency
$15 < H \le 20$	
$20 < H \le 25$	
...	

8 The masses, in grams, of apples in a box were measured as below.

75.3   88.4   99.2   89.6   78.8   77.2   89.1   90.9   82.5   75.6
82.7   90.8   75.1   79.5   87.2   83.8   99.9   88.2   77.5   87.2
88.0   76.4   78.6   89.7   80.7   81.5   81.2   80.3

  a Copy and complete the frequency table.

  b What is the modal class?

  c How many apples were heavier than 90 grams?

Mass, M (grams)	Frequency
$75 < H \le 80$	
$80 < H \le 85$	
...	

Reason mathematically

The times, T seconds, of swimmers in a competition were:

14.3	17.4	18.2	18.6	17.8	16.2	18.1	15.9	12.5	13.6	11.7
12.8	14.0	13.5	16.2	12.8	18.9	17.2	16.5	16.2	17.0	15.4
17.6	18.7	12.7	11.5	11.2	12.3

Explain how you would create a suitable grouped frequency table from this data.

Note the range, from 11.2 seconds to 18.9 seconds. You can divide this range into four suitable classes: $11 < T \le 13$, $13 < T \le 15$, $15 < T \le 17$ and $17 < T \le 19$.

So the table becomes:

Time, T (seconds)	Frequency
$11 < T \le 13$	
$13 < T \le 15$	
$15 < T \le 17$	
$17 < T \le 19$	

9 The lengths, L centimetres, of babies born one week were recorded as:

45.7	58.1	49.3	49.9	58.2	47.4	59.8	50.2	52.7	54.2	52.1
50.9	55.3	54.3	49.9	48.3	49.7	58.1	57.6	57.8	46.7	56.2
53.7	52.8	50.4	51.2	51.9	50.8

   a Explain how you would create a suitable grouped frequency table from this data.

   b Create that suitable table.

10 The times, T seconds, Clara took to swim 200 m were recorded as:

51.7	59.1	60.3	60.9	59.2	58.4	59.8	51.2	53.7	55.2	53.1
51.9	55.3	56.3	50.9	51.3	59.7	57.1	56.6	58.8	50.7	51.2
53.7	52.9	50.4	51.2	50.9	50.9

   a Explain how you would create a suitable grouped frequency table from this data.

   b Create that suitable table.

11 The mass, M grams, of each grape in a bunch was recorded as

7.57	6.71	5.83	9.92	8.23	7.43	9.81	8.25	8.17	7.23	9.18
7.19	6.53	7.93	9.39	8.43	5.73	8.41	7.65	7.89	7.78	9.08
5.17	9.83	7.48	8.26	7.96	6.85

   a Explain how you would create a suitable grouped frequency table from this data.

   b Create that suitable table.

12 In a doctors' surgery, the practice manager told each doctor that the length of most consultations should be more than 5 minutes but less than 10. She monitored the consultation times of the three doctors at the practice throughout one day. These are the results.

Dr Speed (minutes):   6, 8, 11, 5, 8, 5, 8, 10, 12, 4, 3, 6, 8, 4, 3, 15, 9, 2, 3, 5

Dr Bell (minutes):   7, 12, 10, 9, 6, 13, 6, 7, 6, 9, 10, 12, 11, 14

Dr Khan (minutes):   5, 9, 6, 3, 8, 7, 3, 4, 5, 7, 3, 4, 5, 9, 10, 3, 4, 5, 4, 3, 4, 4, 9

a Did any of the doctors manage to follow the practice manager's advice?

b Write a short report about the consultation times of the three doctors.

Solve problems

A nurse recorded the reaction times, T seconds, of patients one morning. These are the results.

1.71	1.33	1.64	1.58	1.42	0.65	1.12	1.38	1.79	1.32	1.70	1.63
1.21	1.54	1.24	1.17	0.82	1.43	1.65	0.99	0.86	0.73	1.25	1.78
1.37	1.18	1.53	1.32	0.92	0.84	1.34					

Using a class size of 0.3 seconds, work out the modal class of the patient reaction times.

The shortest time is 0.65, the longest time is 1.79, so set up a table from 0.6 to 1.8. Dividing into groups of 0.3 will give four classes. Set up a group tally and table to find the results, the smallest being 0.6.

Reaction time T (seconds)	Tally	Frequency
$0.6 < T \le 0.9$	ﬀﬀ	5
$0.9 < T \le 1.2$	ﬀﬀ	5
$1.2 < T \le 1.5$	ﬀﬀ ﬀﬀ /	11
$1.5 < T \le 1.8$	ﬀﬀ ﬀﬀ	10

You can now see that the modal class is $1.2 < T \le 1.5$ seconds.

13 A petrol station owner surveyed a sample of customers to see how many litres of petrol they bought. These are the results.

27.6	31.5	48.7	35.6	44.8	56.7	51.0	39.5	28.8	43.8
47.3	36.6	42.7	45.6	32.4	51.7	55.9	44.6	36.8	49.7
37.4	41.2	38.5	45.9	34.1	54.3	41.3	49.4	38.7	33.2

Using a class size of 5 litres, work out the modal class of the amount of petrol that customers bought.

14 A farmer weighed his bags of potatoes in kilograms, W kg. These are the results.

6.81	6.23	6.74	6.57	6.41	6.15	6.12	6.38	6.19	6.32	6.30	6.63
6.81	6.55	6.34	6.18	6.12	6.45	6.45	6.29	6.16	6.63	6.25	6.48
6.37	6.08	6.53	6.22	6.02	6.54	6.34					

Using a class size of 0.2 kg, work out the modal class of the potato bags.

15 The seasons results of Kath's high jumps were recorded.

0.91	1.33	1.24	1.17	1.21	1.15	1.12	1.18	1.19	1.22	1.20	1.03
1.11	1.15	1.04	1.28	1.22	1.25	1.35	1.29	1.16	1.03	0.95	1.28
1.27	1.18	1.23	1.22	1.02	1.34	1.31					

Using a suitable class size, work out the modal class of Kath's height in the high jump.

16 A weights and measures team measured the amount of lemonade, in litres, contained within lemonade bottles in a supermarket. These are their results.

1.01	1.03	0.94	1.07	0.98	1.05	1.12	0.98	1.09	1.12	0.94	1.03
0.99	1.05	1.04	1.15	0.92	0.95	1.05	1.09	1.14	0.93	1.05	0.98
0.97	1.08	1.03	1.12	1.02	0.94	0.96					

Using a class size of 0.05 litres, work out the modal class of the bottles of lemonade.

Now I can...

work out the size of sectors in pie charts by their angles at the centre	understand grouped frequencies	use grouped frequencies
use a scaling method to draw pie charts	understand continuous data	work with continuous data

CHAPTER 18 Pencil and paper calculations

18.1 Short and long multiplication

- I can choose a written method for multiplying two numbers together
- I can use written methods to carry out multiplications accurately

There are many different methods of multiplying two numbers, including the **grid or box method**, **column method** or **long multiplication** and the **Chinese method**.

Develop fluency

Work out 36×43.

Grid or box method (partitioning)	Long multiplication (expanded working)	Long multiplication (compacted working)	Chinese method
	$\begin{array}{r} 36 \\ \times 43 \\ \hline 18 \end{array}$ (3×6) 90 (3×30) 240 (40×6) 1200 (40×30) $\overline{1548}$	$\begin{array}{r} 36 \\ \times 43 \\ \hline 108 \end{array}$ (3×36) $\underset{1}{~}$ 1440 (40×36) $\underset{12}{~}$ $\overline{1548}$	Note the carried figure, from the addition of 9, 1 and 4.

The answer is 1548.

1 Copy each of these grids and fill in the gaps.

 a 21×6

×	20	1	
6			

 b 37×6

×	30	7	
6			

2 Use the grid method to work these out.

 a 18×6 b 18×8 c 22×9 d 22×7

3 Use long multiplication to work these out.

 a 27×5 b 32×5 c 32×8 d 19×8

4 Use the Chinese method to work these out.

    a 42×4 **b** 42×9 **c** 19×9 **d** 19×3

5 Copy each of these grids and fill in the gaps.

    a 21×16 **b** 58×16

×	20	1
10		
6		

×	50	8
10		
6		

6 Use the grid method to work these out.

    a 17×23 **b** 32×23 **c** 15×23 **d** 56×23

7 Use long multiplication to work these out.

    a 27×15 **b** 32×15 **c** 32×18 **d** 19×18

8 Use the Chinese method to work these out.

    a 42×14 **b** 42×19 **c** 19×19 **d** 19×13

9 Use your favourite method to work these out.

    a 32×24 **b** 32×53 **c** 18×53 **d** 27×53

10 Use any method to work these out.

    a Each day 17 jets fly from London to San Francisco. Each jet can carry up to 348 passengers. How many people can travel from London to San Francisco each day?

    b A van travels 34 miles for every gallon of petrol. How many miles can it go on one tank, given that the petrol tank holds 18 gallons?

Reason mathematically

$12 \times 5 \times 17 = \square \times 15 \times 17$

What number goes in the box to make the equation true?

$12 \times 5 = 60$

$60 \div 15 = 4$

4 goes in the box as $12 \times 5 \times 17 = 1020$ and $4 \times 15 \times 17 = 1020$.

11 $15 \times 16 \times 17 = \square \times 8 \times 17$

What number goes in the box to make the equation true?

12 Follow these steps.
- Write down any three-digit number.
- Multiply the number by 7.
- Multiply your answer by 11.
- Multiply your answer by 13.
- Write down your final answer.

    a What do you notice?

    b Now try to explain why this happens.

13   **a** Work out these multiplications.

       i 34×11 **ii** 71×11 **iii** 26×11 **iv** 45×11

    b What do you notice about the answers?

    c Write down the answer to 16×11.

    d Write down the answer to 85×11.

14 Timothy says that 15×31 gives the same answer as 51×13 because the numbers are the same, just swapped around. Is he correct? Explain your answer.

15 Show how you can use the grid method to work out 7.2×12.7 and explain how you have split up the numbers.

Solve problems

Bilal is having a birthday party at an ice-skating rink. The ice-skating rink charges £17 per guest for the party and requires a deposit of £50, with the rest of the payment due on the day of the party. He has 12 guests coming. How much must he pay for the party on the day?

Work out 17×12.

×	10	7	
10	100	70	**170**
2	20	14	**34**
			204

$204 - 50 = 154$

He must pay £154 on the day.

16 Mr Z's class is going on a residential trip. Each student must pay £35 to go on the trip. There are 29 students in the class. What is the total cost of the trip?

17 Taylor is making cakes for the school cake sale. She is going to make five cakes and has everything she needs except for eggs. Each cake needs three eggs. A package of six eggs costs £2.89. How much will it cost to buy all the eggs to make the cakes?

18   **a** One lunchtime, a burger van sells 35 burger meals at £3.98 each and 17 hot dogs at £2.90 each. How much money did they take in that time?

    b Write down a way you can work out 35×3.98 and 17×2.9, using mental methods rather than short and long multiplication.

18.2 Short and long division

- I can choose a written method for dividing one number by another
- I can use written methods to carry out divisions accurately

There are many different ways to work out a division calculation, including **long division**, **short division** and **repeated subtraction**.

Divisions do not always give whole number answers. There is sometimes a number left over. This is called the **remainder**. You will often need to add some decimal places and continue the division. You may sometimes want your answer with a remainder or written as a fraction.

Develop fluency

Work out $970 \div 8$.

a Give your answer as a remainder and as a fraction.

b Give your answer as a decimal.

a

Repeated subtraction	Short division
970 -800 (100 × 8) 170 160 (20 × 8) 10 $-\ \ 8$ (1 × 8) 2	$8\overline{)9^17^10}$ 1 2 1 r 2

As a remainder, $970 \div 8 = 121$ r2

To write this as a fraction, put the remainder in the top and the number you are dividing by in the bottom:

$121\frac{2}{8} = 121\frac{1}{4}$

b

Short division	Long division
$8\overline{)9^17^10.^20^40}$ 1 2 1 . 2 5	1 2 1 . 2 5 $8\overline{)9 7 0 . 0 0}$ 8 17 16 10 8 20 16 40

To give your answer as a decimal, you need to write zeros after the decimal point in 970 in order to complete the division.

$970 \div 8 = 121.25$

1 Use repeated subtraction to work these out.
   a $84 \div 6$ **b** $126 \div 6$ **c** $168 \div 6$ **d** $312 \div 6$

2 Use long division to work these out.
   a $54 \div 3$ **b** $84 \div 3$ **c** $93 \div 3$ **d** $114 \div 3$

3 Use short division to work these out.
   a $84 \div 7$ **b** $119 \div 7$ **c** $105 \div 7$ **d** $168 \div 7$

4 Use any method to work these out.

    a 200 ÷ 50   **b** 200 ÷ 25   **c** 2000 ÷ 50   **d** 2000 ÷ 250

5 Use long division to work these out. Leave your answer as a remainder.

    a 185 ÷ 12   **b** 212 ÷ 12   **c** 370 ÷ 12   **d** 375 ÷ 12

6 Use long division to work these out. Write the remainder as a fraction in simplest form.

    a 248 ÷ 16   **b** 282 ÷ 16   **c** 500 ÷ 16   **d** 500 ÷ 8

7 Use short division to work these out. Write the remainder as a decimal.

    a 171 ÷ 12   **b** 219 ÷ 12   **c** 438 ÷ 12   **d** 657 ÷ 12

8 Use any method you prefer to work these out. Write the remainder as a decimal.

    a 684 ÷ 16   **b** 692 ÷ 16   **c** 684 ÷ 5   **d** 692 ÷ 12

Reason mathematically

A company has 95 boxes to move by van. The van can carry 8 boxes at a time.

How many trips must the van make to move all the boxes?

Divide 95 ÷ 8.   In this case, there is no need to continue the division into a
decimal because the question is asking how many trips the
van must make. The van must make 12 trips.

$$\begin{array}{r} 11\,r\,7 \\ 8\,\overline{)9\,{}^1 5} \end{array}$$

9 **a** Professional cycling teams enter nine riders into the large races such as the Tour de France, Giro d'Italia and the Spanish Vuelta. Given that there are 198 riders in a race, how many teams have entered?

    b Football teams have 11 players. On one afternoon, 286 players start their matches. How many teams are playing?

10 Electric light bulbs can be packed into boxes of 16 or 24. How can 232 bulbs be packed into full boxes only? Find two possible ways.

11 3000 people go on a journey from Paris to Rome. They travel in 52-seater coaches.

    a How many coaches do they need?

    b Each coach costs €680.

       What is the total cost of the coaches?

    c How much is each person's share of the cost?

Solve problems

A group of 8 friends won £3358 in the lottery and split the money equally amongst themselves.

How much money did each friend get?

Divide £3358 ÷ 8.

$$\begin{array}{r} 0\,4\,1\,9.7\,5 \\ 8\,\overline{)3\,3\,{}^35\,{}^18\,{}^76.{}^40\,0} \end{array}$$

Each friend got £419.75.

12 The mathematics department has printed 500 information sheets about long division. They put them into sets of 30 sheets.

    a How many full sets are there?

    b How many more sheets should be printed to make another full set?

13 To raise money, a running club is doing a relay race of 120 kilometres. Each runner except the last one will run 9 km. The last runner will just run the extra distance to the finish.

    a How many runners are needed to cover the distance?

    b How far does the last person run?

14 The diameter of a 2p coin is 26 mm. A piece of A4 paper is 210 mm wide and 297 mm tall. How many 2p coins would it take to fill up the area of an A4 paper? (Hint: You must work out how many coins will fit across and how many will fit up the page.)

15 A contractor has 13 tonnes of waste to move by van. The van can carry 0.8 tonnes at a time.

    a How many trips must the van make to move all the waste?

    b A different contractor had a lorry that can take 1.1 tonnes at a time. How many fewer trips would this contractor need to shift the 13 tonnes?

18.3 Calculations with measurements

- I can convert between common metric units
- I can use measurements in calculations
- I can recognise and use appropriate metric units

You need to know and use these **metric conversions** for length, capacity and mass. Length is measured in **metres**, m. Capacity is measured in **litres**, L. Mass is measured in **grams**, g.

You can use this diagram to help you convert units. Learn the prefix for each sub-unit. For example, to convert mm (millimetres) to metres (the unit), divide by 1000.

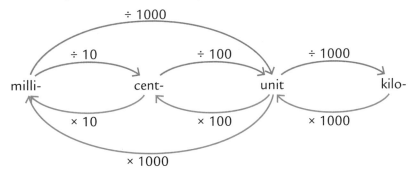

When adding or subtracting metric amounts that are given in different units, you need to convert so that they are in same units first.

You also need to be able to convert times.

| 1 year = 365 days |
| 1 day = 24 hours |

| 1 hour = 60 minutes |
| 1 minute = 60 seconds |

(A **leap year** has 366 days, but usually questions assume 365 days in a year and ignore leap years. Ask your teacher if you are not sure.)

Develop fluency

Add 1.57 m, 23 cm and 0.092 km.

First convert all the lengths to the same units. In this case, metres are a sensible unit.

23 cm = 0.23 m   *23 ÷ 100 = 0.23*

0.092 km = 92 m   *0.092 × 1000 = 92*

1.57 m + 0.23 m + 92 m = 93.8 m

1 Convert each length to centimetres.
   a 60 mm   b 600 mm   c 6 m   d 0.6 km

2 Convert each length to kilometres.
   a 456 m   b 4562 m   c 4 562 000 cm   d 45 620 cm

3 Convert each length to millimetres.
   a 34 cm   b 340 cm   c 3.4 m   d 0.34 km

4 Convert each mass to kilograms.
   a 1259 g   b 125.9 g   c 1259 mg   d 125.9 tonnes

5 Convert each mass to grams.
   a 4.32 kg   b 43.2 kg   c 4320 mg   d 432 000 mg

6 Convert each capacity to litres.
   a 237 cl   b 237 ml   c 2370 ml   d 23.7 ml

7 Convert each capacity to millilitres.
   a 3650 litres   b 356 litres   c 35.6 cl   d 356 cl

8 Convert each capacity to centilitres.
   a 862 litres   b 8.62 litres   c 862 ml   d 86.2 ml

9 Convert each time to hours and minutes.
   a 85 minutes   b 185 minutes   c 108 minutes   d 208 minutes

10 Add together the measurements in each group and give the answer in an appropriate unit.
   a 1.78 m, 39 cm, 0.006 km   b 0.234 kg, 60 g, 0.004 kg
   c 2.3 litres, 46 cl, 726 ml   d 6 hours, 24 minutes, 330 seconds

Reason mathematically

Choose a sensible unit to measure each of these.

a The width of a football field   **b** The length of a pencil

c The mass of a car   **d** A spoonful of medicine

When you are asked to choose a sensible unit, sometimes you will find that there is more than one answer.

a *metres*   **b** *centimetres*

c *kilograms*   **d** *millilitres*

11 Fill in each missing unit.

   a A two-storey house is about 7... high.   **b** Joe's mass is about 47....

   c Ruby lives about 2... from school.   **d** Luka ran a marathon in 3....

12 Asha has these jugs.
How is it possible to use these jugs to measure 100 ml of water?

13 Pierre buys 1 kg of sugar, 750 g of bananas and 1.2 kg of apples.
Is the total mass of the three items more than 3 kg?
Show your working.

14 Jenny knows that she needs 525 minutes of sleep a night in order not to feel tired the following day. She goes to bed at 11:15 p.m. and sets her alarm for 6:45 a.m. the following day. Will she get enough sleep?

Solve problems

Lily ties together three pieces of rope to make one longer piece. The pieces of rope are 62 cm, 1.08 m and 340 mm. She loses 0.14 m of length for each knot. How long is the new piece of rope?

Convert to the same units. Choose a sensible unit. In this case, centimetres is sensible.

1.08 m = 108 cm   1.08 × 100 = 108

340 mm = 34 cm   340 ÷ 10 = 34

0.14 m = 14 cm   0.14 × 100 = 14

Add and subtract the units.

62 + 108 + 34 − (14 × 2) = 176 cm

15 A kitten is born weighing 174 g. The kitten loses 24 050 mg the first day, then gains 7 g on the second day and gains 0.07 562 kg on the third day. How much does the kitten weigh on the third day?

16 The United Arab Emirates produce 21.6 tonnes carbon dioxide per person, per year. Their population is around 9.4 million.

The UK has a larger population, around 66 million, and produces around 5.7 tonnes of carbon dioxide per person, per year.

Which country produces more carbon dioxide in total?

17 This is an extract from the Isle of Man TT (Tourist Trophy) Motorbike races.

Moto GP 2002 Classic Junior

Position	No.	Competitor	Machine	Time	Speed
1	5	Bill Swallow	350 Linton Aermacchi		100.26

Bill Swallow holds the record for the fastest lap on a 350 cc bike. One lap is 37.66 miles and the race was over four laps. Bill was the only racer to average over 100 mph. How many hours and minutes did it take him to win the race, to the nearest minute?

18.4 Multiplication with large and small numbers

● I can multiply with combinations of large and small numbers mentally

You need to be able to complete some multiplications without using a calculator.

Develop fluency

Work these out.

a 4×500 **b** 0.4×0.5 **c** 400×0.05

a $4 \times 5 = 20$ First just multiply the non-zero digits.
$4 \times 500 = 2000$ Because 500 is 5×100 you multiply 20 by 100.

b $4 \times 5 = 20$ First just multiply the non-zero digits.
$0.4 \times 5 = 2$ $0.4 = 4 \div 10$ so divide 20 by 10.
$0.4 \times 0.5 = 0.2$ $0.5 = 5 \div 10$ so divide 2 by 10.

c $4 \times 5 = 20$ First just multiply the non-zero digits.
$400 \times 5 = 2000$ $400 = 4 \times 100$ so multiply 20 by 100.
$400 \times 0.05 = 20$ $0.05 = 5 \div 100$ so divide 2000 by 100.

Work these out.

1 a 300×4 **b** 30×40 **c** 3×4000 **d** 300×400

2 a 0.4×7 **b** 0.4×70 **c** 4×0.7 **d** 40×0.7

3 a 0.2×60 **b** 0.02×6000 **c** 20×0.06 **d** 2000×0.006

4 a 0.6×0.3 **b** 0.6×0.03 **c** 0.6×0.8 **d** 0.06×0.08

5 a 60×9 **b** 600×9 **c** 600×0.9 **d** 600×0.09

6 a 0.3×7 **b** 30×70 **c** 300×0.7 **d** 0.03×0.007

7 a 0.3×0.04 **b** 30×0.4 **c** 30×400 **d** 0.03×4000

Reason mathematically

$17 \times 52 = 884$

Use this fact to work out 170×0.52.

$170 = 17 \times 10$

$0.52 = 52 \div 100$

$170 \times 0.52 = (17 \times 52) \times 10 \div 100$

$170 \times 0.52 = 88.4$

8 $18 \times 35 = 630$

Use this fact to work out:

    a 1.8×35 **b** 18×0.35 **c** 180×350 **d** 0.18×0.35.

9 $23^2 = 529$

Use this fact to work out:

    a 230^2 **b** 2.3^2 **c** 0.23^2.

10 $88^2 = 7744$

Use this fact to work out:

    a 8800^2 **b** 0.88^2 **c** 8.8^2 **d** 0.088^2.

11 $0.32 \times 0.15 = 0.048$

Use this fact to work out:

    a 32×0.15 **b** 3.2×1.5 **c** 32×15 **d** 0.032×0.015.

Solve problems

A piece of paper is cut to be 200 mm wide and 300 mm high.

Work out the area of the piece of paper in cm^2.

There are two methods.

Method 1

Convert each unit to cm and find the area.

$200\,mm = 20\,cm$

$300\,mm = 30\,cm$

$20\,cm \times 30\,cm = 600\,cm^2$

Method 2

Find the area in mm^2 and convert to cm^2.

$1\,cm^2 = 10^2\,mm^2 = 100\,mm^2$

$200\,mm \times 300\,mm = 60\,000\,mm^2$

$60\,000\,mm^2 = 600\,cm^2$

As $60\,000 \div 100 = 600$

Method 2 is sometimes easier depending on the information given in the problem. You can use this method if you are only given an area or volume and not the individual side lengths.

Note that to convert between cubic units, just cube the conversion factor. For example,

$1\,cm^3 = 10^3\,mm^3 = 1000\,mm^3$

12 The lengths of the sides of a large box are 0.4 m, 0.5 m and 0.8 m.

   a Work out the volume, giving your answer in cubic metres (m^3).

   b Work out the volume, giving your answer in cubic centimetres (cm^3).

13 Sound travels about 0.3 km in 1 second. How far does sound travel in the following times? Work in kilometres.

   a 200 seconds   **b** 10 minutes   **c** 0.02 seconds

14 A rectangle measures 60 cm by 80 cm.

   What is the area of the rectangle in: **a** square centimetres?

   b square millimetres?   **c** square metres?   **d** square kilometres?

15 A cuboid measures 30 cm by 40 cm by 50 cm.

   Find the volume of the cuboid in: **a** cubic metres. **b** cubic kilometres.

18.5 Division with large and small numbers

- I can divide combinations of large and small numbers mentally

The number you divide by is called the **divisor**. Changing the divisor to a whole number can make a division easier to do.

Develop fluency

Work these out.

   a $32 \div 0.08$ **b** $8 \div 400$

   a $32 \div 0.08 = 3200 \div 8$ Multiply both numbers by 100 to make the divisor 8.

           $= 400$ $32 \div 8 = 4$ so the answer is 400.

   b $8 \div 400 = 0.08 \div 4$ Divide both numbers by 100 to make the divisor 4.

           $= 0.02$

For questions 1–5, work out the answer.

1 a $6 \div 0.2$ **b** $60 \div 0.2$ **c** $30 \div 0.2$ **d** $3 \div 0.2$

2 a $400 \div 20$ **b** $40 \div 20$ **c** $4 \div 20$ **d** $0.4 \div 20$

3 a $8 \div 20$ **b** $8 \div 200$ **c** $80 \div 200$ **d** $8 \div 0.2$

4 a $0.6 \div 10$ **b** $0.6 \div 20$ **c** $0.6 \div 30$ **d** $0.6 \div 60$

5 a $18 \div 300$ **b** $1.8 \div 30$ **c** $1.8 \div 3$ **d** $1.8 \div 300$

For questions 6–7, work out the answers and then put them in order of size, smallest first.

6 a $21 \div 0.7$ **b** $0.21 \div 0.7$ **c** $21 \div 700$ **d** $0.21 \div 0.07$

7 a $42 \div 0.6$ **b** $420 \div 0.6$ **c** $4.2 \div 0.6$ **d** $42 \div 600$

Reason mathematically

$1425 \div 57 = 25$

Use this fact to work out $14.25 \div 5.7$.

$14.25 \div 5.7 = 142.5 \div 57$ Multiply both numbers by 10 to make the divisor 57.

        $= 2.5$ $1425 \div 57 = 25$, so $142.5 \div 57 = 2.5$

8 $1001 \div 77 = 13$

Use this fact to work out:

  a $100.1 \div 77$ **b** $10.01 \div 77$ **c** $10.01 \div 7.7$ **d** $100.1 \div 0.77$.

9 Which is the odd one out? Give a reason for your answer.

  a $27 \div 0.9$ **b** $270 \div 90$ **c** $2.7 \div 0.09$ **d** $0.27 \div 0.009$

10 Which is the odd one out? Give a reason for your answer.

  a $0.48 \div 1.2$ **b** $0.24 \div 0.6$ **c** $0.16 \div 0.4$ **d** $0.8 \div 0.2$

11 The area of this rectangle is $A \, \text{cm}^2$.

  a Explain why $x = \dfrac{A}{20}$.

  b Work out the value of x when $A = 0.4$.

  c Work out the value of x when $A = 1000$.

Solve problems

The cost to manufacture a single small screw is 0.0002p. In one quarter, the manufacturer spent £15 on manufacturing the screw. How many screws did it make?

Convert £15 to pence, $15 \times 100 = 1500$.

$1500 \div 0.0002 = 15\,000\,000 \div 2$ Multiply both numbers by 10000 to make the divisor a whole number

           $= 7\,500\,000$ $15 \div 2 = 7.5$, so $15\,000\,000 \div 2 = 7\,500\,000$

The manufacturer made 7 500 000 screws.

12 An advertiser pays 0.004p to a website every time it is hit (every time the website receives a visit). In one month, the advertiser pays the website £16. How many hits did the website receive?

13 A website gives 3.6p to charity on every sale it makes. In September, the website gave a mean average of £2.80 per day to charity. How many sales did it make in September?

14 a 3000 bacteria have a mass of 0.000003 g. What is the mass of 12 bacteria?

  b 10 000 000 000 000 atoms have the same mass as one bacterium. What is the mass of 19 atoms? (Bacteria is the plural of bacterium.)

15 The sun is approximately 149 600 000 km from Earth. The distance from the sun to the earth is approximately 3733 times the circumference of the Earth. Given that $1496 \div 3733 \approx 0.40075$, what is the circumference of the Earth in km?

Now I can...

choose a written method for multiplying two numbers together	use written methods to carry out multiplications accurately	recognise and use appropriate metric units
choose a written method for dividing one number by another	use written methods to carry out divisions accurately	multiply with combinations of large and small numbers mentally
convert between common metric units	use measurements in calculations	divide combinations of large and small numbers mentally

CHAPTER 19 Transformations

19.1 Reflections

- I can reflect a shape in a mirror line
- I can use coordinates to reflect shapes in all four quadrants

You can use a mirror line to draw a **reflection**, like this.

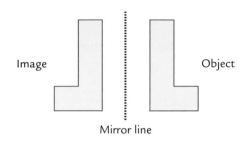

Image   Object

Mirror line

The **object** is reflected in the mirror line to give the **image**.

The mirror line becomes a line of symmetry. So, if the paper is folded along the mirror line, the object will fit exactly over the image.

Note that the image is the same distance from the mirror line as the object is.

Develop fluency

a Reflect this shape in the mirror line.

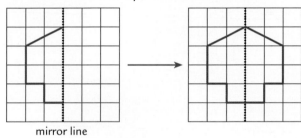

mirror line

Notice that the image is the same size as the object, and that the mirror line becomes a line of symmetry.

b Describe the reflection in this diagram.

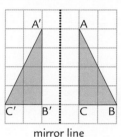

mirror line

Triangle A′B′C′ is the reflection of triangle ABC in the mirror line.

Notice that the line joining A to A′ is at right angles to the mirror line and the three points on the object and image are at the same distance from the mirror line.

1 Copy each diagram onto squared paper and draw its reflection in the dashed mirror line.

  a
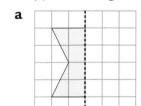
  b
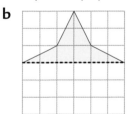
  c
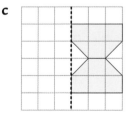
  d
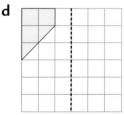

2 Copy and complete the symmetrical shapes.

  a

  b

  c

  d

3 Copy each diagram onto squared paper and draw its reflection in the given mirror line.

a

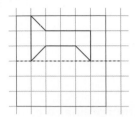

b

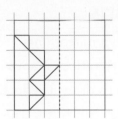

4 Copy each shape onto squared paper and draw its reflection in the dashed mirror line.

a 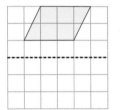 **b** 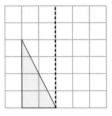 **c** 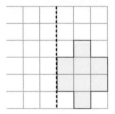 **d**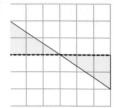

5 Copy and reflect the shape in the two mirror lines.

mirror line

6 Copy and reflect each shape in all the mirror lines.

a **b**

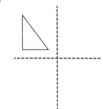

7 a Copy the diagram onto squared paper and reflect the triangle in the series of parallel mirror lines.

b Use a series of parallel mirror lines to make up your own patterns.

8 Which of these shapes could have been formed by a reflection? Copy them onto squared paper and, where possible, draw in the mirror line.

a   **b** 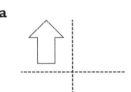 **c**   **d**

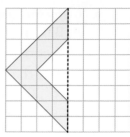

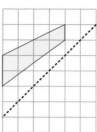

9 Copy each shape onto squared paper and draw its reflection in the mirror line.

a 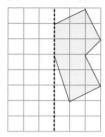   b 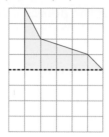   c 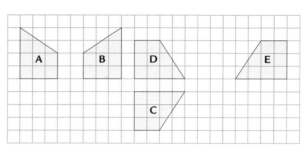   d

10 Jess has produced a pattern by reflecting a shape a number of times. She started with A and reflected it to get B. Then she reflected B to C, C to D and finally D to E. Copy the diagram and draw in the mirror lines that Jess used.

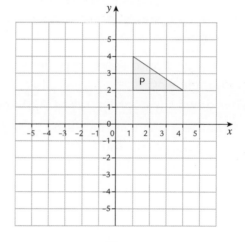

Reason mathematically

a Write down the coordinates of the vertices of triangle P.

b Reflect P in the *x*-axis. Label the new triangle Q.

c Write down the coordinates of the vertices of triangle Q.

d Reflect Q in the *y*-axis. Label the new triangle R.

e Write down the coordinates of the vertices of triangle R.

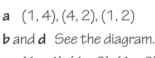

a $(1, 4), (4, 2), (1, 2)$

b and d See the diagram.

c $(1, -4), (4, -2), (1, -2)$

e $(-1, -4), (-4, -2), (-1, -2)$

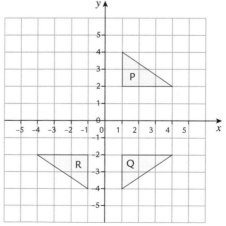

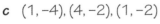

11 The points A(1, 2), B(2, 5), C(4, 4) and D(6, 1) are shown on the grid.

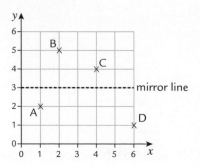

a Copy the grid onto squared paper and plot the points A, B, C and D.
Draw on the mirror line.

b Reflect the points in the mirror line and label them A′, B′, C′ and D′.

c Write down the coordinates of the image points.

d The point E(12, 6) is reflected in the mirror line. What are the coordinates of E′?

12 Look at the points shown on the grid.

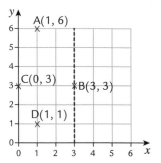

a Copy the grid onto squared paper and plot the points A, B, C and D. Include the coordinates on your diagram. Draw the mirror line.

b Reflect the points in the mirror line and label them A′, B′, C′ and D′.

c Write down the coordinates of the image points.

d The point E′(4, 5) is the reflection of point E. Mark both points E and E′ on your diagram, including the coordinates.

13 a Copy the grid onto squared paper and draw the triangle ABC.
Write down the coordinates of A, B and C.

b Reflect the triangle in the *x*-axis. Label the vertices of the image A′, B′ and C′.
What are the coordinates of A″, B′ and C′?

c Reflect triangle A′B′C′ in the *y*-axis. Label the vertices of this image A″, B″ and C′.
What are the coordinates of A″, B″ and C″?

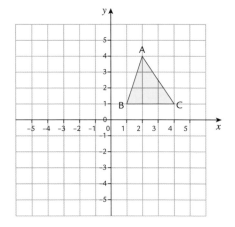

14 a Reflect the shape in the *x*-axis, label it A′B′ C′ and write down the coordinates of the vertices.

b Reflect the original shape in the *y*-axis, label it A″B″C″ and write down the coordinates.

c Complete the diagram with a fourth shape in the lower left quadrant.

d Write down the coordinates of the final triangle and label it appropriately.

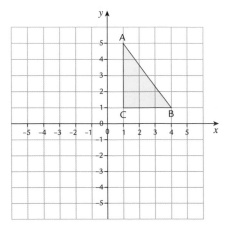

Solve problems

Copy this diagram onto squared paper.

Triangle A is reflected in the x-axis to give triangle B.

a Reflect triangle B in the mirror line $x = -2$ to give triangle C. Draw triangle C.

b Write down the coordinates of the vertices of triangle C.

c Reflect triangle C to give triangle D, which is also a reflection of A. Draw triangle D.

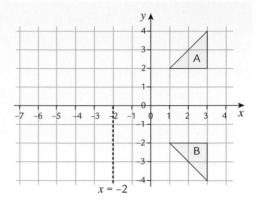

a, c

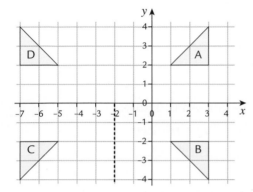

b $(-5, -2), (-7, -2), (-7, -4)$

15 Copy the diagram onto squared paper.
Shape A is reflected in a mirror line to give shape B, and then shape B is reflected to give shape C as shown.

a Draw shape B and the mirror line that has been used to reflect it from A.

b Draw a mirror line that reflects B to C. How would you describe this line?

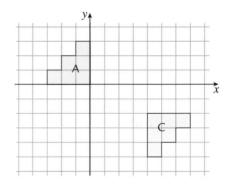

16 Copy this diagram onto squared paper.
Ricardo says: 'I am going to reflect shape A in the y-axis to get shape B. Then, I'll reflect B in the x-axis to get shape C. If I then use the line $y = -x$ as a mirror line, I can reflect C to get back to A.'
On your diagram, follow Ricardo's directions.
Do you get back to the original shape A?

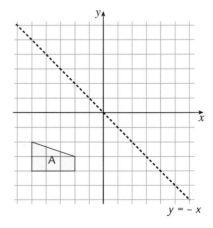

17 Copy this diagram onto squared paper.
    Shape A has been reflected a number of times. First it was
    reflected to give B, then B was reflected to C, then C to D.
    Draw in the mirror lines that achieve these reflections.

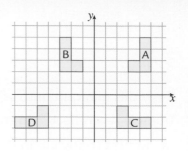

18 Mirrors can be two-way.
    Copy each shape onto squared paper and reflect it in the mirror line.

a 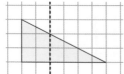   b   c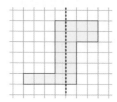

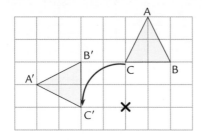

19.2 Rotations

● I can rotate a shape about a point

To describe the **rotation** of a 2D shape accurately, you need to know:

● the **centre of rotation** – the point about which the shape rotates
● the **angle of rotation** – usually 90° (a quarter-turn), 180° (a half-turn) or
   270° (a three-quarter turn)
● the **direction of rotation** – clockwise (to the right) or anticlockwise (to the left).

When you rotate a shape, it is often helpful to use tracing paper.

As with reflections, the original shape is the object and the rotated shape is the image.

Develop fluency

a Describe the rotation of triangle ABC.

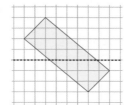

Triangle ABC has been rotated through 90°
anticlockwise, about the point X, onto triangle
A'B'C'.

b Describe the rotation of triangle ABC.

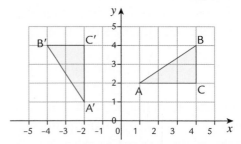

Triangle A'B'C' is the image of triangle ABC
after a rotation of 90° anticlockwise about the
origin, O(0, 0).

The coordinates of the vertices of the object are
A(1, 2), B(4, 4) and C(4, 2).

The coordinates of the vertices of the image are
A'(−2, 1), B'(−4, 4) and C'(−2, 4).

1 Copy each of these shapes and draw the image, after it has been rotated about the point marked X, through the angle indicated. Use tracing paper to help you.

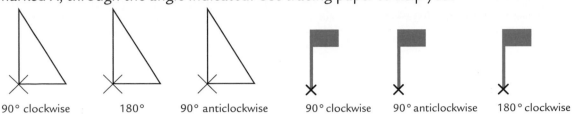

90° clockwise   180°   90° anticlockwise   90° clockwise   90° anticlockwise   180° clockwise

2 Copy each shape onto a square grid.
Draw its image after it has been rotated, about the point marked X, through the angle indicated. Use tracing paper to help you.

a   **b**   **c**   **d**

180° clockwise   90° anticlockwise   180° anticlockwise   270° clockwise

3 Copy each shape onto a square grid.
Draw the image after it has been rotated, about the point marked X, through the angle indicated. Use tracing paper to help.

a **b** **c**
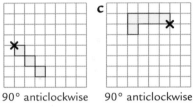

90° clockwise   90° anticlockwise   90° anticlockwise

4 Copy each right-angled triangle ABC onto a coordinate grid, with axes for *x* and *y* both numbered from −5 to 5.

   a Draw the image A′B′C′ of triangle ABC after it has been rotated about the origin, O, through the angle and direction indicated.

   b Write down the coordinates of the vertices of the object.

   c Write down the coordinates of the vertices of the image.

i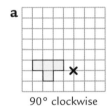

90° anticlockwise

ii

90° clockwise

iii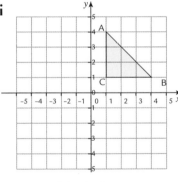

180° anticlockwise

iv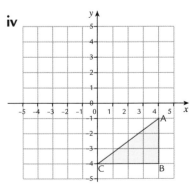

180° clockwise

5 Copy each of these isosceles triangles onto a coordinate grid, with axes for x and y from –5 to 5.

a Draw the image A'B'C' of each triangle after it has been rotated about the origin, O, through the angle and direction indicated.

b Write down the coordinates of the vertices of each object.

c Write down the coordinates of the vertices of each image.

i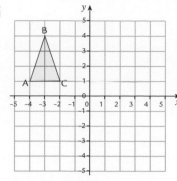

90° clockwise

ii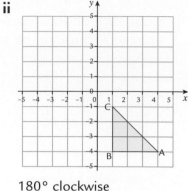

180° clockwise

iii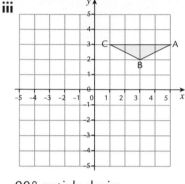

90° anticlockwise

iv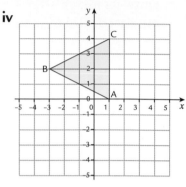

180° anticlockwise

6 Copy the trapezium onto a coordinate grid, with axes for x and y from 0 to 10.

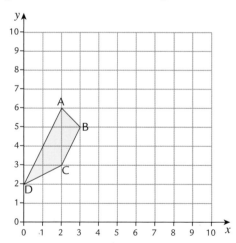

a Rotate the trapezium ABCD through 90° anticlockwise about the point (4, 6) to give the image A'B'C'D'.

b Write down the coordinates of A', B', C' and D'.

7 Copy the diagram onto squared paper.

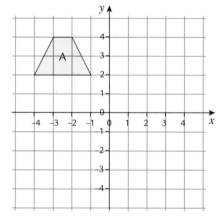

a Rotate shape A 90° clockwise about the point (0, 0) to give image B.

b Rotate shape A 180° anticlockwise about the point (–1, 0) to give image C.

8 Copy each diagram onto squared paper.

a Shape B is the result of rotating shape A. Describe the rotation that has taken place.

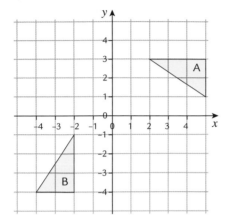

b Shape D is the result of rotating shape C. Describe the rotation that has taken place.

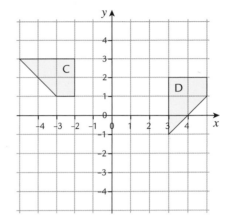

Reason mathematically

a Rotate the triangle ABC through 90° anticlockwise about the point (4, 4) to give the image A′B′C′.

b Write down the coordinates of A′, B′ and C′.

c Which coordinate point remains fixed throughout the rotation?

d Fully describe the rotation that will map the triangle A′B′C′ onto the triangle ABC.

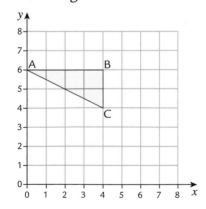

a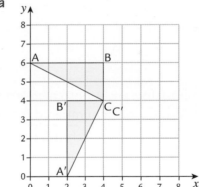

b A′(2, 0), B′(2, 4), C′(4, 4)

c Vertex C(4, 4)

d Rotation 90° clockwise at point (4,4)

9 Copy this rectangle onto a coordinate grid.

a Rotate rectangle ABCD through 90° clockwise about the origin O(0, 0), to give its image A′B′C′D′.

b Write down the coordinates of A′, B′, C′ and D′.

c What rotation will move rectangle A′B′C′D′ onto rectangle ABCD?

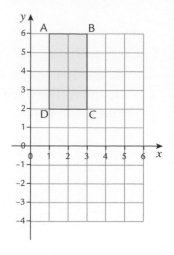

10 Copy the pentagon onto a coordinate grid, with axes for x and y from 0 to 10.

a Rotate the pentagon ABCDE through 90° clockwise about the point (4, 9) to give the image A′B′C′D′E′.

b Write down the coordinates of A′, B′, C′, D′ and E′.

c Write down two different rotations that will move pentagon A′B′C′D′E′ onto pentagon ABCDE.

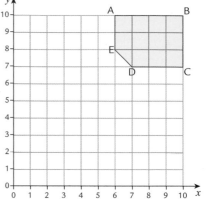

11 a Copy parallelogram A onto a coordinate grid, with axes for x and y both numbered from −5 to 5.

b Rotate parallelogram A through 90° anticlockwise about the point (1, 2). Label the image B.

c Write down two different rotations that will move parallelogram B onto parallelogram A.

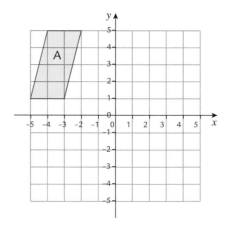

12 Some of these rotations are identical. State which ones are identical and explain why.

90° clockwise   180° clockwise   270° clockwise
270° anticlockwise   450° clockwise   90° anticlockwise
180° anticlockwise

Solve problems

a Triangle A is rotated 90°
clockwise about
(−4, −3) to give triangle B.
Draw triangle B.

b Triangle B is rotated 90°
anticlockwise about (6, 1) to give
triangle C. Draw triangle C.

c Describe the rotation that will
take triangle C back to the initial
position of triangle A.

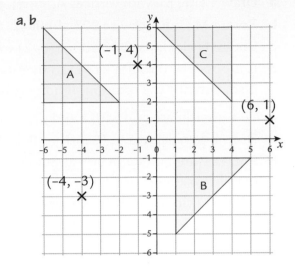

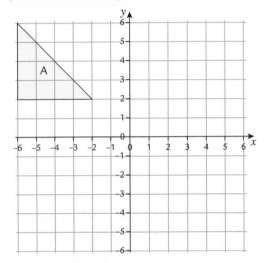

c Rotation 180° about (−1, 4).

---

13 Copy this diagram onto squared paper.

a Shape A is rotated 90° anticlockwise about (8, 8)
to give shape B. Draw shape B.

b Shape B is rotated 90° anticlockwise about (6, 0)
to give shape C. Draw shape C.

c Draw the mirror line that will take shape C back to
shape A.

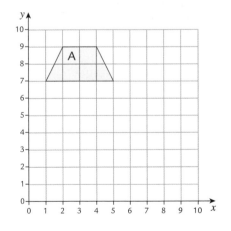

14 Draw a coordinate grid with x and y axes
from 0 to 12.

a Draw a trapezium, X, with vertices A(2, 2), B(2, 4), C(4, 6) and D(4, 2).

b Rotate X clockwise 90° about (8, 4) to give trapezium Y. Draw trapezium Y.

c Rotate Y clockwise 90° about (4, 8) to give trapezium Z. Draw trapezium Z.

d What single rotation would have taken trapezium X to trapezium Z?

15 Draw a coordinate grid with x and y axes from –10 to +10.

   a Draw a kite, M, with vertices A(–6, 5), B(–4, 6), C(–2, 5) and D(–4, 1).

   b Rotate M anticlockwise 90° about (2, 6) to give kite N. Draw kite N.

   c Rotate N clockwise 90° about (8, 2) to give kite O. Draw kite O.

   d Is it possible to rotate kite O so that it becomes a reflection of kite M?
      If so, describe the rotation. If not, explain why this is the case.

16 Draw a coordinate grid with x and y axes from –6 to +6.

   a Draw a parallelogram, P, with vertices A(–6, –4), B(–5, –2), C(–2, –2) and D (–3, –4).

   b Rotate P anticlockwise 90° about the origin to give parallelogram Q. Draw
      parallelogram Q.

   c Rotate Q clockwise 180° about (4, 0) to give parallelogram R. Draw parallelogram R.

   d Is it possible to rotate parallelogram R so that it becomes a reflection of
      parallelogram P?
      If so, describe the rotation. If not, explain why this is the case.

19.3 Translations

● I can translate a shape

A **translation** is a movement of a 2D shape from one position to another, without reflecting it
or rotating it.

The distance and direction of the translation are described by the number of unit squares
moved to the right or left, followed by the number of unit squares moved up or down.

As with reflections and rotations, the original shape is the object and the **translated** shape is
the image.

Develop fluency

a Translate triangle
   A onto triangle B
   by the
   translation 3
   units right and 2
   units up.

b Describe the
   translation of
   rectangle ABCD.

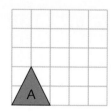

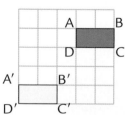

a Points on triangle A
   are translated onto
   corresponding points
   on triangle B, as
   shown by the arrows.

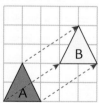

   When an object is
   translated onto its image, every
   point on the object moves the same
   distance, in the same direction.

b The rectangle ABCD has translated onto
   rectangle A′B′C′D′ by the translation 3
   units left and 3 units down.

1 Describe the translation:

a from A to B   b from A to C

c from A to D   d from A to E

e from B to D   f from C to E

g from D to E   h from E to A.

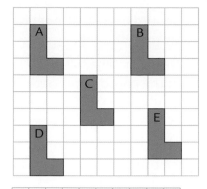

2 Describe each translation.

a A to C   b A to D

c C to B   d D to E

e B to A   f B to D

g D to C

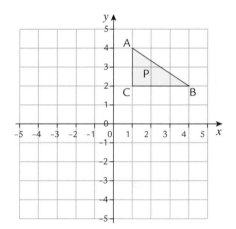

3 Copy triangle ABC onto squared paper, with axes for x and y both numbered from −5 to 5. Label the triangle P.

a Write down the coordinates of the vertices of triangle P.

b Translate triangle P 6 units left and 2 units down. Label the new triangle Q.

c Write down the coordinates of the vertices of triangle Q.

d Translate triangle Q 5 units right and 4 units down. Label the new triangle R.

e Write down the coordinates of the vertices of triangle R.

f Describe the translation that translates triangle R onto triangle P.

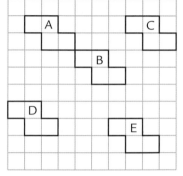

4 Copy the isosceles triangle onto a coordinate grid, with axes for x and y from −5 to 5.

a Plot the image A′B′C′ with A′(2, −1), B′(−1, −1) and C′(−1, 2).

b Describe the translation to translate ABC to A′B′C′.

c Describe the translation to translate A′B′C′ to ABC.

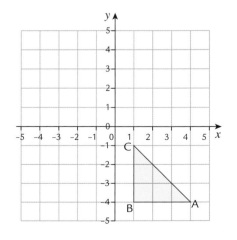

5 Copy the trapezium onto a coordinate grid, with axes for x and y from 0 to 10.

   a Translate ABCD 7 units right and 1 unit up.

   b Write down the coordinates of the image A′B′C′D′.

   c Rotate A′B′C′D′ 90° anticlockwise about the point (7, 3). Label the new image A″B″C″D″.

   d Describe the transformation that translates A″B″C″D″ back onto ABCD.

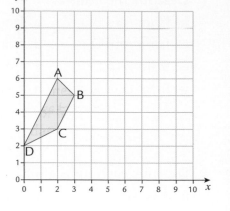

6 Copy the diagram onto a coordinate grid, with axes for x and y from 0 to 10.

   a Translate shape X 5 units left and 3 units down to give shape Y.

   b Translate shape Y 3 units right and 2 units down to give shape Z.

   c Which translation takes shape Z back to shape X?

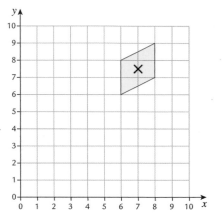

7 Describe the translations of:

   a B to D

   b D to C

   c A to D.

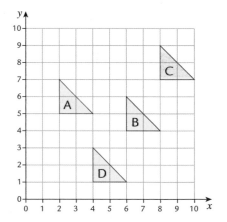

8 Copy the diagram onto a coordinate grid, with axes for x and y from 0 to 10.

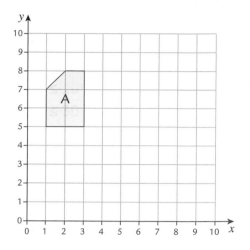

  a Translate pentagon A by 5 units right and 1 unit up to give pentagon B.

  b Translate pentagon B by 2 units left and 5 units down to give pentagon C.

  c Plot the points (8, 1), (8, 4), (9, 4), (10, 3) and (10, 1). Join them up, in order, to form pentagon D.

  d Translate pentagon D by 5 units left and 4 units up to give pentagon E

  e Which transformation connects pentagon A to pentagon E?

Reason mathematically

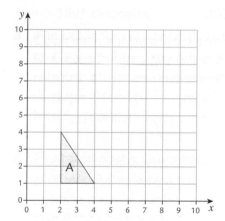

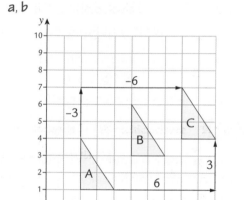

a, b

a Translate shape A 3 units right and 2 units up. Call the image B.

b Translate shape B 3 units right and 1 unit up. Call the image C.

c Which translation will take A to C?

d Which translation will take C to A?

e How could you use the information in parts **a** and **b** to provide an answer to part **c**, without using the diagram?

f What is the connection between your answers to parts **c** and **d**?

c 6 units right, 3 units up

d 6 units left, 3 units down

e By combining the units of movement: 3 right + 3 right = 6 right, 2 up + 1 up = 3 up.

f The movement is reversed: right to left, up to down, but the units are the same.

9 To translate shape A to shape B, A is translated 5 units right and 2 units up.
To translate shape B to shape C, B is translated 3 units right and 7 units up.

   a Describe the translation that translates shape A onto shape C.

   b Describe the translation that translates shape C back onto shape A.

10 Copy trapezium ABCD onto a coordinate grid, with axes for x and y both numbered from –5 to 5.

   a Translate the trapezium 4 units right and 5 units up.

   b Write down the coordinates of the vertices of the image.

   c Describe which properties of the trapezium have changed and which have stayed the same, after this translation.

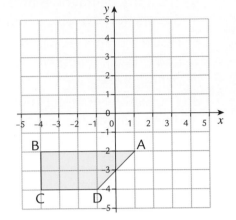

11 Copy the isosceles triangle X onto a coordinate grid, with axes for x and y both numbered from −5 to 5.

   a Rotate the triangle X through 90° anticlockwise about the origin, O. Label the image Y.

   b Now translate triangle Y 4 units right and 1 unit up. Label the image Z.

   c Describe how you would move triangle Z back to triangle X.

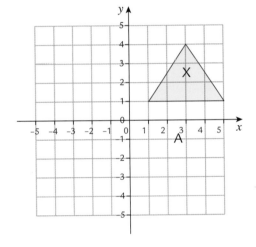

12 The translation from A to B is 4 units right and 2 units up.
The translation from C to A is 6 units left and 1 unit up.
What is the translation from B to C?

Solve problems

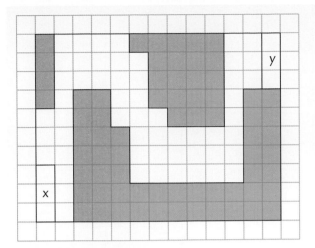

A shipping container is being moved through a storage area. It is at location X and needs to be at location Y. The shaded areas represent other containers.

a List the translations that will move the container from X to Y. Always move horizontally first.

b Does it make any difference to the number of moves if you move vertically first?

a Move 1: right 1, up 7
Move 2: right 3, down 2
Move 3: right 1, down 3
Move 4: right 5, up 5
Move 5: right 2, up 0

b It will still be achieved in 5 moves.

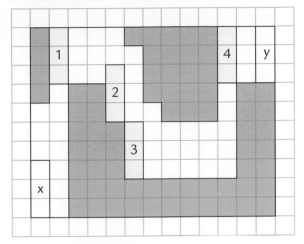

13 Copy the diagram onto a coordinate grid, with axes for x and y from 0 to 13.

A forklift truck driver has to move an isosceles-shaped wooden crate from X to Y. The shaded areas are walls.

List the translations and rotations that will move the crate from X to Y.

For a translation, make the horizontal or vertical moves in either order.

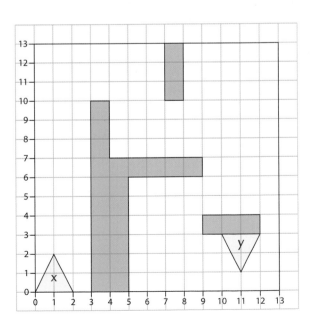

14 Copy the diagram onto a coordinate
grid, with axes for x and y
from −8 to 8.

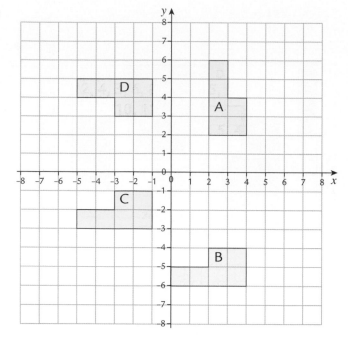

   a Which transformation will
      take A to B?

   b Which transformation will
      take B to C?

   c Draw a mirror line that will
      transform C to D. How would
      you describe this line?

   d A can be moved so that it joins
      C to form a rectangle as:

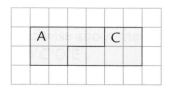

Describe the transformation that achieves this.

15 In a video game a jigsaw piece, A, needs to be placed
in the space A′, at the bottom of the grid. This is
achieved by translations and rotations, but
rotations can only be centred on a vertex of
the shape.
Which three different sets of transformations
would achieve this?

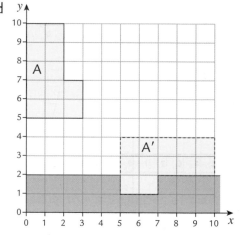

16 Copy the diagram onto squared paper.
A single move (translation) by a knight on a chess board
is made by either:

1 unit up or down and then 2 units to either the right or
left, or

2 units up or down and then 1 unit to either the right
or left

List the series of possible moves that would take the
knight from X to Y.

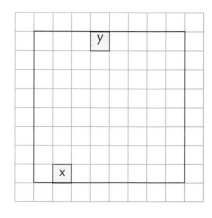

19.4 Tessellations

● I can tessellate shapes

A **tessellation** is a pattern made by fitting the same shapes together without leaving any gaps.

Develop fluency

Tessellate these shapes on square grids.

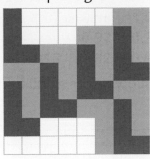
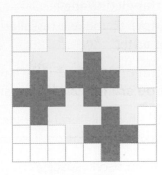

1 Make a tessellation with each shape. Use a square grid.

a    b    c    d

2 Make a tessellation with each shape. Use a triangular grid.

a   b   c   d

3 a Which of these tiling patterns are tessellations of the shaded shape?

i   ii   iii 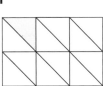   v    vi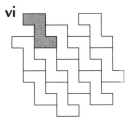

b Two of these patterns are tessellations of two shapes, rather than one. Name the non-shaded shapes that make each tessellation work.

4 Make a tessellation from each of the following shapes, if possible. Use a square grid to help.

a   b 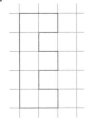   c 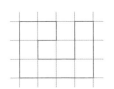   d

5   Draw your own tessellations, using at least six repetitions of each of the shapes below.

a   b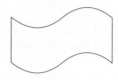

6   The length of the square below is equal to the base and height of the right-angled triangle.
    Combine both shapes to make a tessellation. Use at least four repetitions of each.

7   Draw a regular octagon and a square that have sides of the same length. Combine these shapes to make a tessellation. Use at least four octagons and seven squares.

8   Draw a square and an equilateral triangle that have sides of the same length. Combine both shapes to make a tessellation. Use at least three squares and six triangles.

9   This tree is made from three isosceles triangles placed one above another. Copy the tree and then tessellate. Use at least six trees.

Reason mathematically

Design a tile based on a square that will tessellate and make a pattern. Use at least six of your tiles.

Start with a square and cut a piece from one side. Add the piece you have cut out to the opposite side. This will tessellate.

You can do the same with the other two sides that you have left untouched and it will still tessellate.

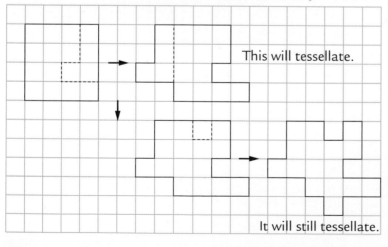

This will tessellate.

It will still tessellate.

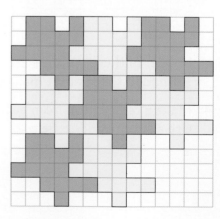

10 This tessellation uses shapes with curves.
Design a different tessellation that uses curves.

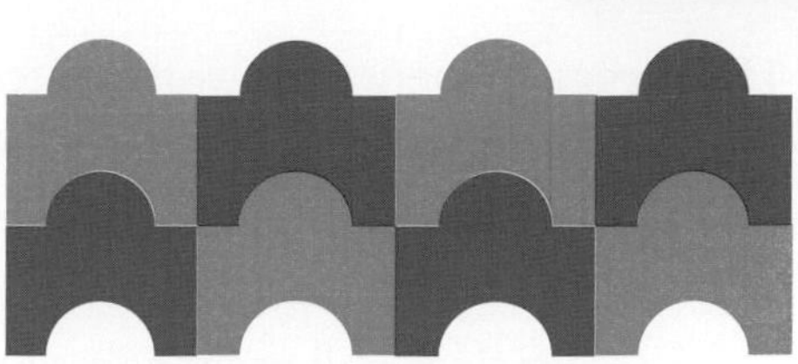

11 Find a way to test whether the shapes below will tessellate, and then create the tessellation patterns.

a

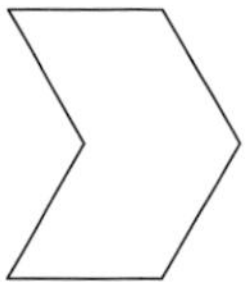

b

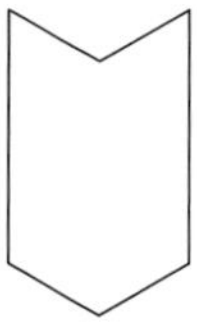

c

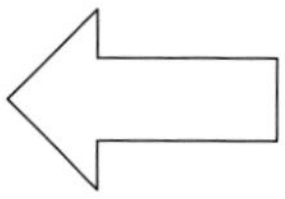

d

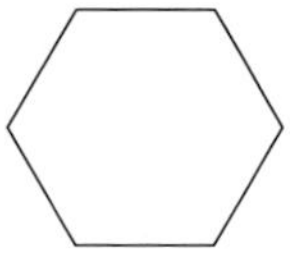

12 **i** Which of these designs will not tessellate?

ii Which of these designs tessellate with a single shape?

iii Which of these designs can only tessellate using two shapes?

a

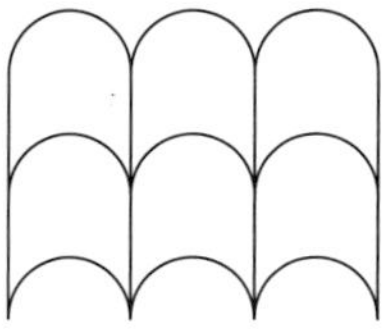

b

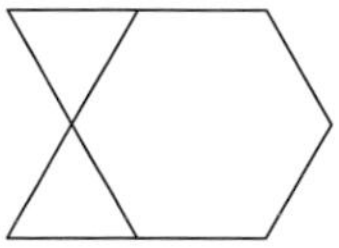

c

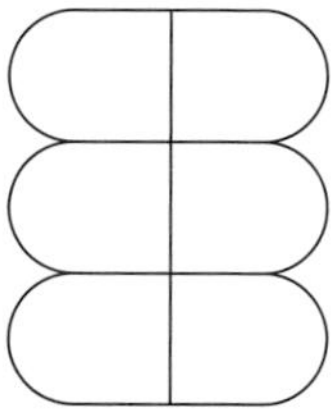

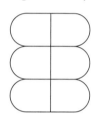

d

13 **a** Show how an equilateral triangle and a regular hexagon can be combined to form a full tessellation.

b What are the angles at the points where the shapes meet?

Solve problems

One of these shapes will not tessellate around a point. Which one? Explain why.
equilateral triangle, square, regular pentagon, regular hexagon.

First work out the interior angles of the shapes.

Equilateral triangle : 60° square: 90°

regular pentagon: 108° regular hexagon: 120°

For a shape to tessellate around a point, its internal angle must be a factor of 360°.

This means that the pentagon will not tessellate around a point.

14 Any quadrilateral will tessellate.
This is an example of an irregular quadrilateral.
Make an irregular quadrilateral tile cut from card.
Then use your tile to show how it tessellates.

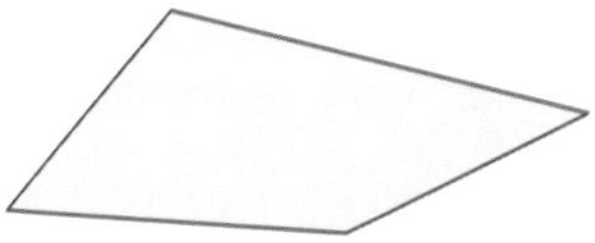

15 A regular pentagon will not tessellate on its own. Investigate what second shape can be used in a tessellation with a pentagon.

16 An irregular pentagon with different angles but sides of the same length can be tessellated. The internal angles need to be adjusted from 108° so that either three or four vertices will meet at a point.

From this information work out what the angles of the pentagon will be and try it out.

17 Circles will not tessellate on their own, a second shape is needed. Draw six identical circles as close together as possible to work out what the second shape will look like.

Now I can...

reflect a shape in a mirror line	use coordinates to reflect shapes in all four quadrants	tessellate shapes
rotate a shape about a point	translate a shape	

CHAPTER 20 Working with numbers

20.1 Powers and roots

● I can use powers and roots

A **power** shows a number being multiplied by itself many times. A power can be any number. It tells you how many 'lots' of the **base number** are being multiplied together.

The inverse process is finding the root: this will usually be the **square** or **cube root**.

Square roots can be positive or negative. This is because multiplying two negative numbers gives a positive number.

A positive cube number can only have a positive cube root and a negative cube number can only have a negative cube root.

Develop fluency

Calculate the value of each number.

a 4^3

b 5.5^2

c $(-3)^4$

a $4^3 = 4 \times 4 \times 4 = 64$

b $5.5^2 = 5.5 \times 5.5 = 30.25$

c $(-3)^4 = -3 \times -3 \times -3 \times -3 = 9 \times 9 = 81$

Be careful when raising negative numbers to a power. A negative number raised to an odd power will have a negative answer and a negative number raised to an even power will have a positive answer. If you are doing this on a calculator, put the negative number in brackets.

Use a calculator to work these out.

a $\sqrt{12.25}$ **b** $\sqrt{33124}$ **c** $\sqrt[3]{2197}$

Depending on your calculator, you might need to select the square root or cube root key before the number. Make sure you know how to use the functions on your calculator.

a $\sqrt{12.25} = 3.5$ **b** $\sqrt{33124} = 182$ **c** $\sqrt[3]{2197} = 13$

1 Use a calculator to find the value of each number.

  a 18^2 **b** 18^3 **c** 12^2 **d** 12^3 **e** 13^2 **f** 13^3

2 The diagrams show cubes made from smaller one-centimetre cubes.
   For each shape, work out the:
   a Area of each face (cm²)
   b Volume of the cube (cm³).

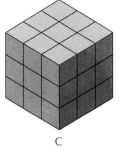

A   B   C

3 Use your answers from question 2 to work out one value for each number.

    a $\sqrt{64}$ **b** $\sqrt{81}$ **c** $\sqrt[3]{729}$ **d** $\sqrt[3]{512}$

4 Use a calculator to find the value of each number.

    a 1.7^2 **b** 1.3^3 **c** $(-2.3)^3$ **d** $(-4.5)^2$ **e** $\sqrt{6.76}$ **f** $\sqrt[3]{9.261}$

5 Use a calculator to find the value of each number.

    a 4^4 **b** 2^5 **c** 3.05^4 **d** 2.32^5 **e** $(-4)^4$ **f** $(-5.05)^4$

6 Without using a calculator, write down the value of each number.

    a 20^2 **b** 30^3 **c** 50^3 **d** 20^5 **e** 70^2 **f** 200^3

7 $10^2 = 100$, $10^3 = 1000$

Copy and complete this table.

Number	100	1000	10 000	100 000	1 000 000	10 000 000
Power of 10	10^2	10^3				

8 Use a calculator to find the value of each number.

    a 17^3 **b** 15^2 **c** 14^2 **d** 15^3

9 Use your answers from question 8 to find the value of each number.

    a 1.7^3 **b** 0.15^2 **c** 140^2 **d** 1.5^3

10 Find two values of x that make the equation true.

    a $x^2 = 36$ **b** $x^2 = 121$ **c** $x^2 = 2.25$ **d** $x^2 = 5.76$ **e** $x^2 = 2.56$ **f** $x^2 = 3600$

Reason mathematically

Given that $13^2 = 169$, find the value of 13^4.

13^4 is $13^2 \times 13^2 = 169 \times 169$

              $= 28561$

11 Given that $3^4 = 81$, find the value of 3^6.

12 **a** Work out the value of each number.

    i 1^2 **ii** 1^3 **iii** 1^4 **iv** 1^5 **v** 1^6

    b Write down the value of 1^{123}.

13 **a** Work out the value of each number.

    i $(-1)^2$ **ii** $(-1)^3$ **iii** $(-1)^4$ **iv** $(-1)^5$ **v** $(-1)^6$

    b Write down the value of: **i** $(-1)^{223}$ **ii** $(-1)^{224}$.

    c Explain why the answers to **b(i)** and **(ii)** are different.

14 Choose any number between 0 and 1, for example, 0.4.

    a Calculate increasing powers of your number, for example, 0.4^2, 0.4^3, 0.4^4.

    b What do you notice about the sizes of your answers as the power increases? Explain why this happens.

Solve problems

Estimate the value of $\sqrt{70}$.

$\sqrt{70}$ is between 8 and 9 because $8 \times 8 = 64$ and $9 \times 9 = 81$.

70 is slightly closer to 64 than it is to 81, so estimate the value to be slightly closer to 8, say 8.4.

The actual value is 8.3666 to 4 dp.

15 Estimate the value of $\sqrt{51}$.

16 64 is a square number (8^2) and a cube (4^3).

    a One other cube number, with a root below 10 (apart from 1), is also a square number. Which is it?

    b Which is the next cube number that is also a square number?

17 Jack is designing a box in the shape of a cube. He needs the box to have a volume of $729\,cm^3$. What side length should he use?

18 Jill is laying square tiles on a square floor. Each tile has a side length of 30 cm. She uses 16 tiles in each direction. What is the area of the room? Give your answer in m^2.

20.2 Powers of 10

● I can multiply and divide by powers of 10

Powers of 10 are numbers that can be written as 10 raised to some power. For example, 100 is 10^2 and 1000 is 10^3.

To multiply by powers of 10, think of the digits moving one place to the left for every power of 10.

To divide by powers of 10, think of the digits moving one place to the right for every power of 10.

Develop fluency

Work these out.

    a 0.937×10 b 2.363×100 c $0.000\,281 \times 10^4$

    a $0.937 \times 10 = 9.37$ Multiplying by 10 moves the digits 1 place to the left.

    b $2.363 \times 100 = 236.3$ Multiplying by 100 (10^2) moves the digits 2 places to the left.

    c $0.000\,281 \times 10^4 = 2.81$ Multiplying by 10^4 moves the digits 4 places to the left.

Work these out.

  a $65 \div 100$ **b** $0.985 \div 10$ **c** $7.8 \div 10^3$

  a $65 \div 100 = 0.65$ Dividing by 100 (10^2) moves the digits 2 places to the right.

  b $0.985 \div 10 = 0.0985$ Dividing by 10 moves the digits 1 place to the right.

  c $7.8 \div 10^3 = 0.0078$ Dividing by 10^3 moves the digits 3 places to the right.

1 Multiply each number by 100.

  a 5.3   **b** 0.79   **c** 24   **d** 5.063   **e** 0.003

2 Divide each number by 100.

  a 83   **b** 4.1   **c** 457   **d** 6.04   **e** 34 781

3 Multiply each number by 10^3.

  a 6.43   **b** 0.685   **c** 35.2   **d** 8.074   **e** 0.0021

4 Divide each number by 10^3.

  a 941   **b** 5.23   **c** 568   **d** 0.715   **e** 45 892

5 Write down the answers.

  a 3.1×10 **b** 6.78×100 **c** $34 \div 1000$ **d** $823 \div 100$ **e** 57.89×100

6 Write down the answers.

  a 4.25×10^3 **b** 5.67×10^2 **c** $23 \div 10^3$ **d** $8.05 \div 10^3$ **e** $68.9 \div 10^2$

7 Multiply each of these numbers by   **i** 10   **ii** 100   **iii** 1000.

  a 2.7   **b** 0.05   **c** 38   **d** 0.008

8 Multiply each of these numbers by   **i** 10^4 **ii** 10^5.

  a 2.7   **b** 0.05   **c** 38   **d** 0.008

9 Divide each of these numbers by   **i** 10^4 **ii** 10^5.

  a 730   **b** 4   **c** 2.8   **d** 35 842

10 Work your way along this chain of calculations for each of these starting numbers.

  a 2000   **b** 7   **c** 0.06

  i

  ii

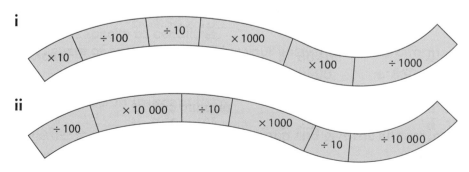

Reason mathematically

The diameter of the Earth is 12 800 km. What is the diameter of the Earth, in m?

To convert from km to m, multiply by 1000.

12 800 × 1000 = 12 800 000   Move the digits three places to the left.

The diameter of Earth is 12 800 000 m.

11 The diameter of Jupiter is 143 000 km. What is the diameter of Jupiter, in m?

12 The diameter of a 50p coin, measured from one vertex to the centre of the opposite side is 27.33 mm. How long would a row of 10 000 50p coins lined up, end to end, be, in m?

13 The distance from the Sun to the Earth is 149 600 000 000 m.

    a Write down how you would find the distance from the sun to the Earth in km using powers of 10.

    b Find the distance from the Sun to the Earth in km.

14 Fill in the squares to make each equation true.

    a $0.678 \, \Box \, 10^2 = 67\,800 \div 10^{\Box}$

    b $253.8 \, \Box \, 10^4 = \Box \div 10^2$

    c $432\,008 \div 10^6 = 4.320\,08 \, \Box \, 10^{\Box}$

    d $3.268 \, \Box \, 10^{23} = 326.8 \, \Box \, 10^{\Box}$

Solve problems

A racing car travels at 100 miles per hour. How long would it take to finish a 160 mile race? Give your answer in hours and minutes.

160 miles ÷ 100 mph = 1.60.   Move the digits two places to the right.

The time taken is 1.6 hours.

0.6 of 1 hour is 0.6 × 60 = 36 minutes

It will take 1 hour and 36 minutes.

15 The closest Mars gets to Earth is approximately 30 000 000 miles.

    a How many hours would it take to travel there if the spacecraft travelled at:

        i 100 miles per hour   **ii** 1000 miles per hour?

    b How many days would it take to travel to Mars if the craft travelled at 3000 mph?

16 A rocket travels at 40 000 miles per hour.

    How far will the rocket travel in:

    a 10 000 hours   **b** 100 hours   **c** 100 weeks?

17 The mass of one electron is 0.000 000 000 000 000 000 000 000 000 911 grams.

    a What is the mass of one million electrons?

    b What is the mass of one billion electrons?

18 The mass of one atom of hydrogen is 0.000 000 000 000 000 000 000 001 673 8 g.

  a How many hydrogen atoms will there be in 16 738 g of hydrogen?

  b Approximately how many hydrogen atoms will there be in 1 kg of hydrogen?

20.3 Rounding large numbers

● I can round large numbers

When you are discussing large quantities, you often only need to use an **approximate** number. To work this out, you decide to which place value you need to **round** and find the number to which the original number is closest. For example, when rounding 137 to the nearest ten, think whether 137 is closer to 130 or closer to 140. In fact, 137 is closer to 140 so it rounds to 140 to the nearest tens.

You can also use rounded numbers to estimate the answers to questions.

Develop fluency

Round these numbers to the nearest hundred.

  a 3612   b 156 258

  a 3612 rounded to the nearest hundred is 3600 because it is closer to 3600 than to 3700.

  Look at the number line. You can either sketch a number line or just visualise it in your head.

  Look at the digit to the right of the hundreds place in 3612. 1 is less than 5 so round down to 3600.

  b 156 258 is 156 300 to the nearest hundred because it is closer to 156 300 than it is to 156 200.

  Using the rules, the digit to the right of the hundreds place in 156 258. 5 is greater than or equal to 5 so round up to 156 300.

1 Round each number to the nearest hundred.

  a 2620   b 861   c 83 717   d 4934

2 Round each number to the nearest ten.

  a 3260   b 1815   c 237 887   d 18 657

3 Round each number to the nearest thousand.

  a 15 873   b 207 984   c 999   d 12 284

4 Round each number to the nearest ten thousand.

  a 3 547 812   b 9 722 106   c 3 045 509   d 15 698 999

5 Round each number to the nearest million.

a 7 247 964　　　b 1 952 599　　　c 645 491　　　d 9 595 902

6 Copy and complete this table.

Number	Nearest ten	Nearest hundred	Nearest thousand	Nearest ten thousand
13 658				
205 843				
16 009 234				
798				

Reason mathematically

The population of Portugal is approximately 10 million. What are the smallest and largest population numbers this could represent?

As the population is given to the nearest million, the actual population could be as much as half a million either way, so the population is between 9.5 million and 10.5 million.

The smallest possible population would be 9 500 000.

The largest possible population would be 10 499 999.

7 At one time, there were 2 452 800 people out of work. The government said: 'Unemployment is just over two million.' The opposition said: 'Unemployment is still nearly three million.' Who is closer to the real figure? Explain your answer.

8 In a show called 'Z-factor', two contestants, Mickey and Jenna, were told: 'Mickey received 8000 votes, Jenna received 7000 votes, but it was so close, there was only one vote between them.'
Explain how this statement can be correct.

9 This table shows the top six UK airports, in 2018, measured by number of passengers.

a Which two of these airports had the same number of passengers, to the nearest million?

b How many passengers used Heathrow, Gatwick or Stansted in 2018? Give your answer to the nearest ten million passengers.

1	Heathrow	80 125 000
2	Gatwick	46 086 000
3	Manchester	28 293 000
4	Stansted	27 996 000
5	Luton	16 770 000
6	Birmingham	14 457 000

10 The president of a society: 'We have about 25 000 members in our society.'
An opponent of the society said that they only had 20 000 members.
State a number of members for which both the president and the opponent are correct and explain how each person rounded it.

Solve problems

a The bar chart shows the annual income for a large company over five years. Estimate the income for each year.

b The company secretary says: 'Income in 2013 was nearly 50 million pounds.' Is the secretary correct?

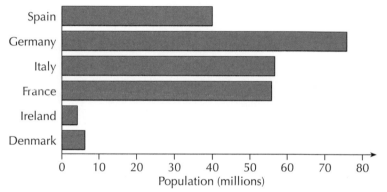

a In 2009 the income was about 19 million pounds.

In 2010 it was about 26 million pounds.

In 2011 it was about 31 million pounds.

In 2012 it was about 39 million pounds.

In 2013 it was about 43 million pounds.

b The secretary is wrong, as in 2013 the income is closer to 40 million pounds.

11 The bar chart shows the populations of some countries in the European Union. Estimate the population of each country.

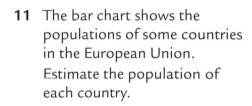

12 Three cruise ships docked in Dubrovnik. The residents were warned that one ship had about 3000 passengers, one ship had about 5000 passengers and one had about 1.5 thousand passengers. All the passengers from all three ships were expected in town that day.

a What was the greatest number of passengers that the residents could expect to have in their town?

b What was the lowest number of passengers that they could expect?

13 These are the populations of England, Scotland, Wales and Northern Ireland, rounded to the nearest thousand.

England   55 620 000
Scotland   5 425 000
Wales   3 125 000
Northern Ireland 1 871 000

a What is the highest the population of the UK could be?

b What is the lowest the population of the UK could be?

14 Use the numbers on these cards to write down two numbers that round to:

a 50 to the nearest ten

b 600 to the nearest hundred

c 4000 to the nearest thousand.

20.4 Significant figures

● I can round to one or more significant figures

In any number:

● the digit with the highest place value is the most **significant figure**
● the digit with the next highest place value is the next most significant, and so on.

To round a number to one significant figure, round the digit with the highest place value. To round to two significant figures (2 sf), use the digit after the one with the highest place value, and so on.

A zero at the end of a whole number is not significant. For example:

Develop fluency

Round:

a 34.87 to one significant figure (1 sf)

b 2 413 589 to four significant figures (4 sf).

a To round 34.87 to 1 sf

The most significant digit is 3.

The figure to the right of it is 4.

The value of this is less than 5, so there is no rounding up. The 3 stays the same.

Put in a zero to preserve the place value of the 3.

Hence 34.87 ≈ 30 (1 sf).

b To round 2 413 589 to 4 sf

The four most significant figures are 2413.

These are followed by 5, so round up and add one to the 2413, which makes 2414.

Replace the 5 and the other digits with zeros to preserve the place value of the digits in the original number.

Hence 2 413 589 ≈ 2 414 000 (4 sf).

1 State the number of significant figures in each number, then round each number to two significant figures.

   a 1.325   **b** 320   **c** 5.24   **d** 0.509   **e** 8 million

2 Round each number to one significant figure.

   a 327   **b** 3760   **c** 60.8   **d** 0.9137   **e** 0.0853

3 Round each number to two significant figures.

   a 5329   **b** 49.7   **c** 9.053   **d** 752.2   **e** 0.082 56

4 Round each number to three significant figures.

   a 2148   **b** 9.5612   **c** 0.265 57   **d** 268.26   **e** 1.879 56

5 Round each number to the given number of significant figures.

  a 3.223 (2 sf) **b** 7.5474 (1 sf) **c** 0.06371 (2 sf) **d** 0.00754 (1 sf) **e** 83.697 (3 sf)

6 Use a calculator to work out each division. Round the answer to the given number of significant figures.

  a $1 \div 7$ (4 sf) **b** $11 \div 7$ (2 sf) **c** $12 \div 13$ (3 sf) **d** $51 \div 13$ (3 sf) **e** $113 \div 37$ (2 sf)

7 State how many significant figures there are in each number.

  a 40 **b** 35 200 **c** 900 000 **d** 470 000

8 Round these numbers to one significant figure.

  a 410 **b** 430 **c** 450 **d** 470

9 State how many significant figures there are in each number.

  a 4.5478 **b** 6 302 000 **c** 0.04923 **d** 5 billion

10 Round each number to one significant figure.

  a 8.265 **b** 6.849 **c** 3.965 **d** 0.095 **e** 4.994 **f** 0.047

Reason mathematically

Rebecca has a monthly salary of £2673.48.

Find her annual salary to four significant figures (4 sf).

£2673.48 × 12 = £32 081.76

Her annual salary is £32 080, to four significant figures (4 sf).

11 Find the annual salary, to four significant figures, of these people, who have a monthly salary of:

  a Tia £4711.50 **b** Zeenat £3188.13 **c** Tiara £4273.76.

12 Find the monthly pay, correct to four significant figures (4 sf), of each person.

Name	Mary	Chadrean	Sheila	Ismail	Sheena
Annual salary (£)	15 750	65 150	55 590	36 470	45 800

13 A football team scored 98 goals in 38 matches. Find, to two significant figures (2 sf), the mean number of goals scored.

14 Four scientists used different scales to weigh the same mass of carbon. Their measurements are 12.031 g, 12.0308 g, 12.0299 g and 120287 mg.

  a To how many significant figures did each scientist record the weights?

  b They need to record the official mass in grams and would like it correct to four significant figures (4 sf). What number should they record?

Solve problems

The population of a city is 2 647 100. Its total land area is 625 600 000 m². Work out, correct to two significant figures (2 sf), how many square metres (m²) there are to each person in this country?

The number of square metres (m²) per person is
625 600 000 ÷ 2 647 100 = 236.334 101 5

To two significant figures, there are 240 m² for each person.

15 Tom painted his sitting room walls. He applied several coats of paint.
Altogether he used 7 litres of paint to cover an area of 75 m².
Work out the area, in square metres (m²), that was covered by each litre.
Give your answer to two significant figures (2 sf).

16 Brant was planning to knock down a tall, circular chimney. He wanted to know approximately how many bricks there were in the chimney.
He counted about 215 bricks in one row, all the way round the chimney.
He estimated there were 47 rows of bricks in the chimney, from bottom to top.
How many bricks would he expect to get after knocking the chimney down?
Give your answer correct to one significant figure (1 sf).

17 A leaking tap drips water at a rate of 1.5 ml every 10 minutes. How much water leaks from the tap in a week? Give your answer to two significant figures (2 sf).

18 A rectangular field has dimensions of 32 m by 28 m. Another field measures 26 m by 18 m. How many times larger is the larger field?
Give your answer to three significant figures (3 sf).

20.5 Large numbers in standard form

● I can write a large number in standard form

There are two things to remember about a number expressed in standard form.

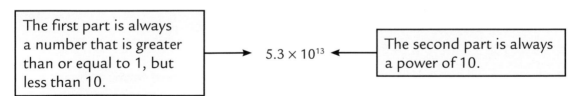

The first part is always a number that is greater than or equal to 1, but less than 10.

5.3×10^{13}

The second part is always a power of 10.

Develop fluency

Express each number in standard form.

  a 760 000   **b** 5420   **c** 36×10^7

  a Count how many places to move the digits to the right, to give a number between 1 and 10. For 760 000 that will be five places. This means that the power of 10 will be 5.

    $760 000 = 7.6 \times 10^5$

  b Count how many places to move the digits to the right, to give a number between 1 and 10. For 5420 that will be three places. This means that the power of 10 will be 3.

    $5420 = 5.42 \times 10^3$

  c Although 36×10^7 is written as a number multiplied by a power of 10, it is not in standard form, as the first part is not a number between 1 and 10.

    Work out the number in standard form, like this:

    $36 = 3.6 \times 10$ The number between 1 and 10 will be 3.6

    $36 \times 10^7 = 3.6 \times 10 \times 10^7 = 3.6 \times 10^8$ Rewrite the number by substituting 3.6×10 for 36.

1 Write each number as a power of 10.

  a 100   **b** 1 000 000   **c** 10 000   **d** 10   **e** 1 000 000 000 000 000

2 State for each number whether it is written in standard form and if not, explain why.

  a 6.8   **b** 0.68×10^{12} **c** 6.8×10^{12} **d** 6.8×9^{12} **e** 680×10^{12}

3 Write each number in standard form.

  a 5690   **b** 1 200 000   **c** 77 800   **d** 396 500 000   **e** 73

4 Write each number in standard form.

  a 3.4 million   **b** 5.6 thousand   **c** 45 thousand   **d** 258 million

5 In all English-speaking nations, a billion is one thousand million. Write each number in standard form.

  a 8 billion   **b** 12 billion   **c** 150 billion   **d** 6.7 billion

6 Write each standard-form number as an ordinary number.

  a 2.3×10^6 **b** 4.56×10^2 **c** 9×10^6 **d** 3.478×10^4

7 Find the square of each number, giving your answer in standard form.

  a 500   **b** 4200   **c** 370   **d** 9000   **e** 650   **f** 30 000

8 Write each number in standard form.

  a 73×10^6 **b** 256×10^2 **c** 259×10^3 **d** 1873×10^7

9 These are the populations of six South American countries in 2013. Write each of the numbers in standard form.

   a Brazil: 201 000 000   **b** Colombia: 47 100 000   **c** Venezuela: 29 760 000

   d Paraguay: 6 800 000   **e** Uruguay: 3 300 000   **f** Guyana: 798 000

10 Write each of the following numbers in standard form.

   a 52×10^7 **b** 0.6×10^{10} **c** 0.007×10^{20} **d** 3600×10^3

Reason mathematically

The volume of water in Loch Ness is 263 billion cubic feet. 1 cubic foot is 28.3168 litres. What is the volume of Loch Ness in litres? Give your answer in standard form correct to three significant figures.

Questions with very large numbers like this are easier if the numbers are converted to standard form first.

263 billion is $263\,000\,000\,000 = 2.63 \times 10^{11}$

$28.3168 = 2.831\,68 \times 10$

$2.63 \times 10^{11} \times 2.831\,68 \times 10$ is $7.447\,318\,4 \times 10^{12} = 7.45 \times 10^{12}$ litres to three significant figures.

11 One ml is 1 cm^3. From the example above, the volume of water in Loch Ness is 7.45×10^{12} litres. Work out the volume of water in Loch Ness, in km^3.

12 $2^3 = 2 \times 2 \times 2 = 8$.

   Find the cube of each of these numbers, giving your answer in standard form.

   a 40   **b** 2 million   **c** 35 thousand

13 Given that $15^2 = 225$, work out the value of each number and write your answer in standard form.

   a $1\,500\,000 \times 0.0015$ **b** $(1.5 \times 10^{12}) \times (0.15 \times 10^2)$ **c** $(2.25 \times 10^6) \div (1.5 \times 10^3)$

14 Write the populations of these countries in standard form, correct to three significant figures, and order them from largest to smallest.

   San Marino 33 683   Germany 824×10^5 Greece 1.1124×10^7

   Russia 144×10^6 Ukraine 46 441 049

Solve problems

There are 5.01833×10^{22} atoms of carbon in one gram of carbon. The average 70 kg human body contains approximately 12.6 kg of carbon. How many atoms of carbon are in the average 70 kg human? Give your answer in standard form to three significant figures (3 sf).

There are 1000 g, or 10^3 g in 1 kg.

$5.01833 \times 10^{22} \times 10^3 = 5.01833 \times 10^{25}$ atoms of carbon in one kg

$5.01833 \times 10^{25} \times 12.6 = 5.01833 \times 12.6 \times 10^{25} = 63.230958 \times 10^{25}$ atoms of carbon in 12.6 kg

So, there are 6.32×10^{26} atoms of carbon in a 70 kg person, in standard form to three significant figures.

15 Together, 1.09×10^{27} electrons will have a total mass of 1 gram. How many electrons will have a total mass of 1 kg? Give your answer in standard form.

16 When someone has a haircut, there will be, on average, approximately one million pieces of human hair left on the floor to be swept up. At Snippers, a popular hair stylist shop, about 300 people come in every week for a haircut. Approximately how many pieces of hair will Snippers sweep up each week? Give your answer in standard form.

17 At the end of 2013, the gross domestic product (GDP) of the United States was $\$1.6 \times 10^7$ and the GDP of the United Kingdom was $\$2.4 \times 10^6$. Approximately how many times larger was the USA GDP than the UK GDP?

18 The distance from the Sun to Mercury is 57.91 million km. Earth is 2.5833 times further from the Sun than Mercury is. How far is Earth from the Sun? Write your answer in standard form correct to four significant figures (4 sf).

Now I can...

use powers and roots	round to one or more significant figures	write a large number in standard form
multiply and divide by powers of 10	round large numbers	

CHAPTER 21 Percentage changes

21.1 Percentage increases and decreases

● I can work out the result of a simple percentage change

In a sale, prices are often reduced by a certain percentage. When the value of something **increases** or **decreases**, the change is often described in terms of a percentage. This is the value of the increase or decrease as a percentage of the original price or value. You add it on for an increase. You subtract it for a decrease. A decrease is often called a **reduction**.

Develop fluency

The original price of a dress is £135.

In the sale the price is reduced by 30%.

What is the sale price?

The reduction is 30% of £135.

30% of 135 is 0.3 × 135 = 40.5.

The reduction is £40.50.

The sale price is £135 – £40.50 = £94.50.

1 Increase each number by 10%.
   a 50   **b** 320   **c** 8   **d** 2500   **e** 35   **f** 123

2 Decrease each number by 25%.
   a 80   **b** 44   **c** 300   **d** 1200   **e** 76   **f** 23

3 Increase each number by 40%.
   a 80   **b** 2110   **c** 190   **d** 780

4 Decrease each number by 35%.
   a 260   **b** 40   **c** 4560   **d** 680

5 Increase each number by 28%.
   a 95   **b** 624   **c** 5810   **d** 6

6 Decrease each number by 31%.
   a 152   **b** 781   **c** 5682   **d** 9

7 Increase each amount by 20%.
   a £621.60   **b** 648.4 km   **c** 62.1 litres   **d** 5.1 tonnes

8 Decrease each amount by 60%.

  a $519.80   **b** 216.7 m   **c** 125.3 kg   **d** 8.4 g

9 In a sale all prices have been reduced by 30%. What is the sale price of these products?

  a A digital camera which cost £268.90

  b A mountain bike which cost £114.99

  c A laptop computer which cost £349.00

10 A farmer uses a new mixture of fertiliser and sees a 14% increase in his produce. What is the new volume of his crops?

  a Barley was 680 tonnes   **b** Wheat was 512 tonnes   **c** Oats were 426 tonnes

Reason mathematically

Farah's house cost her £156 000 in 2010. She sold it to Guanting in 2015 at a 4% profit. In 2018 Guanting sold the house to Daria at a 5% loss.

How much did Daria pay for the house?

4% of £156 000 = 0.04 × 156 000 = £6240

Guanting bought the house for 156 000 + 6240 = £162 240.

5% of £162 240 = 0.05 × 162 240 = £8112

Daria bought the house for 162 240 − 8112 = £154 128.

11 The price of a car is decreased by 3%.
   The original price was £14 800.

  a Work out the decrease in price.   **b** Work out the new price.

12 The population of a village has increased by 45% over the last ten years.
   Ten years ago the population of the village was 360 people.
   Work out the population now.

13 When 200 is increased by 10%, the answer is 220.
   When 220 is decreased by 10%, the answer is 198.
   Why do you not get the answer you started with?

14 A woman earns £36 600 per year.
   She is given a pay rise of 4%.

  a Work out her new salary.

  b She is paid $\frac{1}{12}$ of her salary each month.
   How much extra will she earn each month?

Solve problems

EasyPlanes increases the cost of its flights by 6%.

Jet10 increases the cost of its flights by 4%.

A return ticket from Gatwick to Malaga used to cost £145 with EasyPlanes and £150 with Jet10.

Which company is the cheaper airline now?

EasyPlanes: The increase is 6% of 145 = 0.06 × 145 = £8.70.

The new price is 145 + 8.70 = £153.70.

Jet10: The increase is 4% of 150 = 0.04 × 150 = £6.00.

The new price is 150 + 6.00 = £156.00.

EasyPlanes is still the cheaper airline.

15 Heron Tower is 230 m tall. The Shard is 35% taller than Heron Tower.
broadgate Tower is 47% shorter than The Shard. How tall is Broadgate Tower?

16 a Which of these sales is the best deal?

   b How much would a washing machine with an original price of £400 cost in each of these sales?

17 To calculate the tax paid on an annual salary:
- subtract £4500 from the salary (this amount is not taxed)
- calculate 10% of the next £1500 of the salary
- calculate 25% of the remaining salary.

Calculate the tax due for each of these salaries.

   a £6000   **b** £26 000   **c** £7000

18 A fuel company increases its gas bills by 16% and its electricity bills by 18%.
The Watsons' gas bill last year was £625.
Their electricity bill was £441.

   a Work out the expected bills for the Watsons for gas and electricity after the increases.

   b What is the total increase in their bills for gas and electricity?

21.2 Using a multiplier

● I can use a multiplier to calculate a percentage change

You can solve problems involving percentage increases and decreases by multiplying by a number called a **multiplier**.

Develop fluency

The price of an article before tax is £64.50.

12% tax must be added.

Work out the price including tax.

£64.50 is 100%. You need to add 12%.

100% + 12% = 112% altogether

112% = 1.12

£64.50 × 1.12 = £72.24

The price including the tax is £72.24.

The price of a pair of shoes is £64.50.

In a sale, the price is reduced by 12%.

Work out the sale price.

This time you need to subtract 12%.

100% − 12% = 88% = 0.88

The multiplier this time is 0.88.

£64.50 × 0.88 = £56.76

The sale price is £56.76.

You multiplied by 1.12, which is the multiplier in this example.

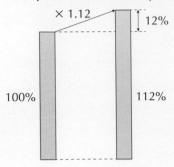

The multiplier this time is 0.88.

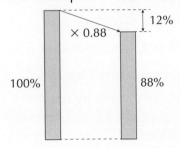

---

1   The price of a coffee machine is £86.00.
    The price is going to increase by 20%.
    a What is the multiplier for a 20% increase?   **b** Work out the price after the increase.

2   Increase each of these prices by 20%.
    a £32.00   **b** £61.00   **c** £184.00   **d** £9.40

3   **a** What is the multiplier for:
        i a 30% increase   **ii** a 36% increase   **iii** a 43% increase   **iv** a 6% increase?
    b Increase £72 by:
        i 30%   **ii** 36%   **iii** 43%   **iv** 6%.

4   Increase £42.00 by:
    a 15%   **b** 35%   **c** 85%   **d** 95%.

5   22% tax must be added to these prices.
    Work out the prices including tax.
    a £13.50   **b** £43.70   **c** £142.00   **d** £385.10

6 Increase 42 kg by:

   a 19%   **b** 39%   **c** 59%   **d** 92%.

7 Increase 270 cm by:

   a 1%   **b** 31%   **c** 51%   **d** 71%.

8 **a** What is the multiplier to decrease an amount by 15%?

   b Decrease these amounts by 15%.

     i £63.00   **ii** £52.50   **iii** £262.00   **iv** £ 59.99

9 Work out the multiplier for a decrease of:

   a 10%   **b** 30%   **c** 37%   **d** 43%   **e** 75%.

10 In a sale these prices are reduced by 40%.
Find the sale price of:

   a a jacket reduced from £115   **b** jeans reduced from £39

   c a shirt reduced from £29.50   **d** boots reduced from £75.80.

Reason mathematically

The number of bacteria in a Petri dish is growing rapidly. At 9 a.m. there are 120 000. By 11a.m. an increase of 28% has been observed. An anti-bacterial agent is introduced and the number of cells is rapidly reduced by 28%. A reporter says that the anti-bacterial agent has taken the colony back to its original size.

He is wrong, it is more effective than that. Explain why and work out the actual number of cells.

The reporter is wrong because 28% of the original cell count is smaller than 28% of the new, larger, cell count. The combined effect is that the reduction is more than the initial increase – the resulting cell count is less than the original.

120 000 increased by 28% is 120 000 × 1.28 = 153 600.

153 600 decreased by 28% is 153 600 × 0.72 = 110 592.

11 The price of a television is £360.
The price is increased by 10%. A month later the price is reduced by 10%. Simon says: 'The price must be £360 again.'
Show that Simon is not correct.

12 A coat costs £96. In a sale, its price is reduced by 20%.

   a Find the sale price of the coat.
      After the sale, the manager increases the price by 20%.

   b Find the new price of the coat.

   c Explain why the new price of the coat is not the same as the original price of £96.

13 Saleh is changing jobs. His salary used to be €35 750 but the company gave everyone a 2% pay cut. His new company said that they would match his current salary plus 2%. Saleh is happy because he thinks that his new salary will be the same as his salary before the pay cut.

   a Why is he wrong to think this?   b What will his new salary be?

14 The population of a town is currently 246 500. A statistician has predicted that over the next 10 years the population will grow by 8%. However, after a further 25 years the population will have fallen, again by 8%. The town mayor says: 'Unfortunately, in 35 years' time the population will be just the same as it is now.' The mayor is wrong.

   a What will the population be?   b Why is the mayor wrong?

Solve problems

A travel agent is offering a 24% discount off all holidays if the total cost of flights, hotel and car hire is more than £1000.

Marco has booked flights for £259, a hotel at £75 per night and a car at £25 per day. He will be in the hotel for 8 nights and needs a car for 7 days. If the hotel is booked for more than 6 nights there is a 10% reduction on additional nights.

How much does Marco pay for his holiday?

Hotel for 8 nights:	6 nights at £75	$6 \times 75 = 450$
	2 nights with 10% off	$2 \times (75 \times 0.9) = 135$
Car hire for 7 days	7 days at £25	$7 \times 25 = 175$
Flights		259
Total	$450 + 135 + 175 + 259$	$= 1019$

Marco can have the price of the holiday reduced by 24%. He will pay 76% or 0.76 of the cost.

The price of 0.76 of Marco's holiday will be $1019 \times 0.76 = £774.44$.

15 Three people each withdrew a certain percentage of their bank balance from their bank account.
   John: 29% of £300   Hans: 75% of £640   Will: 15% of £280
   Which person has the largest remaining bank balance?

16 Marvin's average score on the computer game Space Attack was 256.
   a After a competition against his friends, his average score had increased by 28%. What was his new average score, to the nearest whole number?
   b Marvin went on holiday. When he returned, his average score had decreased by 15%. What was his new average score, to the nearest whole number?

17 Compressing a computer file reduces its size by a certain percentage.

   a Find the smallest file after compressing.

   b Find which file has been reduced by the most kilobytes.

      500 MB file reduced by 17%   740 MB file reduced by 43%

      655 MB file reduced by 23%   1264 MB file reduced by 92%

18 The population of a town ten years ago was 57 000.

   a It has decreased by 4% in the last ten years. Work out the population now.

   b If it had increased by 4% in the last ten years, what would the population be now?

21.3 Calculating a percentage change

● I can work out a change in value as a percentage increase or decrease

Sometimes you will know the increase or decrease in value and you will want to work out what percentage change this is, based on the original value.

original value × multiplier = new value

Develop fluency

The number of students in a college increases from 785 to 834.

Work out the percentage increase.

Use the formula: original value × multiplier = new value

785 × multiplier = 834   Divide by 785.

multiplier = $\dfrac{834}{785}$ The fraction is greater than 1 for an increase.

      = 1.062...

      = 106%   Use a calculator and round the answer.

The percentage increase for an increase of 106% − 100% is 6%.

In a sale, the price of a washing machine is reduced from £429 to £380.

Work out the percentage reduction.

This time:

429 × multiplier = 380   Divide by 429.

multiplier = $\dfrac{380}{429}$ The fraction is less than 1 for a decrease.

      = 0.885...

      = 89%   Round to the nearest whole number.

So this is a reduction of 100% − 89% = 11%.

1 The price of a holiday has been reduced from £460 to £391. What is the percentage reduction?

2   After a series of revision classes a student retakes an exam. His score increases from 56% to 70%. What is the percentage improvement?

3   To fit more comfortable seats into an aircraft its seating capacity is reduced from 320 to 296. What is the percentage reduction in seats?

4   A painting bought ten years ago for $20 400 is sold at auction for $25 704. What is the percentage increase in value?

5   A new product has a recommended price of €140, but a shop has a promotional offer and is selling it for €98. What is the percentage reduction of the promotional offer?

6   A yearly subscription for a magazine used to be £75 but is now £72. What is the percentage reduction?

7   An office worker's salary increases from £27 500 to £28 325. What percentage salary increase was given?

8   The number of monthly visitors to a museum increases from 15 000 to 15 300. What is the percentage increase in visitors?

9   Last season a football team won 30 of the games it played. This season it won 21. What is the percentage change in the number of games the team won?

10   A plant is being monitored for growth. Last week it was 460 mm tall. It is measured today at 621 mm. What is the percentage increase in the height of the plant?

Reason mathematically

The mass of a cat increases from 8 kg to 10 kg.

a   Show that the multiplier is 1.25.

b   What is the percentage increase?

a   Using:

original value × multiplier = new value

8 kg × multiplier = 10 kg

The multiplier is $\frac{10}{8}$ = 1.25.

b   1.25 as a percentage is 125%.

The increase is 125% − 100% = 25%.

11   An actor's following on her social media account has increased from 123 500 to 142 025.

   a   Show that the multiplier is 1.15.

   b   What is the percentage increase?

12   The price of a DVD player increases from £250 to £295.

   a   Show that the multiplier is 1.18.

   b   What is the percentage increase?

13   The length of a race decreases from 1500 m to 1260 m.

   a   Show that the multiplier is 0.84.

   b   What is the percentage decrease?

14 This table shows the numbers of people voting for each party in two local elections. Work out the percentage change for each party. Say whether it is an increase or a decrease.

	2011 election	2014 election
Red Party	1281	1342
White Party	584	260
Blue Party	782	831

Solve problems

In January 2017 one ounce of gold was worth £937. In January 2018 it was worth £969 and then in January 2019, £1016.

In these calculations give your final answers to 2 decimal places.

  a What was the percentage increase each year?

  b What was the change from 2017 to 2019?

  c Why is your answer to part **b** not equal to the sum of your answers in part **a**?

  a The multiplier from 2017 to 2018 is $\dfrac{969}{937} = 1.034\,15\ldots$

    So the percentage increase is 3.42% (to 2dp).

    The multiplier from 2018 to 2019 is $\dfrac{1016}{969} = 1.048\,50\ldots$

    So the percentage increase is 4.85% (to 2dp).

  b The multiplier from 2017 to 2019 is $\dfrac{1016}{937} = 1.084\,31\ldots$

    So the percentage increase is 8.43% (to 2dp).

  c The percentage amount needed to increase from 937 to 1016 is more than the gradual percentage increases from 937 to 969 and then 969 to 1016.

15 The number of people in a health club decreases from 289 to 214.

  a $289 \times$ multiplier $= 214$

    Work out the multiplier. Round your answer to 2 decimal places.

  b What is the percentage decrease?

16 In a sale, prices are reduced by £40.

  Work out the percentage reduction for something marked down:

  a from £65 to £25   **b** from £119 to £79   **c** from £242 to £202.

17 Maria bought a skateboard for £25. Two months later she sold it to Zeenat for £18. Zeenat then decorated the skateboard and sold it to Delaney for £27.

  a What was Maria's percentage loss?

  b What was Zeenat's percentage profit?

  c What was the overall percentage change in the value of the skateboard?

18 An odd job man kept a record of his income and costs for each job.
Copy and complete this table.

Job	Costs (£)	Income (£)	Profit (£)	Percentage profit (%)
4 Down Close	120	162		
High Birches	50			82%
27 Bowden Rd	25		17	
Church hall		34		70%

Now I can...

work out the result of a simple percentage change	use a multiplier to calculate a percentage change	work out a change in value as a percentage increase or decrease

CHAPTER 22 Graphs

22.1 Graphs from linear equations

- I can recognise and draw the graph of a linear equation

A **linear equation** connects two **variables** by a simple rule. Linear equations use any of the four operations: addition, subtraction, multiplication and division. For example in the equation $y = x + 2$, the two variables x and y are connected by a simple rule.

Develop fluency

Draw a graph of the equation $y = 3x + 1$.

First, draw up a table of simple values for x. Then substitute that value of x in the equation to determine the corresponding y-value.

x	−2	−1	0	1	2
3x	−6	−3	0	3	6
y = 3x + 1	−5	−2	1	4	7

The middle line of the table helps you to work towards the final value of $3x + 1$.

Finally, take the pairs of (x, y) coordinates from the table, plot each point on a grid and join up all the points.

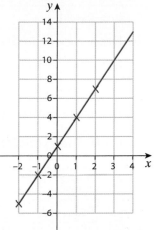

Notice that the line passes through other coordinates, as well. All of these fit the same equation, that is $y = 3x + 1$. Choose any points on the line that have not been plotted in the table and show that this is true.

1 **a** Copy and complete this table for the equation $y = x + 3$.

x	−2	−1	0	1	2	3
y = x + 3			3			

   b Draw a coordinate grid, with an x-axis from −2 to 3 and a y-axis from −1 to 7.

   c On your grid, use values from the table to draw the graph of $y = x + 3$.

2 **a** Copy and complete this table for the equation $y = x - 2$.

x	−2	−1	0	1	2	3
y = x − 2			−2			1

   b Draw a coordinate grid, with an x-axis from −2 to 3 and a y-axis from −4 to 2.

   c On your grid, use values from the table to draw the graph of $y = x - 2$.

3 **a** Copy and complete this table for the equation $y = 3x$.

x	−2	−1	0	1	2	3
y = 3x		−3				0

   b Draw a coordinate grid, with an x-axis from −2 to 3 and a y-axis from −6 to 12.

   c On your grid, use values from the table to draw the graph of $y = 3x$.

4 a Copy and complete this table for the equation $y = 4x + 1$.

x	−2	−1	0	1	2	3
4x		−4	0	4		
y = 4x + 1			1			

b Draw a coordinate grid, with an x-axis from −2 to 3 and a y-axis from −7 to 13.

c On your grid, use values from the table to draw the graph of $y = 4x + 1$.

5 a Copy and complete this table for the equation $y = 4x − 1$.

x	−2	−1	0	1	2	3
4x		−4	0	4		
y = 4x − 1			−1			

b Draw a coordinate grid, with an x-axis from −2 to 3 and a y-axis from −9 to 11.

c On your grid, use values from the table to draw the graph of $y = 4x − 1$.

6 a Draw a coordinate grid, with an x-axis from −2 to 3 and a y-axis from −15 to 15.

b Draw the graph of the equation $y = 3x + 1$.

7 a Copy and complete this table for the equation $y = 0.5x + 2$.

x	−4	−2	0	2	4	6
0.5x	−2			1		
y = 0.5x + 2	0			3		

b Draw a coordinate grid, with an x-axis from −4 to 6 and a y-axis from 0 to 5.

c On your grid, use values from the table to draw the graph of $y = 0.5x + 2$.

8 a Copy and complete this table for the equation $y = 5 − x$.

x	0	1	2	3	4	5	6	7	8	9	10
5	5	5		5							5
y = 5 − x	5	4		2							−5

b Draw a coordinate grid, with an x-axis from 0 to 10 and a y-axis from −5 to 5.

c On your grid, use values from the table to draw the graph of $y = 5 − x$.

Reason mathematically

a Copy and complete the table for the equation $y = 3x + 2$

x	−2	−1	0	1	2	3	4
$3x$	−6		0				
$y = 3x + 2$			2			11	

b Draw a coordinate grid, numbering the *x*-axis from −2 to 4 and the *y*-axis from −4 to 16 and draw the graph of $y = 3x + 2$.

c Use your graph to estimate the solution to $9 = 3x + 2$.

d Why is your answer to **c** an estimate?

a

x	−2	−1	0	1	2	3	4
$3x$	−6	−3	0	3	6	9	12
$y = 3x + 2$	-4	-1	2	5	8	11	14

b

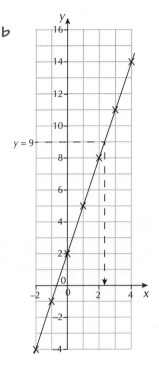

c The question says 'Use your graph', so it is really important to make it very clear on the graph where your answer comes from. In this case draw a line at $y = 9$ to meet the line of the equation and then, from this point, draw a line down to the *x*-axis.

The solution of $9 = 3x + 2$ is approximately 2.4.

d The answer is an estimate because it is read from the graph. Your graph might not be drawn really accurately, or the scale on the *x*-axis may not be detailed enough for better than a guess at something a little less than 2.5.

9 a Copy and complete this table for the given equations.

x	−2	−1	0	1	2	3
$y = x$	−2		0			3
$y = 2x$	−2		0		4	
$y = 4x$			0	4		

b Draw a coordinate grid, numbering the *x*-axis from −2 to 3 and *y*-axis from −10 to 15.

c Draw the graph of each equation in the table.

d What two properties do you notice about each line?

10 Look at the lines $y = 4x$, $y = 4x + 3$ and $y = 4x − 1$ and their graphs. What two things do you notice?

11 a Copy and complete the table for the given equations.

x	-2	-1	0	1	2	3
$y = 2x$		-2	0	2		
$y = 2x + 2$		0	2		6	
$y = 2x - 2$		-4	-2			
$y = 2x - 4$		-6	-4			

b Draw a coordinate grid, numbering the x-axis from −2 to 3 and the y-axis from −8 to 10.

c Draw the graph of each equation in the table.

d What two properties do you notice about each line?

12 a Copy and complete this table for each of the equations.

x	-2	-1	0	1	2	3
$y = x + 2$						
$y = 2x + 2$						
$y = 3x + 2$						
$y = 4x + 2$						

b Draw a coordinate grid, numbering the x-axis from −2 to 3 and y-axis from −10 to 15.

c Draw the graph of each equation in the table.

d What is the same about the lines?

e What is different about the lines?

f Use what you've noticed to sketch the graph of $y = 2.5x + 2$.

Solve problems

a Copy and complete this table for the equations shown.

x	−2	−1	0	1	2	3	4
y = 2x + 1	−3		1				
y = 2x + 2	−2		2			8	
y = 2x + 3			3			9	

b Draw a coordinate grid, numbering the x-axis from −2 to 4 and the y-axis from −4 to 10.

Draw the graph for each equation in the table, on the same coordinate grid.

c Geometrically, what is special about the three lines you have drawn?

d Describe the connection between the point at which a line cuts the y-axis and the equation.

a

x	−2	−1	0	1	2	3	4
y = 2x + 1	−3	-1	1	3	5	7	9
y = 2x + 2	−2	0	2	4	6	8	10
y = 2x + 3	−1	1	3	5	7	9	11

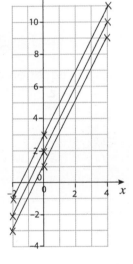
b

c They are parallel.

d The point where the line cuts through the y-axis has the same value as the number added to the multiple of x, in each case.

13 Look back at question 12. Use the properties you have noticed to draw the graphs of these equations.

   a $y = x + 2.5$ **b** $y = x − 1.5$

14 Look back at question 12. Use the properties you have noticed to draw the graphs of these two equations.

   a $y = 2x + 2.5$ **b** $y = 2x − 2.5$

15 a Copy and complete this table for each equation.

x	−2	−1	0	1	2	3
y = 5x − 1						
y = 2x − 4						

   b Write down the coordinates of the point where the lines intersect.

16 Draw the graph of the equation $10y + 4x = 20$ by finding three points that lie on the line.

22.2 Gradient of a straight line

● I can work out the gradient of a graph from a linear equation
● I can work out an equation of the form $y = mx + c$ from a linear graph

The **gradient** of a straight line is its steepness or slope. You can measure it by calculating the increase in value up the y-axis for an increase of 1 in the value along the x-axis.

In a straight-line graph:

· the gradient is the same as the coefficient of x in the equation

· the line cuts the y-axis at the value that is added to the x-term.

The point where the line meets or cuts the y-axis is called the y-**intercept**.

A linear equation can be shown generally in the form $y = mx + c$, where m is the coefficient of x. The number added to the x-term is called the constant and is represented by c. It is the y-intercept.

Develop fluency

What is the equation of this graph?

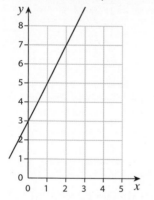

The gradient of the line is 2 and it cuts (intercepts) the y-axis at 3. Its equation is $y = 2x + 3$.

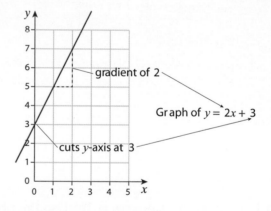

gradient of 2

Graph of $y = 2x + 3$

cuts y-axis at 3

Write down the equation of the straight line that cuts the y-axis at $(0, 5)$ and has a gradient of 3.

The equation of a straight-line graph can be written as $y = mx + c$, where m is the gradient and c is the y-intercept

So the equation of this line is $y = 3x + 5$.

1 State the gradient of each line.

a
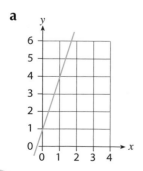

b
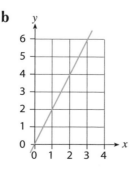

c
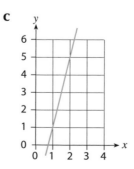

d
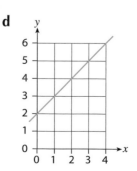

2 For each of these lines, write down:

   i the gradient   ii where it cuts the y-axis.

   a 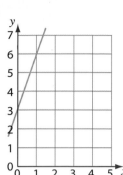   **b** 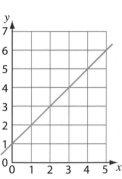   **c** 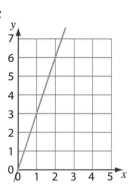   **d**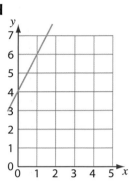

3 Write down the equation of each line in question 2.

4 State the gradient of each straight line.

   a $y = 7x + 2$ **b** $y = 2x - 8$ **c** $y = x + 5$

5 State the equation of the straight line with:

   a a gradient of 3, passing through the y-axis at $(0, 5)$
   b a gradient of 2, passing through the y-axis at $(0, 7)$
   c a gradient of 1, passing through the y-axis at $(0, 4)$
   d a gradient of 7, passing through the y-axis at $(0, 15)$.

6 A straight line passes through the points $(0, 1)$ and $(4, 9)$.

   a What are the coordinates of the point where this line cuts the y-axis?
   b By plotting both points on a coordinate grid, calculate the gradient of the line.
   c What is the equation of the line?

7 For each graph:

   i   find the gradient of the coloured line
   ii   write down the coordinates of where the line crosses the y-axis
   iii write down the equation of the line.

   a 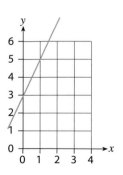   **b** 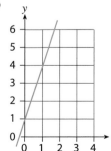   **c** 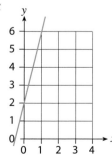   **d**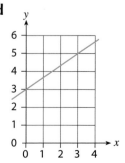

8 A straight line is drawn between the points A $(0, -3)$ and B$(5, 12)$.
   What is the equation of the line AB?

9 A straight line is drawn between the points M(0, 1) and N(6, 13).
   What is the equation of the line MN?

10 A straight line passes through the points (0, 5) and (7, 5).
    What is the equation of the line?

Reason mathematically

Look again at these lines.
Their equations are $y = 2x + 1$, $y = 2x + 2$ and $y = 2x + 3$.

You can see from the diagram that the lines are parallel.

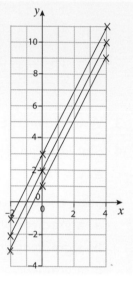

   a Describe the connection between the 3 coefficients of x.

   b Examine the equations $y = 4x + 3$ and $y = 3x + 3$.

      i Without drawing the equations on a coordinate grid, explain
         why these equations will not produce parallel lines

      ii What do these two lines have in common?

   a *The coefficients are the same: this is what makes the lines parallel.*

   b i *The coefficients of x are different, which tells you that they are
          not parallel.*

      ii *The lines cut the y-axis at the same point (0, 3).*

11 Look at this diagram.

    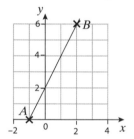

    a Is the straight line $y = 2x - 5$ parallel to the one already drawn?
       How do you know?

    b Does $y = 2x - 5$ cut the y-axis at the same point as the one
       already drawn? How do you know?

12 A straight line passes through the points
    (1, 2) and (2, 5).

    a By plotting both points on a
       coordinate grid, calculate the gradient
       of the line.

    b What are the coordinates of the point
       where this line cuts the y-axis?

    c What is the equation of the line?

13 Think about the line joining (2, 1)
    and (4, 7).

    a By plotting both points on a
       coordinate grid, calculate the gradient
       of this line.

    b What is the coordinate of where this
       line cuts the y-axis?

    c What is the equation of this line?

14 When the lines of the four equations $y = x + 7$, $y = x + 1$, $y = 2x$ and $y = 2x + 10$ are drawn
    on a coordinate grid, they form a quadrilateral.
    Think about the gradients of these lines and, without drawing the graphs, determine
    what special quadrilateral this will be.

Solve problems

What is the equation of the line that passes through the coordinates (2, –5) and (10, 11)?

To calculate the gradient, work out the horizontal and vertical distances between the two sets of coordinates.

y-axis distance = 11 – (–5) = 16 units

x-axis distance = 10 – 2 = 8 units

Gradient $= \dfrac{\text{y-axis distance}}{\text{x-axis distance}} = \dfrac{16}{8} = 2$

To calculate the y-intercept, substitute the coordinates of one of the points, for example, (10, 11), with the calculated gradient (2), in the equation $y = mx + c$.

$11 = 2(10) + c$

$11 = 20 + c$

$c = -9$

The equation is: $y = 2x - 9$

15 Work out the equation of the line that passes through each pair of coordinates.
- **a** (0, 2) and (1, 5)
- **b** (1, 3) and (2, 7)
- **c** (2, 5) and (4, 9)

16 Work out the equations of the lines that pass through each pair of coordinates.
- **a** (2, 8) and (4, 10)
- **b** (2, 1) and (5, 7)
- **c** (1, 5) and (2, 9)

17 The line AB has the equation $y = 3x - 4$. Another line, MN, is parallel to AB and cuts the y-axis at (0, 3).
Which of these points lie on MN?
- **a** (–2, –3)
- **b** (7, 23)
- **c** (4, 15)
- **d** (–6, –18)

18 a Copy and complete the table for these equations.

x	–2	0	4	5
y = x + 2	0			
y = 2x		0		

b Draw the lines for both equations on the same coordinate grid.

c At what coordinate point do the two equations have exactly the same x- and y-values?

22.3 Graphs from quadratic equations

● I can recognise and draw the graph from a simple quadratic equation

You can follow the same techniques for finding coordinates that fit a **quadratic equation** and plotting them on a graph as for linear equations. This time, the lines are not straight!

Develop fluency

Draw the graph of the equation $y = x^2$.

First, draw up a table of values for x. Then substitute each value of x to determine the corresponding y-value.

x	−4	−3	−2	−1	0	1	2	3	4
$y = x^2$	16	9	4	1	0	1	4	9	16

Now take the pairs of (x, y) coordinates from the table and plot each point on a grid.

Join up all the points.

Notice the shape for a quadratic equation is a smooth curve. It is important to draw a quadratic graph very carefully, especially at the bottom of the graph where it needs to be a smooth curve.

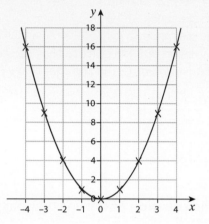

1   **a**   Copy and complete this table for the equation $y = x^2 + 1$.

x	−3	−2	−1	0	1	2	3
x^2	9	4	1	0	1	4	9
$y = x^2 + 1$				1			

   b   Draw a coordinate grid, with an x-axis from −3 to 3 and a y-axis from −1 to 10.

   c   On your grid, use values from the table to draw the graph of $y = x^2 + 1$.

2   **a**   Copy and complete this table for the equation $y = x^2 + 3$.

x	−3	−2	−1	0	1	2	3
x^2							
$y = x^2 + 3$							

   b   Draw a coordinate grid, with an x-axis from −3 to 3 and a y-axis from −1 to 12.

   c   On your grid, use values from the table to draw the graph of $y = x^2 + 3$.

3   **a**   Use your answers to questions 1 and 2 to help you copy and complete this table of values.

x	−3	−2	−1	0	1	2	3
$y = x^2 + 2$							
$y = x^2 + 4$							
$y = x^2 + 5$							

   b   Draw a coordinate grid, with an x-axis from −3 to 3 and a y-axis from −1 to 15.

   c   On your grid, use values from the table to draw the graph of each equation.

4 a Copy and complete this table for the equation $y = 2x^2$.

x	−3	−2	−1	0	1	2	3
x^2	9	4	1	0	1	4	9
$y = 2x^2$	18			0			

b Draw a coordinate grid, with an x-axis from −3 to 3 and a y-axis from −1 to 20.

c On your grid, use values from the table to draw the graph of $y = 2x^2$.

5 a Copy and complete this table for the equation $y = 2x^2 + 1$.

x	−3	−2	−1	0	1	2	3
x^2	9	4	1	0	1	4	9
$y = 2x^2 + 1$	18			1			

b Draw a coordinate grid, with an x-axis from −3 to 3 and a y-axis from 0 to 20.

c On your grid, use values from the table to draw the graph of $y = 2x^2 + 1$.

6 a Copy and complete this table for the equation $y = 0.5x^2$.

x	−3	−2	−1	0	1	2	3
x^2	9	4	1	0	1	4	9
$y = 0.5x^2$	4.5			0			

b Draw a coordinate grid, with an x-axis from −3 to 3 and a y-axis from −1 to 6.

c On your grid, use values from the table to draw the graph of $y = 0.5x^2$.

7 a Copy and complete this table for the equation $y = 0.5x^2 - 3$.

x	−3	−2	−1	0	1	2	3
x^2	9	4	1	0	1	4	9
$y = 0.5x^2 - 3$	1.5			-3			

b Draw a coordinate grid, with an x-axis from −3 to 3 and a y-axis from −4 to 4.

c On your grid, use values from the table to draw the graph of $y = 0.5x^2 - 3$.

Reason mathematically

a Copy and complete this table for the equation $y = x^2 - x$.

x	−3	−2	−1	0	1	2	3
x^2	9					4	
$y = x^2 - x$	12		2		0		6

b Copy and complete this table for the equation $y = x^2 + x$.

x	−3	−2	−1	0	1	2	3
x^2	9					4	
$y = x^2 + x$	6		0		2		12

c Draw a coordinate grid, numbering the x-axis from −3 to 3 and the y-axis from −2 to 14.

d What effect does − x in the first equation and + x in the second have on the position of the graph of x^2?

a

x	−3	−2	−1	0	1	2	3
x^2	9	4	1	0	1	4	9
$y = x^2 - x$	12	6	2	0	0	2	6

b

x	−3	−2	−1	0	1	2	3
x^2	9	4	1	0	1	4	9
$y = x^2 + x$	6	2	0	0	2	6	12

d ±x moves (shifts) the graph horizontally in the opposite direction to the sign.

c

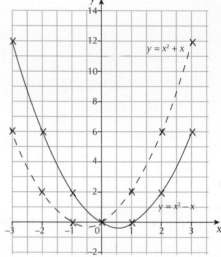

8 Look again at question 3. What do you notice about each line?

9 Consider the graphs of $y = x^2$ and $y = 3x^2$. What do you notice?

10 a Copy and complete the table for $y = x^2 - 1$.

x	−3	−2	−1	0	1	2	3
x^2							
$y = x^2 - 1$							

b Explain why there are two values of x for which $x^2 - 1$ is equal to 3.

11 Rosie and Emma are discussing the equation $y = x^2 + 4$.
Rosie says that when $x = -2$, $y = 0$.
Emma says that when $x = -2$, $y = 8$.
Who is correct and why?

Solve problems

Look again at your solutions to question 3. Use what you have noticed to draw the graphs of these two equations.

  a $y = x^2 + 2.5$ **b** $y = x^2 - 1$

  a For $y = x^2 + 2.5$ the graph of $y = x^2$ is shifted up by 2.5 units.

  b For $y = x^2 - 1$ the graph of $y = x^2$ is shifted down by 1 unit.

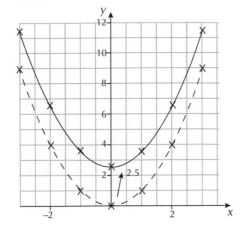   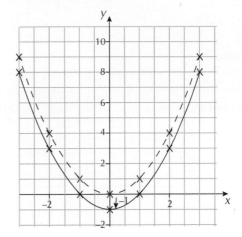

12 Look again at your solution to question 6.
Use what you have noticed to draw the graph of $y = 3x^2$.

13 Imagine that you are standing on a cliff, on a clear day, looking out to sea so that you can see the horizon. The distance (D km) to the horizon is related to the height (H metres) of your line of sight (your eyes) above sea level by the equation $H = D^2 \div 13$.

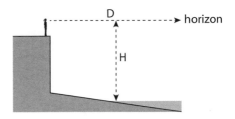

  a Draw a graph to show the heights at which you can see distances from 0 to 50 km out to sea.

  b Find the distance you can see on a clear day from a height of 25 m above sea level.

  c On a clear day, Roy and Hayley were standing on the observation level of the Blackpool tower, which is at a height of 150 m above sea level. How far out to sea could they see?

14 A ball is dropped from a tall building. The distance (D metres) travelled by the ball is related to the time (T seconds) it has been falling by the equation $D = 5T^2$.

  a Draw a graph to show how far the ball has fallen for times from 0 to 4 seconds.

  b The building is 100 m tall. Find the time it takes the ball to fall until it is 40 m above the ground.

15 Use the values in the tables to find the equations of the lines.

x	−3	−2	−1	0	1	2	3
Line 1: $y =$	19	14	11	10	11	14	19
Line 2: $y =$	7	2	−1	−2	−1	2	7
Line 3: $y =$	81	36	9	0	9	36	81
Line 4: $y =$	−1	−6	−9	−10	−9	−6	−1
Line 5: $y =$	31	16	7	4	7	16	31

Now I can...

recognise the graph of a linear equation	draw the graph of a linear equation	work out an equation of the form $y = mx + c$ from a linear graph
work out the gradient of a graph from a linear equation	recognise the graph from a simple quadratic equation	draw the graph from a simple quadratic equation

CHAPTER 23 Correlation

23.1 Scatter graphs and correlation

- I can read scatter graphs
- I can understand correlation

Positive correlation means that the bigger one factor is, the bigger the other is as well.

Negative correlation means that the bigger one factor is, the smaller the other one is.

Develop fluency

Describe the relationship in each scatter graph.

a

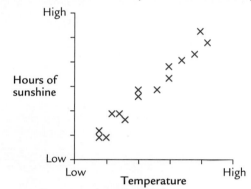

This graph shows that the temperature is higher when there are more hours of sunshine.

The points lie close to a straight line. This is a **strong positive correlation**.

b

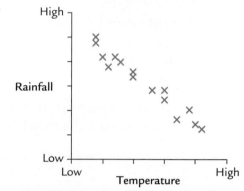

This graph shows that as the temperature increases, the rainfall decreases.

The points lie close to a straight line. This is a **strong negative** correlation.

Use these diagrams to answer questions 1–2.

a 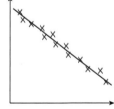   b 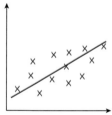   c

d 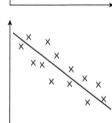   e 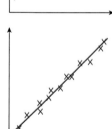   f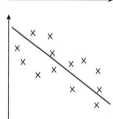

1 Decide which of these correlations is best applied to each diagram.

| strong positive | strong negative | moderate positive |
| moderate negative | weak positive | weak negative | none |

2 Match each of these pairs of axis labels to one of the diagrams.

i Daily temperature,
number of ice creams sold

ii Distance you drive,
amount of fuel left in your tank

iii Age of child, height

iv Age of adult, height

v Daily temperature,
number of coats sold

vi Amount of time spent on revision,
expected exam score

vii Price of perfume, amount of rainfall

Reason mathematically

Josh is asked to draw a scatter graph to show the correlation between the number of sweets that children eat and the number of times they visit the dentist.

Josh thinks that the correlation will be positive. Explain why this might not be true.

Eating more sweets is likely to increase tooth decay, but this won't necessarily mean more visits to the dentist. People who look after their teeth are likely to visit the dentist regularly for check-ups.

3 Look back at your answers for question 2.
Explain how you selected the matching graphs.

Solve problems

The scatter graph shows the scores for students who took Maths and Science tests in the same week.

a How many students did both tests?

b From these results how much would you expect a student to score:

i in Science if they got 68 in Maths

ii in Maths if they got 64 in Science?

c Is it fair to say students who are good at Maths are also good at Science?

d Is this group of pupils generally better at Maths or Science? Explain your answer.

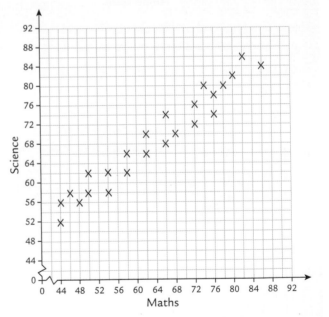

a Count the crosses. 24 students.

b You can draw a line through the middle of the crosses as a guide. It's called a **line of best fit**.

i 72   **ii** 58

c Yes. The graph shows a positive correlation.

d Some of the scores are higher in Science than in Maths, but there is not enough evidence to conclude that the group is generally better at science.

4 The table records the numbers of comedy programmes shown on BBC channels each evening for a week and the number of times a mountain rescue team were called out the same week.

	Mon	Tue	Wed	Thurs	Fri	Sat	Sun
Programmes	7	6	8	5	7	8	6
Call-outs	2	1	3	0	1	2	2

a As you can see, as the number of programmes increased so did the call-outs. Describe the type of correlation shown by the data.

b Does this mean that a higher number of comedies causes more accidents? What does this tell you about making conclusions from this sort of data?

23.2 Creating scatter graphs

● I can create scatter graphs

You need to be able to plot points to make a scatter graph from a table of data.

Develop fluency

Ten people entered a craft competition.

Their displays of work were awarded marks by two different judges.

The table shows the numbers of marks that the two judges gave to each competitor.

Competitor	A	B	C	D	E	F	G	H	I	J
Judge 1	90	35	60	15	95	25	5	100	70	45
Judge 2	75	30	55	20	75	30	10	85	65	40

a Draw a scatter graph to illustrate this information.

b Is there any correlation between the marks awarded by the two judges and, if so, what type?

a Use a scale of 0 to 100 on each axis.

Label the horizontal axis as Judge 1 and the vertical axis as Judge 2.

For each competitor, use the two marks to form a pair of coordinates, (Judge 1, Judge 2).

Plot the points on the grid.

Draw the scatter graph.

b There is positive correlation between the judges' marks.

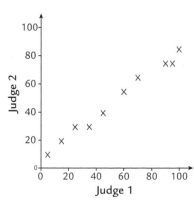

1 The table shows students' scores in their Science and Maths classes.

Student	Jo	Ken	Lim	Tony	Dee	Sam	Pat	Les	Kay	Val	Rod
Science	24	3	13	21	5	15	26	1	12	27	14
Maths	20	5	10	14	9	15	22	7	16	24	11

   a Draw a scatter graph for the data.
      Use the horizontal axis for Science scores, from 0 to 30.
      Use the vertical axis for Maths scores, from 0 to 30.

   b Describe in simple terms what the graph tells you.

2 Language students were given oral and written tests. Use a scatter diagram to find out if there is a relationship between the scores from the two types of test.

Oral test	20	10	14	7	16	8	7	6	14	8
Written test	19	4	8	12	5	12	14	12	10	17

Reason mathematically

Look at the diagram from the previous Worked Example.

Which judge tends to give the higher scores? Explain your answer.

Judge 1 tends to give the higher scores. The gradient of a line through the middle of the crosses is inclined slightly below 45 degrees.

3 **a** Draw a scatter graph of the data in the table. Work out your own scales.

Distance to work (km)	2	5	8	8	9	12	15	16	17	20
Average time taken to get to work (minutes)	6	11	14	29	26	34	37	56	50	55

   b How would you describe the correlation between the time taken to get to work and the distance to work for this group of people: strong, moderate, weak or none at all?

   c Two people live 8 kilometres away from work. Why do you think one usually gets there much more quickly than the other?

Solve problems

One summer's day, a student records the outside temperature every 30 minutes from 6 a.m. until 12 p.m. She finds that there is a strong positive correlation between the time of day and the temperature.

Explain how she could use her results to estimate what the temperature was at 9.15 a.m.

She could plot the values of time and temperature on a scatter graph and draw a line of best fit through the points. Using the line of best fit, she could find the value on the temperature scale that corresponds to a time of 9.15 a.m.

4 A teacher had given her class two Maths tests, but four students were absent for one or the other of them. These are the class results.

Student	Test A	Test B
Andy	69	59
Ben	34	17
Celia	8	absent
Dot	42	43
Eve	77	54
Faye	72	43
Gill	54	38
Harry	40	25
Ida	64	absent

Student	Test A	Test B
Joy	76	61
Kath	85	65
Les	absent	41
Meg	35	30
Ned	14	15
Olly	82	63
Pete	36	35
Quale	20	30
Jess	63	42

Student	Test A	Test B
Robin	48	32
Sophia	71	52
Tom	52	49
Ulla	absent	28
Vera	81	62
Will	41	36
Xanda	58	48
Yin	40	32
Zeb	28	19

Use the test scores to create a scatter graph and so estimate what score the absent students might have been expected to get if they had taken the missing test.

Now I can...

read scatter graphs	understand correlation	create scatter graphs

CHAPTER 24 Congruence and scaling

24.1 Congruent shapes

● I can recognise congruent shapes

If two shapes are exactly the same shape and size, they are **congruent**. Reflections, rotations and translations all produce images that are congruent to the original object. For shapes that are congruent, all the corresponding sides are equal and all the corresponding angles are equal.

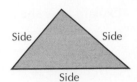

Two triangles are congruent if:

- all three sides are the same lengths in both triangles (SSS)

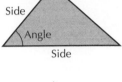

- two sides are the same length and the angle between them is the same size in both triangles (SAS)

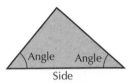

- two angles are the same size and the side between them is the same length in both triangles (ASA).

Develop fluency

Which two shapes below are congruent?

Shapes **b** and **d** are exactly the same shape and size, so they are congruent.

Use tracing paper to check that the two shapes are congruent.

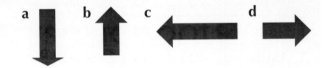

Show that triangle ABC is congruent to triangle XYZ.

The diagram shows that:

∠B = ∠X = 70°

∠C = ∠Y = 56°

BC = XY = 5 cm

So, triangle ABC is congruent to triangle XYZ (ASA).

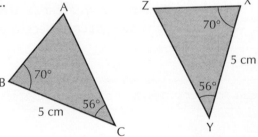

1   Look at the shapes in each pair and state whether they are congruent or not. Use tracing paper to help if you are not sure.

2 Which pairs of shapes on the grid below are congruent?

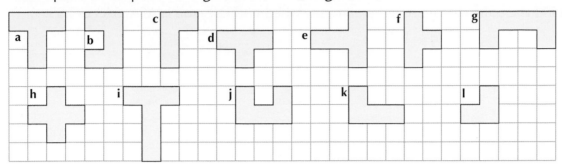

3 Which of these shapes are congruent? Write your answers, in the form A = B = C.

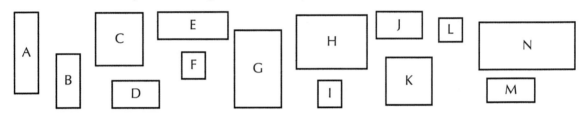

4 Which of these shapes are congruent? Write your answer in the form P = T = W.

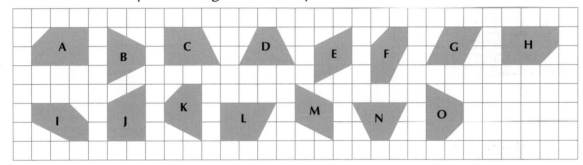

5 Which of the shapes below are congruent?

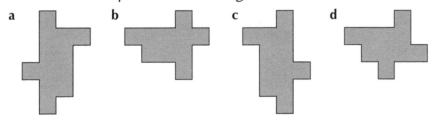

6 Use your ruler to check which of these triangles are congruent.
Write your answer in the form P = Q.

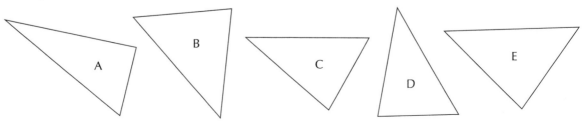

7 There are three pairs of congruent triangles pictured below.
Find the pairs that go together, and say whether the reason is ASA, SAS or SSS.

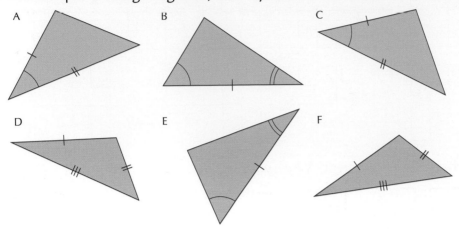

8 Find out whether triangles in each pair are congruent or not.
Give reasons to justify your answer.

a

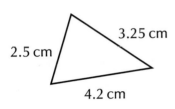

2.5 cm 3.25 cm 4.2 cm

42 mm $3\frac{1}{4}$ cm $2\frac{1}{2}$ cm

b

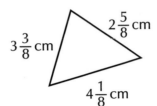

$3\frac{3}{8}$ cm $2\frac{5}{8}$ cm $4\frac{1}{8}$ cm

3.375 cm 2.625 cm 4.125 cm

c

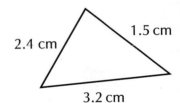

2.4 cm 1.5 cm 3.2 cm

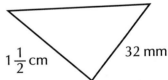

$2\frac{1}{4}$ cm $1\frac{1}{2}$ cm 32 mm

9 Which of these triangles are congruent? Give reasons to justify your answer.

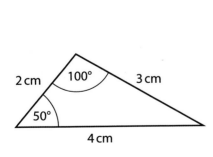
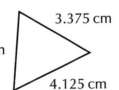
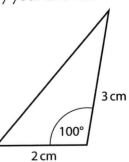

2 cm 100° 3 cm 50° 4 cm

4 cm 50° 30°

3 cm 100° 2 cm

Reason mathematically

Eve draws a triangle with sides 3 cm, 4 cm and 5 cm.

Ollie draws a triangle with sides 3 cm, 4 cm and 6 cm.

Are the two triangles congruent? Explain how you know.

Although the triangles have two sides the same, the third side is different so they cannot satisfy the condition of SSS.

The triangles are not congruent.

10 Show how this right-angled triangle can be split up into four congruent right-angled triangles.

11 Show how this cross can be split into four congruent pentagons.

12 Show how this shape can be split into three congruent shapes.

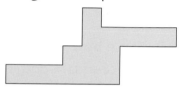

13 Show that the triangles in each pair are congruent. Give reasons for each answer and state which condition of congruency you are using: SSS, SAS or ASA.

a

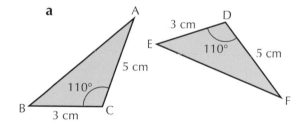

b

c

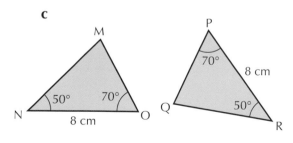

d

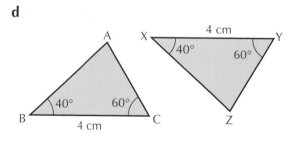

Solve problems

a Draw a rectangle ABCD and draw in the diagonal AC.

b Use a ruler to see if the two triangles you have formed are congruent.

c State how you could have checked without having to have done any measuring.

a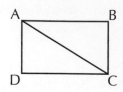

b Measure all the sides, to find matching lengths of triangles ABC and ADC.

c As ABCD is a rectangle, you know that AB = DC and AD = BC. AC is common to both triangles, hence the triangles have sides that satisfy SSS and hence they are congruent.

14 a Draw a parallelogram PQRS and draw in the diagonal SQ.

   b State, without any measuring, if the two triangles SPQ and SRQ are congruent.

15 This 4 by 4 pinboard is divided into two congruent shapes.

   a Use square-dotted paper to show different ways this can be done.

   b Show how you can divide the pinboard into four congruent shapes.

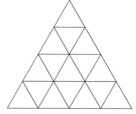

16 In the diagram each small triangle has an area of 1 unit. How many congruent triangles are there with:

   a area 1   **b** area 4   **c** area 9?

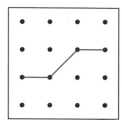

17 Compare this triangle to the eight in the box, on the next page.

Work out which of these triangles:

   a must be congruent to it

   b could be congruent, but more information is needed

   c are definitely not congruent to it.

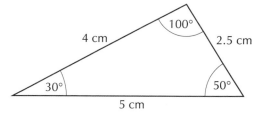

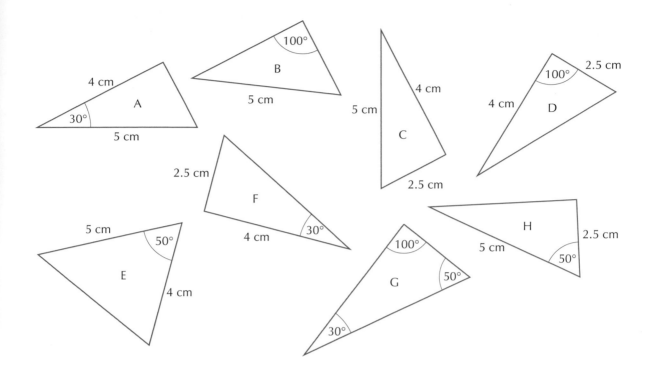

24.2 Enlargements

● I can enlarge a 2D shape by a scale factor

To enlarge a shape, you need a **centre of enlargement** and a **scale factor**.

Develop fluency

Enlarge the triangle XYZ by a scale factor of two about the centre of enlargement O.

Draw rays OX, OY and OZ.

Measure the length of the three rays and multiply each of these lengths by two.

Then extend each of the rays to these new lengths, measured from O, and plot the points X′, Y′ and Z′.

Join X′, Y′ and Z′.

Triangle X′Y′Z′ is the enlargement of triangle XYZ by a scale factor of two about the centre of enlargement O.

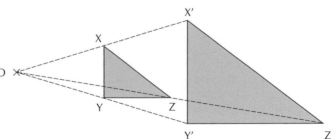

1 Copy or trace each shape and enlarge it, by the given scale factor, about the centre of enlargement O.

a Scale factor 2   **b** Scale factor 3   **c** Scale factor 2   **d** Scale factor 3

(**Note:** ✕ is the centre of square)

O ✕

2 Trace each shape with its centre of enlargement O. Enlarge the shape by the given scale factor.

a

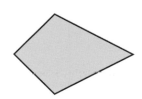

Scale factor 3

b

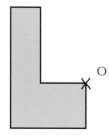

Scale factor 2

3 Copy or trace each shape and enlarge it, by the given scale factor, about the given centre of enlargement.

a   **b**   **c**   **d**

Scale factor 2

Scale factor 3

Scale factor 3

Scale factor 2

4 Copy or trace each shape and enlarge it, by the given scale factor, about the given centre of enlargement

a   **b**   **c**   **d**

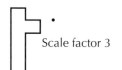

Scale factor 3

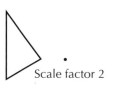

Scale factor 2

Scale factor 4

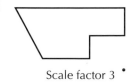

Scale factor 3

5 Copy each diagram onto centimetre-squared paper and enlarge it, by the given scale factor, about the origin O(0, 0).

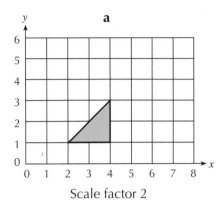

a

Scale factor 2

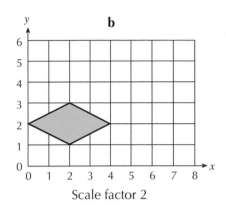

b

Scale factor 2

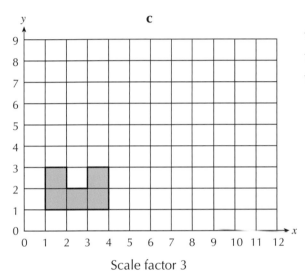

c

Scale factor 3

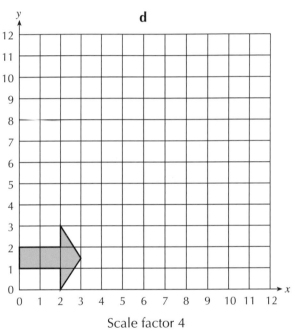

d

Scale factor 4

6 **a** Copy the grid and shape A only. Enlarge shape A by a scale factor of 2, using the origin as the centre of enlargement. Label the image A?.

   b Copy the grid and shape B only. Enlarge shape B by a scale factor of 2, using the point (7, 10) as the centre of enlargement. Label the image B?.

   c Copy the grid and shape C only. Enlarge the shape by a scale factor of 3, using the point (5, 4) as the centre of enlargement. Label the image C?.

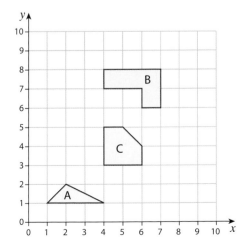

7 Draw a coordinate grid on centimetre-squared paper, labelling the *x*- and *y*- axes from 0 to 10.

Plot the points A(1, 1), B(4, 1), C(5, 3) and D(2, 3) and join them together to form the parallelogram ABCD. Enlarge the parallelogram by a scale factor of 2 about the origin O(0, 0).

8 Draw a coordinate grid on centimetre-squared paper, labelling the *x*- and *y*-axes from 0 to 12.

Plot the points A(4, 6), B(5, 4), C(4, 1) and D(3, 4) and join them together to form the kite ABCD. Enlarge the kite by a scale factor of 2 about the origin O(0, 0).

9 Describe fully these two enlargements.

a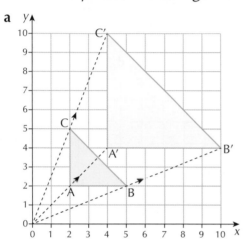

b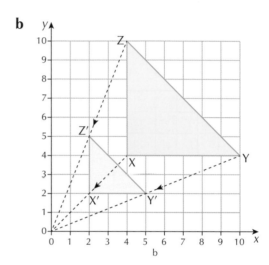

10 Draw a coordinate grid on centimetre-squared paper, labelling the x- and y-axes from 0 to 12.

Plot the points A(1, 1), B(1, 3), C(3, 2) and D(1, 2) and join them together to form a flag ABCD. Enlarge the flag by a scale factor of 3 about the point (0, 1).

Reason mathematically

a Copy the grid and the shape.

b Label the points ABCD and write down their coordinates.

c Enlarge the shape ABCD about (0, 0) with scale factor 2 and label the shape A'B'C'D'.

d Write down the coordinates of the new positions of A'B'C'D'.

e What do you notice about the original coordinates and the new coordinates?

a, b *A(0, 2), B(2, 3), C(4, 2), D(2, 1)*

c, d *A'(0, 4), B'(4, 6), C'(8, 4), D'(4, 2)*

e *The new coordinates are all double the numbers in the original coordinates.*

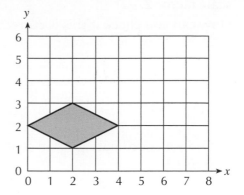

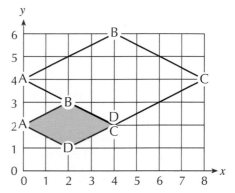

11 Copy the diagram onto centimetre-squared paper.

a Enlarge the square ABCD by a scale factor of two about the point (5, 5). Label the square A'B'C'D'. Write down the coordinates of A', B', C' and D'.

b On the same grid, enlarge the square ABCD by a scale factor of three about the point (5, 5). Label the square A''B''C''D''. Write down the coordinates of A'', B'', C'' and D''.

c On the same grid, enlarge the square ABCD by a scale factor of four about the point (5, 5).

Label the square A'''B'''C'''D'''. Write down the coordinates of A''', B''', C''' and D'''.

d What do you notice about the coordinate points that you have written down?

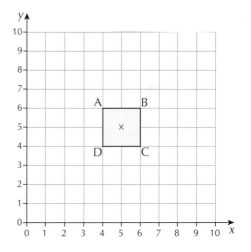

12 Copy the diagram onto centimetre-squared paper.
Brendon said the scale factor of the enlargement was
scale factor 2.
How can you check if this is correct?

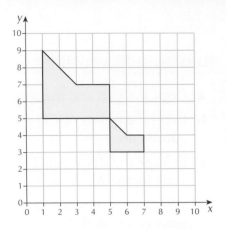

13 a Draw a coordinate grid on centimetre-squared paper, labelling the *x*- and *y*-axes from 0 to 12.
Plot the points A(1, 3), B(3, 3), C(3, 1) and D(1, 1), and then join them together to form the square ABCD.

b Write down the area of the square.

c Enlarge the square ABCD by a scale factor of 2 about the origin.
What is the area of the enlarged square?

d Enlarge the square ABCD by a scale factor of 3 about the origin.
What is the area of the enlarged square?

e Enlarge the square ABCD by a scale factor of 4 about the origin.
What is the area of the enlarged square?

f Write down anything you notice about the increase in area of the enlarged squares.
Try to write down a rule to explain what is happening.

g Repeat the above using your own shapes. Does your rule still work?

14 Draw a coordinate grid on centimetre-squared paper, labelling the *x*- and *y*-axes from 0 to 6.
Plot the points A(0, 0), B(0, 1), C(1.5, 2), D(3, 1) and E(3, 0).

a What is the mathematical name for the shape you have drawn?

b Enlarge the shape with scale factor 2, centred on the origin. Label the new points A′, B′, etc.

c Do the same with scale factor 3. Label these new points A′′, B′′, etc.

d Which point does not move its position? Does this happen with any enlargement?

e Write the new coordinates in a table like this:

Original point	Enlargement scale factor 2	Enlargement scale factor 3
A(0, 0)	A′(,)	A′′(,)
B(0, 1)	B′	
C(1.5, 2)	C′	
D(3, 1)	D′	
E(3, 0)	E′	

Solve problems

A shape ABCD has coordinates A(4, 2), B(8, 5), C(12 ,4) and D(10, 0). It is enlarged by scale factor 0.5 about the origin (0, 0). What are the coordinates of the enlarged shape?

This enlargement is a reduction, making the shape smaller. You still multiply the ordinates by the scale factor to give new coordinates as A'(2, 1), B'(4, 2.5), C'(6, 2), D'(5, 0).

15 Triangle ADE is the image of triangle ABC after an enlargement of scale factor 2 from point A.

   a Write down the length of AD if AB = 5 cm.

   b What is the length of CB if ED is 7 cm?

   c Work out the ratio of the areas of triangle ADE to triangle ABC.

   d What is the size of angle ACB if angle CED is 65°?

   e Name two congruent triangles.

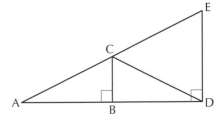

16 Copy the diagram onto centimetre-squared paper.

   a What is the scale factor of the enlargement?

   b By adding suitable rays to your diagram, find the coordinates of the centre of enlargement.

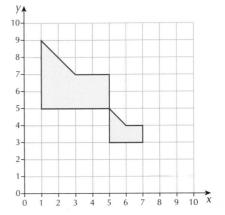

17 Find the coordinates of the centre of enlargement in this diagram.

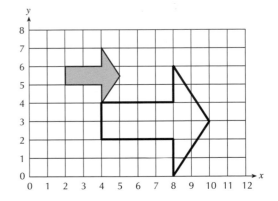

18 a Shape ABC has coordinates A(2, 2), B(3, 5), C(6, 7) and D(7,4). It is enlarged by scale factor 3 about the origin (0, 0). What are the coordinates of the enlarged shape?

   b Shape ABC has coordinates A(4, 2), B(6, 8), C(8, 10), D(12, 6). It is enlarged by scale factor 0.5 about the origin (0, 0). What are coordinates of the enlarged shape?

24.3 Shape and ratio

● I can use ratio to compare lengths, areas and volumes of 2D and 3D shapes

Develop fluency

Work out the ratio of the length of the line AB to the length of the line CD.

A———B   C————————————————D
  12 mm   4.8 cm

First convert the measurements to the same units and then simplify the ratio.

Always use the smaller unit, which here is millimetres (mm).

4.8 cm = 48 mm

Then the ratio is 12 mm : 48 mm = 1 : 4.

Remember that ratios do not have units.

Work out the ratio of the area of rectangle A to the area of rectangle B, giving the answer in its simplest form.

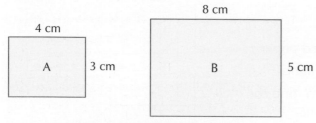

The ratio is 12 cm² : 40 cm² = 3 : 10

Work out the ratio of the volume of the cube to the volume of the cuboid, giving the answer in its simplest form.

The ratio is 8 cm³ : 72 cm³ = 1 : 9

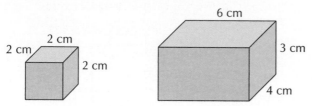

1   Express each ratio in its simplest form.

   a 10 mm : 25 mm   **b** 2 mm : 2 cm   **c** 36 cm : 45 cm

   d 40 cm : 2 m   **e** 500 m : 2 km

2   Express each ratio in its simplest form.

   a 90 cm : 20 cm   **b** 32 mm : 72 mm   **c** 150 cm : 2 m

   d 1.8 cm : 40 mm   **e** 0.65 km : 800 m   **f** 45 cm to 3 m

3 Work out the ratio of the length of the line AB to the length of the line CD.

A ———— B   C ————————————— D
　15 mm　　　　　　4.5 cm

4 Work out the ratio of the area of rectangle A to the area of rectangle B, giving the answer in its simplest form.

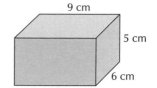

5 cm / A / 4 cm

7.5 cm / B / 6 cm

5 Work out the ratio of the volume of the cube to the volume of the cuboid, giving the answer in its simplest form.

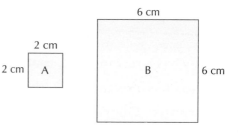

3 cm / 3 cm / 3 cm

9 cm / 5 cm / 6 cm

6 Which of these shapes has sides that are in the same ratio? (They are not drawn to scale.)

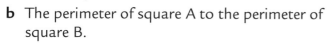

1 cm / A / 3 cm

2 cm / B / 5 cm

2.5 cm / C / 7.5 cm

4.8 cm / F / 1.6 cm

2.5 cm / D / 6.5 cm

1.8 cm / E / 5.2 cm

7 Look at the two squares then work out each ratio, giving your answers in their simplest form.

　a The length of a side of square A to the length of a side of square B.

　b The perimeter of square A to the perimeter of square B.

　c The area of square A to the area of square B.

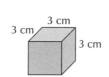

2 cm / 2 cm / A

6 cm / B / 6 cm

8 Look at the two rectangles then work out each ratio, giving your answers in their simplest form.

　a The length of the sides of rectangle X to the length of the sides of rectangle Y.

　b The perimeter of rectangle X to the perimeter of rectangle Y.

　c The area of rectangle X to the area of rectangle Y.

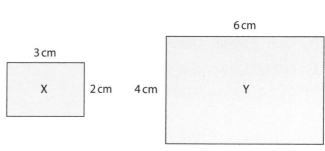

3 cm / X / 2 cm

4 cm

6 cm / Y

9 Look at the two triangles then work out each ratio, giving your answers in their simplest form.

   a The length of the sides of triangle A to the length of the sides of triangle B.

   b The perimeter of triangle A to the perimeter of triangle B.

   c The area of triangle A to the area of triangle B.

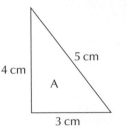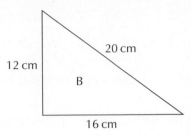

10 a Find the ratio of the base of rectangle A to the base of rectangle B.

   b Find the ratio of the area of rectangle A to the area of rectangle B.

   c What fraction is area A of area B?

   d Find the ratio of the area of rectangle B to the area of rectangle C.

   e Which is greater: area A as a fraction of area B, or area B as a fraction of area C?

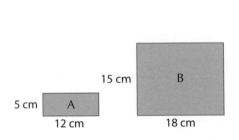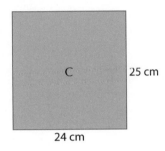

Reason mathematically

Kirsty has a rectangle, K, that measures 3 cm by 4 cm.

Robert has a rectangle, R, that measures 6 cm by 8 cm.

Kirsty says: 'Our rectangles are in the ratio of 1:2.'
Robert says: 'They are in the ratio of 1:4.'

Explain how they are both correct.

The ratio of the lengths of the rectangles are 3:6 or 4:8, both simplifying to 1:2.

The ratio of the areas of the rectangles are 12:48 which simplifies to 1:4.

Hence they are both correct.

11 The dimensions of lawn A and lawn B are shown in the diagram.

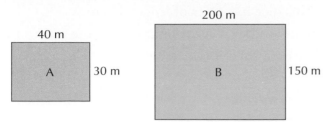

   a Calculate the area of lawn A, giving your answer in square metres.
   b Calculate the area of lawn B, giving your answer in:
      i square metres   **ii** hectares. (1 hectare = 10 000 m²)
   c Work out the ratio of the length of lawn A to the length of lawn B, giving your answer in its simplest form.
   d Work out the ratio of the area of lawn A to the area of lawn B, giving your answer in its simplest form.
   e Express the area of lawn A as a fraction of the area of lawn B.

12 The dimensions of a fish tank are given on the diagram.

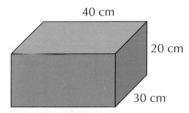

   a Calculate the volume of the fish tank, giving your answer in litres. (1 litre = 1000 cm³)
   b The fish tank is filled with water to a depth that is $\frac{3}{4}$ of the height.

   Calculate the volume of water in the fish tank, giving your answer in litres.
   c Work out the ratio of the volume of water in the fish tank to the total volume of the fish tank, giving your answer in its simplest form.

13 **a** Work out the ratio of the sides of the two cuboids. Give your answer in the form $1:n$.

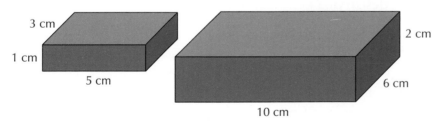

   b Work out the ratio of the total surface area of the two cuboids.
      Give your answer in the form $1:n$.
   c Work out the ratio of the volume of the two cuboids.
      Give your answer in the form $1:n$.
   d The ratio of the sides of two cuboids is $1:3$.
      i Write down the ratio of the total surface area of the two cuboids.
      ii Write down the ratio of the volume of the two cuboids.

14 A school has lots of small dice that are cubes of side length 2 cm, but a teacher wants to use a large foam dice to demonstrate. This dice has sides of length 10 cm.

   a Work out the ratio of the sides of the two cubes. Give your answer in the form $1:n$.

   b Work out the ratio of the total surface areas of the two cubes.
     Give your answer in the form $1:n$.

   c Work out the ratio of the volume of the two cubes. Give your answer in the form $1:n$.

   d The ratio of the sides of two other cubes is $1:4$.

     i Write down the ratio of the total surface area of the two cubes.

     ii Write down the ratio of the volume of the two cubes.

Solve problems

Look at the cuboids A and B.

   a What is the ratio of their lengths?

   b What is the ratio of the surface areas of corresponding faces?

   c What is the ratio of their volumes?

   d Find a connection between the length ratio, the area ratio and the volume ratio.

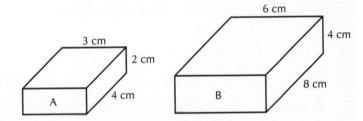

   a Length ratios; $2:4$, $4:8$, $3:6$ all simplify to $1:2$.

   b Area ratios; base $12:48$, front face $6:24$, side face $8:32$ all simplify to $1:4$.

   c Volume ratio $24:192$ simplifies to $1:8$.

   d Area ratio is length ratio squared, volume ratio is length ratio cubed.

15 The diagram shows the design of a new flag.

   a Calculate the ratio of the red area to the green area.
     Use ratios to answer these questions.

   b 60 m² of red cloth is used to make some flags. How much green cloth is needed?

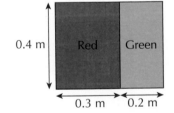

   c The total area of some flags is 200 m².

     i How much red cloth do they contain?

     ii How much green cloth do they contain?

     iii If red cloth costs £8 per square metre, and green cloth costs twice as much, how much would it cost to make these flags?

16 The diagram shows the plan of a garden.

    a Find the ratio of the perimeter of the fence to the perimeter of the pool.

    b Calculate the area of the pool.

    c Calculate the area of the lawn.

    d Find the ratio of the lawn area to the pool area.

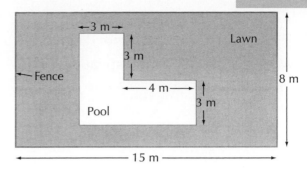

17 The diagram shows a picture, 30 cm by 8 cm that has been put in a frame, 45 cm by 12 cm.

    a Find the ratio of the lengths of the picture to the lengths of the frame.

    b Find the ratio of the picture area to the frame area.

18 Cube A has side length of 2 cm. It is placed on top of cube B with a side length of 3 cm.

    a What is the ratio of the lengths of the cubes?

    b What is the ratio of the areas of the faces on the cubes?

    c What is the ratio of the volumes of the cubes.

    d Find a connection between the ratios of length, area and volumes.

24.4 Scales

- I can understand and use scale drawings
- I can use map ratios

A **scale drawing** is a smaller or larger drawing of an actual object or diagram. A **scale** must always be clearly given by the side of or below the scale drawing.

Maps also are examples of scale drawings. On most maps the scale is given as a **map ratio**.

A map ratio is always written in terms of the same units. So, for example, a scale of 1 cm : 1 km would be 1 : 100 000 in a map ratio.

Develop fluency

The diagram is a scale drawing of Rebecca's room.

What are the real measurements of the room?

- On the scale drawing, the length of the room is 5 cm, so the actual length of the room is 5 m.

- On the scale drawing, the width of the room is 3.5 cm, so the actual width of the room is 3.5 m.

- On the scale drawing, the width of the window is 2 cm, so the actual width of the window is 2 m.

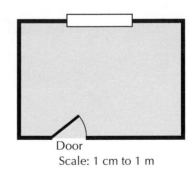

Door
Scale: 1 cm to 1 m

The scale of a map is 1 cm to 5 km.

Work out the scale as a map ratio.

1 m = 100 cm and 1 km = 1000 m

5 km = 5 × 1000 × 100 = 500 000 cm

So, 5 km = 500 000 cm

The map ratio is 1 : 500 000.

1 The lines shown are drawn using a scale of 1 cm to 10 m.
Write down the length each line represents.

a ————————

b ————————————————

c ——————————————

d ——————————————————————

e ————————————————————

2 These objects have been drawn using the scales shown. Find the true lengths of the objects. (In part a, measure the length of the golf club shaft only.)

a   Scale

                                                                 1 cm to 10 cm

b

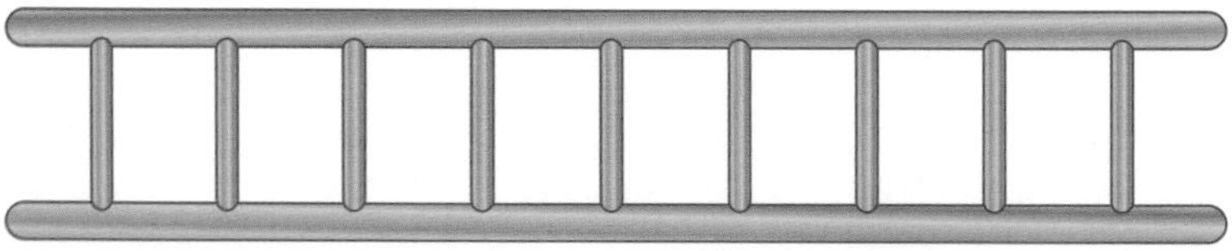

                                                                 2 cm to 1 m

c

                                                                 1 cm to 0.7 m

d

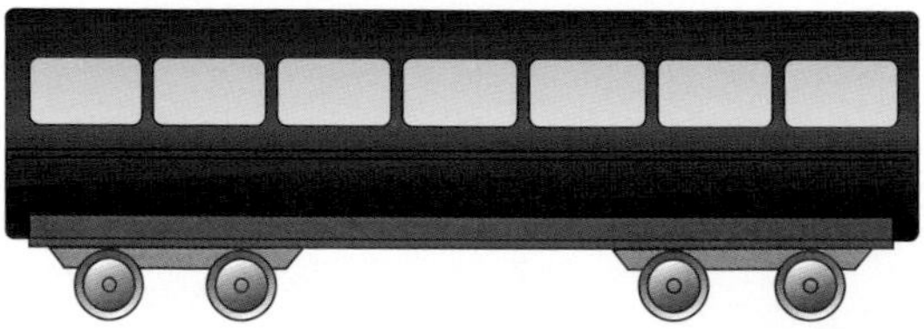

                                                                 2 cm to 3 m

3 The diagram shows a scale drawing for a school hall.
   a Find the actual length of the hall.
   b Find the actual width of the hall.
   c Find the actual distance between the opposite corners of the hall.

Scale: 1 cm to 5 m

4 The real dimensions of some objects are given below. Work out their size on a scale drawing using the scales shown.
   a Bus: length = 11 m, height = 4.5 m, scale 1 cm to 2 m
   b Spade: length = 1.5 m, width of blade = 12 cm, scale 1 cm to 0.5 m
   c Calculator: length = 120 mm, width = 60 mm, scale 1 cm to 40 mm

5 The diagram shows a scale drawing of an aircraft hangar (plan view).
   a Calculate the real length of the hangar.
   b Calculate the real width of the hangar.
   c Calculate the area of the hangar.

Scale: 1 cm to 120 m

6 The diagram shown is Ryan's scale drawing for his mathematics classroom. Nathan notices that Ryan has not put a scale on the drawing, but he knows that the length of the classroom is 8 m.
   a What scale has Ryan used?
   b What is the actual width of the classroom?
   c What is the actual area of the classroom?

7 Copy and complete this table for a scale drawing in which the scale is 4 cm to 1 m.

	Actual length	Length on scale drawing
a	4 m	
b	1.5 m	
c	50 cm	
d		12 cm
e		10 cm
f		4.8 cm

8 Copy and complete the table.

	Scale	Scaled length	Actual length
a	1 cm to 3 m	5 cm	15 m
b	1 cm to 2 m		24 m
c	1 cm to 5 km	9.2 cm	
d		6 cm	42 miles
e	5 cm to 8 m	30 cm	

9 Copy and complete the table.

	Scale	Scaled length	Actual length
a		3 cm	18 m
b		8 cm	24 km
c		5.2 cm	26 km
d		8 cm	40 miles
e		15 cm	90 miles

10 This is a plan for a bungalow.

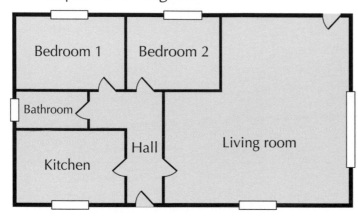

Scale: 1 cm to 2 m

a Find the actual dimensions of:
   i the kitchen   **ii** the bathroom   **iii** bedroom 1   **iv** bedroom 2.
b Calculate the actual area of the living room.

Reason mathematically

James measured a map and saw that a distance of 5 km was represented by 2 cm.

His son said: 'That's a scale of 2 to 5.'

Is he correct? Explain your answer.

No he's not correct. The scale is 2 cm to 5 km, which is simplified to 1 cm to 2.5 km.

11 Joe was told that a distance of 54 miles on his map was represented by 12 cm.
He said: 'That's a scale of 1 cm to 4 miles.'
Is Joe correct? Explain your answer.

12 Helen looks at a map and sees that a distance of 45 km is represented by 15 cm.
She calculates the scale as

$15\,\text{cm} \rightarrow 45\,\text{km}$

$5\,\text{cm} \rightarrow 30\,\text{km}$

$1\,\text{cm} \rightarrow 6\,\text{km}$

Helen has made an error. Correct Helen's error.

13 The diagram shows the plan of a
football pitch. It is not drawn to scale.
Use the measurements on the diagram
to make a scale drawing of the pitch.
Choose your own scale.

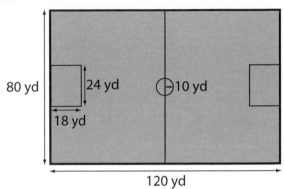

14 The diagram shows a scale drawing of a shop, where 1 cm represents 2 m. Make a table
showing the real-life dimensions and area of each section of the shop.

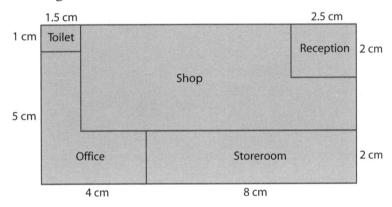

a Work out:

   i the scaled area

   ii the real-life area of each part of the shop.

b **i** Write the scale factor of the dimensions as a ratio in the form $1:n$.

   ii Write the scale factor of the areas as a ratio in the form $1:n$.

Solve problems

David wanted to complete a 30 km walk in the Peak District.

The map ratio of his map was 1 : 200 000.

What length on the map will his walk have to be?

Map ratio of 1 : 200 000 means $1\,\text{cm} \rightarrow 200\,000\,\text{cm}$

$200\,000\,\text{cm} = 2000\,\text{m} = 2\,\text{km}$

So the distance on the map will be $30 \div 2 = 15\,\text{cm}$

15 Joy wanted to set a 12 km walk for her Guides.
She looked at a map with a map ratio of 1 : 300 000.
What length on the map must her walk be?

16 Chris is creating a 45 km hike for the Scouts.
He looked at a map with a map ratio of 1 : 150 000.
What length on the map must his hike be?

17 A 60 km rail journey is planned.
A map with a map ratio of 1 : 500 000 is being used.
What length on the map will this rail journey be?

18 The map shows York city centre.

   a Write the scale in map format.

   b How far is it from the Minster to:

      i the Barbican

      ii the Railway Museum?

   c Some football fans need to get catch a train at 5:30 p.m. to get home. If the match finishes at 4:55 p.m. and their fastest walking pace is 8 m/s, how much time will they have to spare? (Remember they cannot go direct as the black line is the railway, so you will have to find their shortest route by road.)

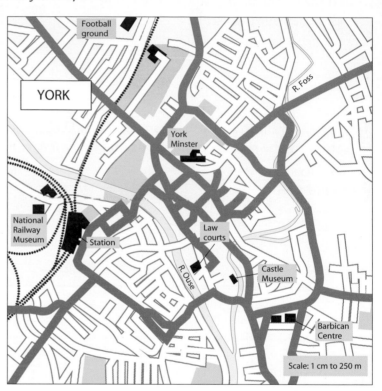

The scale on the map is approximate and for calculation purposes only.

Now I can...

recognise congruent shapes	use ratio to compare lengths, areas and volumes of 2D shapes	use ratio to compare lengths, areas and volumes of 3D shapes
enlarge a 2D shape by a scale factor	use map ratios	
understand and use scale drawings		

CHAPTER 25 Manipulating algebraic expressions

25.1 Algebraic notation

● I can simplify algebraic expressions involving the four basic operations

When you write an algebraic expression you usually leave out the multiplication sign ($\times$), so instead of $4 \times a$ you would write $4a$. You do not usually use the division sign ($\div$) either, so you would write $a \div 4$ as $\frac{a}{4}$ or $\frac{1}{4}a$. You always write a number in front of a letter, for example, write $4a$, never $a4$.

Develop fluency

Write each expression as simply as possible.

a $3 \times t$　　　　**b** $k \times 2 + 4$　　　　**c** $d \times (8 - 2.5)$

d $p \times k \times r$　　　　**e** $2 \times k \times 4$　　　　**f** $f \times \frac{2}{3} \times f$

a $3 \times t = 3t$　　　　You can leave out the × sign.

b $k \times 2 + 4 = 2k + 4$　　　　Put the 2 in front of the k.

c $d \times (8 - 2.5) = d \times 5.5$　　　　Work out the subtraction in the brackets first.

　　$d \times 5.5 = 5.5d$　　　　Put the 5.5 in front of the d.

d $p \times k \times r = pkr$　　　　You do not have to put the letters in alphabetical order.

e $2 \times k \times 4 = 8k$　　　　Multiply the numbers first.

f $f \times \frac{2}{3} \times f = \frac{2}{3}f^2$　　　　Put the $\frac{2}{3}$ in front and then $f \times f = f^2$.

1 Write each expression as simply as possible.

a $4 \times k$　　　　**b** $t \times 3$　　　　**c** $2 \times x$　　　　**d** $y \times 20$

e $a \times b$　　　　**f** $n \times m$　　　　**g** $e \times \frac{1}{2}$　　　　**h** $r \times \frac{1}{4}$

2 Write each expression as simply as possible.

a $2 \times a \times b$　　　　**b** $4 \times 2 \times 5$　　　　**c** $2 \times 5 \times e$　　　　**d** $4 \times 20 \times w$

e $a \times b \times c$　　　　**f** $n \times 6 \times \frac{1}{2}$　　　　**g** $2.5 \times x \times 3$　　　　**h** $b \times \frac{3}{4} \times h$

3 Write each expression as simply as possible.

a $a \times a$　　　　**b** $4 \times x \times x$　　　　**c** $9 \times n \times n$　　　　**d** $t \times t \times 1.4$

e $2 \times a \times 1.6$　　　　**f** $d \times \frac{1}{5} \times d$　　　　**g** $21 \times x \times x$　　　　**h** $\frac{2}{3} \times m \times m$

4 Write each expression without using a × sign.

a $2 \times (n + 1)$　　　　**b** $4 \times (t + 12)$　　　　**c** $8 \times (3 + k)$　　　　**d** $0.5 \times (t - 6.4)$

e $(k + 2) \times 4$　　　　**f** $(12 - y) \times 2$　　　　**g** $(2 + 6) \times (x - 2)$　　　　**h** $(40 - y) \times 3 \times 2$

5 Write each expression as simply as possible.

   a $2 + 3 \times a$ **b** $(5 + 4) \times x$ **c** $w \times (3.5 + 4.2)$ **d** $a \times b - 1.4$

   e $2 + 3 + 4 \times a$ **f** $d \times (10 - 1.5)$ **g** $12 - 7 \times n$ **h** $f \times 13 + 9$

6 Write each expression as simply as possible.

   a $2 \times 3n$ **b** $4 \times 5b$ **c** $4 \times 0.5d$ **d** $0.1 \times 5q$

   e $5 \times 2k$ **f** $4g \times 6$ **g** $8t \times 1.5$ **h** $0.2 \times 5h$

7 Write each expression as simply as possible.

   a $2n \times 2n$ **b** $3d \times 4d$ **c** $4p \times 2p$ **d** $0.5a \times 6a$

8 Write each expression without using a ÷ sign.
   The first one has been done for you.

   a $x \div 3 = \dfrac{x}{3}$ **b** $y \div 4$ **c** $t \div 1.5$ **d** $24 \div n$

9 Write each expression using a fraction. The first one has been done for you.

   a $h \div 5 = \dfrac{h}{5}$ **b** $m \div 3$ **c** $n \div 4$ **d** $ab \div 2$

10 Write each expression as a single fraction, as simply as possible.
   The first one has been done for you.

   a $\dfrac{a \times b}{3} = \dfrac{ab}{3}$ **b** $\dfrac{3c \times d}{2}$ **c** $\dfrac{e \times 5d}{4}$ **d** $\dfrac{2f \times 6g}{7}$

Reason mathematically

Ali has simplified some expressions. State which ones he has answered correctly.

Explain where he has made errors and write the correct answer.

   a $2s \times 3t = 5st$ **b** $4v \times 4v = 16v^2$ **c** $2y + 7y - 3 = 6y$

   a Ali has incorrectly calculated 2 + 3; rather than 2 × 3 correct answer is 2s × 3t = 6st

   b This is correct.

   c Ali has incorrectly calculated 2 + 7 – 3 to get 6, rather than just adding together the y numbers to get 9 and leaving the 3 separate. Correct answer is 2y + 7y – 3 = 9y – 3

11 Rosie has simplified some expressions. State which ones she has answered correctly.
   Explain where she has made errors and write the correct answer.

   a $5c \times 3c = 15c$ **b** $4a \times 3b = 12ab$ **c** $g \div 5 = \dfrac{5}{g}$

   d $2u \times 4v = 6uv$ **e** $8 \times f - 2 = 8f - 2$ **f** $9 - 5 \times y = 4y$

12 Match each expression on the top row with an equivalent expression on the bottom row.

   a $2 \times x + 3 \times y$ **b** $y \times (2 + 3) \times x$ **c** $3y \times 2x$ **d** $x + (2 + 3) \times y$

     $x + 5y$ $6xy$ $3y + 2x$ $5xy$

13 Which of these expressions is the odd one out?
Give a reason for your answer.

$2w + 6h$ $\qquad\qquad$ $4w - 6h$ $\qquad\qquad$ $3w \times 2h$

14 Which of these expressions is the odd one out?
Give a reason for your answer.

$3x \times 6y$ $\qquad$ $4xy + 14xy$ $\qquad$ $36xy \div 2xy$ $\qquad$ $20xy - 2xy$

Solve problems

I am 64 years old.

In four years' time my son will be half my age.

How old is my son now?

Let my son's age now be x years.

In four years' time his age will be $x + 4$ and my age will be $64 + 4 = 68$.

So, $x + 4$ is half of 68.

$x + 4$ is 34, so $x = 30$.

My son is now 30 years old.

15 My friend is 2 years younger than I am.
The sum of our ages is 46.
How old am I?

16 I am thinking of two numbers.
They add up to 40.
The difference of the two numbers is 6.
What are the two numbers I am
thinking of?

17 I am thinking of two numbers.
They have a difference of 2.
The sum of the two numbers is 14.
What is the product of the two numbers I
am thinking of?

18 $x + 7 = 15$
Work out the value of $x - 7$.

25.2 Like terms

● I can simplify algebraic expressions by combining like terms

Look at this expression:

$5a + 4b - 3a + b$

It has four **terms** altogether. $5a$ and $3a$ are called **like terms** and like terms can be combined to simplify an expression. $4b$ and b are also like terms.

$2a$ and $5b$ are not like terms, because they contain different letters. They are **unlike** terms.

The expression simplifies to $2a + 5b$.

Develop fluency

Simplify each expression.

   a $4f - 3f + 6$ **b** $2g - 3h + 2g - 4h$ **c** $x^2 - 4x + x - 5$

   a $4f - 3f + 6 = f + 6$ The first two terms are like terms and can be combined.

                                6 and f are not similar terms and the expression cannot be simplified further.

   b $2g - 3h + 2g - 4h = 2g + 2g - 3h - 4h$ Collect like terms.

                  $= 4g - 7h$ Combine the g-terms and the h-terms.

Note that subtracting $3h$ and then subtracting another $4h$ is the same as subtracting $7h$.

   c $x^2 - 4x + x - 5 = x^2 - 3x - 5$ Only the middle terms are like terms and can be combined.

Note that subtracting $4x$ and then adding $1x$ is the same as subtracting $3x$.

1 Simplify each expression.

   a $5h + 6h$ **b** $4p + p$ **c** $9u - 3u$ **d** $3b - b$

   e $-2j + 7j$ **f** $6pr - pr$ **g** $2k + k + 3k$ **h** $9y^2 - y^2$

   i $6h + 2h + 5g$ **j** $4g - 2g + 8m$ **k** $8f + 7d + 3d$ **l** $4x + 5y + 7x$

   m $6 + 3r - r$ **n** $4 + 5s - 3s$ **o** $c + 2c + 3$ **p** $12b + 7 + 2b$

2 Simplify each expression.

   a $7w - 7 + 7w$ **b** $2bf + 4bf + 5g$ **c** $7d + 5d^2 - 2d^2$ **d** $6t^2 - 2t^2 + 5t$

   e $4s - 7s + 2t$ **f** $5h - 2h^2 - 3h$ **g** $4y - 2w - 7w$

3 Simplify each expression.

   a $9e + 4 + 7e + 2$ **b** $10u - 4 + 9u - 2$ **c** $b + 3b + 5d - 2d$

   d $4 + 5c + 3 + 2c$ **e** $1 + 2g - 3g + 5$ **f** $9h + 4 - 7h - 2$

4 Which of these expressions will not simplify?

   a $3a + 2$ **b** $5a + 6a + 2$ **c** $7j - 2jk + j$ **d** $2x + 3y + 4z$

5 Which of these expressions simplify to the same answer?

   a $5x + 2x$ **b** $-3x + 6x + 4x$ **c** $5x - 2x + 4x$ **d** $x + x + 5x$

6 Simplify each expression, if possible.

   a $2a^2b + 3a^2b$ **b** $2x^2y + 3xy^2$ **c** $2cd^2 + 3cd^2$ **d** $2e^2f + 3e^2f$

7 Simplify each expression.

   a $2a^2b + ab + 3a^2b$ **b** $2x^2y + 4xy^2 + 3xy^2$ **c** $2cd^2 + 3cd^2 + 5cd^2$ **d** $2e^2f + 3e^2f - 5e^2f$

Reason mathematically

Match each expression on the top row with the correct simplified form on the bottom row.

a $5y - 2x - 5y + 4x$ **b** $3x + 2y - y - 2x$ **c** $-2x - 3y + 2y + 3x$ **d** $3y - 5x - 2y + 6x$

$x + y$ $2x$ $x - y$

a $5y - 2x - 5y + 4x = 4x - 2x + 5y - 5y$ *Collect and combine like terms.*
$ = 2x$

b $3x + 2y - y - 2x = 3x - 2x + 2y - y$ *Collect and combine like terms.*
$ = x + y$

c $-2x - 3y + 2y + 3x = 3x - 2x - 3y + 2y$ *Collect and combine like terms.*
$ = x - y$

d $3y - 5x - 2y + 6x = 6x - 5x + 3y - 2y$ *Collect and combine like terms.*
$ = x + y$

8 Match each expression on the top row with the correct simplified form on the bottom row.

a $2y - x - 5y + 7x$ **b** $x + 5x - y - x$ **c** $2x - 3y + 2y + 3x$ **d** $y + 5x - 4y + x$

$5x - y$ $6x - 3y$

9 Match each expression on the top row with the correct simplified form on the bottom row.

a $6x - x - 5y + 4y$ **b** $x + 2y + 2y + x$ **c** $-4x - 2y + 9x + y$ **d** $-y + 5x + 5y - 3x$

$5x - y$ $2x + 4y$

10 Which of these expressions is the odd one out?
Give a reason for your answer.

$3x + 6y - 3x + 6y$ $3x + 6y - 3x - 6y$ $-3x - 6y + 3x + 6y$ $3x - 6y - 3x + 6y$

Solve problems

The operator signs + and − are missing from this four-term expression.

$5a \ldots 7b \ldots a \ldots 3b = 4a - 4b$

Fill in the missing signs to make the equation correct.

$5a - 7b - a + 3b = 5a - a - 7b + 3b$
$ = 4a - 4b$

11 The operator signs + and − are missing from this four-term expression.
$6a \ldots b \ldots 3a \ldots 2b = 3a - b$
Fill in the missing signs to make the equation correct.

12 Fill in the missing terms to make each expression correct.

  a $2a + \ldots + \ldots - 5b = 6a - 2b$ **b** $7a + 4b - \ldots - \ldots = -a - b$

  c $2a - 3b + \ldots + \ldots = 6a + 2b$ **d** $\ldots - 6b + 4a - \ldots = 8a - 11b$

13 Fill in the missing terms to make the expression correct.

   a $4x + 6y - \ldots + \ldots = 7x + 5y$ **b** $2x - 3y + \ldots - \ldots = x + 6y$

   c $x - \ldots + \ldots + y = 2x - 7y$ **d** $-3x - \ldots - \ldots + y = -5x - 9y$

14 Combine two or more of these three expressions to make the given answer.

   $2x + y$ $3x - 2y$ $4x - y$

   You can use each expression more than once in each question part.

   The first one has been done for you.

   a Answer $6x$ $2x + y + 4x - y = 6x$

   b Answer $5x - y$ **c** Answer $8x + y$ **d** Answer $9x - 2y$

25.3 Expanding brackets

● I can remove brackets from an expression

You can remove brackets and write expressions in different ways. For example:

$2(a + 3) = 2a + 6$ This means $2 \times a$ added to 2×3.

So, $2(a + 3)$ and $2a + 6$ are **equivalent** expressions.

$4(c - d) = 4c - 4d$ This means $4 \times c$ take away $4 \times d$.

This is called **multiplying out** or **expanding** an expression with brackets.

Develop fluency

Expand and simplify each expression.

   a $3(t - 5)$ **b** $3(2f + 4)$ **c** $2(a + 3) + 3(a - 4)$

   a $3(t - 5) = 3t - 15$ This means $3 \times t$ minus 3×5.

   b $3(2f + 4) = 6f + 12$ $3 \times 2f = 6f$ and $3 \times 4 = 12$

   c $2(a + 3) = 2a + 6$

      $3(a - 4) = 3a - 12$ Multiply out the brackets separately.

      So, $2(a + 3) + 3(a - 4) = 2a + 6 + 3a - 12$ Put the two expressions together.

                    $= 2a + 3a + 6 - 12$ Collect like terms.

                    $= 5a - 6$ Combine like terms.

1 Expand the brackets.

   a $5(p + 2)$ **b** $4(m - 3)$ **c** $2(t + u)$ **d** $4(d + 2)$

   e $5(b + 5)$ **f** $6(j - 4)$ **g** $2(5 + f)$ **h** $10(1 - n)$

2 Expand the brackets.

   a $2(a + b)$ **b** $3(q - t)$ **c** $4(t + m)$ **d** $5(x - y)$

   e $2(f + g)$ **f** $4(1.5 - f)$ **g** $1.5(h + 10)$ **h** $4(3.5 - f)$

3 Expand the brackets and simplify each expression.

    a $3w + 2(w + 1)$ **b** $5(d + 2) - 2d$ **c** $4h + 5(h + 3)$

    d $2x + 4(3 + x)$ **e** $2(m - 3) - 7$ **f** $16 + 3(q - 4)$

4 Expand the brackets and simplify each expression.

    a $4(a + 1) + 2(a + 2)$ **b** $3(i + 4) + 5(i + 2)$

    c $3(p + 2) + 2(p - 1)$ **d** $5(d - 2) + 3(d + 1)$

    e $4(e + 2) + 2(e - 3)$ **f** $2(x - 2) + 6(x + 1)$

    g $5(m - 3) + 3(m + 4)$ **h** $4(u - 3) + 5(u - 2)$

5 Write these expressions without brackets.

    a $2(2a + 3)$ **b** $3(2x - 5)$ **c** $4(3t + 5)$ **d** $10(5n - 3)$

    e $6(5 + 2a)$ **f** $4(2 - 3y)$ **g** $20(3 + 4r)$ **h** $5(7 - 2m)$

6 Expand the brackets and simplify each expression.

    a $3(2a + 1) + 4(a + 2)$ **b** $4(3c + 2) + 3(c + 4)$

    c $2(2e + 3) + 5(2e + 1)$ **d** $5(f + 5) + 2(4f + 3)$

7 Expand the brackets and simplify each expression.

    a $5(2g + 3) - 4(g + 2)$ **b** $3(3h + 4) - 3(h + 2)$

    c $2(2j + 5) - 4(j + 1)$ **d** $4(3k + 4) - 2(4k + 3)$

8 Expand the brackets and simplify each expression.

    a $4(2m - 3) - 3(m + 2)$ **b** $5(3n + 2) + 3(n - 2)$

    c $3(3p - 5) - 4(p + 3)$ **d** $2(3q + 4) - 2(4q - 3)$

Reason mathematically

Show that $2(3a + 1) + 4(a + 5)$ and $2(5a + 11)$ are equivalent.

$2(3a + 1) + 4(a + 5) = 6a + 2 + 4a + 20$
$= 10a + 22$

$2(5a + 11) = 10a + 22$

So they are equivalent as both give $10a + 22$ when expanded and simplified.

9 Show that $5(a + 6) + 3(a + 2)$ and $4(2a + 9)$ are equivalent.

10 Show that $6(2x + 3) - 3(3x + 2)$ and $3(x + 4)$ are equivalent.

11 Which of these expressions are equivalent?

    a $2(5x + 3) + 2(4x - 1)$ **b** $9(2x - 5) + 3(x + 7)$ **c** $3(5x + 2) + 2(3x - 2)$

    d $7(4x + 1) + 2(3x - 5)$ **e** $4(4x + 3) + 5(x - 2)$ **f** $3(2x + 5) + 4(3x - 4)$

12 **a** Expand and simplify $4(2x + 3) - 3(x - 2)$.

    b Expand and simplify $3(x - 2) - 4(2x + 3)$.

    c What do you notice about your answers to parts **a** and **b**?

Solve problems

$a(4x + 3) + 2(6x - 1) = 24x + b$

Work out the values of a and b.

$a(4x + 3) + 2(6x - 1) = 4ax + 3a + 12x - 2 = 24x + b$

Looking at the terms in x: $\qquad 4a + 12 = 24$

$\qquad\qquad\qquad\qquad 4a = 12 \qquad$ Subtract 12 from both sides.

$\qquad\qquad\qquad\qquad a = 3 \qquad$ Divide both sides by 4.

Looking at the constants (the terms not involving x): $\quad 3a - 2 = b$

$\qquad\qquad\qquad\qquad 3 \times 3 - 2 = b \quad$ Use $a = 3$.

$\qquad\qquad\qquad\qquad 7 = b$

Checking gives: $3(4x + 3) + 2(6x - 1) = 12x + 9 + 12x - 2$
$\qquad\qquad\qquad\qquad\qquad\qquad = 24x + 7$

13 $6(2x + 1) + 3(x + 5) = ax + b$
Work out the values of a and b.

14 $4(3x - 1) + a(5x + b) = 22x$
Work out the values of a and b.

15 Dan is given this question:
Expand and simplify $5(3x + 4) - 2(x - 3)$.
This is his working: $\quad 5(3x + 4) - 2(x - 3) = 8x + 20 - 2x - 6$
$\qquad\qquad\qquad\qquad\qquad\qquad = 6x + 14$

a What are the two mistakes he has made?

b Correct his working to show the correct answer.

25.4 Using algebraic expressions

- I can manipulate algebraic expressions
- I can identify equivalent expressions

In mathematics, and in other subjects, you can use algebraic expressions to show relationships.

Sometimes expressions can be written in different ways. You can **manipulate** expressions to show that they are equivalent.

Develop fluency

In this diagram, all the lengths are in centimetres.

Work out an expression for the perimeter of this shape.

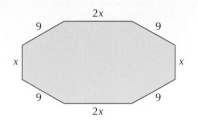

The perimeter is

$2x + 9 + x + 9 + 2x + 9 + x + 9$ — Add the lengths of the 8 sides.

$= 2x + x + 2x + x + 9 + 9 + 9 + 9$ — Collect like terms.

$= 6x + 36$ — Combine like terms.

1 These two pipes are joined end to end.
Write down an expression for the total length
of the joined pipe.

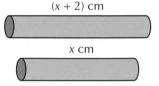

2 A square has side length x cm.
 a Write down an expression for the perimeter.
 b Write down an expression for the area.

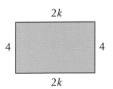

3 Work out an expression for the perimeter of each rectangle.
Write your answer as simply as possible.

 a 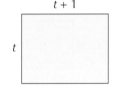 $t+1$, t

 b 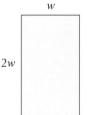 w, $2w$

 c 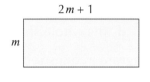 $2m+1$, m

4 Work out an expression for the area of each rectangle in question 3.

5 Work out an expression for the perimeter of each shape.
Write your answer as simply as possible.

 a 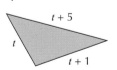 $t+5$, t, $t+1$

 b  $2k$, 4, 4, $2k$

 c c, c, 8, 8, c, c

 d 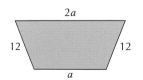 $2a$, 12, 12, a

 e 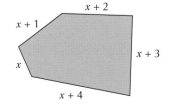 $x+1$, $x+2$, $x+3$, x, $x+4$

6 Work out an expression for the volume of each cuboid.
Write your answer as simply as possible.

a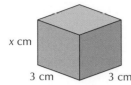

b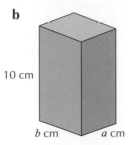

c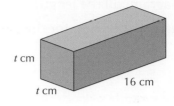

7 In the diagram, all lengths are in centimetres.
Work out an expression for the area, in cm², of:

a **i** rectangle P **ii** rectangle Q **iii** rectangle R **iv** rectangle S.

b Work out the total area of rectangles P, Q, R and S.

8 Look again at the diagram in question 7.
Work out an expression for the perimeter, in cm, of:

a rectangle P **b** rectangle Q **c** rectangle R

d rectangle S **e** the whole shape.

9 In the diagram, all lengths are in centimetres.
Work out an expression for the area, in cm², of:

a rectangle A **b** rectangle B

c rectangle C **d** the whole shape.

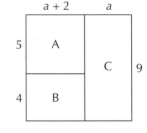

10 Look again at the diagram in question 9.
Work out an expression for the perimeter, in cm, of:

a rectangle A **b** rectangle B **c** rectangle C **d** the whole shape.

Reason mathematically

This is an algebra wall.

The expression in each brick is the sum of the expressions in the two
bricks below it. For example, the sum of $x + 3$ and $2x + 5$ is $3x + 8$.

Show that the expression in the top brick can be written as $6(x + 2)$.

The missing expression in the middle row is $2x + 5 + x - 1 = 3x + 4$.

The missing expression in the top brick is $3x + 8 + 3x + 4 = 6x + 12$.

$6(x + 2) = 6x + 12$ *so the expression in the top brick can be written as $6(x + 2)$.*

11 Show that the expression in the top brick of this algebra wall
can be written as $8(x + 2)$.

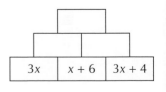

12 Show that the expression in the top brick of this algebra wall can be written as $5(x + 3)$.

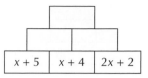

13 Show that the expression in the top brick of this algebra wall can be written as $6(x - 1)$.

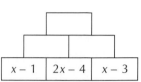

14 Show that the given expressions are equivalent expressions for the perimeters of the shapes in question 5.

   a $3(t + 2)$ **b** $4(k + 2)$ **c** $4(c + 4)$ **d** $3(a + 8)$ **e** $5(x + 2)$

15 This shape has been divided into rectangles in two different ways.

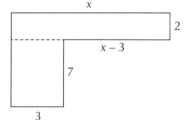 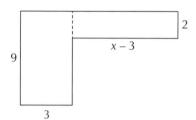

   a Use the first diagram to write an expression for the area of the whole shape.

   b Use the second diagram to write a different expression for the area of the whole shape.

   c Show that the expressions are equivalent.

Solve problems

Work out the expressions missing from the blank bricks.

Working backwards:

In the bottom row, the middle brick is $2x + 5 - (x + 3) = 2x + 5 - x - 3$
$= x + 2$

In the middle row, the right brick is $4x + 11 - (2x + 5) = 4x + 11 - 2x - 5$
$= 2x + 6$

In the bottom row, the right brick is $2x + 6 - (x + 2) = 2x + 6 - x - 2$
$= x + 4$

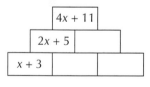

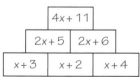

16 Work out the expressions missing from the blank bricks.

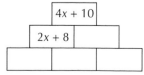

17 Work out the expressions missing from the blank bricks.

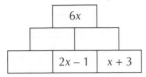

18 For each question, mark any necessary lengths on your diagram.
  a Sketch a rectangle with area abc.
  b Sketch a rectangle with perimeter $6x + 4y$.
  c Sketch a triangle with area $8x$.
  d Sketch a rectangle with area $5x + 10$.

25.5 Using index notation

● I can write algebraic expressions involving powers

You have already used some **powers** such as $x \times x = x^2$ and $y \times y \times y = y^3$.

You can have powers larger than 3.

$a \times a \times a \times a = a^4$

This is a to the power 4. The number 4 is called the **index**.

Develop fluency

Write each expression in index form.

  a $5 \times 5 \times 5 \times 5$ **b** $m \times m \times m$ **c** $t \times t^2$

  d $n \times 2n \times 4n$ **e** $x \times 2y \times x$

  a $5 \times 5 \times 5 \times 5 = 5^4$

  b $m \times m \times m = m^3$

  c $t \times t^2 = t \times t \times t$ Because $t^2 = t \times t$.
           $= t^3$

  d $n \times 2n \times 4n = 2 \times 4 \times n \times n \times n$ You can change the order. Put the numbers together.
                   $= 8n^3$ $2 \times 4 = 8$ and $n \times n \times n = n^3$.

  e $x \times 2y \times x = 2 \times x \times x \times y$ Put the number in front.
           $= 2x^2y$ Leave out the $\times$ signs.

  1 Write each expression in index form.
    a $4 \times 4 \times 4$ **b** $3 \times 3 \times 3 \times 3 \times 3 \times 3$ **c** $10 \times 10 \times 10 \times 10 \times 10$

  2 Write each of these out in full.
    a 2^4 **b** 5^3 **c** 6^5 **d** 9^6

3 Write these numbers as a power of the base 2.
The first one has been done for you.
   a $8 = 2^3$ **b** 16   **c** 32   **d** 64

4 Write each expression in index form.
   a $a \times a \times a \times a$ **b** $r \times r \times r$ **c** $b \times b \times b \times b \times b$
   d $m \times m \times m \times m \times m \times m$ **e** $4a \times 3a$ **f** $p \times 2p$
   g $2g \times 3g \times 2g$ **h** $k \times 4 \times 2k \times k \times 3k$

5 Write each expression as simply as possible.
   a $f + f + f + f + f$ **b** $w \times w \times w \times w$ **c** $c + c + c + c + c + c + c$
   d $k \times k \times k \times k$ **e** $D + D + D + D + D + D$

6 Write each expression as simply as possible.
   a $a \times b \times a$ **b** $x \times y \times y \times 6$ **c** $t \times u \times t \times u$ **d** $d \times 2d \times c$
   e $a \times 2b \times b$ **f** $w \times x \times 2w$ **g** $2t \times 2t \times u$ **h** $21e \times 3f \times 4e$

7 Write each expression as simply as possible.
   a $c^2 \times c$ **b** $5c \times 3c^4$ **c** $c^3 \times c^3$ **d** $9c^3 \times 7c^3$

8 Write each expression as simply as possible.
   a $t \times t^3$ **b** $2t \times 3t^3$ **c** $t^2 \times t^2$ **d** $5t^2 \times 6t^2$

9 Expand the brackets. Write your answer as simply as possible.
   a $4w(5w - w^2)$ **b** $7x^2(x + 2x)$ **c** $8yz(3y^2 - 5z)$

10 Write each expression as simply as possible.
   a $2ab^3 \times 5ab^5$ **b** $(4c^2)^3$ **c** $(2d^2e)^4$

Reason mathematically

Jo says that $5j$ and j^5 are the same.

Is she correct?

Give a reason for your answer.

She is not correct because $5j = j + j + j + j + j$ but $j^5 = j \times j \times j \times j \times j$.

11 Explain the difference between $6w$ and w^6.

12 The formula for the area of a triangle is $A = \frac{1}{2}b \times h$
where b is the base and h is the height.

The area of this triangle is $\frac{1}{4}b^2\,\text{cm}^2$.
What does this tell you about the height?

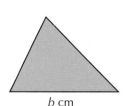

b cm

13 Show that 2^6 has the same value as 4^3.

14 Fill in the missing term.

$2a^4 \times 3a^2 \times 4a \times \ldots = 48a^9$

Solve problems

Work out an expression for the volume of this cube.

Volume is $2x \times 2x \times 2x = 8x^3 \text{ cm}^3$

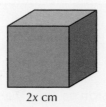

2x cm

15 Work out an expression for the volume of each cuboid.

a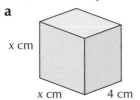

x cm

x cm   4 cm

b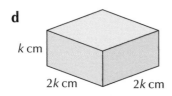

y cm

y cm   y cm

c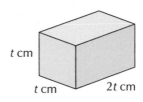

t cm

t cm   2t cm

d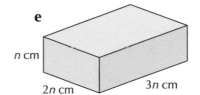

k cm

2k cm   2k cm

e

n cm

2n cm   3n cm

16 You are given that $2^{10} = 1024$.
Work out the value of

a 2^9 **b** 2^{11}

17 Joan says that to simplify $x^a \times x^b$, you just multiply the powers so the answer is x^{ab}.
Give an example, using numbers, to show that she is not correct.

Now I can...

simplify algebraic expressions involving the four basic operations	manipulate algebraic expressions	write algebraic expressions involving powers
simplify algebraic expressions by combining like terms	identify equivalent expressions	
remove brackets from an expression		

CHAPTER 26 Working with fractions

26.1 Adding and subtracting fractions

● I can add or subtract any two mixed numbers

Before you can add or subtract two fractions you need to make sure they are written with the same number in the denominator. Remember to give an answer in its simplest terms.

Develop fluency

Work these out.

a $4\frac{1}{2} + 2\frac{2}{3}$

b $3\frac{1}{4} - \frac{7}{12}$

a $4\frac{1}{2} + 2\frac{2}{3} = \frac{9}{2} + \frac{8}{3}$ Write the mixed numbers as improper fractions.

$\qquad = \frac{27}{6} + \frac{16}{6}$ 6 is a multiple of 2 and 3 so change both fractions to sixths.

$\qquad = \frac{43}{6}$ Add the numerators. The denominator stays the same.

$\qquad = 7\frac{1}{6}$ $43 \div 6 = 7$ remainder 1

b $3\frac{1}{4} - \frac{7}{12} = \frac{13}{4} - \frac{7}{12}$

$\frac{39}{12} - \frac{7}{12} = \frac{32}{12}$ Change $\frac{13}{4}$ into twelfths and then subtract.

$\qquad = \frac{8}{3}$ Divide numerator and denominator by 4 to find an equivalent fraction.

$\qquad = 2\frac{2}{3}$

1 Work these out.

   a $\frac{3}{4} + \frac{1}{8}$ **b** $\frac{3}{4} + \frac{1}{6}$ **c** $\frac{3}{4} - \frac{2}{3}$ **d** $\frac{3}{4} - \frac{7}{10}$

2 Work these out.

   a $1\frac{1}{2} + 2\frac{3}{4}$ **b** $\frac{7}{8} + 2\frac{1}{4}$ **c** $3\frac{3}{4} + 1\frac{5}{6}$ **d** $6\frac{1}{4} + 2\frac{2}{3}$

3 Work these out.

   a $4\frac{1}{2} - 2\frac{3}{4}$ **b** $2\frac{7}{8} - 2\frac{1}{4}$ **c** $4\frac{1}{4} - 1\frac{5}{8}$ **d** $6\frac{3}{4} - 6\frac{2}{3}$

4 Work out **a** the sum **b** the difference of these two fractions. $\frac{7}{12}$ $\frac{3}{8}$

5 Work these out.

a $1\frac{1}{2} + 2\frac{3}{4} + 3\frac{5}{8}$ **b** $2\frac{3}{4} + 1\frac{1}{2} + 3\frac{1}{3}$ **c** $2\frac{2}{3} + 1\frac{1}{6} + 3\frac{5}{9}$

6 Work out the missing numbers in these calculations.

a $4\frac{1}{3} + \dots = 8$ **b** $2\frac{3}{10} + \dots = 5\frac{9}{10}$ **c** $1\frac{2}{3} + \dots = 4\frac{1}{2}$

7 Work these out.

a $1\frac{1}{2} + 2\frac{3}{4} - 1\frac{1}{8}$ **b** $4\frac{3}{5} - 2\frac{1}{2} - 1\frac{1}{10}$ **c** $2\frac{1}{2} + 1\frac{1}{4} + \frac{5}{8}$

8 Work out the missing fraction in each of these calculations.

a $\frac{7}{8} + \frac{4}{5} + ? = 2$ **b** $2\frac{3}{5} + 3\frac{1}{4} + ? = 6$ **c** $1\frac{5}{6} + ? - 2\frac{1}{3} = 3\frac{7}{8}$

9 $x = 3\frac{3}{4}$ and $y = 2\frac{2}{5}$

Work these out.

a $x + y$ **b** $x - y$ **c** $2x + y$ **d** $2x - y$

Reason mathematically

a Work out the lengths marked a and b in this shape.

b Work out the perimeter of the shape

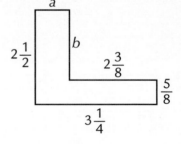

a $a = 3\frac{1}{4} - 2\frac{3}{8} = \frac{7}{8}$ $b = 2\frac{1}{2} - \frac{5}{8} = 1\frac{7}{8}$

b The perimeter is the distance around the edge of the shape.

$\frac{7}{8} + 1\frac{7}{8} + 2\frac{3}{8} + \frac{5}{8} + 3\frac{1}{4} + 2\frac{1}{2} = 11\frac{1}{2}$

10 Work out the perimeter of this triangle.

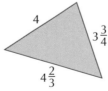

11 The perimeter of this triangle is $13\frac{1}{2}$. What is the length of the base of the triangle?

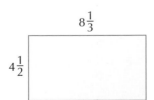

12 **a** Work out the difference between the length and the width of this rectangle.

b Work out the perimeter of the rectangle.

13 A quadrilateral has a perimeter of $12\frac{3}{4}$.

The lengths of three of the sides are $4\frac{1}{2}$, $2\frac{1}{8}$ and $3\frac{1}{4}$.

What is the length of the fourth side?

Solve problems

Look at these two equations.

$$4\frac{5}{16} + x = 7\frac{3}{8}$$

$$x - y = 2\frac{1}{3}$$

Work out the value of y.

In the first equation, $x = 7\frac{3}{8} - 4\frac{5}{16}$

$$= \frac{59}{8} - \frac{69}{16}$$

$$= \frac{118}{16} - \frac{69}{16}$$

$$= \frac{49}{16}$$

$$= 3\frac{1}{16}$$

Using $x = 3\frac{1}{16}$ in the second equation,

$$3\frac{1}{16} - y = 2\frac{1}{3}.$$

So, $y = 3\frac{1}{16} - 2\frac{1}{3}$

$$= \frac{49}{16} - \frac{7}{3}$$

$$= \frac{147}{48} - \frac{112}{48}$$

$$= \frac{35}{48}$$

14 Here are four numbers.

$$2\frac{2}{3} \qquad 2\frac{1}{6} \qquad 2\frac{1}{4} \qquad 2\frac{5}{12}$$

Which two fractions have a **a** sum of $5\frac{1}{2}$? **b** difference of $\frac{1}{12}$?

15 Look at this sum. $3\frac{5}{6} + 1\frac{2}{9} = 5\frac{1}{18}$

Use the sum to write down the answer to each of these calculations.

a $5\frac{1}{18} - 1\frac{2}{9}$ **b** $4\frac{5}{6} + 2\frac{2}{9}$ **c** $1\frac{5}{6} + 6\frac{2}{9}$ **d** $5\frac{1}{18} - 3\frac{2}{9}$

16 A plumber has $1\frac{5}{6}$ m of pipe. He needs to cut a piece measuring $\frac{7}{8}$ m from it.
How much pipe is left over?

17 A jug has $1\frac{3}{4}$ litres of lemonade in it. Semma pours $\frac{5}{8}$ litres into one glass and $\frac{1}{3}$ litre into another glass.
How much lemonade is left in the jug?

26.2 Multiplying fractions

● I can multiply two fractions

When you are multiplying fractions, the order of the multiplication does not matter.

This is the same as when you multiply integers. For example, $2 \times 3 = 6$ and $3 \times 2 = 6$.

Similarly, $\frac{2}{3} \times \frac{2}{5} = \frac{4}{15}$ and $\frac{2}{5} \times \frac{2}{3} = \frac{4}{15}$.

To multiply two fractions, you multiply the numerators and multiply the denominators. There is no need to find a common denominator.

To multiply a fraction by an integer, you multiply the numerator by the integer and leave the

denominator unchanged. For example, $200 \times \frac{3}{4} = \frac{600}{4} = 150$.

Often the word 'of' is used instead of $\times$. So, $\frac{2}{3}$ of $\frac{2}{5}$ and $\frac{2}{3} \times \frac{2}{5}$ mean the same.

Remember always to give an answer in its lowest terms.

Develop fluency

Work these out. **a** $\frac{3}{4}$ of $\frac{1}{2}$ **b** $\frac{2}{3} \times \frac{3}{5}$

a $\frac{3}{4}$ of $\frac{1}{2} = \frac{3}{4} \times \frac{1}{2}$

$= \frac{3 \times 1}{4 \times 2}$ The numerator is 3×1 and the denominator is 4×2.

$= \frac{3}{8}$

b $\frac{2}{3} \times \frac{3}{5} = \frac{6}{15}$ $2 \times 3 = 6$ and $3 \times 5 = 15$

$= \frac{2}{5}$ Simplify the fraction as much as possible.

1 Work these out.

a $2 \times \frac{3}{8}$ **b** $3 \times \frac{3}{4}$ **c** $3 \times \frac{4}{5}$ **d** $4 \times \frac{3}{8}$

e $\frac{1}{5}$ of 3 **f** $\frac{4}{5}$ of 3 **g** $\frac{2}{3}$ of 4 **h** $\frac{5}{6}$ of 2

2 Work these out.

a $\frac{1}{2} \times \frac{1}{3}$ **b** $\frac{1}{2} \times \frac{3}{5}$ **c** $\frac{1}{3} \times \frac{1}{4}$ **d** $\frac{1}{3} \times \frac{2}{3}$

3 Work out $\frac{2}{3}$ of:

a $\frac{1}{2}$ **b** $\frac{3}{4}$ **c** $\frac{2}{3}$ **d** $\frac{4}{5}$ **e** $\frac{1}{8}$ **f** $\frac{5}{6}$

4 Work these out.

a $\dfrac{2}{3} \times \dfrac{1}{4}$ **b** $\dfrac{3}{5} \times \dfrac{3}{4}$ **c** $\dfrac{4}{9} \times \dfrac{3}{8}$ **d** $\dfrac{4}{5} \times \dfrac{5}{12}$

5 Work these out.

a $\dfrac{1}{2} \times \dfrac{1}{2}$ **b** $\dfrac{2}{3} \times \dfrac{2}{3}$ **c** $\dfrac{3}{5} \times \dfrac{3}{5}$ **d** $\dfrac{3}{4} \times \dfrac{3}{4}$

6 Work these out.

a $\dfrac{5}{8} \times \dfrac{4}{5}$ **b** $\dfrac{1}{10} \times \dfrac{5}{6}$ **c** $\dfrac{7}{8} \times \dfrac{2}{3}$ **d** $\dfrac{5}{6} \times \dfrac{3}{4}$

7 Work these out.

a $\left(\dfrac{1}{4}\right)^2$ **b** $\left(\dfrac{3}{4}\right)^2$ **c** $\left(\dfrac{2}{3}\right)^2$ **d** $\left(\dfrac{4}{5}\right)^2$

8 Work these out.

a $\left(\dfrac{1}{2} + \dfrac{1}{4}\right) \times \dfrac{3}{5}$ **b** $\left(\dfrac{2}{3} + \dfrac{1}{6}\right) \times \dfrac{3}{10}$ **c** $\left(\dfrac{1}{2} - \dfrac{1}{6}\right) \times \dfrac{3}{4}$

9 $s = \dfrac{5}{8}$ and $t = \dfrac{2}{3}$

Work these out.

a st **b** ts **c** s^2 **d** t^2

Reason mathematically

Zafar has raised £2400 for some charities. He is going to give $\dfrac{3}{8}$ of the money to charity A, $\dfrac{7}{16}$ to charity B and the remainder of the money to charity C.

How much money does each charity receive?

Charity A: $\dfrac{3}{8}$ of £2400 $= \dfrac{3}{8} \times 2400$

$= \dfrac{3}{8} \times \dfrac{2400}{1}$

$= \dfrac{7200}{8}$

$= £900$

Charity B: $\dfrac{7}{16}$ of £2400 $= \dfrac{7}{16} \times 2400$

$= \dfrac{7}{16} \times \dfrac{2400}{1}$

$= \dfrac{16800}{16}$

$= £1050$

Charity C: £2400 – £900 – £1050 = £450

10 a Belinda uses $\frac{1}{3}$ of a tin of polish every time she cleans her car.

How many tins has she used after cleaning the car 12 times?

b Trevor spends one quarter of an hour every day tidying his room.
How many hours does he spend tidying his room in four weeks?

11 Here is a multiplication. $\frac{5}{12} \times \frac{4}{15} = \frac{1}{9}$

Use this result to work these out.

a $\frac{5}{12} \times \frac{8}{15}$ **b** $\frac{5}{12} \times \frac{2}{15}$ **c** $\frac{5}{6} \times \frac{4}{15}$

12 $\frac{4}{5} \times \frac{3}{4} \times \frac{2}{3} \times \frac{1}{2} = \frac{24}{120} = \frac{1}{5}$ $\frac{5}{6} \times \frac{4}{5} \times \frac{3}{4} \times \frac{2}{3} \times \frac{1}{2} = \frac{120}{720} = \frac{1}{6}$

a Show that if you extend the series by placing $\frac{6}{7} \times$ at the front of the calculation, the answer will be $\frac{1}{7}$.

b Without doing the calculation, what is the answer to the following:

$$\frac{9}{10} \times \frac{8}{9} \times \frac{7}{8} \times \frac{6}{7} \times \frac{5}{6} \times \frac{4}{5} \times \frac{3}{4} \times \frac{2}{3} \times \frac{1}{2}$$

13 Tom has an amount of money that he wants to share among his grandchildren. He gives half of his money to his daughter, Jenny, and the other half to his son, Mike. He asks Jenny and Mike to share the money he has given them evenly among their children. Jenny has three children and Mike has two children. Tom assumes that each of his grandchildren will be given $\frac{1}{5}$ of the original amount.

Explain why Tom is wrong to assume this.

Solve problems

An area of land is being split up into allotments. The land is $\frac{1}{4}$ km wide and $\frac{2}{5}$ km long. Each allotment is $\frac{1}{400}$ of the total area size.

How many square metres is each allotment?

The size of the land is $\frac{1}{4} \times \frac{2}{5} = \frac{2}{20} = \frac{1}{10}$ km².

1 km² = 1000 m × 1000 m = 1 000 000 m²

Each allotment will be $\frac{1}{400} \times 100\,000 = 250$ m²

14 $\frac{1}{2} \times \frac{a}{b} = \frac{3}{8}$

Work out the values of a and b.

15 A poster uses red, blue and yellow inks. Of the ink used, $\frac{2}{9}$ is red and $\frac{1}{6}$ is blue.

    a To print the poster, 36 ml of ink is used. Calculate the amount of each ink used.

    b Another poster uses 6 ml of red ink. How much ink is used altogether?

16 Jamie and Ollie are plastering three walls of a rectangular living room. The area is 30 m² in total.

    a Jamie does $\frac{3}{5}$ of the total and Ollie does the rest.
      How many square metres does each person do?

    b The boss pays them a total of £90 divided in the ratio of the amount they complete. How much does each person get?

    c The area of the fourth wall in the room is 5 m². What fraction of the room did they plaster?

    d What fraction of the whole room did Jamie plaster?

17 1 mile = $1\frac{3}{5}$ km

Lorna is driving from London to Edinburgh, a journey of 405 miles.

    a She passes a road sign which tells her she has 360 miles to go. How many kilometres is she from Edinburgh?

    b What fraction of the journey has she completed?

26.3 Multiplying mixed numbers

● I can multiply one mixed number by another

To multiply two mixed numbers, you need to change them to **improper fractions**. Then you can multiply the numerators and multiply the denominators. Remember to simplify your answers.

Develop fluency

Work these out.

a $\frac{2}{3}$ of $3\frac{1}{2}$

a $\frac{2}{3}$ of $3\frac{1}{2} = \frac{2}{3} \times \frac{7}{2}$ Change $3\frac{1}{2}$ to an improper fraction.

             $= \frac{14}{6}$ $2 \times 7 = 14$ and $3 \times 2 = 6$

             $= \frac{7}{3}$ Simplify the fraction by dividing numerator and denominator by 2.

             $= 2\frac{1}{3}$ Convert the answer to a mixed number.

b $1\frac{3}{4} \times 2\frac{1}{2}$

b $1\frac{3}{4} \times 2\frac{1}{2} = \frac{7}{4} \times \frac{5}{2}$ Change both mixed numbers to improper fractions.

$= \frac{35}{8}$

$= 4\frac{3}{8}$ Convert the answer to a mixed number.

1 Work these out.

  a $\frac{1}{2} \times 1\frac{1}{2}$ **b** $\frac{1}{2} \times 1\frac{1}{4}$ **c** $\frac{1}{4} \times 2\frac{1}{2}$ **d** $\frac{1}{4} \times 3\frac{1}{2}$

2 Work these out.

  a $1\frac{2}{3} \times 2\frac{1}{2}$ **b** $1\frac{1}{4} \times 3\frac{1}{2}$ **c** $2\frac{1}{3} \times 2\frac{1}{2}$ **d** $4\frac{1}{2} \times 2\frac{2}{3}$

3 Multiply $2\frac{2}{3}$ by:

  a $1\frac{2}{3}$ **b** $2\frac{1}{3}$ **c** $1\frac{3}{4}$ **d** $5\frac{1}{2}$

4 Work these out.

  a $\left(\frac{3}{4}\right)^2$ **b** $\left(2\frac{1}{4}\right)^2$ **c** $\left(4\frac{1}{4}\right)^2$ **d** $\left(1\frac{3}{4}\right)^2$

5 Work these out.

  a $2\frac{2}{3} \times 3\frac{3}{4}$ **b** $3\frac{3}{4} \times 3\frac{1}{5}$ **c** $1\frac{2}{5} \times 2\frac{4}{7}$ **d** $2\frac{4}{9} \times 4\frac{1}{3}$

6 Work these out.

  a $2\frac{1}{5} \times 6$ **b** $4\frac{3}{4} \times 7$ **c** $3\frac{2}{3} \times 4$ **d** $5\frac{2}{7} \times 5$

7 Work these out.

  a $\frac{2}{3} \times \frac{3}{5} \times \frac{1}{2}$ **b** $1\frac{2}{3} \times \frac{3}{5} \times 4$ **c** $\left(2\frac{1}{2}\right)^2$ **d** $\left(\frac{2}{3}\right)^3$

Reason mathematically

Peter has been given this calculation to do: $3\frac{3}{5} \times 4\frac{2}{3}$

Here is his working out:

$$3\frac{3}{5} \times 4\frac{2}{3} = 3 \times 4\frac{3}{5} \times \frac{2}{3}$$

$$= 12\frac{6}{15}$$

$$= \frac{72}{15}$$

$$= 4\frac{4}{5}$$

However, the actual answer is $16\frac{4}{5}$.

a Explain how you know, without making the calculation, that Peter has made a mistake.

b What mistake has Peter made?

a Looking at the whole-number parts, 3 and 4, the answer should be at least $3 \times 4 = 12$, so an answer of 4 is obviously wrong.

b Peter has used the wrong method. His method would work if he wanted to add the two fractions. In multiplication calculations, the mixed numbers cannot be separated into whole numbers and fractions.

8 Rana and Ana have each worked out the answer for $4\frac{2}{7} \times 5\frac{1}{4}$.

Rana has an answer of $12\frac{3}{7}$ and Ana $22\frac{13}{28}$.

a Without doing the calculation whose answer is more likely to be correct? Why?

b What is the correct answer?

9 a Work out the value of x^2 when x is equal to:

    i $1\frac{1}{2}$ **ii** $1\frac{1}{3}$ **iii** $1\frac{1}{4}$ **iv** $1\frac{2}{5}$

b Which answer in part a is closest to 2?

10 a Work out the value of x^2 when x is equal to:

    i $1\frac{1}{5}$ **ii** $1\frac{1}{6}$ **iii** $1\frac{1}{7}$ **iv** $1\frac{1}{8}$

b What number are the answers to this series getting closer to?

c Will any value of x^2 be less than 1? Why?

11 Does $\left(\frac{3}{4}\right)^2 \times \left(1\frac{1}{2}\right)^2 \times 2^2$ have the same value as $\left(\frac{3}{4} \times 1\frac{1}{2} \times 2\right)^2$?

Explain your reasoning.

Solve problems

A half marathon is run over a distance of $13\frac{1}{10}$ miles. On one particular course, four drinks stations are located at $\frac{1}{5}$, $\frac{2}{5}$, $\frac{3}{5}$ and $\frac{4}{5}$ of the race distance.
At what distances from the race start are the drinks stations?

Station 1: $13\frac{1}{10} \times \frac{1}{5} = 2\frac{31}{50}$ miles   Station 2: $13\frac{1}{10} \times \frac{2}{5} = 5\frac{6}{25}$ miles

Station 3: $13\frac{1}{10} \times \frac{3}{5} = 7\frac{43}{50}$ miles   Station 4: $13\frac{1}{10} \times \frac{4}{5} = 10\frac{12}{25}$ miles

12 In cricket each session usually lasts 2 hours.

   a A player who bats for $2\frac{3}{4}$ sessions has batted for how many minutes?

   b A bowler is used for a quarter of one session, $\frac{2}{3}$ of the next and half of the final one. How many minutes was he bowling for?

   c Due to bad weather one session is extended to last $1\frac{2}{5}$ of its normal time. How long is that?

13 a Work these out.

   i $\frac{1}{5} \times 1\frac{1}{2}$ ii $\frac{1}{5} \times 2\frac{1}{2}$ iii $\frac{1}{5} \times 3\frac{1}{2}$ iv $\frac{1}{5} \times 4\frac{1}{2}$

   b The multiplications in part **a** follow a pattern.
      Write down and calculate the next two terms in the pattern.

14 Konika is doing some design work. She shades all the squares in her first rectangle purple and then part of a second identical rectangle purple.

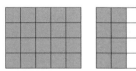

   a Counting one rectangle as a whole one, what fraction of the two rectangles has been shaded?

   b Konika decides to change a quarter of the shaded squares to yellow. What fraction of a rectangle is yellow?

   c What fraction is still purple?

15 a Work these out.

   i $1\frac{1}{2} \times \frac{2}{3}$ ii $1\frac{1}{3} \times \frac{3}{4}$ iii $2\frac{1}{2} \times \frac{2}{5}$ iv $1\frac{3}{4} \times \frac{4}{7}$

   b Write down two more multiplications like those in part **a**.

26.4 Dividing fractions

● I can divide one fraction or mixed number by another

To divide by a fraction, you **invert** it or 'turn it upside down'. This means that you swap the numerator and the denominator. You then multiply by the new fraction.

If the calculation includes mixed numbers, change them to improper fractions first.

Develop fluency

Work out: **a** $3 \div \dfrac{2}{3}$ **b** $\dfrac{2}{3} \div \dfrac{3}{4}$

a $3 \div \dfrac{2}{3} = \dfrac{3}{1} \times \dfrac{3}{2}$

$\phantom{3 \div \dfrac{2}{3}} = \dfrac{3 \times 3}{1 \times 2}$

$\phantom{3 \div \dfrac{2}{3}} = \dfrac{9}{2}$

$\phantom{3 \div \dfrac{2}{3}} = 4\dfrac{1}{2}$

Write 3 as $\dfrac{3}{1}$ and invert $\dfrac{2}{3}$ to get $\dfrac{3}{2}$.

b $\dfrac{2}{3} \div \dfrac{3}{4} = \dfrac{2}{3} \times \dfrac{4}{3}$

$\phantom{\dfrac{2}{3} \div \dfrac{3}{4}} = \dfrac{8}{9}$

Invert $\dfrac{3}{4}$ but do not change $\dfrac{2}{3}$

Work these out.   **a** $2\dfrac{1}{4} \div 1\dfrac{1}{3}$ **b** $1\dfrac{1}{2} \div 3\dfrac{1}{2}$

a $2\dfrac{1}{4} \div 1\dfrac{1}{3} = \dfrac{9}{4} \div \dfrac{4}{3}$

Write the mixed numbers as improper fractions.

$\phantom{2\dfrac{1}{4} \div 1\dfrac{1}{3}} = \dfrac{9}{4} \times \dfrac{3}{4}$

Invert $\dfrac{4}{3}$ and multiply.

$\phantom{2\dfrac{1}{4} \div 1\dfrac{1}{3}} = \dfrac{27}{16}$

$\phantom{2\dfrac{1}{4} \div 1\dfrac{1}{3}} = 1\dfrac{11}{16}$

Write the improper fraction as a mixed number.

b $1\dfrac{1}{2} \div 3\dfrac{1}{2} = \dfrac{3}{2} \div \dfrac{7}{2}$

Write the mixed number as improper fractions.

$\phantom{1\dfrac{1}{2} \div 3\dfrac{1}{2}} = \dfrac{3}{2} \times \dfrac{2}{7}$

Invert $\dfrac{7}{2}$ and multiply.

$\phantom{1\dfrac{1}{2} \div 3\dfrac{1}{2}} = \dfrac{6}{14}$

$\phantom{1\dfrac{1}{2} \div 3\dfrac{1}{2}} = \dfrac{3}{7}$

Simplify the fraction as much as possible

Work these out.

1   **a** $3 \div \frac{1}{2}$ **b** $3 \div \frac{1}{3}$ **c** $3 \div \frac{2}{5}$ **d** $3 \div \frac{3}{5}$

2   **a** $2 \div \frac{1}{4}$ **b** $5 \div \frac{2}{3}$ **c** $12 \div \frac{3}{4}$ **d** $2 \div \frac{5}{12}$

3   **a** $\frac{1}{2} \div \frac{1}{4}$ **b** $\frac{1}{2} \div \frac{1}{3}$ **c** $\frac{1}{2} \div \frac{1}{5}$ **d** $\frac{1}{2} \div \frac{1}{8}$

4   **a** $\frac{1}{2} \div \frac{2}{5}$ **b** $\frac{1}{4} \div \frac{3}{4}$ **c** $\frac{2}{5} \div \frac{1}{10}$ **d** $\frac{1}{6} \div \frac{2}{3}$

5   **a**   **i** $\frac{1}{4} \div \frac{1}{3}$ **ii** $\frac{1}{3} \div \frac{1}{4}$ **b**   **i** $\frac{1}{3} \div \frac{1}{6}$ **ii** $\frac{1}{6} \div \frac{1}{3}$

6   **a** $2\frac{1}{2} \div \frac{1}{2}$ **b** $2\frac{1}{2} \div \frac{1}{4}$ **c** $2\frac{1}{2} \div \frac{3}{4}$ **d** $2\frac{1}{2} \div \frac{5}{8}$

7   **a** $1\frac{1}{2} \div \frac{2}{3}$ **b** $3\frac{1}{4} \div \frac{3}{4}$ **c** $4\frac{1}{2} \div \frac{2}{3}$ **d** $7\frac{1}{2} \div \frac{3}{5}$

8   **a** $4\frac{1}{2} \div \frac{1}{2}$ **b** $6\frac{1}{2} \div 1\frac{1}{2}$ **c** $1\frac{1}{2} \div 4\frac{1}{2}$ **d** $3\frac{1}{2} \div 4\frac{1}{2}$

9   **a** $2\frac{1}{2} \div 1\frac{3}{4}$ **b** $12\frac{1}{2} \div 7\frac{1}{2}$ **c** $3\frac{1}{2} \div 4\frac{2}{3}$ **d** $10\frac{1}{2} \div 4$

10 **a** $\left(\frac{2}{5} \div \frac{1}{2}\right) \div \frac{1}{3}$ **b** $\frac{2}{5} \div \left(\frac{1}{2} \div \frac{1}{3}\right)$

Reason mathematically

  a   Work out: **i** $\frac{2}{5} \div 10$ **ii** $10 \div \frac{2}{5}$

  b   Explain why these answers are different.

  a   **i** $\frac{1}{25}$ **ii**   25

  b   The first asks you to find $\frac{1}{10}$ of $\frac{2}{5}$ (because '÷ 10' means $\frac{1}{10}$), which is $\frac{1}{10} \times \frac{2}{5} = \frac{1}{25}$.

      The second is asking, 'How many $\frac{2}{5}$ are there in 10?', which is 25 because $25 \times \frac{2}{5} = 10$.

11 **a**   Work out

     i $5 \div \frac{1}{2}$ and $\frac{1}{2} \div 5$ **ii** $4 \div \frac{1}{3}$ and $\frac{1}{3} \div 4$ **iii** $8 \div \frac{1}{10}$ and $\frac{1}{10} \div 8$

  b   What did you notice each time?

  c   When you divide an integer by a fraction, is the answer bigger or smaller?

  d   When you divide a fraction by an integer, is the answer bigger or smaller?

12 What do you notice about the pairs of answers in question 5?

13 **a** Calculate $5\frac{1}{4} \div 2\frac{1}{3}$.

   b Check your calculation in part **a** by multiplying the answer by $2\frac{1}{3}$.

Solve problems

Max walks $2\frac{1}{2}$ miles from home to school in 45 minutes.

Kate cycles $3\frac{1}{2}$ miles from home to school in 28 minutes.

   a How many minutes does it take Max to walk 1 mile

   b How many minutes does it take Kate to cycle 1 mile?

   c How many times further away from school is Kate's home than Max's home?

a Max: minutes for 1 mile $= 45 \div 2\frac{1}{2}$

$$= 45 \div \frac{5}{2}$$

$$= 45 \times \frac{2}{5}$$

$$= \frac{90}{5}$$

$$= 18$$

b Kate: minutes for 1 mile $= 28 \div 3\frac{1}{2}$

$$= 28 \div \frac{7}{2}$$

$$= 28 \times \frac{2}{7}$$

$$= \frac{56}{7}$$

$$= 8$$

c Kate's house is $3\frac{1}{2} \div 2\frac{1}{2}$ times further from school than Max's:

$$\frac{7}{2} \div \frac{5}{2} = \frac{7}{2} \times \frac{2}{5}$$

$$= \frac{14}{10}$$

$$= 1\frac{2}{5}$$

14 Ali has just discovered that, at the equator, a point on the Earth's surface is spinning at 1060 mph on average, which is $1\frac{2}{3}$ times faster than it would be in London.

This is because London is further north and has a smaller circle to travel in the same time.

   a How fast are you spinning if you live in London?

   b Kieran says this should be in km/h, so it needs to be divided by $\frac{5}{8}$.

   However, Hussain says that's wrong and you should multiply by $1\frac{3}{5}$.

   Who is right and how fast is London spinning in km/h?

15 A bucket contains $15\frac{1}{2}$ litres of milk. 6 small bottles each holding $\frac{7}{8}$ litre are filled with milk from the bucket. The remaining milk is poured into bottles each holding $1\frac{1}{2}$ litres. How many of the $1\frac{1}{2}$ litre bottles can be filled?

16 A garden patio is being laid. Its length is $5\frac{3}{4}$ m and its width is $2\frac{1}{2}$ m.

The tiles to be used are square with a side length $\frac{1}{4}$ m .

How many tiles will be needed to cover the patio?

Now I can...

add any two mixed numbers	multiply two fractions	divide one fraction or mixed number by another
subtract any two mixed numbers	multiply one mixed number by another	

CHAPTER 27 Circles

27.1 Parts of a circle

● I can define a circle and name its parts

A **circle** is a set of points that are all the same distance from a fixed point, called the centre. The centre of the circle is usually called O.

Develop fluency

a Draw a semicircle with diameter 5 cm.

b Draw a quadrant of a circle with radius 30 mm.

a *Draw a line 5 cm long, for the diameter.*

Mark a dot at the midpoint, for the centre.

Use compasses to draw the semicircle with radius 2.5 cm.

b *Draw a line 3 cm long, for the radius.*

Construct a right angle at one end and draw a 3 cm line, perpendicular to the original line.

Use your compasses to draw the arc with radius 3 cm to create the quadrant.

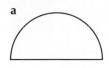

1 a A circle has a radius of 6 cm. What is the length of its diameter?

   b A circle has a diameter of 30 m. What is the length of its radius?

2 i Measure the radius of each circle, giving your answer in centimetres.

   ii Write down the diameter of each circle.

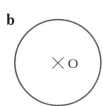
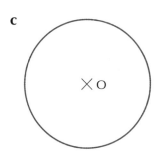

3 Draw circles with these measurements.

   a radius = 2.5 cm   **b** radius = 3.6 cm   **c** diameter = 8 cm   **d** diameter = 6.8 cm

4 a Draw a circle with radius 33 mm.   **b** Draw a circle with diameter 9.2 cm.

   c Draw a semicircle with diameter 8 cm.   **d** Draw a quadrant of a circle with radius 52 mm.

5 Construct these diagrams.

a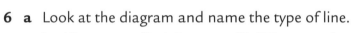

b

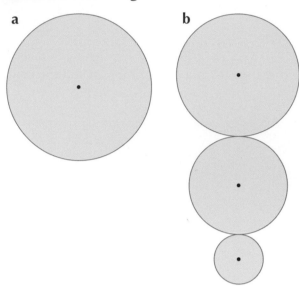

6 a Look at the diagram and name the type of line.

    i AB   ii AC   iii DE   iv CF

b Name the type of shape.

    i ABCA   ii ACFA   iii ABCFA   iv GABCG

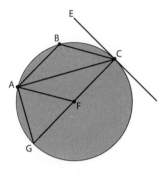

7 Draw each of these shapes accurately. Use a ruler, compasses and a protractor.

a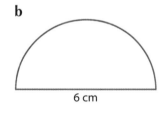

Concentric circles

b

6 cm

Semicircle

c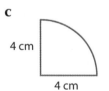

4 cm

4 cm

Quadrant of
a circle

d

60°

Sector of
a circle

8 Draw each of these shapes accurately.

a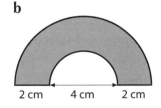

4 cm

4 cm

b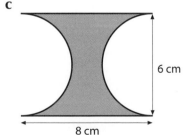

2 cm   4 cm   2 cm

c

6 cm

8 cm

9 Construct these diagrams accurately.

a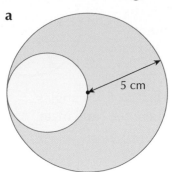

5 cm

b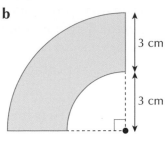

3 cm

3 cm

c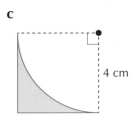

4 cm

10 Match each word on the left with its correct definition on the right

a	Circle	A	The distance from the centre of a circle to its circumference.
b	Arc	B	The distance across a circle, through its centre.
c	Semicircle	C	A set of points all the same distance from a fixed point.
d	Sector	D	A straight line that joins two points on the circumference of a circle.
e	Circumference	E	A straight line that touches a circle at only one point on its circumference.
f	Diameter	F	A portion of a circle that is enclosed by a chord and an arc.
g	Tangent	G	The length round the outside of the circle.
h	Radius	H	A portion of a circle enclosed by two radii and an arc.
i	Segment	I	A part of the circumference of the circle.
j	Chord	J	One half of a circle; either of the parts cut off by a diameter.

Reason mathematically

Draw a circle with centre O and a radius of 5 cm.

Use a protractor to draw three radii that form angles of 120° at the centre of the circle, as shown in the diagram.

Join the three points where the radii meet the circumference, to make a triangle.

a Explain why the triangle is equilateral.

b Explain how you could use a similar method to draw a square.

a The three radii are all the same length, with the same angle, 120° between each radii. This gives us three congruent shapes. Each chord is the same length and so each side of the triangle will be the same, hence the triangle will be equilateral.

b Draw a circle. Then draw four radii that form angles of 90° at the centre of a circle. Now join the four points where the radii meet the circumference to make a square.

11 **a** Draw a circle with centre O and a radius of 4 cm.

Use a protractor to draw six radii that form angles of 60° at the centre of the circle, as shown in the diagram. Join the six points where the radii meet the circumference, to make a regular hexagon.

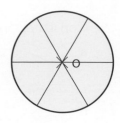

b Use a similar method to draw a regular pentagon.

12 Amber draws a circle. She uses a protractor to draw eight radii that form angles of 45° at the centre of the circle. She joins the eight points where the radii meet the circumference.

a What shape has she drawn with these eight points?

b Explain how you could use a similar method to draw a decagon.

13 Explain how you could use a circle to help you draw a regular twenty-sided polygon.

14 Theo draws a circle, then uses a protractor to draw radii that form angles of 20° at the centre of the circle. He says: 'When I join the points where the radii meet the circumference, I will have drawn a regular polygon with 20 sides.'
Is Theo correct? Explain your answer.

Solve problems

Draw a circle. Then draw two tangents to the circle that meet each other at a point T.

Label the points of the tangents that meet the circle A and B.

Measure the lengths AT and BT. What can you say about these lengths?

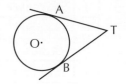

Measuring the two lengths AT and BT shows that they are equal.

15 **a** Draw a circle around a circular object so that you don't know where the centre is.

b Draw a chord AB on the circle.

c Mark a point halfway along AB, label it X.

d Draw a line perpendicular to AB at X so that the line passes though the circle.

e What name can we give to the part of the perpendicular line that passes through the circle?

16 **a** Find a circular object with an unknown centre. Draw a circle around it.

b Draw two chords AB and BD on the circle.

c Mark a point halfway along each chord, label them X and Y.

d Draw perpendicular lines to the chords at X and Y so that the lines pass right across the circle.

e What is special about the point where these two perpendicular lines cut each other?

17 Draw a circle. Then draw two tangents to the circle that meet at a point T.
Label the points of the tangents that meet the circle as A and B.

   a Draw perpendicular lines to the tangents at A and B.

   b Where do these two perpendicular lines meet?

18 Eve draws two tangents to a circle, centre O, that meet point T.
The points of the tangents that meet the circle are A and B.
What can you say about the triangles AOT and BOT?

27.2 Formula for the circumference of a circle

● I can calculate the circumference of a circle

The formula for calculating the circumference, C, of a circle with diameter d is written as $C = \pi d$.

As the diameter is twice the radius, r, the circumference is also given by the formula:

$C = \pi d = \pi \times 2r = 2\pi r$

These formulae include a special number represented by the Greek letter π (pronounced pi). It is impossible to write down the value of π exactly, as a fraction or as a decimal, so you will use approximate values. The most common of these are:

• $\pi = 3.14$ (as a decimal rounded to two decimal places)

• $\pi = 3.142$ (as a decimal rounded to three decimal places)

• $\pi = 3.141\ 592\ 654$ (on a scientific calculator)

• $\pi = \dfrac{22}{7}$ (as a fraction).

Develop fluency

Calculate the circumference of each circle. Give each answer correct to one decimal place.

a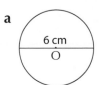

b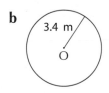

   a The diameter $d = 6$ cm, which gives:

   $C = \pi d = \pi \times 6 = 18.8$ cm (to 1 dp)

   b The radius $r = 3.4$ m, so $d = 6.8$ m.

   This gives $C = \pi d = \pi \times 6.8 = 21.4$ m (to 1 dp)

In this exercise, take $\pi = 3.14$ or use the π key on your calculator.

1 Find the circumference of a circle with:

   a diameter 10 m   b radius 4 m.

2 Calculate the circumference of each circle. Give each answer correct to one decimal place.

a

7 cm
O

b

11 mm
O

c

21 mm
O

d

O
2.4 m

e

1.4 cm
O

3 Calculate the circumference of each wheel.
Write your answers correct to the nearest centimetre.

a

60 cm

b

17.2 cm

4 Calculate the circumference of each coin.
Give each answer to one decimal place.

a

14 mm
1998

b
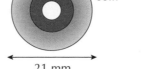
Car wash coin
21 mm

5 Copy and complete the table for each circle (using suitable values of π).

Radius	Diameter	Circumference
7 cm		
	5 cm	
		314

6 Calculate the perimeter, *P*, of this semicircle. Give your answer correct to one decimal place.

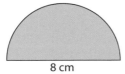
8 cm

7 Calculate the perimeter of this quadrant, taken from a circle radius 4 cm.

O

Give your answer to one decimal place.

8　A circle, centre O, radius 5 cm has three radii drawn as shown.
Calculate the perimeter of the shape AOB, giving your answer to one decimal place.

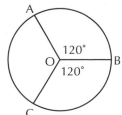

9　Calculate the perimeter of this sector of a circle radius 14 cm.
Give your answer as an integer.

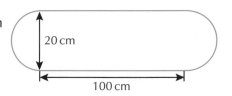

10　The diagram is made up of a rectangle 20 cm by 100 cm with a semicircle of diameter 20 cm at each end.
Calculate the perimeter of the shape, giving your answer to the nearest integer.

Reason mathematically

Mae told Ben the perimeter of her circular table is 66 cm.
Ben said: 'The diameter of the table will be 21 cm.'

Could Ben be correct? Explain what assumption Ben has made.

Yes Ben is correct if he assumed π to be $\frac{22}{7}$, as $\frac{22}{7} \times 21 = 66$

11　The Earth's orbit can be taken as a circle with radius approximately 150 million kilometres.
Calculate the distance the Earth travels in one orbit of the Sun.
Give your answer correct to the nearest million kilometres.

12　The distance round a circular running track is 200 m. Calculate the radius of the track.
Give your answer correct to the nearest metre.

13　The London Eye has a diameter of 120 m.
How far would you travel in one complete revolution of the wheel?
Give your answer correct to the nearest metre.

14　Calculate the total length of the lines in this crop circle diagram.
It has two semicircles, a circle and straight lines.
Write your answers correct to the nearest metre.

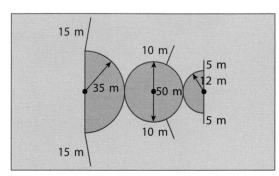

Solve problems

The diagram shows the dimensions of a running track at a sports centre. The bends at the ends are semicircles.

Calculate the distance round the track. Give your answer to the nearest metre.

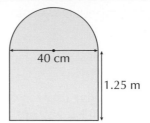

The circumference of circle made up of the semi-circular ends is $\pi \times 51 = 160$ m.

Add the two straight sections to give $160 + 240 = 400$ m

15 Calculate the total perimeter of this arched window.
Give your answer correct to the nearest centimetre.

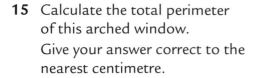

16 The radius of a quadrant is 20 mm. Calculate the perimeter in millimetres. Give your answer correct to three significant figures.

17 The curved parts of this shape are all semicircles. Calculate the perimeter of the shape. Give your answer correct to one decimal place.

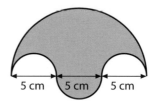

18 The world's tallest Ferris wheel is the Singapore Flyer, opened in 2008.
Its diameter is 150 m and it has 28 capsules, spread equally around the circumference.
Kamile gets into a capsule.
Ewa gets into the next capsule.
Find the distance travelled by Kamile before Ewa gets into her capsule, correct to the nearest metre.

27.3 Formula for area of a circle

● I can calculate the area of a circle

The formula for the area, A, of a circle of radius r is $A = \pi r^2$.

Develop fluency

Calculate the area of each circle. Give your answers correct to one decimal place.

a 3 cm O

b 3.4 m O

a The radius, $r = 3$ cm, which gives:

$A = \pi r^2 = \pi \times 3^2$

$= 9\pi$

$= 28.3 \text{ cm}^2$ (to 1 dp)

When using a calculator, you can use the 'square' key $\boxed{x^2}$.

Simply key in: $\boxed{\pi}$ $\boxed{\times}$ $\boxed{3}$ $\boxed{x^2}$ $\boxed{=}$

b The diameter $d = 3.4$ m, so $r = 1.7$ m. This gives:

$A = \pi r^2 = \pi \times 1.7^2$

$= 2.89\pi$

$= 9.1 \text{ m}^2$ (to 1 dp)

Note that you can also leave your answers in terms of π.

For example, this may be necessary when you are not allowed to use a calculator.

This gives answers of **a** 9π and **b** 2.89π.

In this exercise, take $\pi = 3.14$ or use the π key on your calculator.

1 Find the area of a circle with:

    a diameter 10 m   **b** radius 4 m.

2 Calculate the area of each of the following circles.
Give each answer to one decimal place.

a 8 cm O

b 7 cm O

c 9 cm O

d 3.6 cm O

3 Calculate the area of each circle. Give your answers correct to one decimal place.

a   **b**   **c**   **d**   **e**

1 cm / O   14 mm / O   2.1 m / O   3.5 cm / O   O / 5.5 m

4 Calculate the area of a circular tablemat with a diameter of 21 cm.
Give your answer correct to the nearest square centimetre.

5 A CD has a diameter of 12 cm.
Calculate its circumference and area. Give your answers to one decimal place.

6 The diagram is a quadrant of a circle with radius 5 cm.
Calculate the area of the quadrant. Give your answers to one decimal place.

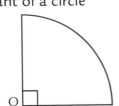

7 This shape is a sector from a circle radius 9 cm.
Calculate the area of the shape. Give your answers to one decimal place.

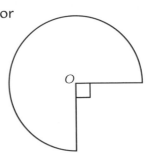

8 Show that the area of this semicircle is 8π cm^2.

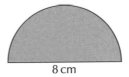

8 cm

9 Calculate the area of a semicircle with radius 10 cm. Give your answer in terms of π.

10 Calculate the area of a quadrant with radius 12 cm. Give your answer in terms of π.

Reason mathematically

Laz has a circular plate with diameter 12 cm. His brother Theo has a square plate of side length 10 cm. Theo says to Laz: 'My square plate has a larger area than your circular plate.'

Is Theo correct? Explain your answer.

Laz's circular plate has area of $\pi \times 6^2 = 3.14 \times 36 = 113.04$ cm^2

Theo's square plate has an area of $10^2 = 100$ cm^2

So, Theo is incorrect.

11 Ollie has a circular table with a diameter of 30 cm.
He says the area of this table is 707 cm.
Is Ollie correct? Explain your answer.

12 Andrew has a pizza with diameter 14 cm.
He says the area of the top is 154 cm² exactly.
What assumption must he make to be correct? Explain your answer.

13 Randal is working out the area of this circle.
This is his working.

Area = π × d = π × 8 = 8π cm

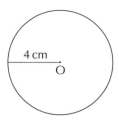

4 cm
O

Explain why Randal's working is wrong.
Write down the correct answer to the problem.

14 Finlay has cut out a circle of radius 5 cm and Jackson has cut out a circle of radius 10 cm.
Jackson says to Finlay, 'My circle is twice the size of yours, so the area of my circle should
be twice the area of yours.'
Calculate the areas of both circles, giving your answers in terms of π.
Is Jackson correct? Give a reason for your answer.

Solve problems

Calculate the area of the shaded part of this shape.
The whole circle has area π × 5² = 3.14 × 25 = 78.5 cm²
The circle in the middle has area π × 1² = 3.14 × 1 = 3.14 cm²
So, the area of the shaded part is 78.5 − 3.14 = 75.36 cm²

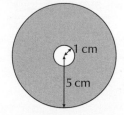

1 cm
5 cm

15 The diagram represents a sports ground.
The curved ends are semicircles. Calculate
the area of the sports ground. Give your
answer correct to the nearest square metre.

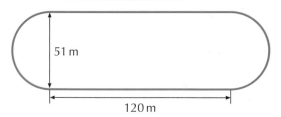

51 m
120 m

16 The length of the minute hand on a clock is 24 cm.
Calculate the area swept by the minute hand in:
a 1 hour   **b** 5 minutes   **c** 1 minute.
Give your answers in terms of π.

17 Calculate the area of this shape.
Write your answer correct to the nearest square centimetre.

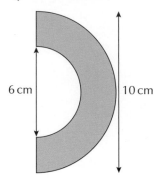

6 cm   10 cm

18 Find the area of this shape. Give your answer in terms of π.

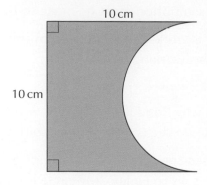

10 cm

10 cm

Now I can...

define a circle	calculate the circumference of a circle	calculate the area of a circle
name the parts of a circle		

CHAPTER 28 Finding probabilities

28.1 Using probability scales

● I can use a probability scale to represent a chance

When you do something such as rolling a dice, this is called an **event**.

The possible results of the event are called its **outcomes**. For example, rolling a dice has six possible outcomes: scoring 1, 2, 3, 4, 5 or 6.

You can use **probability** to decide how likely it is that different outcomes will happen.

Equally likely outcomes are those that all have the same chance of happening. For example, when you roll a dice, there are six different possible outcomes. This is because it could land so that any one of its six numbers shows on top.

The probability of an equally likely outcome is:

$$P(outcome) = \frac{\text{the number of ways that the outcome could occur}}{\text{the total number of possible outcomes}}$$

Probabilities can be written as either fractions, decimals or percentages. They always take values from 0 to 1. The probability of an event happening can be shown on the probability scale.

| 0 | 0.1 | 0.2 | 0.3 | 0.4 | 0.5 | 0.6 | 0.7 | 0.8 | 0.9 | 1 |

Impossible   Even   Certain

If one outcome is the absolute opposite of another outcome, such as 'raining' and 'not raining', then the probabilities of the two outcomes add up to 1.

Similarly, if the outcomes of an event do not overlap the probabilities add up to 1, for example, if the result of a game is win, lose or draw, then P(win) + P(lose) + P (draw) = 1.

Develop fluency

a What is the probability of scoring a number less than 5 when you roll a dice?

b What is the probability of not scoring a number under 5 when you roll a dice?

a There are four possible outcomes that give you a number less than 5: 1, 2, 3 and 4.
There are six different possible outcomes altogether, when you roll a dice: 1, 2, 3, 4, 5 and 6.

So. P(rolling a dice and getting a number less than 5) is $\frac{4}{6} = \frac{2}{3}$.

b These two outcomes are the opposite of each other, so the probabilities add up to 1.

P(number not less than 5) is $1 - \frac{2}{3} = \frac{1}{3}$.

1 A set of cards is numbered from 1 to 50.
One card is picked at random. Give the probability that the number on it:

    a is even   **b** has a 7 in it   **c** has at least one 3 in it

    d is a prime number   **e** is a multiple of 6   **f** is a square number.

2 Joe has 1000 tracks on his phone. He has:

250 tracks of pop   200 tracks of blues   400 tracks of country and western
100 tracks of heavy rock 50 tracks of quiet romantic.

He sets the player to play tracks at random.

What is the probability that the next track to play is:

a pop
b blues
c country and western
d heavy rock
e quiet romantic
f not heavy rock?

3 The probability of outcomes A, B, C and D are shown on the scale. Copy the scale and mark underneath it the probabilities of A, B, C and D not happening.

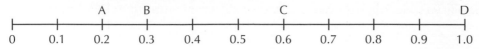

4 Copy and complete each of these tables.

Outcome	Probability of outcome occurring (P)	Probability of event not occurring (1 – P)
A	$\frac{1}{4}$	
B	$\frac{1}{3}$	
C	$\frac{3}{4}$	
D	$\frac{1}{10}$	
E	$\frac{2}{15}$	

Outcome	Probability of outcome occurring (P)	Probability of event not occurring (1 – P)
A	$\frac{2}{3}$	
B	0.35	
C	8%	
D	0.04	
E	$\frac{5}{8}$	
F	0.375	

5 A spinner has four coloured sections: blue, red, green and yellow.

The probability of landing on some of the sections is shown in the table.

Colour	Blue	Red	Green	Yellow
Probability	0.2	0.3	0.1	

Work out the probability of landing on yellow.

6 In a bus station there are 24 red buses, 6 blue buses and 10 green buses. Work out the probability that the next bus to leave is:

a green
b red
c red or blue
d yellow
e not green
f not red
g neither red nor blue
h not yellow.

7 The probability of an egg having a double yolk is 0.009. What is the probability that an egg does not have a double yolk?

8 100 rings are placed in a box. Ten are gold, 20 are silver, 36 are plastic and the rest are copper. A ring is chosen at random. What is the probability that it is:

    a gold   **b** not silver   **c** copper   **d** not plastic?

9 The diagram shows 12 dominoes.
A domino is chosen at random.
Work out the probability that it:

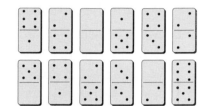

    a has a 4   **b** does not have a 3
    c has a total over 5   **d** totals less than 10
    e has a total of 9   **f** is a double totalling less than 5.

Reason mathematically

A bag contains 32 counters. Some are black and the others are white.

The probability of picking a black counter is $\frac{1}{4}$.

Show that there are 24 white counters in the bag.

The probability of picking a white counter is
$1 - \frac{1}{4} = \frac{3}{4}$.
So, the number of white counters is
$\frac{3}{4}$ of $32 = \frac{3}{4} \times 32 = 24$,

Alternatively:
Number of black counters is $\frac{1}{4}$ of $32 = \frac{1}{4} \times 32 = 8$
So, the number of white counters is $32 - 8 = 24$

10 In a raffle there are 2400 tickets.
The probability that a ticket is a winning ticket is $\frac{1}{100}$.
Show that there are 2376 losing tickets.

11 A child is asked to choose a number at random from:
1 2 3 4 5 6 7 8 9
For each part, compare the probabilities, stating which, if any, is more likely.

    a even number or number more than 6   **b** prime number or odd number
    c multiple of 5 or multiple of 4   **d** triangular number or square number.

12 In a bag there are green, blue, red and yellow discs.
Jack says this means the probability of picking a green disc is $\frac{1}{4}$ because green is one of out of four colours. Is his statement correct? Give a reason for your answer.

13 A box contains blue and black pens.
A pen is picked at random. There are more blue pens than black pens. What does this tell you about the probability a choosing a blue pen?

Solve problems

A bag contains 10 blue and 15 red discs.

a A disc is picked at random and then replaced.
What is the probability the disc is blue?

b More red discs are added to the bag.
The probability of picking a blue disc is now $\frac{1}{3}$.
How many red discs were added to the bag?

a P(blue) is $\frac{10}{25} = \frac{2}{5}$.

b More red discs are added so the total number in the bag will increase so that $\frac{10}{total} = \frac{1}{3}$.
So, the total number must be 30, meaning 5 red discs were added.

14 There are two white counters and one black counter in a bag.

a One counter is picked and replaced.
What is the probability that it is a black counter?

b More black counters are put in the bag.
The probability that a black counter is picked at random is now $\frac{2}{3}$.
How many more black counters were put in the bag?

15 In a box of cereal there is a free gift of a model dinosaur.
There are five animals to make up the set:
Tyrannosaurus Rex, Stegosaurus, Triceratops, Brachiosaurus, Diplodocus.
You cannot tell which animal will be in the box. Each one is equally likely.

a What is the probability that the animal is a diplodocus?

b What is the probability that the animal is a stegosaurus or a triceratops?

The box is opened. The animal is not a brachiosaurus.

c Now what is the probability that the animal is a diplodocus?

d Now what is the probability that the 1 is a stegosaurus or a triceratops?

16 The probability of Bel winning a match is $\frac{1}{2}$.
She says that this means the probability of losing is also $\frac{1}{2}$.
What assumption has she made?

28.2 Mutually exclusive outcomes

● I can recognise mutually exclusive outcomes

Mutually exclusive outcomes cannot occur together. Each excludes the possibility of the other happening. For example, when you roll a dice, throwing a 1 or a 6 are mutually exclusive as you cannot get both results on the same throw of one dice.

However, suppose you have a dice and are trying to throw numbers less than 4, but you also want to score an even number. Which number is common to both outcomes?

- The numbers less than 4 are 1, 2 and 3.
- The even numbers are 2, 4 and 6.

Because the number 2 is in both groups, the outcomes are not mutually exclusive. This means it is possible to achieve both outcomes at the same time if you throw a 2.

You can use a **Venn diagram** to illustrate this.

The two circles show the sets of the possible outcomes: 'less than 4' and 'even'.

In this diagram: ● the set 'less than 4' has 1, 2 and 3 in it
● the set 'even numbers' has 2, 4 and 6 in it.

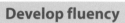

The numbers that satisfy either or both outcomes together are in the **union** of the sets. The numbers that can satisfy either outcome are in the **intersection** of the sets. Because there are some numbers in the intersection, this shows that the outcomes are not mutually exclusive.

Notice that all the possible outcomes of the dice roll are included in the Venn diagram. The number 5 does not form part of either outcome, so it is positioned outside the circles.

Develop fluency

Liz is buying fruit. Here is a list of possible outcomes.

A: She chooses strawberries.   B: She chooses red fruit   C: She chooses green apples
D: She chooses red apples   E: She chooses oranges

She chooses one item only. State which pairs of outcomes are mutually exclusive.

a A and B   **b** A and E   **c** B and C   **d** B and D

a Strawberries are red fruit, so they are not mutually exclusive.

b Strawberries are not oranges, so they are mutually exclusive.

c Green apples are not red fruit, so they are mutually exclusive.

d Red apples are red fruit, so they are not mutually exclusive.

1 These are the home shirt colours of the teams in a football league:
   7 are red, 4 are blue, 3 are white, 1 is blue and white stripes
   and 1 is black and white stripes.

   a Are red shirts and blue shirts mutually exclusive?

   b Are striped shirts and shirts containing blue mutually exclusive?

2 Which of these are mutually exclusive?
   A: Picking an odd number   B: Picking a prime number   C: Picking an even number

3 Which of these are mutually exclusive?
   A: Picking a square number   B: Picking a prime number   C: Picking a triangular number

4 A card is picked at random from a normal pack of cards. A pack has two red suits and two black suits. Each suit contains 13 cards, ranging from values 1 (ace) to 10 and then jack, queen and king, known as picture cards.

   a Is picking a red card and picking a picture card mutually exclusive?

   b Is picking a picture card and picking a number card mutually exclusive?

   c Is picking a number above 6 and picking a black card mutually exclusive?

5 In a game you need to roll a dice and score an odd number larger than 2.

   a Draw a Venn diagram showing the two sets 'odd numbers' and 'numbers larger than 2'.

   b Are the outcomes 'scoring an odd number' and 'scoring a number larger than 2' mutually exclusive?

   c Use your Venn diagram to state the probability of rolling an odd number larger than 2.

6 Soolin has a bag containing ten cards, each showing one of the integers 1 to 10. She is playing a game and needs to select a card at random that shows a prime number smaller than 4.

   a Draw a Venn diagram showing the two sets 'prime numbers' and 'numbers smaller than 4'.

   b Are the outcomes 'selecting a prime number' and 'selecting a number smaller than 4' mutually exclusive?

   c Use your Venn diagram to state the probability of selecting a card showing a prime number smaller than 4.

7 A number square contains the numbers from 1 to 100.
   Numbers are chosen from the number square. Here is a list of outcomes.
   A: The number chosen is greater than 50.
   B: The number chosen is less than 10.
   C: The number chosen is a square number (1, 4, 9, 16, …).
   D: The number chosen is a factor of 100 (1, 2, 5, 10, …).
   E: The number chosen is a prime number (2, 3, 5, 7, …).
   State whether the outcomes in each pair are mutually exclusive or not.
   a A and B   b A and C   c B and C   d B and D   e C and D   f C and E

8 The diagram shows sketches of eight faces.
   Which pairs of outcomes are mutually exclusive?

   a Having a smiling face and a sad face.

   b Smiling and having both eyes shut.

   c Wearing a hat and having both eyes open.

   d Having both eyes open and a sad face.

   e Wearing a hat and having one eye shut.

   f Smiling and having one eye shut.

Reason mathematically

a Draw a Venn diagram with the two sets 'square numbers up to 50' and 'numbers less than 10'.

b Which numbers are in both sets?

c Are the outcomes 'selecting a square number' and 'selecting a number smaller than 10' mutually exclusive? Explain your answer.

a
Square numbers up to 50		Numbers less than 10

16 25 36 49 | 1 4 9 | 2 3 5 6 7 8

b 1, 4 and 9 are in both sets.

c Not mutually exclusive as 1, 4 and 9 are in both sets.

9 In a game you need to roll a dice and get an odd number less than 5 to win.

a Draw a Venn diagram for the two sets 'odd numbers' and 'numbers less than 5'.

b Are the outcomes 'rolling an odd number' and 'rolling a number less than 5' mutually exclusive? Explain your answer.

10 Give an example to show that multiples of 5 and multiples of 7 are not mutually exclusive.

11 a Copy and complete the table to show all the possible pairs of scores if you spin these two spinners.

Spinner 1	Spinner 2	Total score
+2	0	2
+2	−1	1

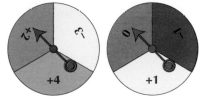

b Which two outcomes are mutually exclusive?
Give a reason for your answer

A: A score of +4 on the 1st spinner   B: A score of −1 on the 2nd spinner

C: A total score of 1   D: A total score of 3

12 These are some outcomes from rolling a dice.

A: Rolling a 5 or a 6   B: Rolling an odd number

C: Rolling a 4 or a 6   D: Rolling a number less than 6

E: Rolling a number greater than 1

a Which two outcomes are mutually exclusive?
Give a reason for your answer.

b For each of the other pairs of outcomes, give a reason why they are not mutually exclusive.

Solve problems

An ordinary six-sided dice is rolled.

One possible outcome is scoring an odd number.

List three different outcomes that are mutually exclusive to getting an odd number.

For example: an even number, a 2, a number greater than 5.

13 Ten cards are numbered 1 to 10
    A card is picked at random. One

| 1 | 2 | 3 | 4 | 5 | 6 | 7 | 8 | 9 | 10 |

possible outcome it to pick the card with 5 on it.
List three different outcomes that are mutually exclusive with picking the card with 5 on it.

14 A spinner has three sections red, blue and yellow. There are four possible outcomes.
    A: Landing on red   B: landing on blue   C: landing on yellow   D: Not landing on red
    How many pairs of these outcomes are mutually exclusive?

15 A dice has six faces numbered 1, 2, 3, 4, 5 and 6. Here are four possible outcomes.
    A: Rolling a 1   B: Rolling a 6   C: Rolling a square number   D: Rolling a prime number
    How many pairs of these outcomes are mutually exclusive?

28.3 Using sample spaces to calculate probabilities

● I can use sample spaces to calculate probabilities

To help you work out the probabilities of outcomes happening together you can use a table or diagram called a **sample space**, which is the set of all possible outcomes from a specific event.

You can use a Venn diagram or a table to illustrate a sample space.

Develop fluency

Find the probability of getting a head and a 6 when you roll a dice and flip a coin at the same time.

This sample space shows all the possible outcomes of flipping a coin and rolling a dice at the same time.

	1	2	3	4	5	6
Head	H, 1	H, 2	H, 3	H, 4	H, 5	H, 6
Tail	T, 1	T, 2	T, 3	T, 4	T, 5	T, 6

You can now work out the probability of getting both a head and a 6.

$$P(\text{outcome}) = \frac{\text{the number of ways that the outcome could occur}}{\text{the total number of possible outcomes}}$$

$$P(\text{head and a six}) = \frac{1}{12}$$

In a class of 32 students, there are 15 boys. Out of the 8 left-handed students in the class, 6 are girls.

What is the probability of choosing a right-handed boy from the class?

You could put this information into a table.

	Boy	Girl	Total
Left-handed	2	6	8
Right-handed	13	11	24
Total	15	17	32

You can also show this in a Venn diagram.

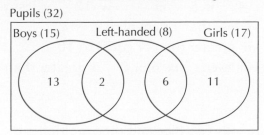

You can now see that:

$P(\text{right-handed boy}) = \dfrac{13}{32}$

1 **a** Draw a sample space to show the results of flipping a coin and rolling a dice at the same time.

   b Use your sample space to find the probability of getting:
   i a 3 and a tail   **ii** a head and an even number   **iii** a number less than 5 and a tail.

2 Two tetrahedral (four-sided) dice, each numbered 1, 2, 3 and 4, are rolled together. Use a sample space diagram to show all the possible totals when the numbers are added.

3 Two children, Kim and Franz, have a bag containing 1p, 2p, 5p, 10p, 20p, 50p and £1 coins. They each write their name on a card and put the cards in the same bag. A card and a coin are taken from the bag at random. The coin is given to the named person.

   a Make a table to show the possible outcomes.

   b Calculate the probability that:
   i Kim receives 20p   **ii** Franz receives less than 10p   **iii** one of them receives 10p
   iv Kim does not receive £1   **v** neither receives more than 20p.

4 A market trader sells jacket potatoes plain, with cheese or with beans. Clyde and Delroy each buy a jacket potato.

   a Copy and complete the sample space table.

   b Write down the probability of:
   i   Clyde choosing plain
   ii   Delroy choosing plain
   iii both choosing plain
   iv   Clyde choosing plain and Delroy choosing beans
   v   Clyde choosing beans and Delroy choosing cheese
   vi   both choosing the same
   vii neither choosing plain
   viii each choosing a different flavour.

Clyde	Delroy
plain	plain
plain	cheese

5 Bret rolls two dice and adds the scores together.
   Copy and complete the sample space of his scores.

   a What is the most likely total?

   b Write down the probability that the total is:
      i 4   ii 5   iii 1   iv 12   v less than 7
      vi less than or equal to 7
      vii greater than or equal to 10
      viii even   ix 6 or 8   x greater than 5.

	1	2	3	4	5	6
1	2	3				
2	3					

6 Bart rolls two dice and then multiplies the results together to give a score.

   a Draw the sample space of his scores.

   b Which is the most likely to occur?

   c What is the probability of rolling a score greater than 17?

7 An ordinary dice is rolled at the same time as this spinner is spun. Their numbers are
   added together to give the total score.

   a Make a list of all the possible outcomes.
      (Axes like the set to the right may help.)

   b How many possible outcomes are there?

   c Find the probability of scoring:
      i 8   ii 1   iii an even number
      iv any score other than 5   v a score higher than 6
      vi a total where the dice score is lower than the spinner score.

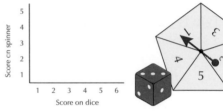

Reason mathematically

Tia throws two coins.

Gary says: 'There are three possible outcomes: 2 heads, head and tail or 2 tails, so
the probability of 2 heads is $\frac{1}{3}$.'

Is he correct?

Give a reason for your answer.

Listing the outcomes in a table shows that there are
four outcomes, so the probability of 2 heads is $\frac{1}{4}$.

He is incorrect.

		2nd coin	
		Head	Tail
1st coin	Head	HH	HT
	Tail	TH	TT

8 Three coins are thrown.

   By listing all possible outcomes show that the probability of three heads is $\frac{1}{8}$.
   The first three have been done for you.
   HHH   HHT   HTH

9 A bag contains apples, bananas and pears. Liam chooses two fruits at random.

   a List all the possible outcomes.

   b You are told that you have more chance of choosing an apple and pear than an apple and a banana. Explain how that could happen.

10 A combination lock has four digits. Each digit is 0, 1, 2, 3, 4, 5, 6, 7, 8 or 9. Mark remembers that:
- all the digits are different
- all the digits are odd
- the first digit is 3 and the last digit is 7.

Show that there are six possible combinations.

Solve problems

At a school fair there is a competition.

   a Draw a sample space diagram to show all the possible outcomes.

   b What is the probability of winning a prize?

Roll a double with 2 dice to win a prize.

a

	1	2	3	4	5	6
1	D	×	×	×	×	×
2	×	D	×	×	×	×
3	×	×	D	×	×	×
4	×	×	×	D	×	×
5	×	×	×	×	D	×
6	×	×	×	×	×	D

b P(double) is $\frac{6}{36} = \frac{1}{6}$

11 Bella spins two spinners together. Each spinner has the numbers 0, 1, 2 and 3 on it. The numbers the spinners land on are added to give a score.

   a Draw a sample space diagram to show all the possible scores.

   b What is the probability that the score is less than 2?

   c What assumption have you made in part **b**?

12 A café makes 100 paninis. All the paninis are either meat or cheese, but 75 of them have salad in them, as well.

30 paninis have cheese with salad.

15 paninis have meat with no salad.

Work out the probability of selecting at random a cheese panini without salad.

13 Two piles of three cards are each numbered 1 to 3. One card from each pile is turned over and the numbers are multiplied to give a score.

Which is more likely; an odd score or an even score?

Now I can...

use a probability scale to represent a chance	recognise mutually exclusive outcomes	use sample spaces to calculate probabilities

CHAPTER 29 Equations and formulae

29.1 Equations with and without brackets

● I can solve equations involving brackets

If an equation involves brackets you can either multiply out the brackets or divide each side first, to remove the need for the brackets.

Develop fluency

Solve these equations.

a $3x - 2 = 15$ **b** $3(x - 2) = 15$ **c** $\frac{1}{3}(x - 2) = 15$

a $3x - 2 = 15$

$\qquad 3x = 17$ Add 2 to both sides. $15 + 2 = 17$

$\qquad x = \frac{17}{3}$ Divide both sides by 3.

$\qquad = 5\frac{2}{3}$ $17 \div 3 = 5$ remainder 2.

Write the answer as a mixed number, because the decimal for $5\frac{2}{3}$ is 5.666 666… and you would need to round it.

b $3(x - 2) = 15$ There are two ways to solve this.

Method 1

$3(x - 2) = 15$

$\qquad x - 2 = 5$ Divide both sides by 3.
$\qquad\qquad\qquad\quad$ $15 \div 3 = 5$

$\qquad\quad x = 7$ Add 2 to both sides.

Method 2

$3(x - 2) = 15$

$\qquad 3x - 6 = 15$ Multiply out the brackets.
$\qquad\qquad\qquad\qquad$ $3(x - 2) = 3x - 6$

$\qquad\quad 3x = 21$ Add 6 to both sides.

$\qquad\quad\: x = 7$ Divide by 3: $21 \div 3 = 7$

Make sure you can use both of these methods.

c $\frac{1}{3}(x - 2) = 15$ Finding $\frac{1}{3}$ is the same as dividing by 3.

$\qquad x - 2 = 45$ Multiply both sides by 3. $15 \times 3 = 45$

$\qquad\quad x = 47$ Add 2 to both sides. $45 + 2 = 47$

The equation in part **c** could also be written as $\frac{x - 2}{3} = 15$ and solved.

Notice that multiplying out the brackets of $\frac{1}{3}(x - 2) = 15$ first would give $\frac{1}{3}x - \frac{2}{3} = 15$. Then you would have to deal with coefficients that are fractions. The method shown above is usually easier.

1 Copy and complete these equations, given that $x = 7$.

   a $4x + 5 = \ldots$ **b** $4(x + 5) = \ldots$ **c** $21(x + 5) = \ldots$ **d** $\dfrac{x + 5}{3} = \ldots$

2 Solve these equations. If the solution is not a whole number, write it as a mixed number.

   a $2x - 12 = 14$ **b** $5y + 4 = 49$ **c** $5k - 11 = 2$ **d** $8n + 3 = 40$

3 Solve these equations.

   a $2(x - 3) = 16$ **b** $4(x + 5) = 32$ **c** $3(y - 6) = 36$ **d** $3(a - 4) = 120$

4 Solve these equations.

   a $\dfrac{1}{4}x = 12$ **b** $\dfrac{1}{4}x + 3 = 12$ **c** $\dfrac{1}{4}(x + 3) = 12$ **d** $\dfrac{x + 4}{12} = 12$

5 Solve these equations.

   a $\dfrac{1}{2}y - 5 = 7$ **b** $\dfrac{1}{2}t + 4 = 13$ **c** $8 = \dfrac{1}{2}(t - 11)$ **d** $\dfrac{1}{8}(x + 10) = 4$

6 Look at this equation: $3(x - 5) = 11$

   a Solve the equation by multiplying out the brackets first.

   b Solve the equation by dividing by 3 first.

7 Solve these equations. Give the answers as mixed numbers.

   a $4x - 12 = 15$ **b** $4(x - 3) = 9$ **c** $2(y - 1\tfrac{1}{4}) = 11$ **d** $10r + 5 = 69$

8 Solve these equations. Write the answers as decimals.

   a $2x + 4.1 = 11.3$ **b** $1.4(w + 6.2) = 12.6$ **c** $\dfrac{t - 1.7}{3.2} = 4.5$ **d** $1.8t + 32 = 144.5$

9 Solve these equations.

   a $\dfrac{a - 2}{4} = 3$ **b** $\dfrac{b + 6}{2} = 5$ **c** $\dfrac{c + 9}{11} = 4$ **d** $\dfrac{d - 4}{7} = 5$

10 Solve these equations. Write the answers as decimals. Do not use a calculator.

   a $3.6p + 2.8 = 17.56$ **b** $\dfrac{2q - 1.4}{3} = 2.8$ **c** $5(w + 3.2) = 29.5$

Reason mathematically

This is Johnny's homework. Explain what is wrong with each of his solutions.

a $8x - 5 = 19$

   $8x = 19 - 5 = 14$

     $x = 14 \div 8$

       $= 1.75$

b $3(5x - 1) = 27$

   $15x - 1 = 27$

     $15x = 30$

       $x = 2$

a Second line error, should be

   $8x = 19 + 5 = 24$

    $x = 24 \div 8 = 3$

b Second line error, should be

   $5x - 1 = 27 \div 3 = 9$

   $5x = 9 + 1 = 10$

    $x = 2$

11 This is Helmut's homework. Find any errors and explain what is wrong.

   a $9x - 2 = 25$
     $9x - 2 = 25 + 2 = 3$

   b $3 + 4y = 19$
     $3 + 4y = 19$
     $4y = 19 + 3 = 22$
     $y = 5.5$

   c $2(2x - 3) = 14$
     $2(2x - 3) = 14$
     $4x - 3 = 14 \Rightarrow 4x = 17$
     $x = 4.25$

12 This is Jos's homework. Find any errors and explain what is wrong.

   a $4x + 3 = 9$
     $4x = 9 + 3 = 12$
     $x = 12 \div 4 = 3$

   b $4(2x + 3) = 24$
     $8x + 8 = 24$
     $8x = 16$
     $x = 2$

13 Ann helped her sister with this question.
Solve the equation $\dfrac{2(b + 2)}{5} = 4$.
Here is her solution $2(b + 2) = 9$
Divide both sides by 2 $b + 2 = 4.5$
                   So $b = 2.5$

Is Ann correct? Explain your answer.

14 Here is Peter's homework. He was asked
to solve the equation $\dfrac{6x - 3}{3} = 3$.
      $6x - 3 = 9$
      $6x = 12$
      $x = 2$

Is Peter correct? Explain your answer.

Solve problems

The perimeter of this rectangle is 15 cm.

   a Write down an equation to show this.

   b Solve the equation to find the value of x.

   a $x + x + 2(x + 1) + 3(x - 5) = 15$
              $7x - 13 = 15$

   b $7x = 28$
      $x = 4$

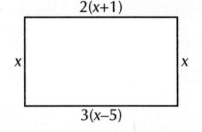

15 The perimeter of this trapezium is $37\frac{1}{2}$ cm.

   a Write down an equation to show this.

   b Solve the equation to find the value of x.

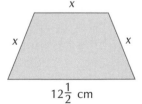

16 The formula for the perimeter of a rectangle is given by $P = 2(a + b)$.
   a and b are different positive integers.

   a Write down all the possible pairs of values for a and b when $P = 16$.

   b Write down all the possible values for P when the area of the rectangle is 30.

   c Find the value of a when $P = 356$ and $b = 29$.

   d Explain why the rectangle cannot have a perimeter of 35.

17 The perimeter of this pentagon is 24 cm.

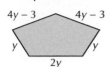

    a Write down an expression to show this.

    b Solve the equation to find the value of y.

18 The perimeter of this triangle is 11 cm.

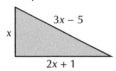

    a Write down an equation to show this.

    b Solve the equation to find the value of x.

29.2 Equations with the variable on both sides

● I can solve equations with the variable on both sides

Some equations have the unknown variable on one side only. It will often happen that the variable appears on both sides of the equation and then you need to bring the variables all to the same side in order to combine them.

Develop fluency

Jake is thinking of a number, 'If I add 12 to my number it is the same as tripling it and subtracting 45.' What is Jake's number?

Call Jake's number x. Then you can write this equation: $x + 12 = 3x - 45$

This equation has x on both sides. When the variable occurs on both sides of the equal sign, remove it from one side by adding or subtracting it on the other side.

    $x + 12 = 3x - 45$

First, subtract x from both sides.

    $12 = 2x - 45$ This removes the x from the left-hand side. On the right, $3x - x = 2x$.

Now x only appears on one side. Solve this in the usual way.

$57 = 2x$	Add 45 to both sides.	$12 + 45 = 57$
$28.5 = x$	Divide by 2.	Half of 57 is 28.5.

Solve these equations. **a** $2x - 4 = 5x - 24$ **b** $4y + 8 = 43 - y$

a $2x - 4 = 5x - 24$ Subtract $2x$ from both sides.

    $-4 = 3x - 24$ You now have -4 on the left-hand side. Add 24 to both sides.

    $20 = 3x$ $-4 + 24 = 20$. Now divide by 3.

    $x = \dfrac{20}{3}$

    $= 6\dfrac{2}{3}$ It is better to leave the answer as a fraction in this case.

b $4y + 8 = 43 - y$ To remove the $-y$ from the right-hand side, add y to both sides.

    $5y + 8 = 43$ $4y + y = 5y$. Solve this new equation in the usual way.

    $5y = 35$ Subtract 8 from both sides. $43 - 8 = 35$

    $y = 7$ Divide by 5 to find the value of y.

You can check this is correct.
When $y = 7$, then $4y + 8 = 4 \times 7 + 8 = 36$
and $43 - y = 43 - 7 = 36$.
They have the same value.

Solve these equations.

1 **a** $2x = x + 15$ **b** $4x = x + 45$ **c** $3t = t + 24$ **d** $5x = x + 44$

2 **a** $4x = 2x + 18$ **b** $5y = 2y + 18$ **c** $6t = 2t + 18$ **d** $6k = 5k + 18$

3 **a** $2y - 8 = y + 12$ **b** $4z - 17 = z + 4$ **c** $3x - 14 = x + 8$ **d** $6x - 13 = 2x + 9$

4 **a** $m = 3m - 16$ **b** $n = 5n - 84$ **c** $2p = 5p - 18$ **d** $4x = 10x - 48$

5 **a** $2x - 15 = x + 3$ **b** $3t - 7 = t + 14$ **c** $5x - 4 = 2x + 6$ **d** $n + 15 = 2n - 19$

  e $3 + 4x = 17 + 2x$ **f** $5d - 27 = 3d - 1$ **g** $x = 5x - 20$ **h** $2.5x + 6 = 3x - 4$

6 **a** $x = 24 - x$ **b** $2x = 24 - x$ **c** $3x = 24 - x$ **d** $5x = 24 - x$

7 **a** $x - 12 = 18 - x$ **b** $2x - 5 = 22 - x$ **c** $4x + 3 = 27 - 2x$ **d** $40 - 2x = 4 + x$

  e $3x - 1 = 19 - 2x$ **f** $x + 28 = 100 - 2x$ **g** $6x - 4 = 10 - x$ **h** $31 - 2\frac{1}{2}x = 16 + \frac{1}{2}x$

8 **a** $x + 11 = 3x + 2$ **b** $x + 2 = 11 - 3x$ **c** $x + 2 = 3x - 11$ **d** $11 - x = 3x - 2$

9 **a** $8y = 6y - 10$ **b** $6k - 6 = 9k$ **c** $-4r = 3r + 21$ **d** $7n - 3 = 3n - 15$

10 **a** $6 + 2x = x - 5$ **b** $6 + x = 5x - 2$ **c** $5x + 2 = 6 - x$ **d** $2 + x = 5 - 6x$

Reason mathematically

This is David's homework. Is he correct? Explain your answer.

Solve the equation $5x + 2 = 3x + 8$.

$5x + 2 = 3x + 8$	The second line is incorrect, it should be $5x - 3x = 8 - 2$
$5x - 3x = 8 + 2$	giving $2x = 6$
$2x = 10$	$x = 3$
$x = 5$	

11 This is Joel's homework. Is Joel correct? Explain your answer.

$4x - 3 = x + 9$

$4x - x = 9 - 3 = 6$

$3x = 6$

$x = 2$

12 This is Amy's homework. Explain what is wrong with each part.

  a $4x = 9 - 2x$ **b** $3(x - 3) = 4x$

  $4x - 2x = 9$ $3x - 9 = 4x$

  $2x = 9$ $9 = x$

13 Kathy helped her brother with this question.
Solve the equation $\frac{b+8}{3} = b$.
Here is her solution:

$b + 8 = 3b$

$b + 3b = 8$

$4b = 8$

$b = 2$

Is she correct? Explain your answer

14 Here is Jack's homework.
Solve $\frac{8x-5}{2} = 6x$.

$8x - 5 = 6x$

$8x - 6x = 5$

$x = 3$

Is Jack correct? Explain your answer.

Solve problems

Find the area of this rectangle.

The top length is equal to the bottom length, leading to the equation:

$5x - 1 = 3x + 4$

$5x - 3x = 4 + 1$

$2x = 5$

$x = 2.5$

Area of rectangle = height × base = $(2 \times 2.5 - 1) \times (3 \times 2.5 + 4)$

$= 4 \times 11.5 = 46$ square units

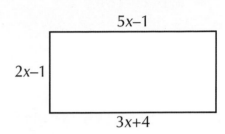

15 The perimeters of these shapes are all the same length.
a Write an equation to show this.
b Solve the equation.
c Work out the perimeter of each shape.

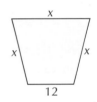

16 Simrath gets £8 pocket money per week. Ruby gets £10 pocket money per week.
Simrath spent £x of her pocket money on stamps.
a Write an expression for the amount of money Simrath had after she bought her stamps.
Ruby bought five times as many stamps as Simrath.
b Write an expression for the amount of money Ruby had after she bought her stamps.
Simrath and Ruby each had the same amount of money left after buying their stamps.
c Write and solve an equation to find out how much a stamp cost.
d How much money did Ruby have left after buying her stamps?

17 a Find the area of this rectangle.
b Find the area of this square.

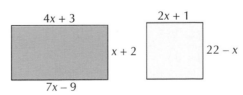

29.3 More complex equations

- I can solve equations with fractional coefficients
- I can solve equations with brackets and fractions

When you have brackets in an equation where the variable occurs more than once, it is usually best to multiply out the brackets first. When there is a fraction in front of a bracketed term, you can get rid of it by multiplying by the denominator.

Develop fluency

Solve the equation $3(t - 11) = 2(t + 8)$.

$3(t - 11) = 2(t + 8)$	There are two sets of brackets. Multiply out both of them.
$3t - 33 = 2t + 16$	Now subtract $2t$ from both sides.
$t - 33 = 16$	By subtracting $2t$ you remove the t-term from one side.
$t = 49$	Add 33 to both sides. $16 + 33 = 49$

You can check this is correct.

$3(49 - 11) = 3 \times 38 = 114$ and $2(49 + 8) = 2 \times 57 = 114$ so they are the same.

Solve the equation $\frac{2}{3}(x + 8) = x$.

$\frac{2}{3}(x + 8) = x$	First, multiply by 3.
$2(x + 8) = 3x$	Now multiply out the brackets.
$2x + 16 = 3x$	Now subtract $2x$ from both sides.
$16 = x$	$3x - 2x = x$

Solve these equations.

1 a $2(x - 4) = 10$ **b** $2(x - 4) = x$ **c** $2(x - 4) = x + 10$ **d** $2(x + 4) = x + 10$

2 a $3(y - 5) = 2y + 9$ **b** $a + 15 = 3(a - 1)$ **c** $2(t + 12) = 5t$
d $18 + x = 3(x - 8)$ **e** $2(n + 6) = 3n - 10$ **f** $3x = 2(20 - x)$

3 a $5x + 2 = 20$ **b** $5(x + 2) = 20$ **c** $5(x + 2) = x + 20$ **d** $5(x - 2) = 20 - x$

4 a $3(x + 2) = 2x + 8$ **b** $3(2g + 3) = 4g + 17$ **c** $8w - 12 = 3(3w - 1)$

5 a $3(x - 6) = 2(x + 3)$ **b** $2(a + 9) = 3(a - 1)$ **c** $4(t - 3) = 2(t + 8)$
d $4(10 - p) = 2(5 + p)$ **e** $10(x + 4) = 2(x + 20)$ **f** $5(x - 4) = 3(x + 7)$

6 The answers are the whole numbers from 1 to 8. Each answer should be used once.
a $2(a + 6) = 3(a + 3)$ **b** $4(b - 1) = 2(b + 6)$ **c** $5(c + 2) = 3(c + 6)$
d $2(d + 2) = 4(d - 2)$ **e** $6(8 - e) = 12(e + 1)$ **f** $11(f + 3) = 22(f - 2)$

7 a $6(x + 3) = 10(x + 1)$ **b** $2(x - 1) = 5(x - 7)$ **c** $2(13 - x) = 4(8 - x)$

d $3(x + 6) = 4(x + 3)$ **e** $8(x + 10) = 12(x + 8)$ **f** $5(x + 11) = 10(x + 7)$

8 a $\frac{2}{5}(w - 4) = 6$ **b** $\frac{x - 5}{4} = x - 11$ **c** $\frac{1}{2}(y + 9) = 2(y - 6)$

9 a $\frac{x - 5}{3} = 2$ **b** $\frac{y + 8}{4} = 5$ **c** $\frac{t + 8}{2} = 6$

d $\frac{x + 1}{2} = 12$ **e** $\frac{n - 4}{4} = 7$ **f** $\frac{x - 12}{8} = 2$

10 a $\frac{3x - 2}{2} = x$ **b** $\frac{3y + 1}{2} = y$ **c** $\frac{5t - 3}{2} = t$

Reason mathematically

a Copy and complete this table.

x	1	2	3	4	5
$4(x + 6)$					
$8(9 - x)$					

b Use the table to solve the equation $4(x + 6) = 8(9 - x)$.

c Solve the equation algebraically and check that you get the same answer.

a

x	1	2	3	4	5
$4(x + 6)$	28	32	36	40	44
$8(9 - x)$	64	56	48	40	32

b From the table, both expressions have the same value when $x = 4$.

c Multiply both brackets out first.

$4x + 24 = 72 - 8x$

$4x + 8x + 24 = 72$ Add $8x$ to both sides

$12x = 48$ Subtract 24 from both sides and combine the xs.

So $x = 4$, which is the same solution we had from the table.

11 a Copy and complete this table.

x	5	6	7	8	9
$2(x + 1)$	12				
$3(x - 2)$	9				

b Use the table to solve the equation $2(x + 1) = 3(x - 2)$.

c Solve the equation algebraically and check that you get the same answer.

12 Each of the sides of a hexagon is $(y + 6)$ metres long.
Each of the sides of a pentagon is $(y + 8)$ metres long.
The hexagon and pentagon have the same perimeter.
   a Write expressions for the perimeter of each shape.
   b Solve an equation to find the value of y.
   c Work out the perimeter of the hexagon.

13 Pair up the equations that have the same answers.
   a $3(w + 6) = 9(w - 2)$ **b** $3(w + 3) = 6(w - 6)$ **c** $6(w + 2) = 2(w + 11)$
   d $5(10 - w) = 10(w - 4)$ **e** $9(w - 1) = 3(w + 2)$ **f** $10(22 - w) = 2(w + 20)$

Solve problems

Ciara is c years old. She is six years older than her sister.

In two years' time she will be three times as old as her sister.

How old is Ciara?

Ciara's sister is $(c - 6)$ years old.

Write what you know about, in two years' time, as an equation.

$c + 2 = 3(c - 6)$

$c + 2 = 3c - 18$

Subtract c from both sides: $2 = 2c - 18$

Add 18 to both sides: $20 = 2c$

So $c = 10$. Ciara is ten years old.

14 Here are two straight lines.
The first line is divided into five equal parts,
each is $(x - 4)$ cm long.
The second line is divided into three equal parts,
each is $(x + 2)$ cm long.

   a The lines are the same length. Write an equation to show this.
   b Solve the equation.
   c Work out the length of each line.

15 a Write down an expression for the length of the perimeter of the triangle.
   b Write down an expression for the length of the
      perimeter of the square.
   c The perimeters of the shapes are the same
      length. Write down an equation to show this.
   d Solve the equation.
   e Work out the lengths of the sides of each shape.

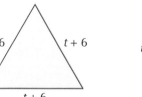
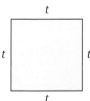

16 a Write down an expression for the area of the first rectangle.

  b Write down an expression for the area of the second rectangle.

  c The shapes have the same area. Write down an equation to show this.

  d Solve the equation.

  e Work out the area of each shape.

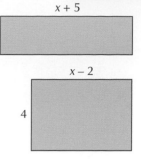

17 Alex is a years old.

  a Write down an expression for Alex's age in six years' time.

  b In six years' time, Alex's age divided by 4 will be 12. Write down an equation to show this.

  c Solve the equation. How old is Alex?

29.4 Rearranging formulae

● I can change the subject of a formula

This is a formula that is used in science.

$v = u + at$

The **subject** of the formula is v.

The formula shows you how to work out the value of v if you know the values of u, a and t.

Develop fluency

Rearrange the formula $v = u + at$ to make a the subject.

$v = u + at$	You need to get a on its own, on one side of the equals sign.
$v - u = at$	First, subtract u from both sides.
$\dfrac{v - u}{t} = a$	Now divide by t. Remember that at means $a \times t$.
The required formula is $a = \dfrac{v - u}{t}$	Write it the other way round, so that a is on the left.

1 Make t the subject of each formula.

  a $s = t + 25$ **b** $w = t - 6.5$ **c** $a = 4t$ **d** $m = 21t$

2 Make n the subject of each formula.

  a $T = m + nb$ $q = n + t - 12$ **c** $y = 3n + a$ **d** $y = 3(n + 1)$

3 Rearrange each formula to make m the subject.

  a $r = m - 3$ **b** $r = 4(m - 3)$ **c** $r = 4m - 3$ **d** $r = \dfrac{1}{4}m - 3$

4 Rearrange each formula to make b the subject.

  a $a = b + c$ **b** $a = b - c$ **c** $a = bc$ **d** $a = \dfrac{b}{c}$

5 Rearrange each formula to make x the subject.

   a $y = x + 9$ **b** $y = 4x$ **c** $y = 5x - 1$

   d $y = 6(x + 5)$ **e** $y = \frac{1}{3}(x + 1)$ **f** $y + 3x = 8$

6 Rewrite each formula as indicated.

   a $A = 9p$ Make p the subject of the formula.

   b $y = x + 5$ Make x the subject of the formula.

7 The equation of a straight line is a formula connecting x and y.

   Make x the subject of each of these equations of straight lines.

   a $y = 5x + 12$ **b** $y = 31x - 2$ **c** $x + y = 50$

   d $y = 45 + 25x$ **e** $x + 3y = 20$ **f** $4x + y = 18$

8 Look at this formula.

   $k = a + 3b - 1$

   a Rearrange the formula to make a the subject.

   b Rearrange the formula to make b the subject.

9 **a** Rearrange the perimeter formula to make a the subject.

   b Rearrange the perimeter formula to make b the subject.

   c Rearrange the area formula to make a the subject.

   d Rearrange the area formula to make b the subject.

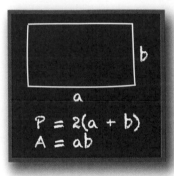

$P = 2(a + b)$
$A = ab$

10 Here is a formula.

   $t = 10p + 20q$

   a Rearrange the formula to make p the subject.

   b Rearrange the formula to make q the subject.

Reason mathematically

The perimeter of a rectangle is given by the formula $P = 2(a + b)$.

   a Show that the formula can be rearranged as $a = \dfrac{P - 2b}{2}$

   b Rearrange the formula to make b the subject.

   a From $P = 2(a + b)$, multiply out the bracket to give $P = 2a + 2b$.

      Subtracting $2b$ from both sides gives $P - 2b = 2a$

      Divide both sides by 2 to give $\dfrac{P - 2b}{2} = a$

   b From $P = 2a + 2b$

      Subtracting $2a$ from both sides gives $P - 2a = 2b$

      Divide both sides by 2 to give $\dfrac{P - 2a}{2} = b$

11 This is the formula for the mean, m, of two numbers, x and y.

$$m = \frac{x+y}{2}$$

   a Work out the value of m, given that $x = 3.5$ and $y = 12.5$.

   b Work out the value of m, given that $x = 124$ and $y = 138$.

   c Show that the formula can be rearranged as $x = 2m - y$.

   d Use the formula in part c to find the value of x, given that $m = 41$ and $y = 48$.

   e Rearrange the formula to make y the subject.

   f Use your formula from part e to find the value of y, given that $m = 8.5$ and $x = 6.3$.

12 The formula $D = \dfrac{M}{V}$ is used in science.

   a Find the value of D when $M = 30$ and $V = 6$.

   b Rearrange the formula to make M the subject.

   c Find the value of M when $D = 12$ and $V = 4$.

   d Rearrange the formula to make V the subject.

   e Find the value of V when $M = 40$ and $D = 5$.

13 The area of a trapezium is given by the formula $A = (a + b)h$.

   a Show that the formula can be rearranged as $a = \dfrac{2a}{h} - b$.

   b Rearrange the formula to make b the subject.

14 Mart was given the equation $w = 3(x + 5y)$ to rearrange to make y the subject.

   His solution was $\dfrac{w}{3} = x + 5y$

$$\frac{w - x}{3} = 5y$$

$$\frac{w - x}{15} = y$$

   Mart has made errors. Correct his solution.

Solve problems

This is a rectangle.

Show that the formula for the perimeter, P, is $P = 6x + 5(y + 1)$.

The perimeter is the sum of all the sides, giving

$P = 2y + 3 + 4x - 1 + 3y - 2 + 2x + 5$

$\quad = 6x + 5y + 5$

$\quad = 6x + 5(y + 1)$

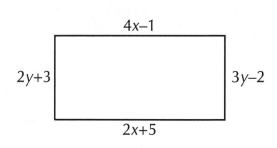

15 a Show that the formula for the perimeter, P, of this isosceles
       triangle is $P = 2x + y$.

    b Work out the value of P when $x = 14$ and $y = 10$.

    c Rearrange the formula to make y the subject.

    d Rearrange the formula to make x the subject.

16 a Show that the formula for the area, a, of this shape is $a = 4x + 30$.

    b Work out the value of a when:

        i $x = 7$ ii $x = 20$.

    c Rearrange the formula to make x the subject.

    d Work out the value of x, given that:

        i $a = 50$ ii $a = 70$.

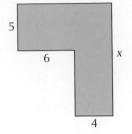

17 a Show that the formula for the perimeter, P, of this heptagon
       is $P = 5u + 18$.

    b Rearrange the formula to make u the subject.

    c Find the value of u when the heptagon has a perimeter of 173.

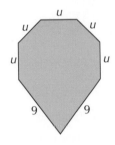

Now I can...

solve equations involving brackets	solve equations with fractional coefficients	change the subject of a formula
solve equations with the variable on both sides	solve equations with brackets and fractions	

CHAPTER 30 Proportion

30.1 Direct proportion

- I can understand the meaning of direct proportion
- I can find missing values in problems involving proportion

Two variables (such as the number of litres and the total price you pay) are in direct proportion if, when you multiply one by a number (such as 2 or 3 or 0.5) you need to multiply the other by the same number.

Here are some examples of pairs of variables that are in **direct proportion**.

- The distance travelled by a car moving at 100 km/hour and the time taken.

- The volume of water flowing out of a tap and the time, in seconds, for which the water flows.

Develop fluency

At a petrol station, 15 litres of petrol costs £20.40.

Work out the cost of: **a** 30 litres **b** 60 litres **c** 5 litres.

a It helps to put the numbers in a table.

Petrol (litres)	15	30	60	5
Cost (£)	20.40			

$30 = 15 \times 2$ The number of litres (15) is multiplied by 2, so the cost (£20.40) is also multiplied by 2.

The cost of 30 litres is £20.40 × 2 = £40.80.

If you multiply or divide the number of litres by any number, you must multiply or divide the cost by the same number.

b $60 = 15 \times 4$ The number of litres is multiplied by 4.

The cost of 60 litres is £20.40 × 4 = £81.60.

c $5 = 15 \div 3$ The number of litres is divided by 3. Do the same to the cost.

The cost of 5 litres is £20.40 ÷ 3 = £6.80.

> You say that the number of litres and the cost in pounds are directly **proportional**.
>
> You can leave out the word 'directly' and just say they are proportional.

1 A train is travelling at a constant speed.
The distance travelled is proportional to the time taken.
In 5 minutes the train travels 13 kilometres.
Copy and complete this table.

Time taken (minutes)	5	10	20	30	45
Distance (km)	13				

2 Jacob buys 300 g of carrots and they cost 84 pence in total.
   Work out the cost of:
   a 600 g of carrots   **b** 900 g of carrots   **c** 150 g of carrots   **d** 100 g of carrots.

3 250 ml of cola contains 27 g of sugar.
   Work out the amount of sugar in:
   a 500 ml of cola   **b** 1 litre of cola   **c** 2 litres of cola   **d** 125 ml of cola.

4 Paulo knows that 1.5 kg of flour is enough to make four small loaves.
   a How much flour will he need to make 16 small loaves?
   b How many small loaves can he make from 9 kg of flour?

5 A distance of 5 miles is approximately the same as 8 kilometres.
   Copy and complete this table to show equivalent distances.

Miles	5	40	100			
Kilometres	8			24	40	200

6 The perimeter of a circle is called the circumference.
   The circumference of a circle is proportional to the diameter of the circle.
   A circle with a diameter of 3.5 m has a circumference of 11 m.
   Work out the circumference of a circle with a diameter of:
   a 7 m   **b** 10.5 m   **c** 35 m   **d** 1.75 m.

7 Water is dripping from a tap at a steady rate.
   In 15 minutes there are 80 drips.
   a Work out the number of drips in one hour.
   b Work out the time taken for 800 drips to come from the tap.

8 In a shop, 100 g of sweets cost 64 pence.
   a Work out the cost of:
      i 300 g of sweets   **ii** 500 g of sweets   **iii** 25 g of sweets.
   b What mass of sweets can you buy for:
      i £1.28   **ii** £6.40   **iii** 32p?

9 The pressure of a car tyre can be measured in two different units, bars or psi (pounds per square inch).
   A pressure of 2.1 bars is the same as 30 psi.
   Copy and complete this table to show conversions between the two units.

bar	2.1			8.4	12.6
psi	30	10	20		

10 The mass of a steel cable is proportional to its length.

    a A five-metre length of a particular cable has a mass of 8.2 kg.

    b Work out the mass of 20 metres of the cable.

    c Another length of cable of the same type has a mass of 49.2 kg. How long is it?

Reason mathematically

A wholesaler sells paint at £10 per litre.

Work out the missing values in this table.

Volume (litres)	4.7	11.8	
Cost (£)	47		73.60

In this case, it is easier to compare pairs of values vertically rather than horizontally.

Note that 4.7 × 10 = 47

Now, 11.8 × 10 = 118 so the cost of 11.8 litres of paint is £118.

Reversing the process for the last pair of values:

73.60 ÷ 10 = 7.36 so 7.36 litres of paint may be purchased for £73.60.

11 The heights of some fence posts and the lengths of their shadows are found to be in proportion.

    Copy and complete the table.

Post height (cm)	128	134	175	
Shadow length (cm)	64			104.5

12 The exchange rate between pounds (£) and US dollars (US$) is £42 = US$63.

    Copy and complete this table.

Pounds (£)	42	66	210	228	
US dollars (US$)	63				450

13 Energy content on food labels is given in two different units, kilocalories (kcal) and kilojoules (kJ).

    Here is part of a conversion table.

Kilocalories (kcal)	38	
Kilojoules (kJ)	160	800

    a Work out the missing value.

    b Show that the ratio of the two amounts of kilojoules is the same as the ratio of the two amounts of kilocalories.

14 **a** 8 light bulbs cost £7.20. How much does a box of 20 light bulbs cost?

    b How many light bulbs can you buy for £9?

Solve problems

The table shows how the length of a spring changes with its tension. The extension of a spring is the increase in its length due to the tension in newtons (N).

Tension (N)	10	20	40
Length (cm)	30	35	45

a Is the length of the spring proportional to its tension? Explain.

b Calculate the length of the spring when the tension is zero.

c Calculate the tension required to make the spring double its length.

d Copy and complete the table.

Tension (N)	10	20	40	50
Extension (cm)				

e Is the tension proportional to the extension? Explain.

a *No. The first two columns of the table show that when the tension doubles, the length does not. This means that the length of the spring is not proportional to its tension.*

b *The length of the spring reduces by 5 cm as the tension reduces by 10 N. When the tension is 0 N the length will be 30 cm – 5 cm = 25 cm.*

c *The spring will have doubled its length when its length is 50 cm. The tension will then be 10 N more than when its length was 45 cm which is 40 N + 10 N = 50 N.*

d

Tension (N)	10	20	40	50
Extension (cm)	5	10	20	25

e *Yes. For example, the extension is 5 cm for every 10 N of tension.*

15 Temperature can be measured in degrees Celsius (°C) or degrees Fahrenheit (°F). Here is a table of values.

Degrees Celsius (°C)	20	30	50	100
Degrees Fahrenheit (°F)	68	86	122	212

Is temperature in degrees Celsius proportional to temperature in degrees Fahrenheit? Give a reason for your answer.

16 A car uses 20 litres of petrol in travelling 140 km.

a How much would be used on a journey of: i 35 km   ii 210 km?

b a full tank of petrol is 60 litres. At the end of a journey there were 15 litres left in the car. How long was the journey?

17 This table shows the exchange rate between pounds (£) and New Zealand dollars (NZ$).

Pounds (£)	50	150
New Zealand dollars (NZ$)	96	

   a Work out the missing value.

   b Work out the ratio of the two amounts of pounds.

   c Work out the ratio of the two amounts of dollars.

30.2 Graphs and direct proportion

● I can represent direct proportion graphically and algebraically

When two variables are in direct proportion, you can plot them on a graph and join them with a line.

A graph of values of two variables that are in direct proportion always has these properties:

·   The points are in a straight line.

·   The line passes through the origin.

You can write a formula linking the two variables in the form $y = mx$ where m is a constant.

Develop fluency

Ribbon is sold by the metre. Lucy buys 6 metres and the total cost is £5.04.

  a Find a formula for the cost, y pence, of x metres of ribbon.

  b Draw a graph to show the cost of different lengths of ribbon.

  a The cost is proportional to the length of ribbon, so the formula is $y = mx$.

    You need to work out the value of m.

    You know that when x is 6 then y is 504. Notice that y is the cost, in pence.

    $504 = m \times 6$

    $m = 504 \div 6 = 84$

    The formula is $y = 84x$.

  b Use the formula to find the costs of different lengths.

    Multiply the length by 84 to find the cost.

    Choose some values for the length.

Length (x metres)	1	2	3	5	7	10
Cost (y pence)	84	168	252	420	588	840

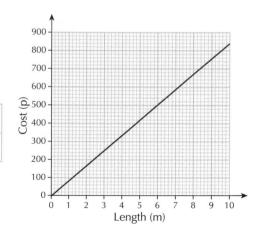

Plot the points on a graph, then join them up.

The points should be in a straight line. The line should go to the origin.

1 The perimeter (y cm) of a square, of side x cm, is given by the formula $y = 4x$.

  a Copy and complete this table to show values of x and y.

Side (x cm)	2	5	7	8	10
Perimeter (y cm)					

  b Draw a graph to show the values in your table. Label the axes.

2 A 200 ml glass of a fizzy drink contains 20 g of sugar.

  a Copy and complete this table.

Drink (x ml)	200	100	500	1000
Sugar (y g)	20			

  b You are told that x ml of fizzy drink contains y g of sugar.
    Show that the formula is $y = 0.1x$.

  c Draw a graph to show the figures in your table.

3 A helicopter is travelling at a constant speed. The distance travelled is proportional to the time taken. It travels 7 km every 5 minutes.

  a Copy and complete this table.

Time taken (minutes)	5	10	15	20	25
Distance (km)	7				

  b What is the scale factor as a decimal?

  c Write down the formula for distance, D, compared to time, T.

4 Some perimeters (y cm) of a rhombus of side x cm are shown.

  a Copy and complete this table to show values of x and y.

Side (x cm)		4		9	
Perimeter (y cm)	10	16	24	36	60

  b Find the formula.

5 This table shows the price of different masses of potatoes.

Mass (x kg)	0.5	1	1.5	2	3
Price (y pence)		48		96	

  a Work out the missing values of y.

  b What do you multiply the mass by, to find the price?

  c Work out a formula for y in terms of x.

  d Use the formula to find the cost of 7.5 kg of potatoes.

  e Draw a graph to show the cost of potatoes.

6 This graph shows the exchange rate between pounds (£) and Hong Kong dollars (HK$).

a Use the graph to complete this table.

Pounds (£x)	25	50	75	100
Hong Kong dollars (HK$y)				

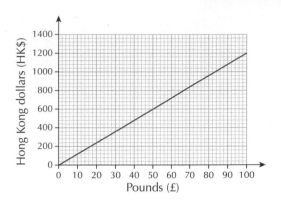

b Work out a formula for y in terms of x.

c Use the formula to change £1270 into Hong Kong dollars.

7 The speed of a car can be measured in metres per second (m/s) or in kilometres per hour (km/h).

A speed of 5 m/s is the same as a speed of 18 km/h.

a Write 10 metres per second in kilometres per hour.

b What do you multiply a speed in metres per second by, to find the speed in kilometres per hour?

c a speed of y km/h is the same as a speed of x m/s.
Write down a formula for y in terms of x.

8 The lengths of the side and the diagonal of a square are proportional.

a This table shows possible values of x and y. Fill in the missing values.

Side (x mm)	5	10	15	20	25
Diagonal (y mm)	7				

b Work out a formula for y in terms of x.

c Use your formula to calculate the diagonal of a square, if the length of the side is:

i 12 mm   ii 19 mm   iii 31 mm.

Reason mathematically

The table shows measurements of some off-cuts of carpet and their cost.

Measurement	1 m × 2 m	1 m × 3 m	2 m × 2 m	2 m × 3 m
Cost (£)	10	15	20	

a Use proportion to calculate the missing value in the table.

b Find a formula for the cost £C of an off-cut measuring x m × y m.

a *The cost of an off-cut is proportional to its area.*
A 2 m × 3 m off-cut has an area of 6 m². Its cost is £(5 × 6) = £30.

b *The area of an off-cut measuring x m × y m is xy m². So, the cost is given by $C = 5xy$.*

9   The angles of this triangle are 30°, 60° and 90°.
    The lengths of AB and AC are proportional.
    This table shows some possible values of x and y.

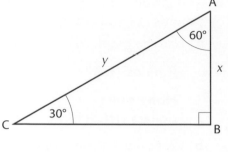

AB (x cm)	4.1	5.2	12.9		
AC (y cm)	8.2			18.8	6.4

   a   Write down a formula for y in terms of x.

   b   Work out the missing values in the table.

   c   Draw your own triangle with angles of 30°, 60° and 90°, like the one in the diagram.
       Measure x and y and check that they agree with your formula.

10 1 inch is almost exactly 2.5 centimetres.

   a   What is the formula to represent this, with x as inches and y as centimetres?

   b   Draw a table to show the conversion from inches to centimetres for the first 6 inches.

   c   Draw a graph of this conversion.

11 Hannah has measured the areas of some circles of different radii.
    Her results are shown in the table.

Radius (cm)	3	6	9	12
Radius^2 (cm^2)				
Area (cm^2)	28.3	113.1	254.5	

   a   Is the area proportional to the radius? Explain.

   b   Copy the table and complete the middle row.

   c   Compare the ratio of two values of the radius^2 with the ratio of the
       corresponding areas.

   d   Copy and complete the statement 'The area of a circle is proportional to … '

   e   Calculate the final missing value in the table to the nearest whole number.

Solve problems

The distance that can be seen to the horizon at sea is proportional to the square root
of the height of the viewing point above sea level.

Height above sea level (m)	1	9	49
Distance to horizon (km)	3.7		

   a   Find a formula for the distance, d km, that can
       be seen to the horizon from a point at sea h m
       above sea level.

   b   Calculate the missing values in the table.

   a   Looking at the first pair of values in the table, $d = 3.7\sqrt{1}$.
       The formula is $d = 3.7\sqrt{h}$.

   b   When $h = 9$, $d = 3.7\sqrt{9} = 3.7 \times 3 = 11.1$ km
       When $h = 49$, $d = 3.7\sqrt{49} = 3.7 \times 7 = 25.9$ km

12 This graph shows the exchange rate between pounds and Canadian dollars. £1 = $1.8

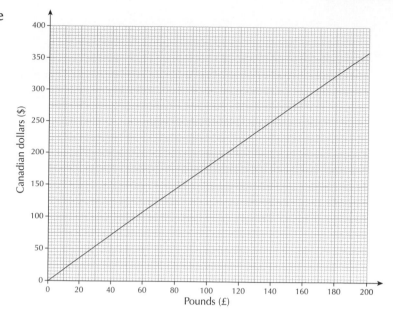

Canadian dollars ($) / Pounds (£)

   a Mark visits his sister in Toronto and needs some money for when he gets there.
   How many dollars will he get for £150?

   b He has $36 left at the end of his visit. How many pounds will he get back?

   c What is the formula connecting the two currencies, using p for pounds and d for dollars?

13 The graph shows how much diesel fuel is used by a car on a 300 mile journey.
Copy the table shown and fill in any gaps.

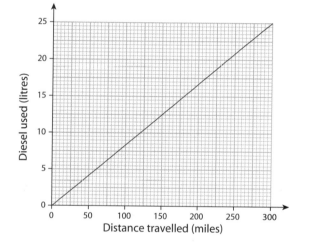

Diesel used (litres) / Distance travelled (miles)

Distance (x miles)		30		96	
Diesel (y litres)	2		6		12

   a What is the formula linking diesel usage with miles travelled?

   b How much diesel would be used up after 60 miles?

   c The car began the journey with 20 litres in the tank. Will the driver have to refuel before he finishes the journey?

   d With diesel priced at £1.40 per litre, how much will the journey have cost?

14 The diagram shows the graphs of some equations.

   a Which graph shows that y is directly proportional to x? Explain your answer.

   b What is its equation?

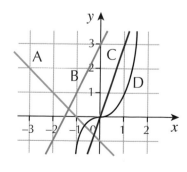

30.3 Inverse proportion

- I can understand what inverse proportion is
- I can use graphical and algebraic representations of inverse proportion

With **inverse proportion**, as one variable increases, the other variable decreases in such a way that their product remains constant.

Develop fluency

Show that the values in this table vary in inverse proportion and find a formula connecting x and y.

x	6	12	15	24
y	20	10	8	5

In this table, as the x values increase from 6 to 24, the y values decrease from 20 to 5.

Each pair of values will multiply to make 120.

x and y vary in inverse proportion and $xy = 120$.

This table shows the journey times for a car travelling a distance of 120 km at various speeds.

Speed (x km/h)	20	30	40	50	60
Time (y hours)	6	4		2.4	

Complete the table and show that the speed and time are inversely proportional.

If you multiply the speed by any number, you must divide the time by the same number.

For example:

- $20 \times 2 = 40$ and $6 \div 2 = 3$
- $40 \times 1.5 = 60$ and $3 \div 1.5 = 2$.

The speed (x) and time (y) are in **inverse proportion**. When you multiply one by a number you divide the other by the same number.

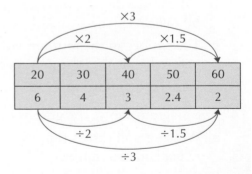

In the example above, if you multiply the speed by the time the answer is always 120, because:

speed × time = distance

If the speed is x km/h and the time is y hours, you can write this as a formula:

$xy = 120$

1 The area of a rectangular field is 2400 m².
The sides of the field are a metres and b metres long.
This table shows possible lengths for the sides of the field.

a metres	40	50	
b metres			80

   a Calculate the missing values.

   b Work out a formula connecting a and b.

2 y is inversely proportional to x. When $y = 3$, $x = 12$.

   a What does xy equal?
      b What will x be when $y = 4$?

   c What will y be when $x = 1$?
      d What other whole number combinations are possible?

3 Four equal-sized pipes can fill a tank in 70 minutes. How long will it take to fill the tank using 7 equal-sized pipes?

4 It takes 8 people 12 hours to dig a hole.
How long would it take the following numbers of people?

   a 4 people
      b 2 people
      c 6 people

5 A train is travelling a distance of 600 km.

   a How long does the train take if it travels at a speed of 100 km/h?

   b If the train travels at 150 km/h, how long will it take for the same journey?

   c Copy and complete this table.

Speed (x km/h)	100	150	120	200	300
Time (y hours)					

   d Show that x and y are inversely proportional.

   e Write down a formula connecting x and y.

   f Draw a pair of axes like this. Plot the points from the table in part **c** and join them with a smooth curve.

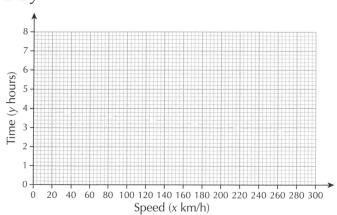

6 A teacher has £1000 to spend on books.

  a Some books cost £10 each. How many can the teacher buy?

  b Other books cost £5 each. How many can the teacher buy?

  c Copy and complete this table.

Cost of a book (£x)	2	2.50	5	10	20	25
Number bought (y)						

  d Show that x and y are inversely proportional.

  e Write down a formula connecting x and y.

  f Draw a pair of axes like this.
Draw a graph to show the information in the table.

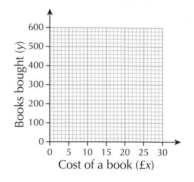

7 The graph shows the time taken by an aeroplane to travel between two airports, at different speeds.

  a Use the graph to find the time taken when the aeroplane flies at a speed of 800 km/h.

  b Use the graph the find the speed when the journey takes 4 hours.

  c The time (y hours) is inversely proportional to the speed (x km/h).
Use your answers to parts **a** and **b** to find a formula connecting x and y.

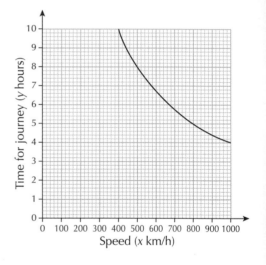

8 Some children are measuring the lengths of their paces and how many paces they take to walk 12 metres.

  a If the length of a pace is 0.5 metres, how many paces will they take to walk 12 metres?

  b Copy this table and fill in the missing values.

Length of pace (p metres)	0.5	0.6	1	1.2
Number of paces (n)				

  c Show that p and n are inversely proportional.

  d Write down a formula connecting p and n.

Reason mathematically

Meryl wants to make a rectangular flower bed with an area of 24 m².

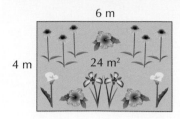

a Draw up a table to show these and other possible length and width measurements of the flower bed.

b Do the measurements have to be whole numbers? Explain your answer.

c In order to sum up what happens to the area, fill in these blanks:

As the length of one side _____, the _____ has to be halved for the _____ to stay the same.

a

Length (m)	6	8	12	24
Width (m)	4	3	2	1

b No. For example, a length of 16 m and a width of 1.5 m gives an area of 24 m².

c As the length of one side doubles, the other has to be halved for the area to stay the same.

9 An isosceles triangle has an area of 100 cm².
The length of the base (b cm) is inversely proportional to the height (h cm).
When the base is 20 cm then the height is 10 cm.

a Copy and complete this formula: $bh = ...$

b Copy and complete this table to show possible values of b and h.

Base (b cm)	20	16	12.5	10	8
Height (h cm)	10				

c Draw a graph to show how the height varies with the base.

d Use your graph to find the height, when the base is 15 cm.

e Use your formula to check the answer to part d.

10 Imran bought 40 toys at £12 each. How many toys can Imran buy at £8 each if he spends the same total amount?

Solve problems

A village held a sponsored 10-mile road race. The graph shows how long it took to complete it, depending on people's average speed.

Use the graph to answer these questions.

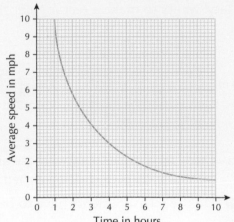

a The winner finished after 75 minutes. What was his average speed?

b The youngest entrant averaged 2 mph. How long did it take her to finish?

c The run started at 11 a.m. About what time did someone finish whose average speed was 7 mph?

a 75 minutes = $1\frac{1}{4}$ hours. Starting at $1\frac{1}{4}$ hours on the time axis, moving up to the curve and across, the winner's average speed was 8 mph.

b Reading across from the average speed axis at 2 mph and down gives the time as 5 hours.

c Starting at 7 mph on the average speed axis and reading across to the curve and down gives a time of about 1 hour 25 minutes, so they finished at about 12.25 p.m.

11 The table shows the time taken to run 100 metres for various running speeds.

Speed (metres/second)	4	5	8	10	20
Time taken (seconds)	25	20	12.33	10	5

a Most Olympic athletes can now run 100 metres in less than 10 seconds, which is 10 metres/second. What is 10 metres/second in kilometres per hour?

b To the nearest whole number, how fast would you have to run to get below 9 seconds? A graph may help, but try to find coordinates for 6, 7 and 9 seconds to get an accurate curve.

c A PE teacher reckons he can run 100 metres in 15 seconds. How fast does he need to run?

12 Some families want to buy play equipment for their local park.
The total cost is £30 000.
The families agree to share the cost among themselves, equally.

  a Work out the cost for each family, when there are 20 families.

  b Work out the cost for each family, when there are 30 families.

  c Copy and complete this table of values.

Number of families (n)	10	20	30	40	50	60
Cost for each family (£c)						

  d Is the cost for each family inversely proportional to the number of families?
Justify your answer.

  e Draw a graph to show how the cost varies with the number of families.

  f They decide that each family should not pay more than £800. Use your graph to work out the smallest number of families that need to take part.

  g Write down a formula connecting n and c. Use it to check your answer to part **f**.

13 When $x = 1.6$, $y = 16$, but when $x = 4$, $y = 6.4$.

  a Work out the formula for this relationship.

  b Draw up a table for integer values of x from 1 to 8. Round your answer to three decimal places.

  c Draw a graph that allows answers to be read accurately to one decimal place.

  d What is y when $x = 3\frac{1}{3}$?

30.4 Comparing direct and inverse proportion

● I can recognise direct and inverse proportion and work out missing values

For direct proportion:

- As one variable increases, so does the other.
- The ratio of the two variables is constant.
- A graph of the two variables is a straight line, through the origin.

For inverse proportion:

- As one variable increases, the other decreases.
- The product of the two variables is constant.

Develop fluency

Here are the values of two variables, p and q.

p	20	50
q	90	

a Find the missing value, given that q is directly proportional to p.

b Find the missing value, given that q is inversely proportional to p.

a Look at the values in the first column.

$q \div p = 90 \div 20 = 4.5$

$q = 4.5p$ Because they are in direct proportion.

When $p = 50$ then $q = 4.5 \times 50 = 225$.

$q = 225$

b Look at the values in the first column.

$pq = 20 \times 90 = 1800$

$pq = 1800$ Because they are in inverse proportion.

When $pq = 1800$ and $p = 50$ then $50 \times q = 1800 \rightarrow q = 1800 \div 50 = 36$.

1 Which of these graphs shows   **a** direct proportion   **b** inverse proportion?

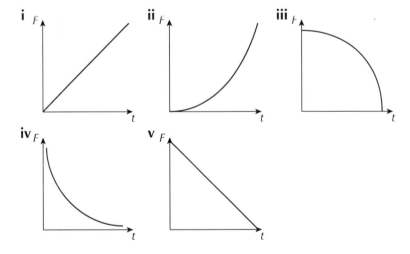

2 Decide which of these you think is direct or inverse proportion:

  a distance travelled and fuel used

  b length of dress and material used

  c running speed and time taken to complete a race

  d number of windows in a house and time taken to clean them

  e diameter of a tube and time taken for water to get through

  f hours worked and amount in pay packet

3 George is walking at a constant speed.
  In 5 minutes he walks 400 metres.
  a Is the distance travelled (*d* metres) proportional to the time taken (*t* minutes)?
  b Write down a formula for *d* in terms of *t*.
  c Use the formula to work out how far George walks in 8.5 minutes.

4 Anne is doing a sponsored walk of 20 km.
  a How long will it take if she walks at 5 km/h?
  b How long will it take if she walks at 8 km/h?
  c Explain why the time taken (*t* hours) is inversely proportional to her walking speed (*w* km/h).
  d Write down a formula connecting *t* and *w*.

5 **a** Two variables, *x* and *y*, are in direct proportion.
    When $x = 40$, $y = 10$.
    Work out a formula for *y* in terms of *x*.
  b Two more variables, *x* and *y*, are in inverse proportion.
    When $x = 40$, $y = 10$.
    Work out a formula connecting *x* and *y*.

6 This graph shows two variables that are in direct proportion.
  a Write down the coordinates of two points on the line.
  b Work out a formula for *y* in terms of *x*.

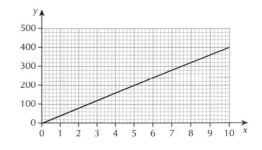

7 This graph shows two variables that are in inverse proportion.
  a Write down the coordinates of three points on the line.
  b Work out a formula connecting *x* and *y*.

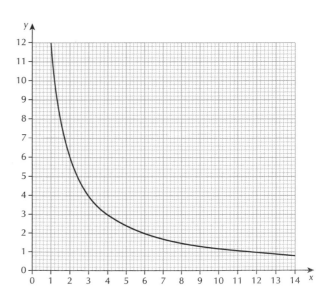

8 What does y equal, in terms of x, for each of these sets of values?

a

x	1	2	3	4	5
y	13	26	39	52	65

b

x	2	4	6	8	10
y	7	14	21	28	35

c

x	5	10	15	20	25
y	20	40	60	80	100

9 What is the formula for each of these tables?

a

x	1	2	3	4	5
y	60	30	20	15	12

b

x	1	3	5	7.5
y	75	15	25	10

c

x	2	3	4	6	8
y	48	32	24	16	12

Reason mathematically

Here are some tables of values. Say whether they show direct proportion, inverse proportion or neither of these.

If the variables are directly or inversely proportional, work out the formula.

a

x	12	17	5	14
y	36	51	20	42

b

c	18	30	12	1.5
d	5	3	7.5	60

a $36 = 12 \times 3$ and $51 = 17 \times 3$, but $20 \neq 5 \times 3$ so the first table shows neither.

b Each pair of values has a product of 90, so they are inversely proportional.
The formula in this case is $cd = 90$.

10 Sketch a graph to show what each of these might look like.
   a Distance travelled and time taken
   b Number of workers doing a repair and time taken to complete the job
   c Number of texts on a pay-as-you-go mobile phone and amount of credit remaining
   d Depth of bath water and time taken for it to empty

11 These are some tables of values. Say whether they show direct proportion, inverse proportion or neither of these.
   If the variables are directly or inversely proportional, work out the formula.

a

f	5.6	9.4	63.8	3.6
r	75.6	126.9	861.3	48.6

b

u	12	16	14.8	46.25
w	12	9	10	3.2

12 How can you tell that this graph does not show direct proportion?

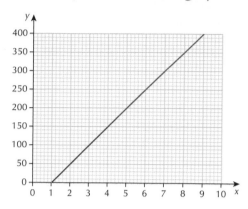

Solve problems

A bath is to be filled with water. If only the cold water tap is used, it will take 10 minutes to fill the bath. If only the hot water tap is used, it will take 15 minutes to fill the bath.

How long will it take to fill the bath if both taps are used?

The LCM of 10 and 15 is 30.

In 30 minutes the cold tap runs enough water to fill the bath 3 times.

Also, in 30 minutes, the hot tap runs enough water to fill the bath twice.

With both taps running, there will be enough water to fill the bath 5 times in 30 minutes.

To fill the bath once takes 30 ÷ 5 = 6 minutes.

13 Alice and Bob are painters. Alice can paint a house in 20 hours. Bob will take 24 hours to do the same job. How long will it take them to paint a house if they work together? Give your answer to the nearest hour.

14 Colin takes x minutes to complete a task. Ali takes y minutes to complete the same task. How long will it take them to complete the task if they work together?

Now I can...

understand the meaning of direct proportion	find missing values in problems involving proportion	recognise direct and inverse proportion and work out missing values
represent direct proportion graphically and algebraically	use graphical and algebraic representations of inverse proportion	
understand what inverse proportion is		

CHAPTER 31 Applications of graphs

31.1 Step graphs

- I can interpret step graphs

A **step graph** is a special type of line graph that is made up of horizontal lines in several intervals or steps.

You would generally use a step graph to represent situations that involve sudden jumps across intervals, such as cost of postage, car parking fees or telephone rates.

Develop fluency

This step graph shows a car park's charges.

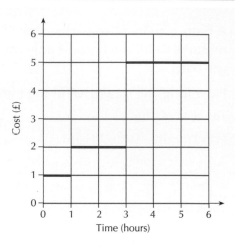

a What is the charge for parking a car for:

   i 30 minutes   **ii** less than 1 hour

   iii 2 hours   **iv** 2 hours 59 minutes

   v 3 hours 30 minutes   **vi** 6 hours?

b How long can I park for:

   i £1   **ii** £2   **iii** £5?

You can read the answers from the graph. The value at the end of each step is usually included.

a **i** £1   **ii** £1   **iii** £2

    iv £2   **v** £5   **vi** £5

b **i** Up to 1 hour   **ii** Up to 3 hours   **iii** Up to 6 hours

1 This step graph shows how the cost of a single ticket on a transport system varies.
What is the price of a ticket to travel:

  a 3 miles   **b** 10 miles

  c 32 miles   **d** 50 miles?

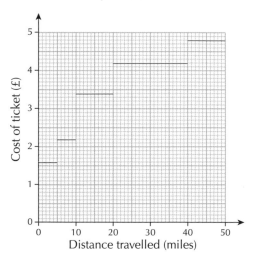

2 The step graph shows how the cost of a ticket on an underground railway network varies.

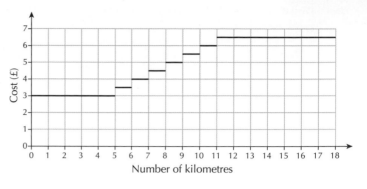

  a What is the cheapest ticket you can buy?

  b What is the fare for a journey of

    i 4.5 km   **ii** 6.5 km

    iii 8.5 km   **iv** 10.5 km

    v 12.5 km?

3 The step graph shows how much a library charges in fines for overdue books.

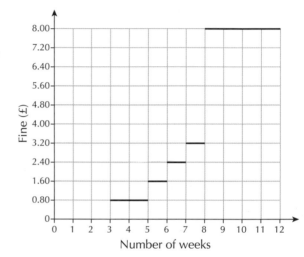

  a What is the longest time you can borrow a book for free?

  b What is the fine for returning a book after

    i four weeks

    ii six weeks and two days

    iii eleven weeks?

  c What is the minimum time a book has been overdue for if the fine is

    i £1.60   **ii** £3.20   **iii** £8?

4 The step graph shows how the taxi fare varies with the distance travelled.

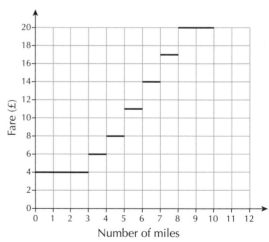

  a What is the minimum fare?

  b What is the price of

    i a 1 mile journey

    ii a 9 mile journey

    iii a 3.5 mile journey and a 5.5 mile journey?

  c What is the furthest you can travel for £11?

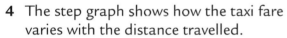

5 The table shows a country's parcel post costs.

Mass, m (kg)	Home country	Abroad
$0 < m \leq 0.5$	£1.40	£3.50
$0.5 < m \leq 1$	£2.50	£4.60
$1 < m \leq 2$	£3.20	£5.90
$2 < m \leq 3$	£4.60	£7.00
$3 < m \leq 5$	£5.50	£9.50
$5 < m \leq 10$	£6.00	£12.00
$10 < m \leq 20$	£8.00	£15.00

Draw step graphs to show charges against mass for:

a the home country b abroad.

6 The table shows the costs of sending a large letter by first-class and second-class post.

Mass of letter (g)	First-class post (£)	Second-class post (£)
0–100	0.90	0.69
101–250	1.20	1.10
251–500	1.60	1.40
501–750	2.30	1.90

Draw step graphs to show cost against mass for:

a first-class post

b second-class post.

Reason mathematically

The step graph shows how much it costs to post a parcel within the same country or abroad.

a What is the difference in price between sending a 4.8 kg parcel to a home country and sending it abroad?

b Show that the cost of sending two 2.5 kg parcels abroad and one 5.7 kg parcel to a home country is less than £15.

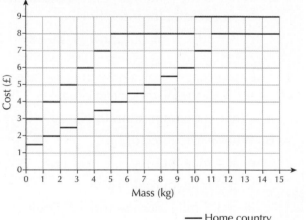

—— Home country

—— Abroad

a Cost to send abroad is £7.

Cost to send to home country is £3.50, so difference is £7 – £3.50 = £3.50

b Cost to send one 2.5 kg parcel abroad is £5, so two parcels will cost £10.

Cost to send a 5.7 kg parcel to a home country is £4.

So total cost is £10 + £4 = £14 which is less than £15.

7 Look again at the step graph in question 2.
   a How much more does a 12 km journey cost than a 4 km journey?
   b Theo goes on a journey. The ticket costs twice as much as the cheapest ticket.
      Show that he can travel over 10 km using this ticket.

8 Look again at the step graph in question 4.
   a What is the difference in price between a 3.2 mile journey and a 6.4 mile journey?
   b Show that the price of the 6.4 mile journey is more than double the price of the
      3.2 mile journey.

9 A town has two car parks, one for short stays
   and one for long stays. The step graph shows
   the different charges at each one.
   a Which car park is cheaper for someone to
      park for 2 hours?
   b Marie thinks she will be parked for
      between $1\frac{1}{2}$ and 3 hours.
      Which car park should she use?
      Give a reason for your answer.

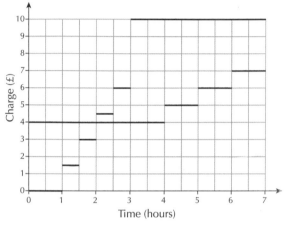

Time (hours)

— Short stay car park   — Long stay car park

10 Match the four graphs to these situations.
   a The amount John gets paid against the number of
      hours he works.
   b The temperature of an oven against the time it is
      switched on.
   c The amount of tea in a cup as it is drunk.
   d The cost of posting a letter compared to the mass.

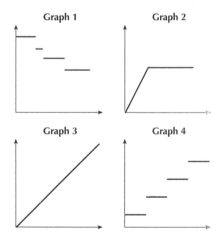

Graph 1   Graph 2

Graph 3   Graph 4

Solve problems

A taxi's meter reads £2 at the start of every journey.

As soon as the taxi has travelled two miles, £3 is added to the fare.

The reading on the meter then increases in steps of £3 for each whole mile covered, up to five miles.

For journeys over five miles, an extra £1 is added per mile.

a Complete a table showing the prices for journeys up to 8 miles.

b Show the results on a step graph.

a

Distance (miles)	0	1	2	3	4	5	6	7	8
Cost	£2	£2	£5	£8	£11	£14	£15	£16	£17

b

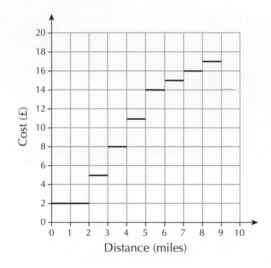

11 An online retailer charges an extra amount for postage and packaging. The amount charged for postage and packing depends on amount spent on goods.

Amount spent	Postage and packaging
£0.01–£4.99	£1.50
£5.00–£9.99	£2.00
£10.00–£16.99	£2.50
£17.00–£24.99	£3.00
£25.00–£59.99	£3.50

a Draw a step graph to show the cost of postage and packaging for the retailer

b DVDs cost £9. Work out how much can be saved on postage and packing by ordering three DVDs together instead of separately.

12 A telecommunications company charges 20p for the first minute of a telephone call and then 10p for each subsequent minute or part of a minute.

   a How much would a 4 minute 18 second call cost?

   b Draw a step graph to show the cost of calls up to 10 minutes in length.

13 An internet provider charges £10 for the first 10 hours spent online. Then for every additional full hour spent online £1.50 is added to the charge.

   a How much does it cost to go online for 7 hours?

   b How much does it cost to go online for 12 hours?

   c What is the maximum time that can be spent online for £20?

31.2 Distance–time graphs

● I can draw graphs from real-life situations to illustrate the relationship between two variables

Graphs are used everywhere; in newspapers, advertisements, on TV and the internet.

Most of these graphs show a relationship between two variables. One variable is shown on one axis and the other is shown on the other axis.

You can use a **distance–time graph**, like the one here, to describe a journey.

The vertical axis shows distance travelled and the horizontal one shows time: this one shows the time of day.

This graph shows that:

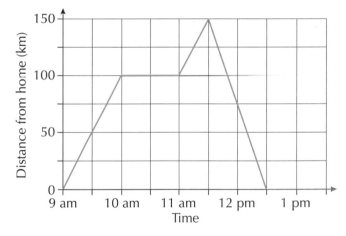

- 100 km was travelled in the first hour

- no distance at all was travelled for an hour (the line is flat)

- the journey continued for half an hour, from 11:00 a.m. to 11:30 a.m.

- the return journey started at 11:30 a.m.

Note that the line shows the return journey by moving in the opposite direction. It goes down instead of up.

You can also use the graph to work out the average speed for a journey. In the example, above, you can see that the first 100 km were covered in 1 hour. This represents an **average speed** of 100 km/h. In general:

$$\text{average speed} = \frac{\text{total distance covered}}{\text{total time taken}}$$

Develop fluency

Sheila set off from home to the vet's surgery. She travelled 90 km in $1\frac{1}{2}$ hours.

Sheila spent 30 minutes at the vet's surgery. She then travelled back home in $2\frac{1}{4}$ hours. Draw a distance–time graph of the journey.

Work out the key coordinates (time, distance from home).

- The start from home is represented by $(0, 0)$.
- After $1\frac{1}{2}$ hours Sheila has travelled 90 km so she arrives at the vet's surgery at $(1\frac{1}{2}, 90)$.
- Sheila spends 30 minutes at the vet's surgery without covering any distance, then sets off home at $(2, 90)$.
- Sheila travels 90 km home in $2\frac{1}{4}$ hours and arrives at $(4\frac{1}{4}, 0)$.

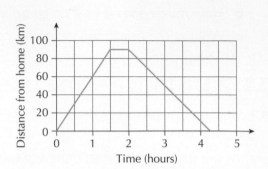

Now plot the points and draw the graph.

1 Mrs Jay had to travel to a job interview at Penford, which was 120 miles away.

She caught the 09:00 train from Shobton and travelled to Deely, 30 miles away. This train journey took 1 hour.

There she had to wait 30 minutes for a connecting train to Penford.

This train took 1 hour.

Her job interview at Penford lasted 2 hours.

Her return journey to Shobton lasted $1\frac{1}{2}$ hours.

a Copy this grid, using a scale of 2 centimetres to 1 hour and 1 centimetre to 10 miles.

b Draw, on the grid, a distance–time graph for Mrs Jay's journey.

c Label Deely on the vertical axis.

d How far was Mrs Jay from Shobton at 11:30?

e At which times was Mrs Jay 60 miles from Shobton?

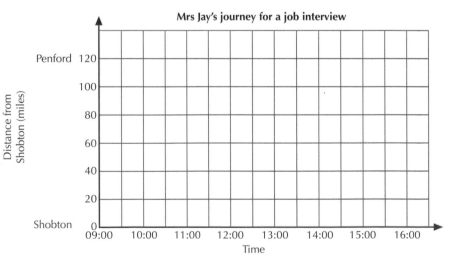

2   A petrol tanker made this journey along a motorway.

Filled up at petrol depot, then drove 10 miles in 30 minutes to Dibley Service Station. Spent 30 minutes filling the pumps, then drove a further 20 miles in 1 hour to Penton Service Station.

Spent 30 minutes filling the pumps and took 30 minutes for a tea break.

Drove a further 30 miles in 1 hour to Hillview Service Station and spent 30 minutes filling the pumps. Then returned to the depot in 90 minutes.

a Draw a grid, using a scale of 1 centimetre to 30 minutes, from 0 to 6 hours, on the horizontal (time) axis and 1 centimetre to 5 miles, from 0 to 60 miles, on the vertical (distance) axis. Use it to draw a distance–time graph for the journey.

b Label the places of delivery on the vertical axis.

c How far was the tanker from the depot after:

i 90 minutes   **ii** 3.5 hours   **iii** 5 hours?

3   A car transporter left the factory and took 30 minutes to travel 15 miles to a car dealer in Harton. It took 30 minutes to unload three of the cars. The transporter then travelled 12.5 miles in the next hour and made a delivery at Glimp. This delivery and lunch took an hour. A final delivery was made 30 minutes later at Unwich after a 10-mile drive. This delivery took 30 minutes. The transporter returned to the factory, taking 2 hours to get back.

Draw a grid, using a scale of 1 centimetre to 30 minutes, from 0 to 6 hours, on the horizontal (time) axis and 1 centimetre to 5 miles, from 0 to 50 miles, on the vertical (distance from factory) axis. Use it to draw a distance–time graph for this journey. Mark the places of delivery on the vertical axis.

4   Marco travelled from home to Manchester Airport, a distance of 100 km.

It took him 2 hours.

He stopped there for 30 minutes, then picked up his sister and took her straight back home, taking $2\frac{1}{2}$ hours.

Draw a grid, using a scale of 1 cm to 30 minutes, from 0 to 5 hours, on the horizontal (time) axis and 1 cm to 20 km, from 0 to 100 km, on the vertical (distance from home) axis. Use it to draw the travel graph for this journey.

5   Sarah travelled from home to her office, 60 km away. She left home at 11:00 am and travelled the first 40 km in 1 hour. She stopped for 30 minutes and then completed her journey in 20 minutes.

Draw a grid, using a scale of 1 cm to 20 minutes, from 0 to 2 hours, on the horizontal (time) axis and 1 cm to 10 km, from 0 to 60 km, on the vertical (distance) axis. Use it to draw the travel graph for this journey.

6 A swimming pool, 2 m deep, was filled with water from a hose. The pool was empty at the start and the depth of water in the pool increased at the rate of 4 cm/minute.

   a Copy and complete this table to show the depth of water after various times.

Time (minutes)	0	10	25	40	50
Depth (cm)			100		

   b Draw a grid, using a scale of 1 cm to 5 minutes, from 0 to 60 minutes, on the horizontal (time) axis and 2 cm to 50 cm. from 0 to 200 cm, on the vertical (depth) axis. Use it to draw a graph to show the increase in depth of water against time.

7 This distance–time graph shows a car journey between two towns.

   a How many times does the car stop on the journey?

   b What is the total distance between the towns?

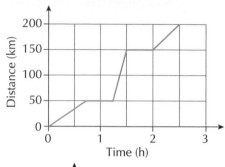

8 This distance–time graph shows Tony's journey from his home to the shop.

   a How far is his home from the shop?

   b How long did he spend at the shop?

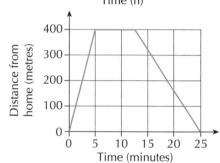

9 This distance–time graph shows a cycle journey.

   a How far has the cyclist travelled after 2 minutes?

   b How long did she stop for?

   c How far was the whole journey?

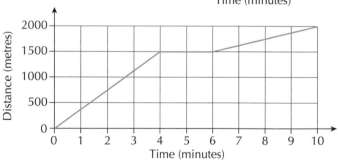

10 Pat walks to the post box and back home. This is his distance–time graph.

   a What time does he set off?

   b How far does he walk altogether?

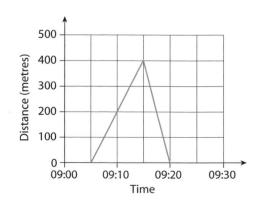

Reason mathematically

Look again at this distance–time graph.

Work out the speed:

a for the first part of the journey.

b for the last part of the journey.

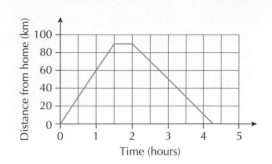

a Distance travelled is 90 km in 1.5 hours

$$\text{speed} = \frac{\text{distance}}{\text{time}}$$

So, speed $= \dfrac{90}{1.5} = 60$ km/h

b Distance travelled is 90 km in 2.25 hours

So, speed $= \dfrac{90}{2.25} = 40$ km/h

11 Look again at your distance–time graph in question 1.
Work out the speed for the return journey to Shobton.

12 Look again at your distance–time graph in question 4.
Work out Marco's speed **a** to the airport **b** from the airport.

13 Look again at your distance–time graph in question 5.
Explain how to find Sarah's average speed over the last 20 minutes.

14 Look again at the distance–time graph in question 8.
Which part of the walk is faster?
Give a reason for your answer.

Solve problems

Station A and station B are 180 miles apart.

A train leaves station A and travels to station B at 90 mph.

Another train leaves station B and travels to station A at 60 mph.

Both trains set off at the same time.

Work out how long it takes before the trains pass each other.

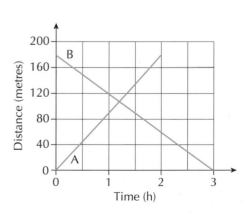

The train leaving station A takes 2 hours for the whole journey.

The train leaving station B takes 3 hours for the whole journey.

Draw a distance–time graph and then read off where they cross.

The trains cross after 1.2 hours or 1 hour 12 minutes.

15 Two service stations, A and B, are 40 miles apart.
One car travels from A to B at 60 mph and
another car travels from B to A at 40 mph.
Both cars leave at the same time.
Work out how long it takes before the cars pass
each other.
Use a copy of this grid to work it out.

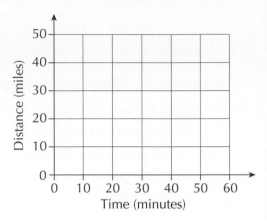

16 Look again at the graph in question 9.
How much greater would the average speed be
without the stop?

17 The water from a swimming pool was pumped out at the rate of 32 litres/minute. It took
about 5 hours for the pool to be emptied.

a Copy and complete this table to show how much water there is in the pool at
various times.

Time (minutes)	0	50	100	150	200	250	300
Water left (litres)	9000						

b Draw a graph to show the amount of water left in the pool, against time.

c How long did it actually take to empty the pool?

31.3 More time graphs

● I can interpret and draw time graphs

A **time graph** is used to show a relationship between a variable and time, for example,

· how a distance changes with time, usually referred to as a distance–time graph

· how temperature changes with time

· how populations change with time.

Develop fluency

This time graph shows how the temperature of water increases with time, as it is heated.

a Estimate the temperature of the water after:

i $4\frac{1}{2}$ minutes **ii** $9\frac{1}{2}$ minutes.

b Estimate how long it takes for the water to boil (when the temperature reaches 100 °C).

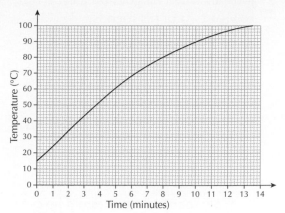

a See the red lines drawn on the graph.

i After $4\frac{1}{2}$ minutes the temperature is about 56 °C.

ii After $9\frac{1}{2}$ minutes the temperature is about 88 °C.

b See the green lines drawn on the graph.

It takes $13\frac{1}{2}$ minutes for the water to boil.

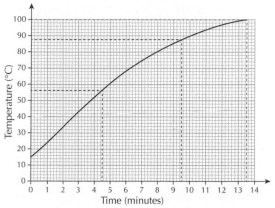

1 The graph shows how a liquid cools down over time.

a How long does it take the liquid to cool from 80°C to 20°C?

b Estimate the temperature of the liquid 10 minutes after it starts to cool.

c Estimate how long it takes the liquid to cool from 60°C to 40°C.

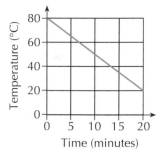

2 This distance–time graph shows the journey of a jogger on a 5-mile run. At one point she stopped to admire the view and at another point she ran up a steep hill.

a For how long did she stop to admire the view?

b What distance into her run was the start of the hill?

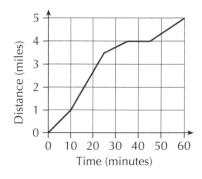

3 The time–distance graph shows Ari's journey when he goes on a cycle ride to visit two of his friends.

a At what time did he arrive at the first friend's house?

b How long did he stay at the first friend's house?

c At what time did he arrive at the second friend's house?

d How long did he stay at the second friend's house?

e How many kilometres did he cycle in total?

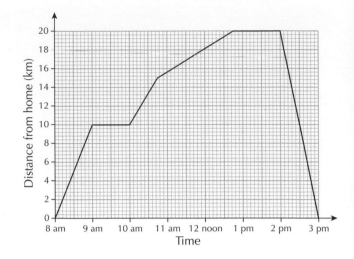

4 The graph shows the journeys of two coaches.

a Which coach travelled further?

b Which coach was travelling faster at the beginning of its journey?

c How long did Coach B stop for, in total?

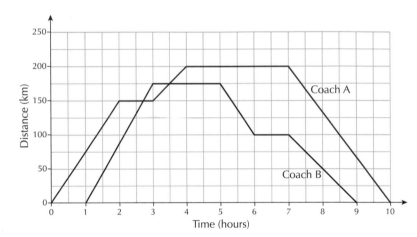

5 This distance–time graph shows how two rockets flew during a test flight. Rocket D flew higher than Rocket E.

a Estimate the maximum height reached by Rocket D.

b Estimate how much higher Rocket D reached than Rocket E reached.

c How long after the launch were both rockets at the same height?

d For how long was each rocket higher than 150 m?

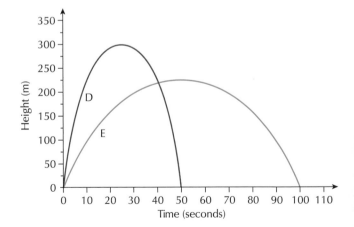

6 The graph shows how the population of a small town changed over 20 years.

  a What was the population at the start of the 20 years?

  b How much did the population increase in the first 10 years?

  c What was the population at the end of the 20 years?

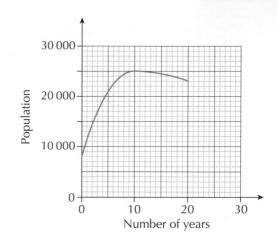

7 This graph illustrates the amount of water in a bath after it has started to be filled.

  a How long did the bath take to fill?

  b When was the plug pulled out, for the bath to start emptying?

  c How long did the bath take to empty?

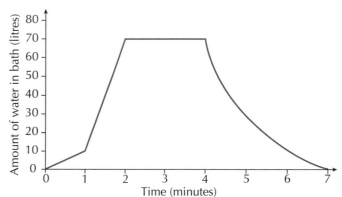

8 Ben and Gurjit left their homes at the same time and travelled to each other's home. This graph shows their journeys.

  a What is the distance between the two houses?

  b What time did they meet?

  c How long were they together?

  d What time did Gurjit arrive at Ben's house?

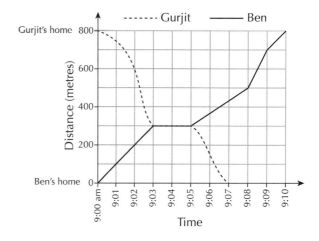

Reason mathematically

Match each description to its graph.

a The temperature of the desert over a 24-hour period.

b The temperature of a kitchen over a 24-hour period.

c The temperature of a cup of tea as it cools down.

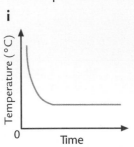
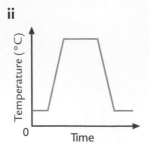
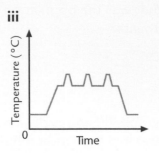

a This matches **ii** as temperature is low at night and high during the day.

b This matches **iii** as temperature increases quickly, probably cooking three meals in the day.

c This matches **i** as the temperature does not rise at any point.

9 Match these distance–time sketch graphs to the situations described.

Graph 1   **Graph 2**   **Graph 3**

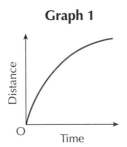
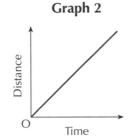
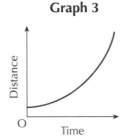

a The distance travelled by a train moving at a constant speed.

b The distance travelled by a motorbike accelerating to overtake.

c The distance travelled by an old car, which gradually slows down.

10 Match each graph with the correct description.

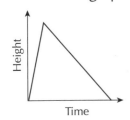
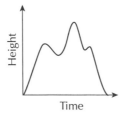
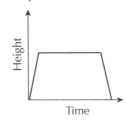
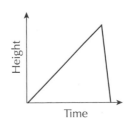

a A hovercraft journey.

b A rocket launch and parachute landing.

c A flight of a radio-controlled model aeroplane.

d A model hot air balloon that takes off and catches fire.

11 This distance–time graph illustrates three people running in a race.
Describe the race.

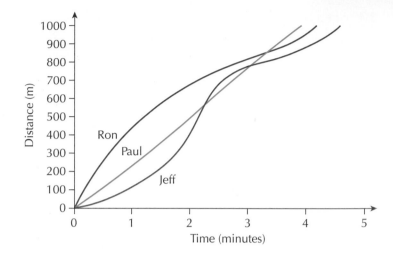

Solve problems

Each of these candles is 25 cm tall and burns to the ground in 20 hours.

For each candle, sketch a graph showing its height over time.

a

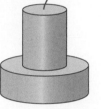

b

c

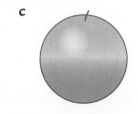

a

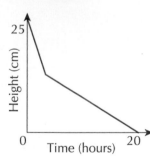

b

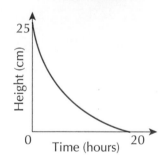

c

12 Water drips steadily into the container shown.

The sketch graph shows how the depth of water varies with time.

Make similar sketch graphs for bottles with these shapes.

a

b

c

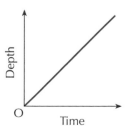

d

13 Sketch a graph to show:

   a the height of a space shuttle that takes off, circles the Earth, then lands

   b the mass of a woman 3 months before she is pregnant, during her pregnancy and 3 months after she gives birth.

14 Sketch a graph to illustrate:

   a the height of a person from birth to age 30 years

   b the temperature in summer from midnight to midnight the next day

   c the height of water in a WC cistern from before it is flushed to afterwards. Estimate any necessary measurements.

31.4 Graphs showing growth

● I can interpret and draw exponential growth graphs

Graphs that continuously increase at a fixed rate are known as **exponential growth graphs**.

Graphs that show an increase in population or the increase in investments at a bank are examples of exponential growth graphs.

Develop fluency

The population of a village doubles every five years.

The table shows the increase in the population over 20 years.

Number of years	0	5	10	15	20
Population	500	1000	2000	4000	8000

Estimate the number of years it will take for the population of the village to reach 5000.

Plot the points on a graph.

Draw a smooth curve passing through all the points.

Then draw suitable lines on the graph, as shown, to estimate the number of years for the population of the village to reach 5000.

Start with a horizontal line from 5000, then draw a vertical line down, from where the first line meets the curve.

This is $16\frac{1}{2}$ years.

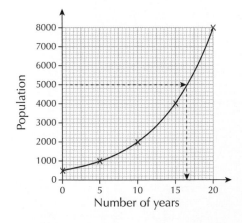

1   The graph shows the population
    growth over time (measured in
    seconds) of a particular strain
    of bacteria.

    a How many bacteria are alive
       after one minute?

    b How many bacteria are alive
       after two minutes?

    c How many bacteria are alive
       after three minutes?

    d How long did it take the
       population of bacteria to
       reach 30 000?

    e How long did it take the
       population of bacteria to
       reach 70 000?

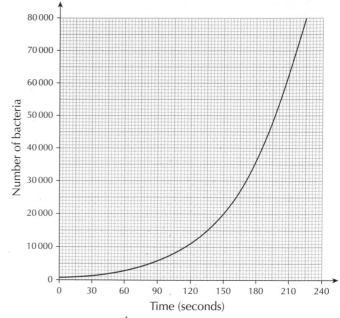

2   The graph shows how an investment has
    grown exponentially over a 30-year period.

    a What was the initial investment?

    b How long did it take for the investment
       value to grow to £20 000?

    c How long did it take for the investment
       value to grow to £50 000?

    d What was the value of the investment after
       20 years?

    e What was the value of the investment after
       30 years?

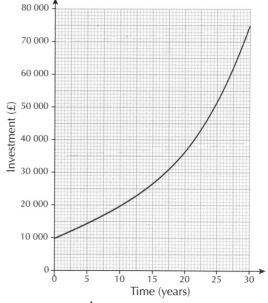

3   The graph shows how fast a truck is travelling as it
    increases its speed.

    a What is the truck's maximum speed?

    b What is the speed of the truck after 11 seconds?

    c How many seconds does it take the truck to reach
       a speed of 50 km/h?

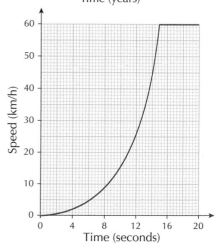

4 The graph shows the average growth rate of black sea turtles.

    a Estimate the length of a black sea turtle when it is born.

    b Estimate the length of a black sea turtle when it is 40 years old.

    c Estimate the age of a black sea turtle when it has grown to 40 cm.

    d Estimate the age of a black sea turtle when it has grown to 80 cm.

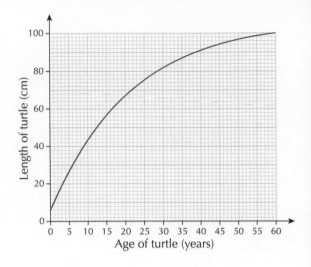

5 The population growth over time (measured in years) of a village is shown on the graph.

    a How many villagers were there initially?

    b How many villagers were there after 10 years?

    c How many years did it take for the population to reach 2000?

    d How much did the population increase by in the first 30 years?

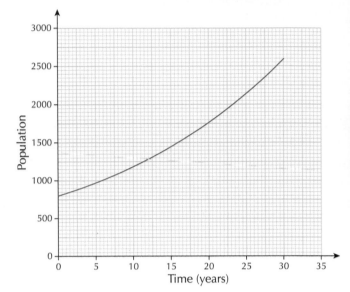

6 The graph shows the population growth over time (measured in seconds) of a strain of bacteria.

    a How many bacteria are alive after 1 minute?

    b How many bacteria are alive after $2\frac{1}{2}$ minutes?

    c How long did it take for the population to reach 10 000?

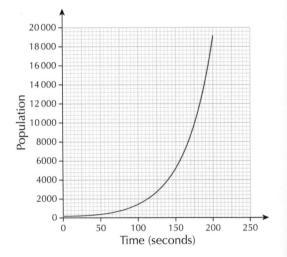

7 Bethany invested £3500 in a bank account offering 7% compound interest.
The graph shows how the amount in the account grew exponentially over 50 years.

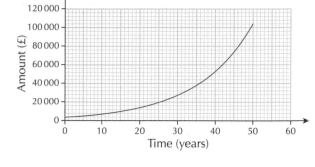

a What was the value of the investment after 20 years?

b What was the value of the investment after 40 years?

c After how many years did Bethany have £80 000 in the account?

8 In order to spread, a virus needs to find hosts to infect. The graph shows how many hosts have been infected by a particular virus.

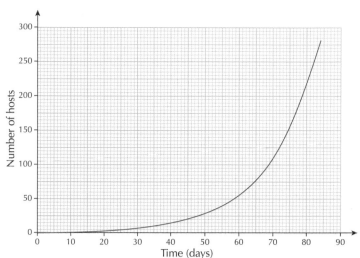

a How many hosts had been infected after 3 weeks?

b How many hosts had been infected after 6 weeks?

c How many hosts had been infected after 9 weeks?

d How many hosts had been infected after 12 weeks?

e After how many days had the virus infected at least 100 hosts?

9 The graph shows the temperature of a cup of tea as it cools.

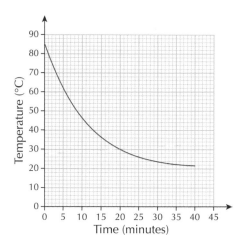

a What was the initial temperature of the cup of tea?

b What was the temperature of the cup of tea after 10 minutes?

c What was the temperature of the cup of tea after half an hour?

d After how many minutes had the tea cooled to 30°C?

e After how many minutes had the tea cooled by 20°C?

Reason mathematically

A population of rabbits is introduced into a new habitat. The table shows the growth of the population over a five-year period.

Year	0	1	2	3	4	5
Number of rabbits	100	150	225	338	506	759

   a Draw a graph to show the increase in the rabbit population.

   b How many rabbits were originally introduced?

   c Describe how the rabbit population is increasing.

   d Estimate how many rabbits there will be after six years, given that the rabbit population growth continues at the same rate.

a

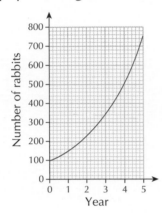

b 100

c Population increases by 50% each year, for example
100 × 1.5 = 150 or 150 ÷ 100 = 1.5.

d 759 × 1.5 is approximately 1140.
This could be obtained by extending the graph.

10 The table shows the exponential growth of the population of a town.

Year	1990	1995	2000	2005	2010	2015
Population	24 000	36 000	54 000	81 000	121 500	182 250

   a How many times greater was the population in 1995 than in 1990?

   b How many times greater was the population in 2000 than in 1995?

   c Show that the population increased by the same rate every five years.

11 Look again at the investment graph in question 7.

   a How many years did it take for the investment to gain £16 500 interest?

   b Estimate the number of extra years until the investment is worth over £200 000.

12 Look again at the graph in question 8.
Estimate the number of hosts that will have been infected by the virus after 15 weeks.

13 Look again at the graph in question 9.
The temperature of the tea approaches room temperature as it cools.
Estimate the temperature of the room.

Solve problems

The number of fish in a pond is decreasing at a fixed rate.

The table shows how the number of fish has changed over 3 years.

Year	2017	2018	2019
Number of fish	800	600	450

a Draw a graph to show this information.

b The trend is expected to continue for 2 more years. Use the graph to estimate the number of fish in the pond in 2021.

a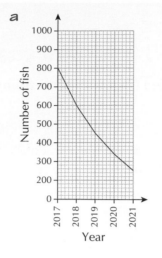

b *Continuing the curve and reading off at 2021 gives about 250 fish.*

14 The value of an investment increases exponentially over time.

The table shows the growth of the investment over a three-year period.

Year	2016	2017	2018	2019
Value (£)	10 400.00	10 920.00	11 466.00	12 039.30

a Draw a graph to show the value of the investment.

b The investment continues to grow at the same rate. Estimate the value of the investment in 2020.

15 A curve has the exponential equation $y = 3 \times 2^x$.

    a i Work out the value of y when $x = 0$.   **ii** Work out the value of y when $x = 3$.

    b Work out the value of x when $y = 96$.

16 A curve has the exponential equation $y = ab^x$.

The curve passes through the points with coordinates $(0, 4)$, $(1, 20)$ and $(2, c)$.

Work out the value of   **a** a **b** b **c** c

Now I can...

interpret step graphs	draw graphs from real-life situations to illustrate the relationship between two variables	draw exponential growth graphs
interpret time graphs	draw time graphs	
interpret exponential growth graphs		

CHAPTER 32 Comparing sets of data

32.1 Grouped frequency tables

- I can create a grouped frequency table from raw data

Develop fluency

You can draw up a frequency table to record how many times each value in a set of data occurs. The number of times a data value occurs is its **frequency**.

If the data is about specific events it is easy to see a pattern, for example, when people vote for different political candidates in an election there is always a limited selection of choices to be made. This is called **discrete** data.

Often, data has a wide range of possible values; examples include masses or heights of all the pupils in a class. This is called **continuous** data and you must group it together, to see any pattern. In a **grouped frequency table**, you arrange information into **classes** or groups of data to do this. You can create frequency diagrams from grouped frequency tables to illustrate the data.

These are the journey times, in minutes, for a group of 16 railway travellers.

25, 47, 12, 32, 28, 17, 20, 43, 15, 34, 45, 22, 19, 36, 44, 17

Construct a frequency table to represent the data.

Looking at the data, 10 minutes is a sensible class interval size.

Write the class intervals in the form $10 < T \leq 20$.

$10 < T \leq 20$ is a short way of writing the time interval of 10 minutes to 20 minutes.

The possible values for T include 20 minutes but not 10 minutes.

There are six times in the group $10 < T \leq 20$: 12, 17, 20, 15, 19, 17.

There are three times in the group $20 < T \leq 30$: 25, 28, –22.

There are three times in the group $30 < T \leq 40$: 32, 34, 36.

There are four times in the group $40 < T \leq 50$: 47, 43, 45, 44.

Put all this information into a table.

Time, T (minutes)	Frequency
$10 < T \leq 20$	6
$20 < T \leq 30$	3
$30 < T \leq 40$	3
$40 < T \leq 50$	4

1 This table shows the lengths of time 25 customers spent in a shop.

a One of the customers was in the shop for exactly 20 minutes. In which class was this customer recorded?

b Which is the modal class?

Time, T (minutes)	Frequency
$0 < T \leq 10$	12
$10 < T \leq 20$	7
$20 < T \leq 30$	6

2 The table shows the lengths (*L* metres) of 30 snakes.

  a One of the snakes is 4.5 metres long. Which class contains this length?

  b How many snakes are shorter than 5 metres?

  c How many snakes have a length of 5.5 metres or more?

  d How many snakes are between 4 and 6 metres long?

  e How many snakes could be exactly 5 metres long?

Length of snake (*L* metres)	Frequency
$3.5 \leq L < 4.0$	2
$4.0 \leq L < 4.5$	4
$4.5 \leq L < 5.0$	7
$5.0 \leq L < 5.5$	9
$5.5 \leq L < 6.0$	5
$6.0 \leq L < 6.5$	3

3 These are the heights (*h* metres) of 20 people.

1.65, 1.53, 1.71, 1.72, 1.48, 1.74, 1.56, 1.55, 1.80, 1.85, 1.58, 1.61, 1.82, 1.67, 1.47, 1.76, 1.79, 1.66, 1.68, 1.73

  a Copy and complete the frequency table.

  b What is the modal class?

Height, *h* (metres)	Frequency
$1.40 < h \leq 1.50$	
$1.50 < h \leq 1.60$	
$1.60 < h \leq 1.70$	
$1.70 < h \leq 1.80$	
$1.80 < h \leq 1.90$	

4 These are the masses (*M* kilograms) of fish caught in one day by an angler.

0.3 5.6 3.2 0.4 0.6 1.1 2.4 4.8 0.5 1.6 5.1 4.3 3.7 3.5

  a Copy and complete the frequency table.

  b What is the modal class?

  c The angler's daughter arrived with some lunch just as he was catching a fish. What is the probability that this fish was in the range of $3 < M \leq 4$?

Mass, *M* (kilograms)	Frequency
$0 < M \leq 1$	
$1 < M \leq 2$	
...	

5 These are the temperatures (in °C) of sixteen towns in Britain on one day.

12, 10, 9, 13, 12, 14, 17, 16, 18, 10, 12, 11, 15, 15, 12, 13

  a Copy and complete the frequency table.

  b What is the modal class?

  c What is the range of temperatures in these cities?

Temperature, *T* (°C)	Frequency
$8 < T \leq 10$	
$10 < T \leq 12$	
...	

6 The volumes (Vcl) of liquid contained in twenty coconuts are shown below.

12.2, 11.1, 10.5, 12.8, 12.0, 10.1, 11.8, 12.3, 10.7, 12.7, 10.0, 11.6, 12.1, 10.5, 10.8, 12.6, 10.7, 11.4, 12.8, 11.3

Volume of liquid (Vcl)	Tally	Number of coconuts
$10 \leq V < 10.5$		
$10.5 \leq V < 11$		
...		

a Copy and complete the table.

b What is the range of the volumes of liquid?

c What is the modal class?

d What proportion of coconuts contain less than 11 cl?

e What is the probability that a coconut chosen at random contains 12 cl or more?

7 These are the results of a maths exam of 30 students.

33, 48, 33, 50, 33, 46, 35, 41, 36, 45, 36, 47, 36, 37, 45, 37, 39, 31, 41, 41, 43, 45, 46, 47, 48, 39, 41, 49 50, 47

Results (R marks)	Tally	Number of students
$30 \leq R < 35$		
...		

a Copy and complete the table.

b What is the range of the results?

c What is the modal class?

d What proportion of students achieved more than 40 marks in the exam?

e What is the probability that a student chosen at random achived a result more or equal to 35, but less than 45?

8 Mr Smith recorded the number of errors in 20 different maths books. Here are the results.

20, 8, 11, 8, 16, 8, 15, 8, 17, 8, 9, 10, 11, 13, 15, 16, 17, 20, 20, 13

a Draw a grouped frequency table for the data. Use a class interval of 4.

b What is the modal class?

c What proportion of the books had less than 16 typing errors?

d What is the probability that a book chosen at random has 20 typing errors?

9 Jason recorded how many copies of a daily magazine he sold each month. Here are his results.

54, 56, 59, 58, 43, 51, 54, 49, 42, 61, 65, 58, 50, 53, 62, 61, 63, 54, 65, 57, 60, 67, 52, 52, 49, 55, 59, 61, 45, 68

a Draw a grouped frequency table for the data.

b What is the modal class?

c For what percentage of the time did Jason sell less than 50 magazines?

10 Here are the heights of 20 students.
170, 165, 175, 182, 196, 152, 162, 154, 163, 178,
188, 165, 171, 168, 167, 170, 161, 174, 180, 190

   a Draw a grouped frequency table for the data.
   b Write down the modal class interval.
   c A student is chosen at random. What is the probability that the student is 180 cm tall or taller?

Reason mathematically

24 boys run a 100 m race. These are the times they take to complete the race.

10, 11, 11, 12, 12, 13, 13, 13, 13, 14, 14, 14,
15, 15, 15, 15, 15, 15, 16, 17, 18, 18, 19, 19

The grouped frequency table shows information based on the data.

   a Find two things that are wrong with this table.
   b Make a new grouped frequency table to show the data.

Time (t seconds)	Frequency
$10 \le t \le 12$	4
$12 \le t \le 14$	7
$14 \le t \le 16$	6
$16 \le t \le 18$	2
$18 \le t \le 20$	4

a The time intervals are incorrect. For example, 12 should not be included in first interval.

The frequencies are incorrect. The total for the frequencies should be 24.

b

Time (t seconds)	Frequency
$10 \le t < 12$	3
$12 \le t < 14$	6
$14 \le t < 16$	9
$16 \le t < 18$	2
$18 \le t < 20$	4

11 Twenty people at a party take part in a music quiz. They were played clips from 25 songs and had to write down what they were and who sang them. Their scores are shown below:

4, 5, 8, 8, 8, 11, 12, 12, 14, 15, 15, 16, 17, 18, 18, 20, 21, 21, 21, 23

   a Put these results into a grouped frequency table. Use a class interval of 5.
   b What was the range of their scores?
   c How many separate modes are there?
   d What was the modal class?
   e Are the modal scores inside the modal group? Explain what has happened here.
   f What proportion of players got less than half correct?

12 The table shows the number of pieces of fruit and vegetables a group of children eat each day.

Look at the statements. Are they true or false? Explain your reasoning. If the statement is false, correct it.

Number of pieces	Number of children
$0 \leq n < 2$	2
$2 \leq n < 4$	4
$4 \leq n < 6$	3
$6 \leq n < 8$	2
$8 \leq n < 10$	3

a There were five children in the group.

b The mode of the data is four.

c A child was picked up at random.
The probability of the child eating more than 8 pieces of fruit is the same as the probability of the child eating less than 6 pieces of fruit.

13 Adam collects 20 leaves. These are their lengths.
41, 41.5, 43.3, 44, 44.6, 45.3, 47.4, 50.3, 53.8, 54.1, 54.7, 55.6, 55.8, 57.1, 60.5, 61, 61.2, 63.3, 63.9, 64
He designs a frequency table to show the information.

Length (L mm)	Frequency
40–44	
45–49	
50–54	
55–59	
More than 60	

a There are two things wrong with this table. What are they?

b Make a new grouped frequency table to show the data.

c What is the modal class?

Solve problems

Lola collected 30 pebbles. She weighed each of them and recorded the information in the table.

Mass (w grams)	Frequency
$40 \leq w < 50$	x
$50 \leq w < 60$	3
$60 \leq w < 70$	$2x$
$70 \leq w < 80$	9
$80 \leq w < 90$	6

a How many pebbles had a mass between 40 and 50 grams?

b How many pebbles had a mass between 60 and 70 grams?

c What is the modal class?

d What percentage of the pebbles have a mass of 70 grams or more?

a $x + 3 + 2x + 9 + 6 = 30$
$3x + 18 = 30$
$3x = 12$
$x = 4$

b $2x = 2 \times 4 = 8$ pebbles.

c Modal class = $70 \leq w < 80$

d $\dfrac{9+6}{30} = \dfrac{15}{30} \times 100 = 50\%$

14 Twenty people at a party play a memory game. They are shown 20 objects for a minute before they are removed. They then must write down as many as they can remember. The table shows how they did.

Objects remembered	Number of people
1–5	0
6–10	2
11–15	12
16–20	6

   a What was the modal class?

   b How many people remembered more than half the objects?

   c Could anyone have remembered them all?

   d What percentage only remembered half or less?

   e What is the minimum possible range?

   f What is the maximum possible range?

15 Twenty girls compare how much money they received for their birthdays, in pounds:
20, 25, 32, 35, 40, 45, 50, 50, 52, 55, 60, 60, 70, 70, 70, 75, 75, 80, 85, 90

Money (£)	Tally	Frequency
$0 < M \leq 20$		
$20 < M \leq 40$		
$40 < M \leq 60$		
$60 < M \leq 80$		
$80 < M \leq 100$		

   a Copy and complete the frequency table of this data.

   b Without knowing the actual data what could the maximum range be? What is the actual range?

   c Why is the actual mode different to the modal class?

   d What proportion of the girls received more than £50?

16 A petrol station owner surveyed a sample of customers to see how many litres of petrol they bought. These are the results.
27.6, 31.5, 48.7, 35.6, 44.8, 56.7, 51.0, 39.5, 28.8, 43.8, 47.3, 36.6, 42.7, 45.6, 32.4, 51.7, 55.9, 44.6, 36.8, 49.7, 37.4, 41.2, 38.5, 45.9, 34.1, 54.3, 41.3, 49.4, 38.7, 33.2
Using a class size of 5 litres, work out the modal class of the amount of petrol that customers bought.

17 In a doctors' surgery, the practice manager told each doctor that the length of most consultations should be more than 5 minutes but less than 10. She monitored the consultation times of the three doctors at the practice throughout one day. These are the results.
Dr Speed (minutes): 6, 8, 11, 5, 8, 5, 8, 10, 12, 4, 3, 6, 8, 4, 3, 15, 9, 2, 3, 5
Dr Bell (minutes): 7, 12, 10, 9, 6, 13, 6, 7, 6, 9, 10, 12, 11, 14
Dr Khan (minutes): 5, 9, 6, 3, 8, 7, 3, 4, 5, 7, 3, 4, 5, 9, 10, 3, 4, 5, 4, 3, 4, 4, 9

   a Did any of the doctors manage to follow the practice manager's advice?

   b Write a short report about the consultation times of the three doctors.

32.2 Drawing frequency diagrams

- I can interpret frequency diagrams
- I can draw a frequency diagram from a grouped frequency table

Develop fluency

Construct a frequency diagram as a block graph and a line graph for this data about runners' race times.

Race times, t (minutes)	Frequency
$0 < t \leq 15$	4
$15 < t \leq 30$	5
$30 < t \leq 45$	10
$45 < t \leq 60$	6

It is important that the diagram has a title and labels, as shown here.

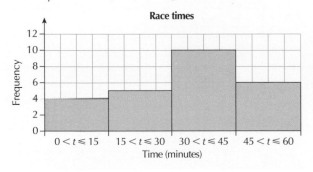 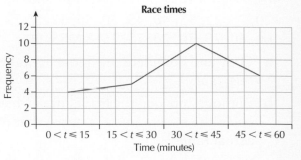

1   Construct a frequency diagram using the data from each frequency table.

a   Aircraft flight times

Time, T (hours)	Frequency
$0 < T \leq 1$	3
$1 < T \leq 2$	6
$2 < T \leq 3$	8
$3 < T \leq 4$	7
$4 < T \leq 5$	4

b   Temperatures of European capital cities

Temperature, T (°C)	Frequency
$0 < T \leq 5$	2
$5 < T \leq 10$	6
$10 < T \leq 15$	11
$15 < T \leq 20$	12
$20 < T \leq 25$	7

c   Lengths of metal rods

Length, l (centimetres)	Frequency
$0 < l \leq 10$	9
$10 < l \leq 20$	12
$20 < l \leq 30$	6
$30 < l \leq 40$	3

d   Masses of animals on a farm

Mass, M (kg)	Frequency
$0 < M \leq 20$	15
$20 < M \leq 40$	23
$40 < M \leq 60$	32
$60 < M \leq 80$	12
$80 < M \leq 100$	6

2 Shop prices for the same beach ball are as follows. Draw a bar chart to illustrate the data.

Price of beach ball (£P)	Number of shops
$2.80 < P \le 3.00$	27
$3.00 < P \le 3.20$	51
$3.20 < P \le 3.40$	30
$3.40 < P \le 3.60$	20
$3.60 < P \le 3.80$	13
$3.80 < P \le 4.00$	9

3 Amounts of coffee served in 500 ml cups from a survey of cafes are shown in the table below.

a Draw a bar chart for the data.

b The table shows that some cafes are cheating their customers by giving short measures and others are being a little generous. Using percentages, show whether more cafes are giving too little or too much.

Amount of coffee (V ml)	Number of cups
$485 \le V < 490$	3
$490 \le V < 495$	7
$495 \le V < 500$	13
$500 \le V < 505$	16
$505 \le V < 510$	11

4 This graph shows the mean monthly temperature for two cities.

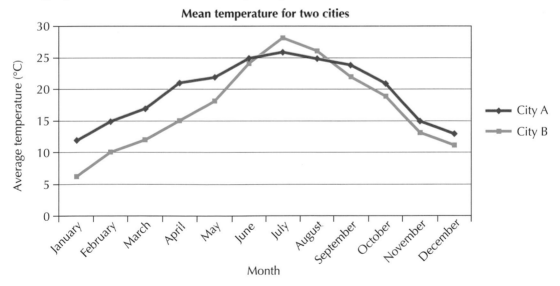

Mean temperature for two cities

a Which city has the highest mean monthly temperature?

b Which city has the lowest mean monthly temperature?

c How many months of the year is the temperature higher in City A than in City B?

d What is the difference in average temperature between the two cities in February?

5 The table shows the pulse rate of forty students immediately after exercising.

   a Draw a block graph for the data.

   b Which is the modal class?

Pulse rate	Frequency
$110 \leq P < 115$	2
$115 \leq P < 120$	8
$120 \leq P < 125$	6
$125 \leq P < 130$	14
$130 \leq P < 135$	7
$135 \leq P < 140$	3

6 Look at these frequency polygons. They show the number of students taking part in the orchestra and choir.

   a Which activity has the highest number of Year 7 students?

   b Which activity was the most popular?

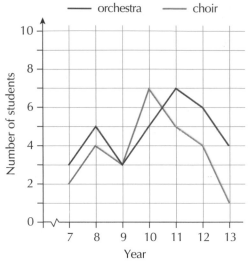

7 The bar chart shows the amount of time a group of students spend on their homework. Draw a frequency table for the data based on this bar chart.

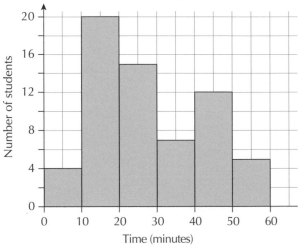

Reason mathematically

A shop manager collected data about the daily sales of ice cream every month throughout the year. He worked out the average (mean) daily sales figure for each month and put it on the graph below. In which month were average sales at their highest? Give a reason why you think this happened.

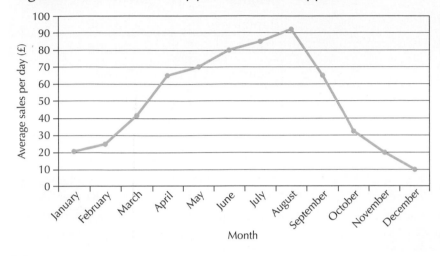

The highest average sales were in August (£92 per day). This was probably because the weather was warmer, as people tend to buy ice cream in warm weather.

8 The table shows the average height of sweetcorn plants after being sprayed with different amounts of a new fertilizer.

a Plot a graph. Use the following scales:
   x-axis (amount of fertilizer): 1 cm to 10 ml
   y-axis (height of plant): 2 cm to 0.1 m

b Estimate the level of fertilizer that would give plants a height of 1.5 m.

c Estimate the height of a plant sprayed with 35 ml of fertilizer.

d Which level of fertilizer would you advise the farmer to use? Explain your answer.

e At which levels did the fertilizer not improve growth?

Amount of fertilizer (A ml)	Height of plant (h m)
0	1.40
10	1.40
20	1.45
30	1.55
40	1.70
50	1.70
60	1.60
70	1.50
80	1.40
90	1.30
100	1.30
110	1.30

9 The table shows the age and price of 15 cars.

Age (years)	Price (£)
0	17 000
1	16 200
2	13 800
3	11 500
4	9900
5	7600
6	6500
7	4800
8	3200
9	2500
10	1000
11	500

a Plot a graph. Use the following scales:
   x-axis (age of cars): 1 cm to 1 year
   y-axis (price of car): 2 cm to 2500

b Estimate the price of the car after 8 years.

c Estimate the number of years that must pass for the car to have a price of £12 000.

d Why would it not be sensible to use the graph to estimate the price of the car after 13 years?

10 A shop manager collected data about the sales of sun cream throughout the year. He worked out the average (mean) daily sales figure for each month. Here is the information he collected.

Month	Number of sun cream
January	17 000
February	16 200
March	13 800
April	11 500
May	9900
June	7600
July	6500
August	4800
September	3200
October	2500
November	1000
December	500

a Plot a graph.

b In which month were average sales at their lowest? Give a reason why you think this happened.

c What do you think the graph will look like if the shop manager lived in Australia? Why?

Solve problems

A company makes and sells birthday cards in a shop and online. The table shows the sales of cards each month from January to August 2019.

Month	Jan	Feb	March	April	May	June	July	August
Sales in the shop	100	260	126	102	128	125	95	80
Sales online	130	140	134	120	132	128	100	95

a Find the difference in the total of online and shop sales for the first 8 months of the year.

b Which month had the most sales?

c Estimate the minimum and the maximum number of cards that can be sold in September.

a Total shop sales = 100 + 260 + 126 + 102 + 128 + 125 + 95 + 80 = 1016 cards.
Total online sales = 130 + 140 + 134 + 120 + 132 + 128 + 100 + 95 = 979 cards.
1016 – 979 = 37 cards

b The total sales are shown in the table

Month	Jan	Feb	March	April	May	June	July	August
Total Sales	230	400	260	222	260	253	195	175

February had the most sales.

c The minimum number of cards sold is 175 (sold in August). The maximum number of cards is 400 (sold in February). It can be estimated that the number of cards in September will be between 175 and 400 cards.

11 The three classes of a year group collected sponsor money for charity. These are the amounts collected by individual pupils.

Class A (£): 3.00, 9.60, 5.50, 8.45, 8.00, 7.35, 1.55, 15.50, 19.00, 14.00, 12.75, 13.50, 11.85, 12.00, 14.36, 15.40, 17.00, 6.55, 7.40, 8.00, 6.32, 1.00, 3.65, 16.50, 14.00, 19.55, 18.00, 16.46, 19.00

Class B (£): 5.00, 11.50, 7.40, 10.55, 10.00, 9.15, 3.55, 18.70, 21.00, 1.16, 14.95, 15.70, 13.95, 14.00, 16.30, 17.50, 19.00, 8.75, 9.20, 10.00, 8.12, 3.00, 5.85, 18.60, 16.00, 21.95, 20.00, 18.26, 21.00, 14.35, 5.47, 13.55

Class C (£): 4.00, 8.50, 6.70, 7.35, 9.00, 6.25, 2.75, 14.30, 20.00, 13.00, 11.65, 14.70, 10.45, 13.00, 13.38, 14.30, 18.00, 5.45, 8.70, 7.00, 7.42, 2.00, 2.15, 17.30, 13.00, 20.65, 17.00, 17.56, 18.00, 17.55, 19.65, 17.36

Create a diagram that the head of year can use, to show how well the pupils had done in raising sponsorship money, so that they can encourage the school to collect for charity.

12 A patient's temperature was taken every hour to check on their wellbeing. 36.8°C is the normal temperature for a human, and anywhere over 37.5°C is classed as fever.
The results are as shown:

a Draw a line graph to show these changes in temperature.

b As accurately as possible, when would you say the fever started and finished?

c What could you draw on your graph to help you answer part b?

d Did the fever last for more or less than a quarter of a day?

e At about what time do you think the medication started to reduce the fever?

Time	Temperature
9:00 a.m.	36.9°C
10:00 a.m.	37.0°C
11:00 a.m.	37.3°C
12:00 p.m.	37.6°C
1:00 p.m.	37.8°C
2:00 p.m.	37.9°C
3:00 p.m.	37.9°C
4:00 p.m.	37.7°C
5:00 p.m.	37.6°C
6:00 p.m.	37.4°C
7:00 p.m.	37.2°C
8:00 p.m.	37.0°C
9:00 p.m.	36.8°C

13 Two brands of batteries are compared by measuring how long they can keep a torch alight before they fail. The time is measured to the nearest hour. The table shows the results.

	1 hour	2 hours	3 hours	4 hours	5 hours	6 hours
Brand A	13	10	16	8	7	6
Brand B	6	7	5	15	19	8

   a Draw a diagram to show the number of hours each battery lasts.

   b Which brand is more likely to last for 4 hours. Why?

   c Which brand is more likely to fail in 3 hours or less?

   d Which brand is more reliable? Why?

32.3 Comparing data

● I can use mean and range to compare data from two sources

Develop fluency

It is often important to know the range of results as this can show how consistent they are. A golfer whose shots ranged over 20 m to 150 m would be performing less consistently than one whose shots ranged from 80 m to 150 m.

The table shows the mean and range of two teams' basketball scores.

Compare the mean and range and explain what they tell you.

	Team A	Team B
Mean	75	84
Range	20	10

The means tell you that the average score for Team B is higher than that for Team A, so they have higher scores generally.

The range compares the differences between the lowest and highest scores. As this is higher for Team A, there is more variation in their scores. You could say that they are less consistent.

1 A factory worker recorded the start and finish times of a series of jobs.

Job number	1	2	3	4	5
Start time	9:00 a.m.	9:20 a.m.	9:50 a.m.	10:10 a.m.	10:20 a.m.
Finish time	9:15 a.m.	9:45 a.m.	10:06 a.m.	10:18 a.m.	10:38 a.m.

   a Work out the range of the times taken for the jobs.

   b Calculate the mean length of time taken for the five jobs.

2 These are the minimum and maximum temperatures for four English counties, in April.

County	Northumberland	Leicestershire	Oxfordshire	Surrey
Minimum (°C)	2	4	4	4.5
Maximum (°C)	12	15	16.5	17.5

a Find the range of the temperatures for each county.

b Comment on any differences you notice, explaining why these differences might occur.

3 The table shows the mean and range of a set of test scores for Jon and Matt.

Compare the mean and range and explain what they tell you.

	Jon	Matt
Mean	64	71
Range	35	23

4 For each set of data, decide whether the range is a suitable measure of spread or not.

a 30, 60, 100, 120, 150, 200 b 13, 45, 48, 52, 66

b 5, 10, 15, 20, 25, 25, 25, 25, 25

5 QuickDrive and Ground Works are two companies that lay drives. The numbers of days each company takes to complete nine drives are shown below.

QuickDrive   2, 5, 3, 3, 6, 2, 1, 8, 3

Ground Works   3, 2, 3, 1, 4, 3, 2, 2, 1

a Calculate the mode, mean and range for each company.

b Comment on the differences between the averages.

c Comment on the difference between the ranges.

6 Mark and Philippa are competitive runners. The table summarises their times in the last 10 km races.

a Comment on the differences between the means.

b Comment on the difference between the ranges.

	Mean (minutes)	Range (minutes)
Mark	45.3	3.8
Philippa	44.9	7.6

7 Margaret Wix school and Beaumont school take part in a dance competition. The table shows the median and range of points they have scored in each of their past 10 competitions.

a Comment on the differences between the medians.

b Comment on the difference between the ranges.

	Median score	Range of scores
Margaret Wix	72.5	12.1
Beaumont	64.8	13.0

8 These are the heights in centimetres of a group of children from forms 8A and 9A.

8A   152, 155, 153, 153, 156, 162, 161, 158, 163

9A   163, 163, 163, 161, 164, 167, 172, 172, 171

　a Calculate the mode, mean and range for each group.

　b Comment on the differences between the averages.

　c Comment on the difference between the ranges.

9 A teacher recorded the start and finish times of children putting their books away at different times during the day in a primary school.

Day	Monday	Tuesday	Wednesday	Thursday	Friday
Start time	9:45 a.m.	10:20 a.m.	11:50 a.m.	13:40 a.m.	14:10 a.m.
Finish time	10:00 a.m.	10:45 a.m.	12:06 a.m.	13:58 a.m.	14:38 a.m.

　a Work out the range of the times taken for each day. Comment on your findings.

　b Calculate the mean length of time taken to put the books away.

10 These are the minimum and maximum number of people going to cinema during the four weeks of February.

	Week 1	Week 2	Week 3	Week 4
Minimum (number of people)	218	206	180	128
Maximum (number of people)	860	815	770	720

　a Find the range of the number of people for each week.

　b Comment on any differences you notice, explaining why these differences might occur.

Reason mathematically

A manager records the number of mobile phones Vivi and Henry sell each week. The table shows the mode, the mean and range for the number of of mobile phones they sold each week.

	Vivi	Henry
Mode	40	30
Mean	38	39
Range	42	50

The manager wants to give a bonus to either Vivi or Henry.

　a Vivi thinks she should get the bonus. Explain why she might think that.

　b The manager gives the bonus to Henry. Explain the manager's justification for that.

　a Vivi thinks she should get the bonus because the mode of her results is 40, whilst the mode of Henry's results is 30. Vivi thinks that she has, in most weeks, sold a higher number of mobile phones than Henry did.

　b The manager gives the bonus to Henry because his mean is higher than Vivi's which means that the total number of mobile phones that Henry sold overall is higher than the total number of mobile phones that Viv sold overall.

11 a Find the mean and range of 21, 24, 25, 25, 26 and 29.

    b Find the mean and range of 7.7, 7.3, 7.9, 7.1, 7.8, 7.2 and 7.5.

    c Find the mean and range of 582, 534, 518, 566 and 550.

    d Can you see a quick way of finding the mean for each of the questions above?

12 Fiona recorded how long, to the nearest hour, Everlast, Powercell and Electro batteries lasted in her MP3 player. She did five trials of each make of battery. The table shows her results.

    a Find the mean and range of the lifetime of each make of battery.

    b Explain which two types of battery you might buy.

Everlast (£1.00 each)	Powercell (50p each)	Electro (£1.50 each)
6	4	9
5	6	8
6	3	9
6	4	9
7	4	9

13 The weekly wages of four workers are £320, £290, £420 and £370.

    a Calculate the mean and range for the wages.

    b The employer hires a fifth worker. She wants the mean wage to be £340. What should she pay the fifth worker?

    c All workers receive a pay rise of £30 per week. How will this affect the mean and range? (Do not recalculate them.)

    d The following year the workers receive a 10% pay rise. How do you think the mean and range will be affected?

14 The table shows the mean and range for the average weekly rainfall (mm) in two holiday resorts.

    Explain the advantages of each island's climate using the mean and range.

	Larmidor	Tutu Island
Mean	6.5	5
Range	33	62

Solve problems

There are four number cards.

Use the information given in the table to find the cards.

Mean	Mode	Median	Range
8	8	8	2

The mode is 8, hence there will be at least 2 cards that are 8. The median is 8, hence the second and third cards must be equal to 8.

The mean = 8, hence the sum of all the 4 cards must be equal to $8 \times 4 = 32$.

The range = 2. The last card is 2 more than the first card.

Let the first card = x

The last card = $x + 2$

$x + 8 + 8 + x + 2 = 32$

$2x + 16 = 30$

$2x = 14$

$x = 7$

First card = 7, last card = $7 + 2 = 9$

The cards are 7, 8, 8, 9.

15 This table shows the average daily maximum temperatures for Cardiff, Edinburgh and London over one year.

Month	Jan	Feb	Mar	Apr	May	Jun	Jul	Aug	Sep	Oct	Nov	Dec
Cardiff (°C)	7	7	10	13	16	19	20	21	18	14	10	8
Edinburgh (°C)	6	6	8	11	14	17	18	18	16	12	9	7
London (°C)	6	7	10	13	17	20	22	21	19	14	10	7

Using the means and ranges of these maximum temperatures, write a report comparing the temperatures in the three cities.

16 **a** Write down two numbers with a range of 4 and a mean of 12.

b Why is there only one possible solution to part **a**?

c Write down three numbers with a mean of 10 and a range of 6.

d Why are there only three possible solutions to part **c**?

e What is crucial about the total of the three numbers in part **c**?

17 **a** Find the mean and range of 2, 4, 6, 8 and 10.

b Add 5 to each number in part **a**.　How are the mean and range affected?

c Double each number in part **a**.　What happens to the mean and range now?

d Add x to each number in part **a**.　Work out the new mean and range.

18 Joanna is thinking of four numbers.
The range of the numbers is 8, the mode is 7 and the median is 8. What could the numbers be?

32.4 Which average to use

● I can decide when each different type of average is most useful

Develop fluency

This table will help you decide which type of average to use for a set of data.

Type of average	Advantages	Disadvantages	Example
Mean	Uses every piece of data. Probably the most used average.	May not be representative when the data contains an extreme value.	1, 1, 1, 2, 4, 15 Mean = 4 which is a higher value than most of the data.

Median	Only looks at the middle values, so it is a better average to use if the data contains extreme values.	Not all values are considered so could be misleading.	1, 1, 3, 5, 10, 15, 20 Median = 4th value = 5 Note that above the median the numbers are a long way from the median but below the median they are very close.
Mode	It is the most common value.	If the mode is an extreme value it is misleading to use it as an average.	Weekly wages of a boss and his four staff: £150, £150, £150, £150, £1000. Mode is £150 but mean is £320.
Modal class for continuous data	This is the class with the greatest frequency.	The actual values may not be centrally placed in the class.	(see table below)
Range	It measures how spread out the data is.	It only looks at the two extreme values, which may not represent the spread of the rest of the data.	1, 2, 5, 7, 9, 40 The range is 40 − 1 = 39 but without the last value (40), the range would be only 8.

Modal class example:

Time (T) minutes	Frequency
$0 < T \le 5$	2
$5 < T \le 10$	3
$10 < T \le 15$	6
$15 < T \le 20$	1

The modal class is $10 < T \le 15$, but all six values may be close to 15.

The list shows the number of cars sold at a garage in the last 10 days.

0, 9, 7, 8, 8, 9, 7, 10, 7, 9

a Calculate: i The mean ii mode iii median

b Explain why the median is a suitable measure for this data.

a i mean = 7.4

   ii There are two modes: 7 and 9

   iii arrange numbers in ascending order: 0, 7, 7, 7, 8, 8, 9, 9, 9, 10
      median = 8

b There are two modes. The mean is affected by the one extreme value of 0. The median is not affected by the extreme value.

1   **i** Calculate the given average for each set of data.

    ii Explain whether the given average is a sensible one to use for that set of data.

    a The mean of 2, 3, 5, 7, 8, 10   **b** The mode of 0, 1, 2, 2, 2, 4, 6

    c The median of 1, 4, 7, 8, 10, 11, 12   **d** The mode of 2, 3, 6, 7, 10, 10, 10

    e The median of 2, 2, 2, 2, 4, 6, 8   **f** The mean of 1, 2, 4, 6, 9, 30

2 These are the times (in seconds) that 15 pupils took to complete a short task.

10.1, 11.2, 11.5, 12.1, 12.3, 12.8, 13.6, 14.4, 14.5, 14.7, 14.9, 15.4, 15.9, 16.6, 17.1

Time, T (seconds)	Tally	Frequency
$10 < T \leq 12$		
$12 < T \leq 14$		
$14 < T \leq 16$		
$16 < T \leq 18$		

    a Copy and complete the frequency table and find the modal class.

    b Explain why the mode is unsuitable for the ungrouped data, but the modal class is suitable for the grouped data.

3   **i** Calculate the range for each set of data.

    ii Decide whether it is a suitable average to represent the data. Explain your answer.

    a 1, 2, 4, 7, 9, 10   **b** 1, 1, 1, 7, 10, 10, 10   **c** 2, 5, 8, 10, 14

4   **i** Calculate the mean for each set of data.

    ii Decide whether it is a suitable average to represent the data. Explain your answer.

    a 12, 11, 17, 12, 18   **b** 1, 78, 79, 90, 75   **c** 32, 35, 30, 29, 89

5   **i** Calculate the mode for each set of data.

    ii Decide whether it is a suitable average to represent the data. Explain your answer.

    a 10, 8, 9, 11, 11   **b** 4, 4, 78, 78, 79   **c** 22, 25, 23, 27, 25

6   **i** Calculate the median and mean for each set of data.

    ii Decide which is a more suitable average to represent the data. Explain your answer.

    a 21, 22, 24, 27, 20   **b** 4, 3, 4, 3, 11   **c** 13, 22, 22, 22, 10

7 A rugby team plays 11 games. Here are the number of points they scored.

27, 16, 37, 16, 12, 19, 12, 8, 10, 24, 12

    a Write down the mode, median, range and mean.

    b Decide which is a more suitable average to represent the data.

8 Here are the lengths, in minutes, of the number of phone calls of five customer advisors in a shopping centre.

46, 14, 13, 15, 18

    a Calculate the mean, median and range for this data set.

    b Explain whether each of them is a sensible one to use for this set of data.

9 Here are the number of people that fail the driving test in the local driving centre during the week.

Monday	Tuesday	Wednesday	Thursday	Friday
30	28	29	35	10

   a Explain why the mode is unsuitable for this set of data.

   b Calculate: **i** the mean   **ii** the median.

   c Anton thinks that mean is more suitable for the data. Explain why he might think that.

Reason mathematically

Here are the annual salaries of nine people working a factory.

£32 000, £18 000, £17 950, £17 900, £17 200, £17 100, £16 800, £16 800, £16 800

  a Which number do you think represents the manager's salary?

  b The workers say, 'We all earn less than the average salary in this company.'
What average do the workers calculate?

  c The manager says, 'All the workers in this company earn the average salary or above the average salary.'
What average is the manager using?

  a £32 000

  b The workers calculate the mean = £18 950

  c The manager calculates the mode = £16 800

10 A group returning from a trip abroad were asked to donate their spare change to charity. This is what they were able to donate.

$0.80, $3.50, $1.60, $5.22, $0.42, $0.06, $3.15, $4.38, $2.72, $0.70, $2.90, $5.45, $1.32, $2.05, $0.67, $5.21, $4.57, $2.30, $4.18, $0.19, $0.01, $5.86, $3.17, $5.08, $3.76, $3.14, $2.19

  a Copy and complete the frequency table for these figures.

  b What is the modal class?

  c Find the median amount donated.

  d Explain why the modal amount donated is a better average to use than the median.

Donation, M ($)	Tally	Frequency
$0 < M \leq 1$		
$1 < M \leq 2$		
$2 < M \leq 3$		
$3 < M \leq 4$		
$4 < M \leq 5$		
$5 < M \leq 6$		

11 A factory employs 500 people. Of these, 490 are workers who each earn less than £250 a week and 10 are managers earning more than £900 a week each.

   a Which average would you use to argue that pay at the factory was low?

   b Which average would you use to argue that pay at the factory was reasonable?

   c In discussions about average pay, which average would be used by:

     i the worker   **ii** the manager   **iii** the owners of the factory?

   Give reasons for your answers.

12 A snooker team needs a replacement player for an important match. The captain decides to look at the last nine scores of two possible replacements. These are their scores.

Joe	48	79	53	88	75	64	72	49	65
Jimmy	110	30	36	119	25	31	28	101	41

   a Imagine you were Joe. Give your reasons for being chosen.

   b Imagine you were Jimmy. Give your reasons for being chosen.

   c If you were the captain of the snooker team, whom would you choose? Why?

Solve problems

Doctor Foster wants to compare the amount of time in hours that people, of different ages, spend watching TV in one week. She asks 100 people from each group, calculates the mean of their results and constructs a table to show her findings.
Her data is in the table.

Age	Time spent watching TV (hours)						
	Mon	Tues	Weds	Thur	Fri	Sat	Sun
17–19	1.3	1.5	1.3	1.6	0.6	4.3	1.6
20–49	1.4	1.6	1.7	1.2	1.3	3.2	1.8
50 and over	1.1	1.8	2.3	3.2	2.2	5.5	2.7

   a Use the median, mean and range, to compare the average amount of television watched by people in the three age groups.

   b Dina watches, on average, 4 hours 28 minutes of television daily. Which age group is she most likely to be in? Give a reason for your answer.

   c Look at the table. How can Dr Foster improve the reliability of her data?

a	17–19	Median = 1.5, mean = 1.74, range = 4.3 – 0.6 = 3.7
	20–49	Median = 1.6, mean = 1.74, range = 2.0
	50 and over	Median = 2.3, mean = 2.69, range = 4.4

On average the 50 and over group watch more TV than the other two groups. The findings from the first two groups are similar. They have the same mean and similar medians though the range of the data for the first group is much higher than that of the second group.

b The mean for the 50 and over group is the highest, so Dina is likely to be in this age group.

c The age interval of each group is not consistent. Doctor Foster can amend the age intervals of the groups e.g. 16–19-year olds, 20–24-year olds, 25–29-year olds and so on.

13 Here are Mike's and Bev's marks in their Science exams (as a percentage of the total score)
Mike: 35%, 82%, 79%, 85%, 27%, 80%   Bev: 78%, 79%, 80%, 78%, 77%, 79%
   a Use the mean score, as well as the range, to compare their results.
   b Their friend Jessica needs sums help with her homework. Who would be the better person to help? Explain your answer.

14 Here are the quarterly expenses figures for two ice cream shops.

	1st quarter	2nd quarter	3rd quarter	4th quarter
Shop A	£2500	£3700	£6900	£3750
Shop B	£2900	£3500	£4500	£9850

   a For each business, work out the mean and the range for their expenses.
   b Write two sentences comparing the shops.
   c One of the shops has recently bought new machinery to make ice-cream. Which one do you think is it?

15 Two shops offer the following sizes of dresses.
   Periwinkle   12, 14, 16, 18, 30   Jenny's 10, 12, 14, 16, 18, 20, 22
   a Which shop has the greater range of sizes?
   b Do you think this is a suitable indication of sizes available? Explain your answer.
   c Would any of the averages be more helpful to potential customers? If so, which one?

Now I can...

create a grouped frequency table from raw data	draw a frequency diagram from a grouped frequency table	decide when each different type of average is most useful
interpret frequency diagrams	use mean and range to compare data from two sources	

CHAPTER 33 Percentage changes

33.1 Simple interest

- I can understand what simple interest is
- I can solve problems involving simple interest

If you take out a loan you usually have to pay interest to the **lender**. This is the person or organisation lending you the money.

One type of interest is called **simple interest**. This is calculated as a percentage. As long as you still have the loan, you will pay the lender a percentage of the loan, at regular intervals.

Develop fluency

Wayne takes out a loan of £3270 to buy a car. He pays simple interest of 1.5% per month. Calculate the amount of interest he pays in two years.

1.5% = 1.5 ÷ 100 = 0.015   Use 0.015 as the multiplier.

1.5% of £3270 = 0.015 × 3270 = £49.05   Use a calculator. This is the amount of interest paid each month.

2 years = 2 × 12 = 24 monthly payments.

In two years, he pays £49.05 × 24 = £1177.20.

He pays £1177.20 in interest.

1 Jim takes out a loan for £15 000 to pay for a new kitchen. He repays the loan over 30 months at 1.3% per month simple interest. How much interest does Jim pay?

2 Jo takes out a loan for £4300. She pays simple interest of 1.8% per month for a year. How much interest does she pay?

3 Sam takes out a loan of £400.
   He pays simple interest of 4% per month for nine months.
   a Work out the amount of interest Sam pays each month.
   b Work out the total amount of interest Sam pays.

4 Harry pays 2.5% monthly interest on a loan of £3800.
   He pays interest for one year.
   a Work out the amount of interest he pays each month.
   b Work out the total interest he has paid after one year.

5 Aaron takes out a loan of £730.
   He pays simple interest of 0.7% per month for eight months.
   Calculate the total amount of interest Aaron pays.

Sharon takes out a loan of £570 for a washing machine. She repays the loan at 2.1% per month simple interest over 12 months. How much does Sharon pay in total?

2.1% = 2.1 ÷ 100 = 0.021	Use 0.021 as the multiplier.
0.021 × £570 = £11.97.	This is the amount of interest she pays each month.
£11.97 × 12 = £143.64	This is the total interest paid.
£143.64 + £570 = £713.64	The original loan amount must be repaid.

The total amount Sharon repays is £713.64.

6 Jason takes out a loan of £460 to pay for a new bike. He repays the loan at 2.3% per month simple interest over two years. How much does Jason repay in total?

7 Zoe takes out a loan of £7300 to pay for her new company website. She repays the loan at 1.7% simple interest per month over two years. How much does she repay in total?

8 Phil takes out a loan of £2600 to pay for a new bathroom. He repays the loan at 1.9% per month simple interest for 18 months. How much does he repay in total?

9 A company takes out a loan of £28 000 to pay for some software development. The loan is repaid over five years at 0.9% simple interest per month. Work out the total repaid.

Reason mathematically

Lucy takes out a loan of £750. She pays interest of £25.50 per month. What is the rate of simple interest?

$\frac{25.50}{750} = 0.034$	Divide the interest by the amount of the loan.
0.034 × 100 = 3.4%	Multiply the decimal by 100 to change it to a percentage.

10 Graham pays £10.80 simple interest per month on a loan of £600. Work out the rate of interest.

11 Amy pays £56 monthly interest on a loan of £4000. Work out the rate of interest.

12 Myra pays £11.52 simple interest per month on a loan of £640. Work out the rate of interest.

Solve problems

Peggy pays £9.35 weekly interest on a loan of £550.

Ciara pays £8.93 weekly interest on a loan of £470.

Work out who pays the higher rate of interest and by how much.

$\frac{9.35}{550} \times 100\% = 1.7\%$ $\frac{8.93}{470} \times 100\% = 1.9\%$

Peggy pays 1.7% interest per week.   Ciara pays 1.9% interest per week.

1.9% – 1.7% = 0.2%

Ciara pays the higher rate of interest by 0.2% per week.

13 Cameron takes out a loan of £5400 and pays 1.4% simple interest every month for six months.

Mary takes out a loan of £3600 and pays 2.8% interest every month for nine months. Who pays more interest, Cameron or Mary? Justify your answer.

14 Jack pays 1.6% simple interest monthly on a loan of £800 for one year.

    a Work out his total interest payment over the year.

    b Work out his total interest payment as a percentage of the original loan.

15 Valerie takes out a loan of £2800. Shannon takes out a loan of £4600. Ruby takes out a loan of £5400.

All three pay 7% simple interest each year for six years.

    a Calculate how much interest Valerie has to pay in total.

    b Calculate how much more interest Ruby must pay in total than Shannon.

16 Rebecca takes out a loan of £630 and pays 2.1% simple interest every week for 12 weeks.

Ntuse takes out a loan of £720 and pays 1.7% simple interest every week for 14 weeks.

Kezia takes out a loan of £790 and pays 1.4% simple interest every week for 15 weeks.

Work out who pays the most interest.

33.2 Percentage increases and decreases

- I can calculate the result of a percentage increase or decrease
- I can choose the most appropriate method to calculate a percentage change

A percentage change may be an **increase** if the new value is larger than the original value or a **decrease** if the new value is smaller than the original value.

There are several methods for calculating the result of a percentage change. The **multiplier** method is often the most efficient. Just multiply the original value by an appropriate number, to calculate the result of the percentage change.

If you use the multiplier method, the multiplier will be larger than 1 for an increase and smaller than 1 for a decrease.

The multiplier may be a fraction or a decimal.

Develop fluency

Alicia buys a bicycle for £350. A year later the value of the bicycle has fallen by 20%.

Calculate the new value.

Take the original value as 100%.

You need to find 100% − 20% = 80% of the original price.

80% = $\frac{4}{5}$ or 0.8. This is the multiplier.

The new value is 0.8 × £350 = £280 You could also work out $\frac{4}{5}$ × 350 to get the answer.

Shaun puts £438 in a bank account. After one year it has earned 3% interest.

Work out the total amount in the account.

The interest must be added to the original.

You need to find 100% + 3% = 103%.

103% = 103 ÷ 100 = 1.03   *Multiply by 1.03.*

The new total is £438 × 1.03 = £451.14.   *For this problem, it is sensible to use a calculator.*

1 Peter has £400 in a savings account.
How much will he have if his savings:
  a increase by 25%   **b** decrease by 25%   **c** decrease by 14%?

2 Decrease each price by 15%.
  a Jacket, £160   **b** Shoes, £90   **c** Handbag, £75   **d** Jeans, £48

3 Increase each bill by 9%.
  a Electricity, £325.40   **b** Gas, £216.53   **c** Rent, £475.50

4 The price of a cooker was £585.
Work out the new price after:
  a an increase of 3%   **b** an increase of 23%   **c** a decrease of 2.3%.

5 The value of an antique painting is £5200.
Work out the new value if it increases by:
  a 20%   **b** 90%   **c** 110%   **d** 135%.

6 Last year, Danny ran a 100-m race in 15.0 seconds.
She has reduced her time by 12%.
How long does it take Danny to run 100 m now?

7 Puppies increase in mass by 25% between the ages
of 8 weeks and 10 weeks.
Copy and complete the table.

Age	Mass (kg)		
8 weeks	2	5	6
10 weeks			

8 A new car will typically lose about 53% of its value over three years.
Work out the expected value after three years on these new prices.
  a £28 800   **b** £19 600   **c** £47 400

9 The expected cost of a large building project was £18 million.
The true cost was 36% higher.
What was the true cost of the project?

10 The volume of Venus is 13.4% less than the volume of Earth.
The volume of Earth is 1083 billion cubic km.
What is the volume of Venus, to the nearest billion cubic km?

Reason mathematically

a When Kelly was calculating a percentage change she multiplied the original value by 1.4.
State whether this was an increase or a decrease and what the percentage change was.

b When Elaine was calculating a percentage change, she multiplied the original value by 0.56.
State whether this was an increase or a decrease and what the percentage change was.

a 1.4 is larger than 1, so it represents an increase.
$1.4 = 1.4 \times 100\% = 140\%$
$140\% - 100\% = 40\%$
The percentage increase was 40%

b 0.56 is less than 1, so it represents a reduction.
$0.56 = 0.56 \times 100\% = 56\%$
$100\% - 56\% = 44\%$
The percentage reduction was 44%.

11 a When Sam was calculating a percentage change she multiplied the original value by 1.07.
State whether this was an increase or a decrease and what the percentage change was.

b When Phil was calculating a percentage change he multiplied the original value by 0.7.
State whether this was an increase or a decrease and what the percentage change was.

c When Simrath was calculating a percentage change she multiplied the original value by 0.13.
State whether this was an increase or a decrease and what the percentage change was.

d When Suzie was calculating a percentage change, she multiplied the original value by 2.8.
State whether this was an increase or a decrease and what the percentage change was.

12 Peter has a rectangular piece of card with sides of length 16 cm and 20 cm.
He cuts off a 2 cm border all around the edge.
Peter says: 'The length and the width are both reduced by 20%.'
Is Peter correct? Justify your answer.

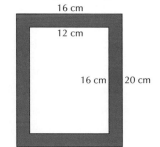

16 cm
12 cm
16 cm   20 cm

13 Work out how much each person now earns.
a Magisha earned £460 per week and then received a 4% pay rise.
b Fiona earned £385 per week and then received a 2% pay cut.
c Gloria earned £514 per week and then received a 9.1% pay cut.
d Anna earned £193.28 per week and then received a 22% pay rise.

14 This table shows how the populations of some towns and villages changed between 1980 and 2000. Copy and complete the table.

Place	Percentage change	Population in 1980	Population in 2000
Smallville	30% decrease		1300
Lansbury	8% increase	46 000	
Gravelton	2% decrease		19 800
Smithchurch	48% increase		144 000
Deanton		28 000	33 320
Tanwich		680	646

Solve problems

An electricity bill increases from £285.79 to £299.42.

Calculate the percentage increase.

$\dfrac{\text{new bill}}{\text{original bill}} = \dfrac{299.42}{285.79}$ Write the new bill as a fraction of the original bill.

$= 1.0476\ldots$ This is $299.42 \div 285.79$

$= 104.8\%$ The new bill is 104.8% of the original bill.

The increase is 104.8% – 100% = 4.8%.

The price of a car is decreased from £8490 to £7750.

Calculate the percentage decrease.

$\dfrac{\text{new price}}{\text{original price}} = \dfrac{7750}{8490}$ Write the new price as a fraction of the original price.

$= 0.9128\ldots$ Change to a decimal. This is $7750 \div 8490$.

$= 91.3\%$ The new price is 91.3% of the original price.

The reduction is 100% – 91.3% = 8.7%.

15 Tracy used a multiplier to increase the amount she charged for a haircut from £9.50 to £11.78.
   a What multiplier did Tracy use?   b What was the percentage increase in price?

16 Ben used a multiplier to reduce the cost of a coat in a sale from £74 to £62.90.
   a What multiplier did Ben use?   b What was the percentage reduction in the sale.

17 The rent for a flat increases from £320 to £336.
   Work out the percentage increase.

18 Work out the percentage change in each price.
   a An increase from £280 to £305   b An increase from £975 to £1125
   c An increase from £76.50 to £105.50   d An increase from £37 to £87

33.3 Calculating the original value

● I can calculate the original value given the result of a percentage change

To find the original value, you need to undo the percentage change.

Original amount $\xrightarrow{\times \text{multiplier}}$ Result of
$\xleftarrow{\div \text{multiplier}}$ percentage change

Do this by dividing by the multiplier.

Develop fluency

Last year, the number of daily visitors to Hastings Museum and Art Gallery increased by 30% to 1066.

What was the number of visitors before the increase?

The multiplier for the increase is 1.3.   100% + 30% = 130% = 1.3

Original number × 1.3 = 1066   The result of the increase is 1066.

Original number = $\dfrac{1066}{1.3}$ Find the original number, by dividing by 1.3.

= 820   1066 ÷ 1.3 = 820

In a survey, the number of butterflies seen in a wood was 150. This was a 40% reduction on the previous year. How many butterflies were there last year?

The multiplier for a 40% reduction is 0.6.   100% – 40% = 60% = 0.6

Last year's number × 0.6 = 150

Last year's number = $\dfrac{150}{0.6}$ = 250   Divide 150 by 0.6.

1 An amount is increased by 25% to make £80. What was the amount?

2 An amount was reduced by 36% to make £144. What was the amount?

3 a After a 50% increase, a price is now £30. Work out the original price.
   b After a 50% decrease, a price is now £30. Work out the original price.

4 Write down the multiplier for each percentage increase.
   a 22%   b 48%   c 6%   d 8.5%   e 120%   f 333%

5 Write down the multiplier for each percentage decrease.
   a 22%   b 48%   c 6%   d 8.5%

6 a After a 10% increase, a price is now £275. Work out the original price.
   b After a 20% increase, a price is now £540. Work out the original price.
   c After a 30% increase, a price is now £234. Work out the original price.
   d After a 5% increase, a price is now £756. Work out the original price.

7 **a** A price is £105 after a 25% increase. Work out the original price.

   b A price is £243 after a 25% decrease. Work out the original price.

   c A price is £234 after a 4% increase. Work out the original price.

   d A price is £648 after a 4% decrease. Work out the original price.

8 In a sale, the cost of a coat is reduced by 20% to £38.40.
   What was the price of the coat before the reduction?

9 Olivia had a pay rise of 5% which took her annual salary to £19 110.
   What was her annual salary before the pay rise?

Reason mathematically

The price of a car repair, including 20% VAT, is £283.68.

Work out the price excluding VAT.

The multiplier for a 20% increase is 1.2.

$100\% + 20\% = 120\% = 1.2$

Price excluding VAT $\times 1.2 = £283.68$

Price excluding VAT $= \dfrac{£283.68}{1.2} = £236.40$.

10 Here are some prices, including VAT at 20%.
    Work out the price excluding VAT in each case.

    a Restaurant bill, £67.68   **b** Garage bill, £548.16   **c** Computer, £1054.80

11 The rate of VAT on energy bills is 8%. These are some energy bills, including VAT.
    Work out the bills before VAT was added.

    a Electricity, £58.64   **b** Gas, £155.30   **c** Electricity, £304.05

12 The standard rate of VAT in Ireland is 23%.
    Here are the prices, including VAT, of some goods in Ireland.
    Work out the prices excluding VAT.

    a Camera, €350.55   **b** Printer, €102.64   **c** Furniture, €2287.80

13 The rent on a house increases by 12% and is now £728.
    Calculate the rent before the increase.

Solve problems

The cost of a laptop computer is £756 including VAT at 20%.

How much VAT is included in the price?

The cost of the laptop before VAT is added is the initial cost.

120% of initial cost = £756.

$20\% = 120\% \div 6$

20% of initial cost = £756 ÷ 6 = £126

The amount of VAT included in the price is £126.

14 The cost of a concert ticket is £330 including VAT at 20%.
How much VAT is included in the price?

15 The value of Sarah's house has increased by 25% since she bought it.
Her house is now valued at £105750.
How much has the value of her house increased?

16 Work out the values of the letters A to F and find the matching pairs.
When A is increased by 75% the answer is 490.
When B is increased by 50% the answer is 435.
When C is decreased by 30% the answer is 196.
When D is increased by 70% the answer is 510.
When E is decreased by 15% the answer is 255.
When F is decreased by 40% the answer is 174.

17 Emily's mass is now 70.3 kg.
This is a 5% decrease from six months ago.
What was Emily's mass six months ago?

33.4 Using percentages

● I can choose the correct calculation to work out a percentage

A pair of numbers may be compared in different ways using percentages.

Develop fluency

The ratio of men to women in a keep-fit class is $4:3$.

a Work out the percentage of the keep-fit class that is made up of men.

b What is the number of women, as a percentage of the number of men?

You can represent the situation with a diagram.

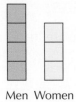

Men Women

a Four out of every seven of the people are men. $4 + 3 = 7$

$\frac{4}{7}$ are men.

$\frac{4}{7} = \frac{4}{7} \times 100\% = 57\%$ $4 \div 7 = 0.5714\ldots$

b There are three women to every four men.

There are $\frac{3}{4}$ as many women as there are men.

$\frac{3}{4} = 75\%$

In a music examination, if you pass you may also get a distinction.

One year 256 pupils passed and 62 of them got a distinction.

109 pupils failed.

What percentage got a distinction:

  a out of those who passed   **b** out of those who took the examination?

  a *62 got a distinction, and 256 passed.*

$$62 \text{ out of } 256 = \frac{62}{256} \times 100\%$$
$$= 24.2\% \qquad 62 \div 256 = 0.2421\ldots$$

  b *365 pupils took the examination.*   *That is 256 + 109.*

$$62 \text{ out of } 365 = \frac{62}{365} \times 100\% = 17.0\% \quad 62 \div 365 = 0.1698\ldots$$

1 **a** One week last year, 93 people took a driving test and 37 passed.
Work out the percentage that passed.

  b The following week the number who passed decreased by 16%.
How many passed in that week?

2 In one year, there were 204 600 applications for exam papers to be re-marked across all exam boards at GCSE and A level. The total number of scripts was 15.1 million. As a result, 38 520 grades were changed.

  a What percentage of the total number of scripts were re-marked?

  b What percentage of the re-marked scripts were awarded a different grade?

3 In an audience there are 80 men, 90 women, 60 girls and 20 boys.

  a What percentage of the children are girls?

  b What percentage of the adults are women?

  c What percentage of the whole audience is female?

4 The number of visitors to a website each day increased from 4600 to 6400.
Work out the percentage increase.

5 In 2000 there were 64 elephants in a wildlife reserve.
In 2010 the number had increased by 25%.

  a Work out the number of elephants in 2010.

  b What percentage of the 2010 number is the 2000 number?

  c What percentage of the 2000 number is the 2010 number?

6 Mike and Joe play in a football team.
In three seasons, Mike scored 40 goals and Joe scored 32 goals.

  a Work out Joe's goals as a percentage of the total number of goals the two scored.

  b Work out the number of goals Joe scored as a percentage of the number Mike scored.

  c Work out Mike's total goals as a percentage of Joe's total goals.

7   These are the numbers of votes cast for four candidates in an election.

Candidate	Alan	Kirsty	Matt	Wendy
Votes	50	65	75	40

   a   What percentage of the votes did Kirsty get?

   b   What percentage of the votes that went to females did Wendy get?

   c   Write the number of votes that Alan got as a percentage of the number of votes that Kirsty got.

8   Some college students are either running or swimming for charity.
    The ratio of runners to swimmers is 3 : 2.

   a   What percentage of the students are runners?

   b   There are 144 runners. How many students are there in total?

   c   What is the number of swimmers, as a percentage of the number of runners?

9   1340 people visited an amusement park on Friday.

   a   On Saturday there were 30% more visitors than on Friday.
       How many visited on Saturday?

   b   Of the total for Friday and Saturday, what percentage visited on Saturday?

   c   On Friday there were 10% more than there were on Thursday.
       How many visited on Thursday?

10   The table shows the numbers of members of a club by gender and age.

   a   What percentage of the members are females under 25?

   b   What percentage of the members are male?

   c   What percentage of the members are 25 or over?

Age group	Male	Female
Under 25	27	45
25 or over	72	36

Reason mathematically

In a sale, the price of a pair of jeans was reduced from £27 to £23.

What was the percentage reduction?

The multiplier is $\frac{23}{27} = 0.85185\ldots$

$$= 0.85185\ldots \times 100\%$$

$$= 85.185\ldots\%$$

The reduction is $100\% - 85.185\% = 14.815\ldots$

$$= 14.8\% \text{ to one decimal place.}$$

11 a The price of a laptop was £185. In a sale the price was reduced to £169.
What was the percentage decrease?

b The price of a television was originally £479. In the sale there was a 15% reduction.
What was the sale price?

c After a 30% reduction, the price of a radio was £58.80.
Work out the original price.

12 Here are some prices. The rate of VAT is 15%.
Work out the missing numbers.

	Item	Price before VAT (£)	Price including VAT (£)
	Fit gas cooker	85.00	97.75
a	Fix broken window	44.00	
b	Install computer		34.27
c	Service boiler	69.40	
d	Replace radiator		164.22

13 There are 145 pupils in year 9 at a school. This is 18% of the total number of pupils in the school.
How many pupils are there at the school, to the nearest 10?

14 216 of the cushions made in a factory each day are red.
This is 45% of the total number of cushions made.

a How many cushions are made in total in one day?
One third of the cushions are blue and the rest are yellow.

b How many yellow cushions are made in five days?
The factory makes £6 profit for each red cushion, £8 profit for each blue cushion and £10 profit for each yellow cushion.

c What percentage of the factory's profits comes from red cushions?

Solve problems

At a tennis tournament, the ratio of sunny days to cloudy days was $9:5$.

a What percentage of the days were sunny?

b Write the number of cloudy days as a percentage of the number of sunny days.

a The fraction of days that were sunny was $\frac{9}{14} = 0.6428...$
As a percentage this is $0.6428 \times 100\% = 64.3\%$ to one decimal place.

b The number of cloudy days as a fraction of the number of sunny days is $\frac{5}{9} = 0.5555...$
As a percentage, this is $0.5555... \times 100\% = 55.6\%$ to one decimal place.

15 Abigail runs a website that sells theatre tickets.

She calculates that sales for child and adult tickets in March were in the ratio 3 : 5.

  a What percentage of the ticket sales were adult tickets?

  b Work out the number of child tickets sold as a percentage of adult tickets sold.

In March, 261 child tickets were sold.

  c How many adult tickets were sold?

Child tickets cost £9 each and adult tickets cost £14 each.

  d Find the total ticket sales on Abigail's website in March.

16 The diameter of the planet Mercury is 38% of the diameter of Earth.
The diameter of Earth is 12 756 km.

  a Round 38% to one significant figure.

  b Round 12 756 km to one significant figure.

  c Use your approximations to estimate the diameter of Mercury.

17 A garden centre is holding a sale in which the prices of all items have been reduced by the same percentage.
Sarah buys a wheelbarrow for £28.50 which has been reduced in price from £37.50.

  a Find the percentage reduction.
    Simi buys a lawnmower which cost £67.50 before the sale.

  b Find the sale price of the lawnmower.
    Izzy buys a bag of compost for £17.10.

  c Find the price of the bag of compost before the sale.

18 There are 60 children and teenagers living in a new housing estate. This is 20% of the total population.

  a How many people live in the estate in total?

  b Work out the ratio of children and teenagers to adults.

Now I can...

understand what simple interest is	solve problems involving simple interest	choose the correct calculation to work out a percentage
calculate the result of a percentage increase or decrease	choose the most appropriate method to calculate a percentage change	
calculate the original value, given the result of a percentage change		

CHAPTER 34 Polygons

34.1 Angles in polygons

- I can work out the sum of the interior angles of a polygon
- I can work out exterior angles of polygons

A **polygon** is a 2D shape that has straight sides. In a **regular polygon**, all the sides are equal and all the angles are equal. In **irregular polygons**, such as the one in the diagram, the sides are not all equal and the angles are not all equal.

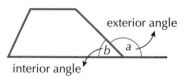

The angles at the vertices, inside a polygon, are its **interior angles**. If a side of a polygon is extended, the angle formed outside the polygon is an **exterior angle**.

At any vertex of a polygon the interior angle plus the exterior angle = 180° (angles on a straight line).

Hence, on the diagram: $a + b = 180°$.

This is true for any polygon.

The sum of the exterior angles for any polygon is 360°.

Develop fluency

a Draw a pentagon and divide into triangles by drawing diagonals from the same vertex.

b Calculate the sum of its interior angles.

c How many exterior angles are there in a pentagon?

d Work out the sum of the exterior angles.

a

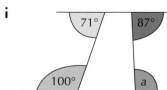

b The diagram shows how a pentagon can be split into three triangles from one of its vertices. The sum of the interior angles for each triangle is 180°.

So, the sum of the interior angles of a pentagon is given by:

$3 \times 180 = 540°$

c There are five exterior angles in a pentagon.

d The sum of all the exterior angles is 360°.

1 a What is the sum of all the exterior angles in **i** a quadrilateral **ii** a pentagon?
b Find the size of the unknown angles.

i

71°   87°

100°   a

ii

65°   b

70°

73°

38°   59°

Hint: What is the sum of the exterior angles?

2 Calculate the unknown angle in each diagram.

a

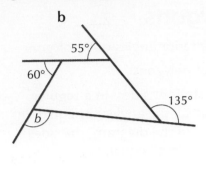

b

c

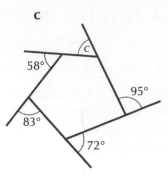

3 Calculate the value of *x* in each pentagon.

a

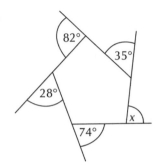

b

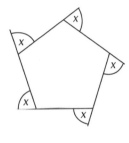

4 a Sketch a hexagon and an octagon.

  b By splitting the above shapes into triangles using diagonals from the same vertex, find the sum of their interior angles.

5 a What is the sum of the interior angles of a pentagon?

  b Four interior angles of a pentagon are 100°, 80°, 76° and 112°. What is the size of the fifth interior angle?

6 Four interior angles of a hexagon are 90°, 128°, 160° and 110°. The other two interior angles are equal. What is the size of the other two interior angles?

7 Calculate the size of the unknown angle in each polygon.

a

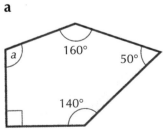

b

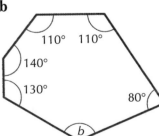

c

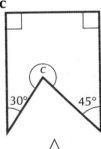

8 a What is the sum of the angles in this polygon?

  b Calculate the value of *x* .

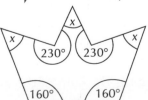

Reason mathematically

Valma thinks that the unknown angle m is $360° - (105° + 75° + 70°) = 360° - 250° = 110°$.

Is Valma correct? Explain why.

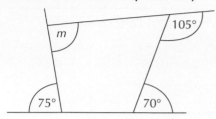

Valma is wrong. She has worked out the size of the exterior angle to m. m is $180° - 110° = 70°$.

9 Calculate the size of each unknown angle, marked with a letter.
For each angle explain how you found the value.

a

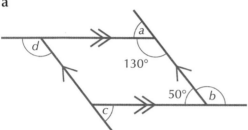

b

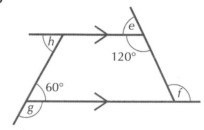

For each diagram, check that the sum of the exterior angles is 360°.

10 Which of these could not be the sum of the interior angles of a polygon? Explain how you know.

a 1180°   b 6660°   c 3060°   d 6488°

11 To calculate the sum of the interior angles of a quadrilateral and a pentagon, Amil divides each polygon into several triangles by connecting a point inside the polygon to each of its vertices.

Amil draws this table.

	Quadrilateral	Pentagon	Hexagon	Octagon
Number of triangles	4	5		
Sum of the interior angles	$4 \times 180° - 360° = 360°$	$5 \times 180° - 360° = 540°$		

a Why has Amil subtracted 360° each time?

b Use Amil's method to calculate the sum of the interior angles of a hexagon and an octagon.

12 The sum of the interior angles of a polygon is 2340°. To find how many sides the polygon has, Jessie and Ben do these calculations.

Jessie's working	Ben's workings
$2340 \div 180 = 13$	$2340 - 2 = 2338$
$13 - 2 = 11$ sides	$2338 \times 180 = 420\,840$ sides

Both Jessie and Ben are incorrect. Find their mistakes. What should the answer be?

Solve problems

The sum of the interior angles of a polygon is 3240°. How many sides does the polygon have?

Sum of interior angles $= (n-2) \times 180°$

$(n-2) \times 180 = 3240°$

$n - 2 = 3240 \div 180°$

$n - 2 = 18$

$n = 18 + 2$

$n = 20$ sides

13 The four interior angles of a quadrilateral are $3x + 80$, $5x + 10$, $3x - 20$ and $4x - 10$. Calculate the size of each interior angle of the quadrilateral.

14 The five interior angles of a pentagon are $6x - 2$, $7x + 1$, $4x + 5$, $2x + 80$ and $8x - 3$. Calculate the size of each interior angle of the pentagon.

15 The sum of the interior angles of a polygon is 1980°. How many sides does the polygon have?

16 A polygon has 50 sides. All but one of its angles is 170°. What is the size of the other angle?

34.2 Constructions

● I can make accurate geometric constructions

It is important that you how to make accurate **constructions**. These are useful because they produce exact measurements and are therefore used by architects and in design and technology.

You will need a sharp pencil, a ruler and compasses. Always leave all your construction lines on your diagrams.

Develop fluency

Construct the perpendicular from point P to the line segment AB.

×P

A ——————————— B

Set the compasses to any suitable radius.

Draw arcs from P to intersect AB at X and Y.

With the compasses still set at the same radius, draw arcs centred on X and Y to intersect at Z below AB.

Join PZ.

PZ is perpendicular to AB and intersects AB at a point C.

CP is the shortest distance from P to the line AB.

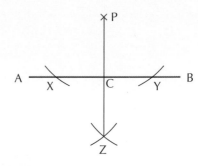

Construct the perpendicular at point Q on the line segment XY.

X ——————×—————— Y
　　　　　Q

Set the compasses to a radius that is less than half the length of XY.

With the centre at Q, draw arcs on either side of Q to intersect XY at A and B.

Set the compasses to a radius that is greater than half the length of XY and, with centre at A and then B, draw arcs above and below XY to intersect at C and D.

Join CD.

CD is the perpendicular from the point Q.

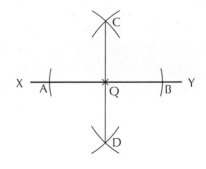

1 **a** Copy the diagram and construct the perpendicular from the point X to the line AB.

　　b Copy the diagram and construct the perpendicular from point X to CD.

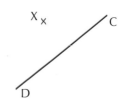

2 **a** Copy the diagram and construct the perpendicular
from the point Z on the line segment XY.

b Copy the diagram and construct the perpendicular
from the point C on the line segment AB.

3 Draw a line 10 cm long. Mark a point
4 cm from one end.
Construct the perpendicular from
this point.

4 Draw a line 10 cm long. Mark a point
about 5 cm from the line.
Construct the perpendicular from the
point to the line.

5 Draw a circle of radius 6 cm and centre O. Draw a line AB of any
length across the circle, as in the diagram. (AB is called a chord.)
Construct the perpendicular from O to the line AB. Extend the
perpendicular to make a diameter of the circle.

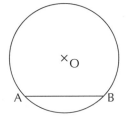

6 Draw a circle. Draw the diameter in the circle. Bisect the diameter to form a
perpendicular diameter. Join the ends of the diameter together to form a square.

7 Construct each triangle. Remember to label all the sides.

a
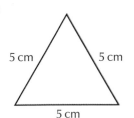
5 cm   5 cm
5 cm

b
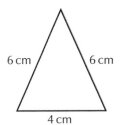
6 cm   6 cm
4 cm

c
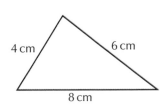
4 cm   6 cm
8 cm

8 Construct triangles with these sides:

a 6 cm, 7 cm and 8 cm   **b** 5 cm, 6 cm and 9 cm

9 Construct each right-angled triangle. Remember to label all the sides.

a

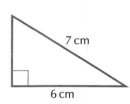
7 cm
6 cm

b

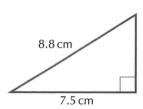
8.8 cm
7.5 cm

c

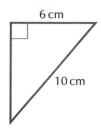
6 cm
10 cm

10 Construct right-angled triangles with these sides:

a hypotenuse: 13 cm, side 12 cm　　　**b** hypotenuse: 7.5 cm, side 6 cm

Reason mathematically

Is it possible to draw this triangle accurately? Explain your answer.

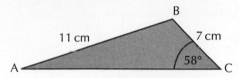
11 cm
B
7 cm
58°
A　　　C

Yes it is.

Draw a horizontal line and construct the angle of 58° at the right-hand end.

Measure BC = 7 cm.

Draw an arc 11 cm long centred on B.

A is the point where the arc crosses the original line drawn for angle C.

Join A to B to make triangle ABC.

11 Which of these shapes cannot be drawn accurately? Explain why.

a

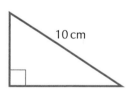
10 cm

b

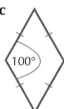
8 cm　61°
7 cm

c

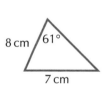
100°

12 Spot the mistakes. Correct them.

a Constructing a perpendicular from a point to a line.

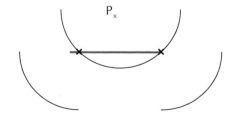
P
x

b Constructing a perpendicular from a point on the line.

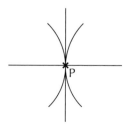
P

13 Lira attempts to draw the perpendicular from a point to line. She thinks that the centre of a triangle is the point where the perpendiculars from each vertex to the opposite side meet.

Selim thinks that the centre of the triangle is the point where the perpendicular bisectors of the three sides meet.

Construct an equilateral triangle and use Lira's and Selim's methods to find the centre of the triangle. What do you notice?

14 Helen and Jake construct a perpendicular from a point to the line.

Helen thinks that she must keep the setting of the compasses the same size for all the arcs that she draws. Tom thinks that he can change the setting of the compasses for each of the arcs that he draws. Jake thinks that after drawing the first arc, he can change the setting of the compasses before he draws the other two arcs, but he needs to ensure that the arcs intersect.

State who is correct and explain why.

Solve problems

Describe how you can draw this quadrilateral using a ruler, protractor and compasses.

Hint: Sketch the quadrilateral first.

Draw a straight line, AB = 6 cm, and construct the angle of 51° at A.

Measure AD = 7 cm.

The length of DC is 6 cm and the length of BC is 7 cm (opposite sides of a parallelogram)

Set the compasses to 6 cm and draw an arc from D.

Set the compasses to 7 cm and draw an arc from B.

The arcs cross at C. Join BC and CD.

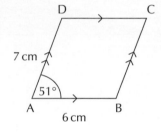

15 There are other ways to draw the parallelogram in the worked example. Describe one.

16 Construct a triangle with side lengths 7 cm, 8 cm and 8 cm as shown. Use a suitable construction to work out the shortest distance from A to BC.

17 ABCD is a rectangle. X and Y are points on AB and AD respectively.

Use a suitable construction to work out the shortest distance from C to XY.

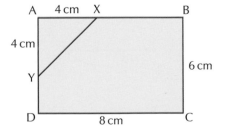

18 Use your knowledge of triangles to draw an angle of 60°. Is there another way you could draw an angle of 60°?

34.3 Angles in regular polygons

- I can work out the exterior angles of a regular polygon
- I can work out the interior angles of a regular polygon

In a **regular polygon**, all the interior angles are equal, and all the sides have the same length.

Develop fluency

Work out the size of each exterior and each interior angle in a regular pentagon.

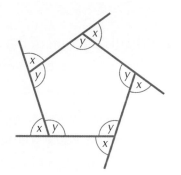

A regular pentagon has five equal exterior angles.

Let the size of each exterior angle be x, as shown on the diagram.

The sum of all the exterior angles is 360°. This gives:

$5x = 360$

$x = \dfrac{360}{5} = 72°$

The regular pentagon has five equal interior angles.

Let the size of each interior angle be y, as shown on the diagram.

The sum of an interior angle and an exterior angle is 180°. This gives:

$y + 72 = 180$

$y = 180 - 72 = 108°$

1 Copy and complete the table below for regular polygons.

Regular polygon	Number of sides	Sum of exterior angles	Size of each exterior angle	Size of each interior angle
Equilateral triangle	3	360°	120°	60°
Square	4	360°	90°	90°
Pentagon	5	360°	72°	108°
Hexagon				
Octagon				
Nonagon				
Decagon				

2 Find the missing angles, marked with a letter, in these shapes.

a
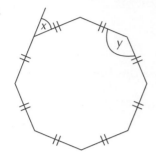

b
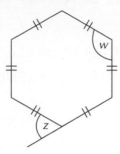

3 **a** What is the sum of the angles in a quadrilateral?
   b What is the special name given to a regular quadrilateral?
   c What is the size of each interior angle in a regular quadrilateral?

4 A regular nonagon has nine sides.
   a How many equal triangles can this polygon be divided into?
   b What is the sum of its interior angles?
   c What is the size of each interior angle?

5 A regular polygon has 12 sides.
   a How many equal triangles can the polygon be divided into?
   b What is the sum of its interior angles?
   c What is the size of each interior angle?

6 Find the sum of the exterior angles of a regular polygon with 13 sides.

7 A regular dodecagon is a polygon with 12 sides. All 12 interior angles are equal, and all 12 sides have the same length.
   a Work out the size of each exterior angle.
   b Work out the size of each interior angle.
   c Calculate the sum of the interior angles.

8 Find the number of sides of a regular polygon with an interior angle of
   a 170°   **b** 179°   **c** 179.9°   **d** 179.999°

9 Calculate the number of sides for a polygon whose:
   a exterior angle is 4°   **b** interior angle is 171°   **c** interior angles add up to 1980°.

10 The exterior angle of a regular polygon is 3°. Calculate:
   a its interior angle   **b** the sum of the interior angles   **c** the number of sides.

Reason mathematically

Tom draws a regular polygon. He says that the exterior angle of the polygon is 60°.
Is he correct?

Yes, 60 divides into 360 exactly, which gives the number of sides of the polygon.

11 True or false? Explain why.

   a A regular quadrilateral is called a rhombus.

   b The size of the exterior angle of a regular triangle is 60°.

   c The sum of the interior angles of a regular polygon with n sides = one interior angle $\times n$.

   d The size of the exterior angle of a regular decagon $= \dfrac{180 - 10}{10}$.

12 Joanne draws a regular polygon. She says that the exterior angle of the polygon is 25°. Ben draws a different regular polygon. He says that the interior angle of his polygon is 30°. Explain why Joanne and Ben are wrong.

13 The centre of the regular pentagon has been joined to its vertices. To calculate the angle x in the centre Stuart does the following calculation:

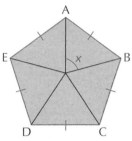

$$x = \frac{360°}{5} = 72°$$

Use Stuart's method to find the missing angles in these regular polygons:

a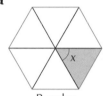
   Regular hexagon

b
   Regular octagon

c
   Regular decagon

d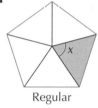
   Regular pentagon

14 ABCDE is a regular pentagon.

   a Triangle ADE is an isosceles triangle. Explain why.

   b Show that angle CAD = 36°. Include reasons for your answer.

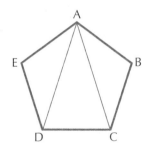

Solve problems

ABCDE is a regular pentagon.

The sides BC and ED are extended to meet at X.

Calculate the size of angle CXD.

Angles XDC and XCD are the exterior angles of the pentagon.

Exterior angle of a pentagon = 360 ÷ 5 = 72°

Angle XDC = angle XCD = 72°

Angle DXC = 180 − 72 − 72

       = 36°

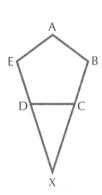

15 ABCDEFGH is a regular octagon. Calculate the size of angle AFB.

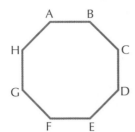

16 a Calculate the marked angles for this regular hexagon.

b Prove that ABDE is a rectangle.
Hint: Show that its angles are all 90°.

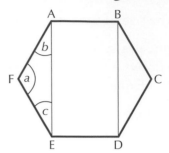

17 Calculate the marked angles for this regular octagon. Give reasons for your answer.

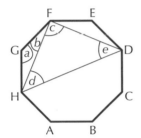

18 This diagram shows a regular pentagon and a regular hexagon joined together. Calculate the size of angle *a*.

34.4 Regular polygons and tessellations

● I can work out which regular polygons tessellate

A **tessellation** is a repeating pattern made from identical 2D shapes that fit together exactly, leaving no gaps. A shape can be tessellated if it can be used to make a repeating pattern with no overlaps or gaps.

Any triangle or quadrilateral can be made to tessellate. A regular polygon can only tessellate if its interior angle is a factor of 360°.

To show how a shape tessellates, draw up to about ten repeating shapes.

Develop fluency

Draw diagrams to show how equilateral triangles and squares tessellate.

Equilateral triangles tessellate like this.

Squares tessellate like this.

1 Show how a regular hexagon tessellates. Use an isometric grid.

2 Trace this regular pentagon onto card and cut it out to make a template.

   a Use your template to show that a regular pentagon does not tessellate.

   b Explain why a regular pentagon does not tessellate.

3 Trace this regular octagon onto card and cut it out to make a template.

   a Use your template to show that a regular octagon does not tessellate.

   b Explain why a regular octagon does not tessellate.

4 Draw six more copies of each shape on squared paper and show how each could tessellate.

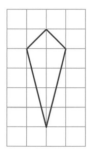
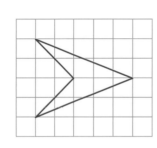

5 Draw a tessellation for each shape.

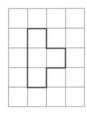

6 Copy on a squared paper. Add six of each shape to each tessellation

7 Polygons can be combined to form a **semi-tessellation**. Here is an example.

Copy on a squared paper. Add six of each shape to draw a tessellation.

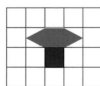
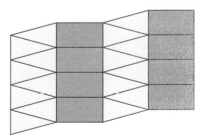

8 Use squared paper or isometric paper.

Make a tessellation using octagons and squares. Is there a different way?

Reason mathematically

Hannah tries to tesselate this isosceles triangle.

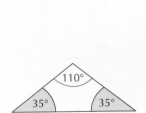

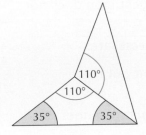

Hannah will not be able to continue her pattern to make a tessellation. Explain why.

If Hannah continues her pattern, there will be a gap, hence she will not be able to tessellate this way. 360 is not a multiple of 110.

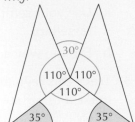

9   Peter tries to tessellate this rhombus. He thinks he needs to place two yellow angles and two black angles together at a point. Is he correct? Explain why.

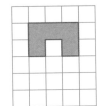

10 **a** Copy each quadrilateral on to squared paper. Does each quadrilateral tessellate?

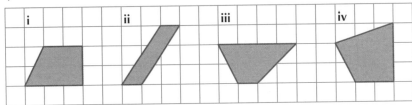

b Draw a quadrilateral of your own on squared paper (not a special quadrilateral). Does it tessellate?

c Does a quadrilateral always tessellate? Explain your answer.

11 Eliza says that a regular octagon does not tessellate, but some irregular octagons can tessellate. Eliza is correct. Explain why.

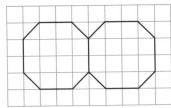

12 **a** Copy and complete the table below for regular polygons.

b Use your table to explain which regular polygons tessellate.

Regular polygon	Size of each interior angle	Does this polygon tessellate?
Equilateral triangle		
Square		
Regular pentagon		
Regular hexagon		
Regular octagon		

Solve problems

This is a scalene triangle.

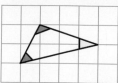

Show that this triangle can be tessellated.

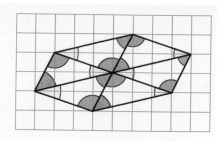

13 Lauren makes a pattern by tessellating five squares.
Find all the different shapes that she makes. Ensure that they are not a reflection or rotation of another.

14 Holly tessellates the small shape to make a rectangle.

George tessellates the small shape to make a bigger similar shape

What is the minimum number of shapes they can use?

15 George uses tiles shaped like this to cover different parts of the patio.
How many of these tiles can you fit into each of the following shapes?

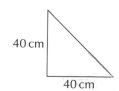
40 cm
40 cm

a

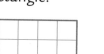
40 cm
40 cm
1.2 m

b

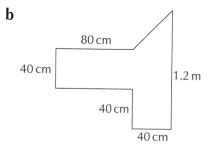
80 cm
40 cm
40 cm
40 cm
1.2 m

16 Ben is buying tiles that are 40 cm wide and 30 cm tall. One of these tiles costs £2.50.
Ben has a budget of £25 to cover an area that is 1.2 m wide and 1 m tall with tiles.
How does he do it?

Now I can...

work out the sum of the interior angles of a polygon	work out exterior angles of polygons	work out which regular polygons tessellate
make accurate geometric constructions	work out the interior angles of a regular polygon	
work out the exterior angles of a regular polygon		

CHAPTER 35 Expressions and equations

35.1 Multiplying out brackets

● I can multiply out brackets

You have seen before how to multiply out a term that includes brackets. If there is a number outside the brackets you multiply each term inside the brackets by that number.

This is also called **expanding** brackets.

When you have more than one set of brackets you can remove them and then simplify the remaining expression by collecting like terms.

Develop fluency

Multiply out the brackets in these expressions. Collect like terms.

  a $x(x - 5)$ **b** $-2(t - 3)$ **c** $2(t - 3) + 3(t - 4)$ **d** $d(d + 1) - d(d - 2)$

  a $x(x - 5) = x^2 - 5x$ $x \times x = x^2$ and $x \times -5 = -5x$

  b $-2(t - 3) = -2t + 6$ -2 is outside the brackets.

                                         $-2 \times t = -2t$ and $-2 \times -3 = 6$

  c $2(t - 3) + 3(t - 4) = 2t - 6 + 3t - 12$ Multiply out each set of brackets separately.

                      $= 5t - 18$ $2t + 3t = 5t$ and $-6 - 12 = -18$

  d $d(d + 1) - d(d - 2) = d^2 + d - d^2 + 2d$ $-d \times d = -d^2$ and $-d \times -2 = 2d$

                        $= 3d$ $d^2 - d^2 = 0$ and $d + 2d = 3d$

1 Simplify:

  a $2 \times 3a$ **b** $5 \times 4b$ **c** $6 \times 6c$ **d** $10d \times 5$.

2 Multiply out the brackets.

  a $2(a + 5)$ **b** $4(t + 7)$ **c** $x(5 - x)$ **d** $y(y + 4)$

3 Multiply out the brackets.

  a $4(2n + 1)$ **b** $3(2t + 7)$ **c** $3(5k - 2)$ **d** $10(2x - 5)$

  e $1.5(2c + 9)$ **f** $2(2z + 3)$ **g** $5(6 - 3y)$ **h** $0.2(15 - 5x)$

4 Multiply out the brackets.

  a $-3(a + 2)$ **b** $-3(m + 5)$ **c** $f(8 - f)$ **d** $-x(4 - x)$

5 Expand the brackets and simplify these expressions by collecting like terms.

  a $4(t + 2) + 3$ **b** $8(g + 1) - 5$ **c** $5(t - 2) - 3$ **d** $7(m - 3) + 13$

  e $6(n - 2) - 5n$ **f** $3(k - 2) + 2k$ **g** $5 + 2(x + 1)$ **h** $9 + 3(a - 2)$

6 Simplify these and then find the odd one out.

    a $5t - 3t + 2$ **b** $4t - 2(t + 1)$ **c** $4t - 2(t - 1)$ **d** $6t - 2(2t - 1)$

7 Simplify each expression as much as possible.

    a $2(x + 1) + 2(x + 3)$ **b** $x(x - 1) + x(x + 3)$ **c** $3(y - 1) + 4(y + 1)$

8 Expand and simplify the expressions.

    a $4(x + 5) - 2x$ **b** $5(2y + 1) + 2y - 3$ **c** $-3(t + 4) + 7t$

    d $10 - 5(x - 1)$ **e** $9g - 2(g + 3)$ **f** $p - 4(2 - p)$

9 Expand both brackets and simplify the expressions.

    a $4(x + 3) + 2(x - 2)$ **b** $2(i - 1) + 3(2i + 1)$ **c** $5(2n - 3) + 3(4n + 1)$

    d $4(r + 2) - 2(r - 3)$ **e** $2(3 - 2f) + 5(4 + 3f)$ **f** $3(6 + u) - 4(3 - 5u)$

Reason mathematically

These brackets have not been expanded correctly.
Spot the mistakes and correct them.

    a $5(2x + 3) = 7x + 15$ **b** $a(a + 3) = 2a + 3a = 5a$ **c** $x(2x + 3) = x(5x) = 5x^2$

    a $5(2x + 3) = 10x + 15$ $5 \times 2x = 10x$ not $7x$

    b $a(a + 3) = a^2 + 3a$ $a \times a = a^2$ not $2a$

    c $x(2x + 3) = 2x^2 + 3x$ You can't add $2x + 3$. $x \times 2x = 2x^2$.

10 Are these statements always, sometimes or never true? Explain your answers.

    a $4y + 5 = 5 + 4y$ **b** $3(p + 2) = 3p + 2$ **c** $10 - a = a - 10$ **d** $m + m = m^2$

11 The brackets have not been expanded correctly. Spot the mistakes and correct them.

    a $3(2y + 3) = 5y + 9$ **b** $x(x + 3) = 3x^2$ **c** $2x(8x - 3) = 10x + 6x = 16x$

12 Sam calculates the perimeter of the rectangle to be $7x + 5$.
Eamon calculates the perimeter of the rectangles to be $10x^2 + 25$.
Explain the mistakes that they may have made.
What should the answer be?

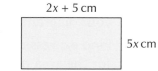

2x + 5 cm

5x cm

13 Write four different expressions that can be simplified to give $12x^2 + 10x$.

Solve problems

The diagram shows a cuboid.

Write an expression for:

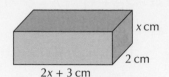

x cm

2 cm

2x + 3 cm

a the surface area of the cuboid

b the volume of the cuboid.

a The cuboid has six faces.

There are two faces $(2x + 3)$ cm long and x cm tall.

There are two faces $(2x + 3)$ cm long and 2 cm wide.

There are two faces 2 cm wide and x cm tall.

Surface area $= 2 \times 2 \times x + 2 \times 2(2x + 3) + 2 \times x(2x + 3)$

$= 4x + 4(2x + 3) + 2x(2x + 3)$

$= 4x + 8x + 12 + 4x^2 + 6x$

$= (4x^2 + 18x + 12)\, cm^2$

b Volume $=$ width $\times$ length $\times$ height

$= 2 \times x \times (2x + 3)$

$= 2x(2x + 3)$

$= (4x^2 + 6x)\, cm^3$

14 The sides of a square all measure $(3s + 5)$ cm.

   a Write an expression for its perimeter.

   b If $s = 7$ cm, what is the perimeter of the square?

15 The sides of a triangle are $(5s - 3)$ cm, $(2s + 7)$ cm and $(3s + 4)$ cm.
The perimeter of the triangle is 38 cm. Work out the length of each side.

16 Make a sketch of this flag.

   a Find an
   expression for:
   i a and **ii** b, and
   mark them on
   your diagram.

   b Find expressions
   for the area of:

   i the yellow part
   ii the white part
   iii the blue part.

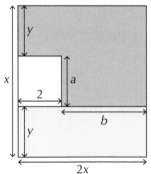

y

x

2

a

b

y

2x

17 In this square, each row, column and diagonal has the same sum. Find the missing expressions.

?	3a + 5	?
a − 1	5a + 1	9a + 3
6a + 4	?	2a + 2

35.2 Factorising algebraic expressions

● I can factorise expressions

Develop fluency

To **factorise** an expression, you do the opposite of multiplying out the brackets. You need to look for a factor that is common to all the terms and take it outside the brackets.

- If you multiply out $4(x - 5)$ you get $4x - 20$.

- If you factorise $4x - 20$ you get $4(x - 5)$.

In this case, 4 is a common factor of 4 and 20 so it can be put outside the brackets.

You could take out 2 as a factor and write $4x - 20 = 2(2x - 10)$.

This is not factorised completely because 2 and -10 have a common factor of 2. To factorise the expression completely you must take out 2 as a factor, as well.

$4x - 20 = 2(2x - 10) = 4(x - 5)$

This is the same answer. Always check that you have factorised as much as possible.

Remember that multiplying out brackets and factorising are inverse operations.

multiply out brackets

$4(x - 5)$ $4x - 20$

factorise

Factorise each expression.

  a $12x + 18$ **b** $9a - 12b - 15$

  a   The HCF of 12 and 18 is 6.   *2, 3 and 6 are factors. Always choose the highest.*

       $12x + 18 = 6(2x + 3)$ *6 goes outside the brackets.*

                                 The numbers inside are $12 \div 6 = 2$ and $18 \div 6 = 3$.

  b   A factor of 9, 12 and 15 is 3.   *3 is their only common factor, apart from 1.*

       $9a - 12b - 15 = 3(3a - 4b - 5)$ *Divide 9, 12 and 15 by 3.*

1   Work out the highest common factor of the numbers in each pair.

    a   15 and 10   **b**   4 and 20   **c**   18 and 24   **d**   12 and 20

2   Copy and complete each factorisation.

    a $2x + 6 = ...(x + 3)$ **b** $3d - 12 = ...(d - ...)$ **c** $4y + 8 = ...(... + 2)$ **d** $5e - 20 = ...(e - ...)$

3   Work out the missing terms.

    a $5t \times ... = 20t$ **b** $7 \times ... = 14a$ **c** $5x \times ... = 10xy$ **d** $... \times 3 = 9xy$

4   Factorise each expression.

    a $5m + 20$ **b** $3x - 21$ **c** $6n - 48$ **d** $8 + 4c$

5 Factorise

   a $4x + 6$ **b** $9x - 6$ **c** $12y + 16$ **d** $10y - 15$

   e $18 - 12x$ **f** $30 - 18t$ **g** $25 + 20t$ **h** $45m + 27.$

6 Work out the highest common factor of each pair of terms.

   a $3x$ and $6x$ **b** $5b$ and $10a$ **c** $10cd$ and $5c$ **d** $4xy$ and $12xy$

7 Factorise each expression.

   a $15x + 10$ **b** $4y - 20$ **c** $18a + 24b$ **d** $12p - 20q$

8 Factorise each expression as much as possible.

   a $2a + 4b$ **b** $6a + 2b$ **c** $5cd + 20d$ **d** $18cd - 12cd$

   e $24x + 30y$ **f** $6x - 8y$ **g** $-27a - 18b$ **h** $-24f - 16g$

9 Factorise each expression fully. Which one is the odd one out?

   a $8m - 12n$ **b** $2m^2 - 3mn$ **c** $10m^2 - 3mn$ **d** $-6m + 9n$

10 Simplify each expression by collecting like terms. Then factorise it.

   a $2a + 5 + 3a + 10$ **b** $9x - 8 + x - 2$ **c** $15x + 2 - 11x - 10$

Reason mathematically

The area of a rectangle is $(10x + 20xy)\,\text{cm}^2$.

Write expressions for its possible width and length.

   area = width × length

   Work out possible factor pairs

   width = x, length = $10 + 20y$

   width = $5x$, length = $2 + 4y$

   width = $10x$, length = $1 + 2y$

11 The perimeter of an equilateral triangle measures $9x + 15$.
    What is the length of each side?

12 Four integers are n, $n + 6$, $n + 12$ and $n + 18$. Mel and Beth add the numbers.
    Mel says that the sum of the numbers is $4(n + 9)$. Beth says that the sum of the numbers is $2(2n + 18)$. Who is correct? Explain why.

13 Ben has factorised these expressions.
    Explain the mistakes he has made and correct them.

   a $2x + xy = x(x + y)$ **b** $10a + 12b = 10(a + 2b)$ **c** $5a - 10b = 5ab(1 - 2)$

14 Three numbers are a, $a + 3$ and $a + 12$.

   a Write down the value of the three numbers, when $a = 10$.

   b Write down the value of the three numbers, when $a = -10$.

   c Write down an expression for the sum of the three numbers.
      Simplify your expression and factorise it as much as possible.

   d Work out the value of the sum of the three numbers, when $a = 45$.

Solve problems

The expression at the top of the pyramid is the product of the two expressions below it. Work out the missing expression.

$16x - 20xy = 4x(4 - 5y)$

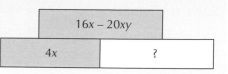

15 For each shape, write an expression for the perimeter. Then simplify it as much as possible. Factorise it if you can.

a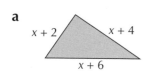

b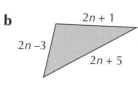

c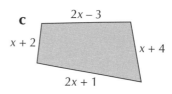

16 For each shape, write an expression for the perimeter. Factorise the answer fully.

a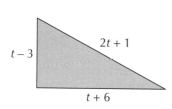

b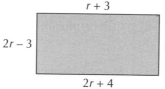

c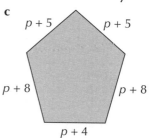

17 The expression at the top of the pyramid is the product of the two expressions below it. Work out the missing expressions.

18 a If the units digit of a number is a, give an expression for:

  i the tens column   **ii** the hundreds column.

b Find an expression for any two-digit number where the tens digit is twice the units digit, e.g. 63.

c Find more two-digit numbers with this property.

d 63 is divisible by 21. Simplify your expression to show that numbers where the tens digit is twice the units digit are always divisible by 21.

35.3 Equations with brackets

● I can solve equations with one or more sets of brackets

You have already learnt how to solve simple equations with brackets. You can use either method in the worked examples.

If there is more than one set of brackets, it is usually easier to multiply them both out first.

Develop fluency

Solve the equation $3(x - 4) = 16$

a by first multiplying out the brackets

b by first dividing by 3.

a $3(x - 4) = 16$

$\quad 3x - 12 = 16$ Multiply x and 4 by 3.

$\qquad 3x = 28$ Add 12 to both sides.

$\qquad x = 9\frac{1}{3}$ Divide 28 by 3.

b $3(x - 4) = 16$

$\quad x - 4 = 5\frac{1}{3}$ Divide 16 by 3.

$\qquad x = 9\frac{1}{3}$ Add 4 to $5\frac{1}{3}$.

1 a Solve these equations.

    i $2x + 6 = 18$ **ii** $2(x + 6) = 18$ **iii** $4y - 2 = 20$ **iv** $4(y - 2) = 20$

  b Solve **ii** and **iv** in a different way. (Hint: Look at the example above.)

2 Solve these equations. Show your method.

  a $2(y - 8) = 20$ **b** $2(y - 8) = 10$ **c** $40 = 5(f - 17)$ **d** $20 = 4(w - 9)$

3 Solve these equations. Give your answers in fraction form.

  a $3(x - 2) = 5$ **b** $4(y + 1) = 11$ **c** $6(t - 4) = 19$ **d** $8(c + 3) = 25$

4 a Simplify $4(x + 2) + x - 5$.   **b** Solve the equation $4(x + 2) + x - 5 = 38$.

5 a Simplify $2(x - 1) + 6(x - 2)$.   **b** Solve the equation $2(x - 1) + 6(x - 2) = 30$.

6 a Simplify $7(x - 1) + 6(x - 2)$.   **b** Now solve the equation $7(x - 1) + 6(x - 2) = 30$.

7 Solve these equations.

  a $2(x + 3) + 2(x + 4) = 28$ **b** $3(x - 3) + 2(x + 4) = 34$ **c** $4(t - 3) - (t - 5) = 20$

8 Solve each of these equations. Start by multiplying out the brackets.

  a $5(x - 1) = 4(x + 1)$ **b** $4(x - 2) = 5(x - 3)$ **c** $6(x - 1) = 4(x + 5)$

9 Solve these equations. Start by dividing both sides by 2.

  a $2(3 + 2v) = 2(v + 7)$ **b** $2(2d + 4) = 2(d - 7)$ **c** $2(5 - 2k) = 4(1 + 2k)$

10 Solve these equations and show your working.

  a $52 - 4(m + 2) = 3(m - 1)$ **b** $7(d + 2) = 21 + 2(2d + 1)$ **c** $3(3 + 2b) = 33 - 4(2 - b)$

Reason mathematically

This solution contains an error. Describe the error and solve the equation correctly.

$2x + 5 = 12$

$\quad 2x = 17$

$\qquad x = 8.5$

Step 2 is incorrect. Instead of adding 5, subtract 5.

$2x = 12 - 5$

$2x = 7$

$\quad x = 3.5$

11 Each equation has been solved incorrectly. Spot the mistakes and solve the equations correctly.

a $3(x + 5) = 12$
$x + 5 = 9$
$x = 4$

b $5y - 11 = 2y + 25$
$7y = 25 - 11$
$7y = 14$
$y = 2$

c $4(z + 3) = 8z + 10$
$z + 3 = 2z + 10$
$z = 7$

12 Thomas solves this equation, as shown.

$18 - 4(x + 1) = 2(x + 12)$

$18 - 4x - 4 = 2x + 24$	Expand brackets.
$14 - 4x = 2x + 24$	Simplify.
$4x = 2x + 10$	Subtract 14 from both sides.
$2x = 10$	Subtract $2x$ from both sides.
$x = 5$	Divide both sides by 2.

Is Thomas correct? Explain your answer

13 Johnny solves this equation:

$5(x - 3) - 15(x - 2) = 10$
$x - 3 - 3(x - 2) = 2$

a Explain his first step.
b Finish Johnny's solution to find x.
c Use Johnny's method to solve $6(x + 1) + 12(x - 2) = 0$

14 a Show that the expression $5(x - 3) - 4(x - 2)$ simplifies to $x - 7$.
b Use the result in part **a** to find a quick solution to the equation $5(x - 3) - 4(x - 2) = 10$.
c Solve the equation $5(x - 3) - 4(x - 2) = -7$.

Solve problems

Form and solve an equation to work out the value of x.

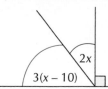

$3(x - 10) + 2x + 90 = 180$	The angles in a straight line add up to 180°.
$3x - 30 + 2x = 180 - 90$	Expand the brackets and subtract 90 from both sides.
$5x - 30 = 90$	Simplify the equation.
$5x = 90 + 30 = 120$	Add 30 to both sides.
$x = 120 \div 5$	Divide both sides by 5.
$x = 24°$	

15 a Explain why a regular hexagon whose perimeter is $12h + 6$ must have sides all equal to $2h + 1$.

b $h = 3.5\,$cm. What are the sides and perimeter of the hexagon?

c If the hexagon's perimeter is 30 cm, what is the value of h?

16 The pentagon has the same perimeter as the square.
Set up an equation and calculate the value of x.

 2(x − 4) cm   3(x − 5) cm

17 I am thinking of a number n. I double it and subtract 11. Then I multiply the answer by 5. I get 35. Construct an equation and solve it to find the number I was thinking of.

18 The sum of three consecutive integers is 54.

a Construct an equation and solve it to find the integers.

b The sum of three consecutive **odd** integers is 63. Construct an equation and solve it to find the integers.

35.4 Equations with fractions

● I can solve equations involving fractions

If there is a fraction in an equation you can remove it by multiplying the whole equation by the denominator of the fraction.

When you cannot write the answer exactly as a decimal, it is better to leave it as a fraction.

Equations with fractions can be written in different ways. Look at these equations.

$\frac{3}{5}x = 8$ could be written as $\frac{3x}{5} = 8$.

$\frac{2}{3}(a - 4) = 5$ could be written as $\frac{2(a - 4)}{3} = 5$.

Develop fluency

Solve the equations:

a $\dfrac{2}{3}(a - 4) = 5$

b $\dfrac{12}{x + 1} = 5.$

a $\dfrac{2}{3}(a - 4) = 5$

$\qquad 2(a - 4) = 15 \qquad$ Multiply both sides by 3.

$\qquad\qquad\qquad\qquad$ 15 is 3×5 and you now have an integer in front of the bracketed term.

$\qquad\quad a - 4 = 7.5 \qquad$ Divide by 2.

$\qquad\qquad\quad a = 11.5 \qquad$ Add 4.

b $\dfrac{12}{x + 1} = 5 \qquad$ $x + 1$ is in the denominator so multiply by it to remove the fraction.

$\qquad 12 = 5(x + 1) \qquad$ Solve this in the usual way. Expand the term with brackets first.

$\qquad 12 = 5x + 5 \qquad$ Now subtract 5.

$\qquad\quad 7 = 5x \qquad$ Then divide by 5.

$\qquad\quad x = 1\dfrac{2}{5}$ or 1.4

Solve these equations.

1 a $\dfrac{x}{5} = 6$ $\qquad$ **b** $\dfrac{y}{3} = 7$ $\qquad$ **c** $\dfrac{t}{4} = 4$ $\qquad$ **d** $\dfrac{n}{2} = 15$ $\qquad$ **e** $\dfrac{m}{15} = 2$

2 a $\dfrac{1}{5}x = 4$ $\qquad$ **b** $\dfrac{2}{5}x = 4$ $\qquad$ **c** $\dfrac{3}{5}x = 4$ $\qquad$ **d** $\dfrac{4}{5}x = 4$

3 a $\dfrac{20}{x} = 5$ $\qquad$ **b** $\dfrac{20}{x} = 10$ $\qquad$ **c** $\dfrac{15}{y} = 6$ $\qquad$ **d** $\dfrac{30}{t} = 5$

4 a $\dfrac{3}{4}a = 6$ $\qquad$ **b** $\dfrac{3}{8}b = 6$ $\qquad$ **c** $\dfrac{5}{8}c = 2$ $\qquad$ **d** $\dfrac{2}{7}k = 4$ $\qquad$ **e** $\dfrac{3}{5}t = 9$

5 a $\dfrac{1}{4}(x + 2) = 5$ $\quad$ **b** $\dfrac{1}{3}(x - 4) = 2$ $\quad$ **c** $\dfrac{1}{6}(x + 13) = 4$ $\enspace$ **d** $\dfrac{1}{8}(x - 9) = 3$

6 a $\dfrac{3d}{5} = 6$ $\qquad$ **b** $\dfrac{4u}{3} = -5$ $\qquad$ **c** $-\dfrac{3m}{5} = 9$ $\qquad$ **d** $\dfrac{2t}{-3} = -5$

7 a $\dfrac{16}{x + 2} = 8$ $\qquad$ **b** $\dfrac{48}{y - 1} = 8$ $\qquad$ **c** $\dfrac{30}{k - 5} = 5$ $\qquad$ **d** $\dfrac{36}{d + 3} = 4$

8 a $\dfrac{2(x + 5)}{3} = 16$ $\quad$ **b** $\dfrac{3(t - 2)}{4} = 4$ $\quad$ **c** $\dfrac{3(2 + x)}{10} = 4$ $\quad$ **d** $\dfrac{5(c - 10)}{9} = 2$

Reason mathematically

Jake and Isabella solved $\dfrac{6(r-8)}{7} = 12$ in two different ways. Explain their methods.

Jake's method

$\dfrac{r-8}{7} = 2$

$r - 8 = 14$

$r = 14 + 8$

$r = 22$

Isabella's method

$6(r-8) = 12 \times 7$

$6r - 48 = 84$

$6r = 132$

$r = 22$

Jake divides both sides by 6, then solves the equation.

Isabella multiplies both sides by 7, so that she removes the fraction first, then solves the equation.

9 Solve each equation in two different ways. Explain your methods.

a $\dfrac{4(x+1)}{3} = 16$ **b** $\dfrac{3(t-3)}{4} = 9$ **c** $\dfrac{6(2+x)}{5} = 4$ **d** $\dfrac{15(c-1)}{9} = 10$

10 The width of this rectangle is 5 cm and the area is $3n + 4$ cm².

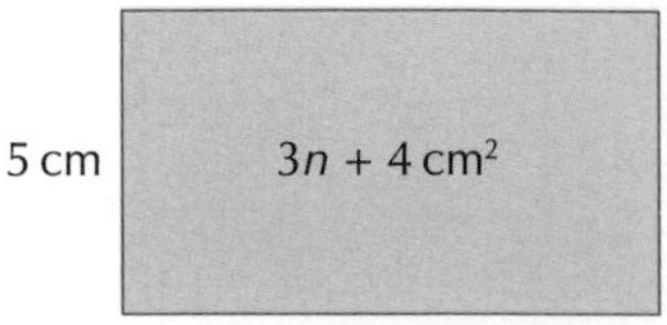

a Explain why the length of the rectangle is $\dfrac{3n+4}{5}$ cm.

b The length of the rectangle is 13 cm.

Write down an equation and solve it to work out the value of n.

5 cm | 3n + 4 cm²

11 The width of this rectangle is 6 cm and the area is $2n + 3$ cm².

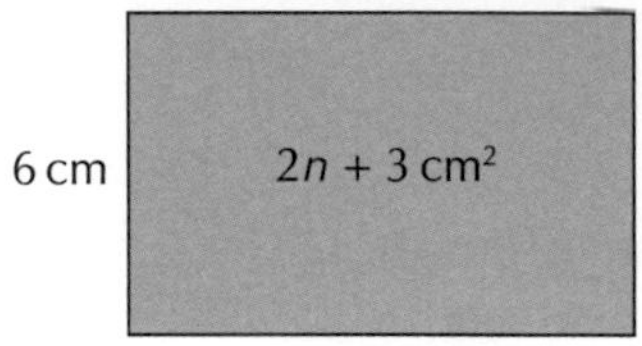

a Explain why the length of the rectangle is $\dfrac{2n+3}{6}$ cm.

b The length of the rectangle is 9 cm.

Write down an equation and solve it to work out the value of n.

6 cm | 2n + 3 cm²

12 The width of a triangle is $n + 5$ cm. The area is 17 cm².

a Explain why the height of the triangle is $\dfrac{34}{n+5}$ cm.

b The width of the triangle is 3 cm. Write down an equation and solve it to work out the value of n.

Solve problems

Three numbers are $2x$, $x + 4$, $x - 3$. The mean of the numbers is $x + 5$.
Work out the numbers.

Mean $= \dfrac{2x + x + 4 + x - 3}{3} = x + 5$	Construct an equation.
$\dfrac{4x + 1}{3} = x + 5$	Simplify the equation.
$4x + 1 = 3 \times x + 3 \times 5 = 3x + 15$	Multiply both sides by 3
$4x + 1 = 3x + 15$	Solve the equation: subtract $3x$ from both sides; subtract 1 from both sides.
$x = 14$	
$2x = 28$	Work out the three numbers.
$x + 4 = 18$	
$x - 3 = 11$	

The numbers are 28, 18 and 11.

13 Four numbers are x, $x + 4$, $x - 3$, and $2x$.
   The mean of the four numbers is 11.

   a Write down an equation and solve it to work out the value of x.

   b Work out the four numbers for the value of x you found in part b.

14 Asma is thinking of a number. She doubles her number and adds 1.
   She then calculates $\frac{2}{3}$ of the answer and gets 6. Set up an equation and solve it to find Asma's number.

15 Work out the value of x by constructing and solving an equation.

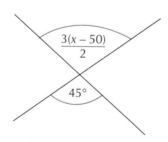

$\dfrac{3(x - 50)}{2}$

$45°$

16 Owen has saved some money. He gives £2 to his brother and spends $\frac{3}{5}$ of what is left at the school fair. Owen spends £9 at the school fair. Construct an equation and find how much money Owen had saved.

Now I can...

multiply out brackets	solve equations with one or more sets of brackets	solve equations involving fractions
factorise expressions		

CHAPTER 36 Prisms and cylinders

36.1 Metric units for area and volume

● I can convert from one metric unit to another

You need to know the metric units for area, volume and capacity.

They are listed in the table, which also gives the **conversions** between these units.

Area	Volume	Capacity
$10\,000\,m^2 = 1$ hectare (ha) $10\,000\,cm^2 = 1\,m^2$ $1\,m^2 = 1\,000\,000\,mm^2$ $100\,mm^2 = 1\,cm^2$	$1\,000\,000\,cm^3 = 1\,m^3$ $1000\,mm^3 = 1\,cm^3$	$1\,m^3 = 1000$ litres (*l*) $1000\,cm^3 = 1$ litre $1\,cm^3 = 1$ millilitre (ml) 10 millilitres = 1 centilitre (cl) 1000 millilitres = 100 centilitres = 1 litre

The unit symbol for litres is the letter l, which is often written as *l*, as shown in the table.

To avoid confusion with the digit 1 (one), it is also common to use the full unit name instead of the symbol.

Remember:

- to convert large units to smaller units, always multiply by the **conversion factor**
- to convert small units to larger units, always divide by the conversion factor.

Develop fluency

Convert each of these units as indicated.

a $72\,000\,cm^2$ to m^2 **b** $0.3\,cm^3$ to mm^3 **c** $4500\,cm^3$ to litres

a Hint: You are converting smaller units to larger units, so divide by the conversion factor 10 000.

$72\,000\,cm^2 = 72\,000 \div 10\,000 = 7.2\,m^2$

b You are converting larger units to smaller units, so multiply by the conversion factor 1000.

$0.3\,cm^3 = 0.3 \times 1000 = 300\,mm^3$

c You are converting smaller units to larger units, so divide by the conversion factor 1000.

$4\,500\,cm^3 = 4500 \div 1000 = 4.5$ litres

1 Express each of these in square centimetres (cm^2).

     a $4\,m^2$ **b** $7\,m^2$ **c** $20\,m^2$ **d** $3.5\,m^2$ **e** $0.8\,m^2$ **f** $540\,mm^2$ **g** $60\,mm^2$

2 Express each of these in square millimetres (mm^2).

     a $2\,cm^2$ **b** $5\,cm^2$ **c** $8.5\,cm^2$ **d** $36\,cm^2$ **e** $0.4\,cm^2$

3 Express each of these in square metres (m²).
 a 20 000 cm²   **b** 85 000 cm²   **c** 270 000 cm²   **d** 18 600 cm²   **e** 3480 cm²

4 Express each of these in cubic millimetres (mm³).
 a 3 cm³   **b** 10 cm³   **c** 6.8 cm³   **d** 0.3 cm³   **e** 0.48 cm³

5 Express each of these in cubic metres (m³).
 a 5 000 000 cm³   **b** 7 500 000 cm³   **c** 12 000 000 cm³

6 Express each of these in litres.
 a 8000 cm³   **b** 17 000 cm³   **c** 500 cm³   **d** 3 m³   **e** 7.2 m³

7 Express each measure as indicated.
 a 85 ml in cl   **b** 1.2 litres in cl   **c** 8.4 cl in ml

8 Convert these measurements to cubic centimetres.
 a 2 m³   **b** 0.5 m³   **c** 78 mm³   **d** 9300 mm³

9 Convert these quantities (remember that cubic centimetres are the same as cm³).
 a 8400 cm³ to litres   **b** 65 cubic centimetres to litres   **c** 4.8 ml to cm³
 d 200 ml to litres   **e** 9 litres to cm³   **f** 3.75 litres to ml

Reason mathematically

25 litres of water flows into a pond every second. The pond is in the shape of a cuboid with dimensions 50 m by 2 m by 3 m. How many hours and minutes does it take to fill the pond from empty?

1 m³ = 1 000 000 cm³

25 litres = 25 000 cm³

Volume of pond = 50 m × 2 m × 3 m = 300 m³

300 m³ = 300 000 000 cm³

Time to fill pond = 300 000 000 ÷ 25 000 = 12 000 seconds = 200 minutes
= 3 hours 20 minutes.

10 a Explain, using areas of squares, why 1 cm² = 100 mm².
 b Explain, using volumes of cubes, why 1000 mm³ = 1 cm³.

11 Spot the mistakes in these conversions. Correct them.
 a 5000 cm² = 5 m²   **b** 3500 mm² = 3.5 cm²   **c** 7.5 m² = 750 000 mm²

12 How many lead cubes of side 2 cm can be cast from 4 litres of molten lead?

13 The volume of a cough-medicine bottle is 240 cm³. The prescription reads: Two 5 ml spoonfuls to be taken four times a day. How many days will the cough medicine last?

Solve problems

A garden path is made from 350 rectangular paving stones, each 30 cm long and 15 cm wide.

  a What is the area of the path?

  b A company is hired to clean the path. They charge £2.75p/m² or £50 to clean the whole path. Which method is cheaper?

  a *30 cm = 0.3 m, 15 cm = 0.15 m*
      Area of one stone = 0.3 × 0.15
                     = 0.045 m².
      Area of path = 350 × 0.045 = 15.75 m².

  b *Cost of the path = 2.75 × 15.75*
                        = £43.31.

    It is cheaper to pay per square metre.

14 How many square paving slabs, each with sides of 50 cm, are needed to cover a rectangular yard measuring 8 m by 5 m?

15 A football pitch measures 120 m by 90 m. It costs £600/month to maintain the pitch. What is the maintenance rate per m²? Give your answer correct to the nearest penny.

16 **a** A car has a 2.3 litre engine. What is this in cubic centimetres (cc)?

      b A car engine has a cylinder capacity of 1487 cc but when advertised this is rounded to the nearest 100 and converted to litres. What size will it be advertised as?

17 A hockey pitch is 80 m long and 45 m wide. The annual maintenance costs for the pitch are £1000/hectare.

    a Work out its area in: **i** square metres   **ii** hectares.

    b Work out the annual cost of maintaining the pitch.

36.2 Volume of a prism

● I can calculate the volume of a prism

A **prism** is a three-dimensional (3D) shape that has exactly the same two-dimensional (2D) shape running all the way through it, whenever it is cut across, perpendicular to its length.

This 2D shape is called the **cross-section** of the prism.

The shape of the cross-section depends on the type of prism, but it is always the same for a particular prism.

You can work out the volume, *V*, of a prism by multiplying the area, *A*, of its cross-section by the length, *l*, of the prism or its height, *h*, if it stands on one end.

Develop fluency

Calculate the volume of this triangular prism.

The cross-section is a right-angled triangle with an area of $\frac{6 \times 8}{2} = 24\,cm^2$.

So, the volume is given by:

$V = $ area of cross-section $\times$ length $= 24 \times 15 = 360\,cm^3$

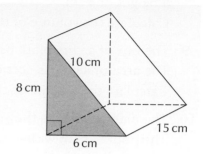

1   Calculate the volume of each prism.

a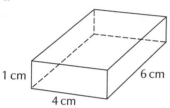

b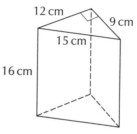

c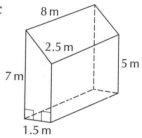

2   Calculate the volume of each cuboid, in the most appropriate units.

a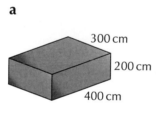

b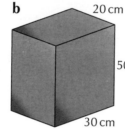

c

3   Calculate the capacity (in litres or ml as appropriate) of each cuboid in question 2.

4   Calculate the volume and capacity of a cube of side 15 cm.

5   A cube has a volume of 8000 cm³. What are the lengths of its sides?

6   The volume of a biscuit tin is 3 litres. The area of the lid is 375 cm². How deep is the tin?

7   The cross-section of a pencil is a hexagonal prism with an area of 50 mm².
    The length of the pencil is 160 mm.

   a   Calculate the volume of the pencil, in cubic millimetres.

   b   Write down the volume of the pencil, in cubic centimetres.

8   A biscuit tin is an octagonal prism with a cross-sectional area of 350 cm² and a height of 9 cm.
    Calculate the volume of the tin.

9   The box of a chocolate bar is an equilateral triangular prism.
    The area of its cross-section is 15.5 cm2 and the length of the prism is 30 cm.
    Calculate the volume of the box.

10 The diagram shows the cross-section of a swimming pool, along its length.
    The pool is 15 m wide.

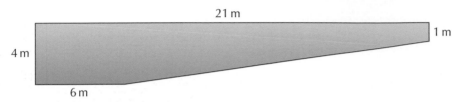

    a   Calculate the area of the cross-section of the pool.

    b   Work out the volume of the pool.

    c   How many litres of water does the pool hold when it is full?

Reason mathematically

A rectangular tank measuring 12 cm by 10 cm by 14 cm contains 1.12 litres of water.

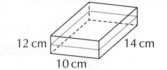

    a   Sasha says that the tank is two-thirds full. Is she correct?

    b   How much more water is needed to fill in the tank completely?
        Give your answer in millilitres.

    a   Volume of water = 1.12 litres = 1.12 × 1000 = 1120 cm³

        Height of water level = $\frac{1120}{14 \times 10}$ = 8 cm

        $\frac{8}{12} = \frac{2}{3}$, so Sasha is correct.

    b   Capacity of tank = 12 cm × 10 cm × 14 cm = 1680 cm³ = 1680 ml

        Amount of water needed = 1680 ml – 1120 ml = 560 ml

11 The cross-section of a block of wood is a trapezium.
    The volume of the block is 9600 cm³.
    Tom thinks that the height of the trapezium is 12 cm.
    Is Tom correct?

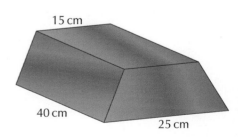

12 Leroy is making a solid concrete ramp for wheelchair access to his house.
The dimensions of the ramp are shown on the diagram.

a Calculate the volume of the ramp, giving your
answer in cubic centimetres.

b One cubic metre of cement weighs 2.4 tonnes.
What is the mass of concrete that Leroy uses?

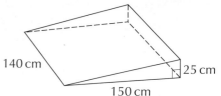

13 Krispies are sold in three sizes: mini, medium and giant. The boxes are filled to the top.

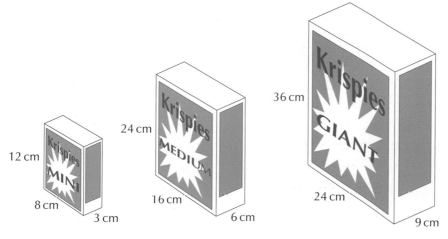

a Write the ratio of the heights of the three boxes in its simplest form.

b Calculate the volume of each box.

c Write the volumes as a ratio in its simplest form.

d Can you see a link between the ratios in parts **a** and **c**?

e How many mini boxes could be fitted into: **i** a medium box   **ii** a giant box?

f The answers in part **e** are the same as which other answers you have found?
Why is that?

g The prices of these packets are 45p, £2.70 and £6.75 respectively. Express these as a
simple ratio.

h Why do you think the ratio of the prices is different to the ratio of the volumes?

14 An empty regular container has a square base of side 10 cm and a height of 20 cm.

a Ben says that it is filled with 2.1 litres of water. Show that Ben can't be right.

b Jordan says that the container is filled with 1.1 litres of water and the height of the
water level is 11 cm. Show that Jordan is correct.

Solve problems

A rectangular tank 30 cm by 20 cm by 12 cm is half full. How much more water is needed to make the tank three-quarters full?

Height of water $= \frac{1}{2} \times 20 = 10$ cm

Height of water when $\frac{3}{4}$ full $= \frac{3}{4} \times 20$ cm $= 15$ cm

15 cm $- 10$ cm $= 5$ cm

Extra volume of water needed $= 30 \times 12 \times 5 = 1800$ cm^3 = 1.8 litres

15 A water tank is in the shape of a trapezium as shown. An overflow pipe is fixed 20 cm below the top of the tank. What is the greatest volume of water the tank can hold?

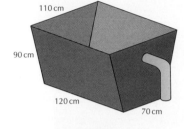

16 The box below is filled with packets of tea. The packets are 5 cm square and 12 cm high. When full, the box contains 8 dozen packets of tea which are stacked upright in the box.

  a What volume of tea is in each packet, in cm^3?

  b How many packets of tea fit into the base of the box?

  c Work out how many layers of packets there must be in the box and then find the height of the box.

  d What is the total volume of tea in the box, in cm^3?

  e The packets cost 75p to produce, and are sold at £1.90 each. How much profit is made on the whole box?

17 A rectangular tank has a square base of sides 20 cm. It contains water to the depth of 5 cm. When 8.4 litres of water is added, the tank becomes half-filled with water. What is the height of the tank?

18 A cubical tank of edge 15 cm is $\frac{1}{3}$ filled with water. The water is then poured into an empty rectangular tank that has a base 25 cm by 6 cm. The water fills up three-quarters of the rectangular tank. Work out:

  a the height of the water in the rectangular tank

  b the height of the tank.

36.3 Surface area of a prism

● I can calculate the surface area of a prism

You can find the **surface area** of a prism by calculating the sum of the areas of its faces.

For example, in the prism shown here, its **total surface area** is made up of the two end pentagons plus five rectangles.

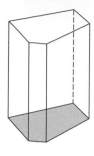

Develop fluency

Calculate the total surface area of this triangular prism.

The cross-section is a right-angled triangle with an area of
$\frac{6 \times 8}{2} = 24\,cm^2$.

The total area of the two end triangles is $48\,cm^2$.

The sum of the areas of the three rectangles is:

$(6 \times 15) + (8 \times 15) + (10 \times 15) = 360\,cm^2$

So, the total surface area is: $48 + 360 = 408\,cm^2$

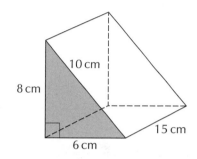

1 Calculate the total surface area of each prism.

a
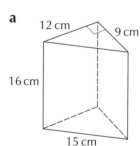

12 cm   9 cm
16 cm
15 cm

b
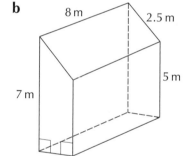

8 m   2.5 m
5 m
7 m
1.5 m

c
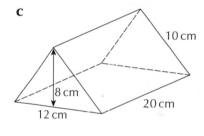

10 cm
8 cm
12 cm
20 cm

2 Calculate the surface area of each prism.

a
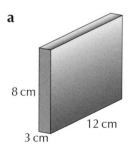

8 cm
12 cm
3 cm

b
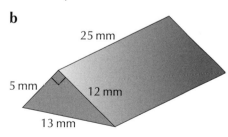

25 mm
5 mm   12 mm
13 mm

3 Convert your answers to question 2 to square metres.

4 A fish tank with glass walls has dimensions 60 cm wide by 30 cm deep by 20 cm high with a solid base and no top.
How much glass is needed?

5 The cross-section of this prism is a trapezium.

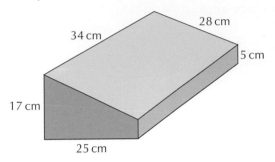

a Calculate the area of the trapezium.
b Calculate the total surface area of the prism.

6 The diagram shows a petrol tank in the shape of a prism.

a Calculate the area of the cross-section.
b Calculate the volume of the tank.
c Calculate the capacity of the tank in litres.
d Calculate the surface area of the tank in square metres.

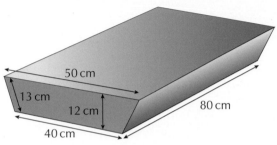

7 This regular octagonal prism has a cross-sectional area of 96 cm^2 and a length of 32 cm.
Each edge of the octagon is 6 cm long.
Calculate the total surface area of the prism.

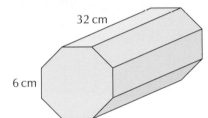

8 A tent is in the shape of a triangular prism. Its length is 2.4 m, its height is 1.6 m, the width of the triangular end is 2.4 m and the length of the sloping side of the triangular end is 2 m.
Calculate the surface area of the outside of the tent.

9 This gift box is in the form of an equilateral triangular prism. The cross-sectional area is 27.7 cm^2.

a Calculate the area of the net needed to make the box. Do not include the area of the tabs.
b Calculate the volume of the box.

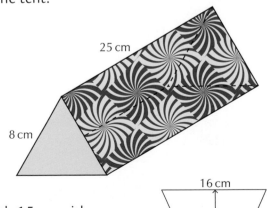

10 Calculate the surface area of a prism of length 15 cm with this isosceles trapezium as its cross section.

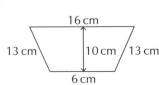

Reason mathematically

Jordan says: 'To work out the surface area of the prism shown, I need to split the shape into two cuboids, find the surface area of each of them and add them up.'

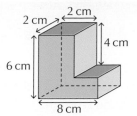

a Is Jordan correct?

b What is the surface area of the shape?

a Jordan is not correct. He only needs to calculate the surface areas of the shape's faces. This shape has two L-shaped faces and six rectangular faces.

b Surface area of the L shape $= 2 \times 6 + 2 \times 6 = 24\,cm^2$

Total surface area $= 2 \times 24 + 8 \times 2 + 6 \times 2 + 2 \times 4 + 6 \times 2 + 2 \times 2 \times 2 = 104\,cm^2$

11 The cross-sectional area of the prism shown is made from five equal squares, each with side length of 4 cm.
The prism is 12 cm long.

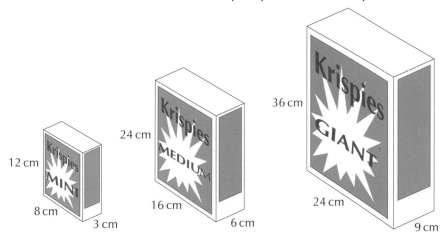

a Calculate the area of the cross-section.

b Calculate the total surface area of the prism.

c Calculate the volume of the prism.

12 Look at the dimensions of the Krispies packets carefully.

a Work out the surface area of each packet.

b Write these areas as a ratio in its simplest form.

c There is a quick way to simplify their ratios from their dimensions. Explain the method.

d Are these boxes similar shapes or not? Explain your answer.

13 Are these statements true or false? Explain your answers.

a When you cut a prism in half, its surface area is halved as well.

b Two prisms can have different dimensions, but equal surface areas.

14 The surface area of a prism is 348 cm². Sketch three different prisms that could have this surface area.

Solve problems

A cuboid has dimensions 6 cm by 10 cm by 2 cm. It has the same surface area as the triangular prism shown. Work out the height of the triangle.

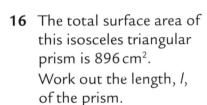

Surface area of cuboid = $2(6 \times 10 + 6 \times 2 + 2 \times 10) = 184$ cm²

Surface area of triangular prism = 184 cm²

$2 \times 10 \times 5 + 6 \times 10 + 2 \times$ area of triangle = 184

Area of triangle = 12 cm²

$\dfrac{6 \times x}{2} = 12$

$x = 4$ cm

15 A cuboid has a surface area of 100 m². Its width is 2 m, and its height is 7 m. Calculate its length.

16 The total surface area of this isosceles triangular prism is 896 cm². Work out the length, l, of the prism.

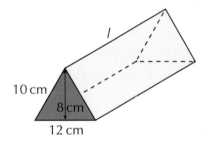

17 Tom has a small tent which is 1 m wide, 1.2 m high, 2.4 m long and has a sewn-in groundsheet. The sloping height of the tent is 1.6 m.

a Work out how much material is needed to make this tent.

When Tom camps with Scouts they have bigger tents that allow more height before the sloping roof begins. The extra vertical bits of canvas are 50 cm high, joining to a sloping section of 1.5 m. They are 1.8 m wide and 2.8 m long but have no groundsheets.

The height of the tent is 1.7 m.

b How much material is used to make these tents?

c How much extra space do you get inside the Scout tent compared to Tom's own tent?

18 Zahara is finding an expression for the surface area of the prism below. She writes:

Area of triangle = $6x$

Total surface area = $6x + 6x + 6 \times 11 + x \times 11 + 10 + 11$
$$= 23x + 176$$

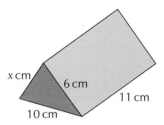

a Is Zahara correct? Explain your answer.

b The total surface area of the prism is 312 cm². What is the value of x?

36.4 Volume of a cylinder

● I can calculate the volume of a cylinder

A cylinder is a circular prism.

The cross-section of a cylinder is a circle of radius r.

The area of the cross-section is A.

$A = \pi r^2$

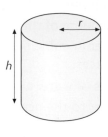

If the height of the cylinder is h, then the volume, V, of the cylinder is given by:

$V = \pi r^2 \times h = \pi r^2 h$

If the length of the cylinder is l, then the volume, V, of the cylinder is given by:

$V = \pi r^2 \times l = \pi r^2 l$

Develop fluency

Calculate the volume of each cylinder, giving your answers correct to one decimal place.

a

2 cm — 3 cm

b

15 m — 3 m

a $V = \pi r^2 h = \pi \times 3^2 \times 2 = 56.5\,\text{cm}^3$ **b** $V = \pi r^2 l = \pi \times 1.5^2 \times 15 = 106.0\,\text{m}^3$

In this exercise, take $\pi = 3.14$ or use the π key on your calculator.

1 Work out the volume of these cylinders:
   a cross-sectional area 3 cm², height 15 cm   **b** cross-sectional area 1.5 m², length 4 m
   c cross-sectional area 5.4 cm², width 8 mm

2 Find the volume of these cylinders. Give your answers to one decimal place.
   a radius 4 cm, length 10 cm   **b** radius 2.5 m, height 6 m
   c diameter 12 cm, length 15 cm   **d** diameter 8.4 m, height 7.5 m

3 Calculate the volume of each cylinder. Give your answers correct to one decimal place.

a 6 cm, 10 cm   **b** 3 cm, 8 cm   **c** 4 m, 1.5 m   **d** 2 m, 12 m   **e** 5 cm, 0.5 cm

4 The diameter of a 2p coin is 26 mm and its thickness is 2 mm. Calculate its volume. Give your answer correct to the nearest cubic millimetre.

5 The diagram shows the internal measurements of a cylindrical paddling pool.

50 cm

2 m

   a Calculate the volume of the pool.
      Give your answer in cubic metres, giving your answer to two decimal places.
   b How many litres of water are there in the pool when it is three-quarters full?
      Give your answer to the nearest litre.

6 A tea urn is a cylinder. The inner diameter of the cylinder is 30 cm and the inner height is 50 cm.
   Calculate the volume of the inner cylinder.
   Give your answer in litres, correct to one decimal place.

7 A winners' podium is going to be built at an athletics stadium.
   The diagram shows the dimensions of the podium.

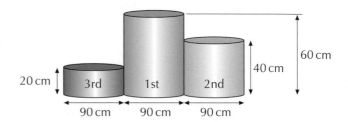

20 cm   3rd   1st   2nd   40 cm   60 cm

90 cm   90 cm   90 cm

   a Calculate the volume of each cylinder.
      Give your answers correct to the nearest cubic centimetre.
   b What is the total volume of the podium?
      Give your answer in cubic metres, correct to two decimal places.

8 The cross-section of this plastic bench has an area of 620 cm². Calculate its volume in:
   a cubic centimetres
   b cubic metres.

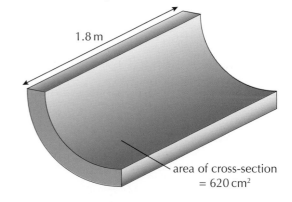

1.8 m

area of cross-section = 620 cm²

9 Work out the volume of each cylinder. Give your answer in terms of π.
   a radius 8 cm, length = 10 cm
   b diameter 8 cm, length 10 cm
   c diameter 16 m, length 10 m
   d radius 16 m, length 10 m

10 Calculate the volume of each cylinder.
   a radius 10 cm, length 12 cm
   b diameter 60 cm, length 1 m
   c diameter 10 m, length 10 m
   d radius 80 cm, length 4 m

Reason mathematically

A cylinder has a height of 5 cm. The volume of the cylinder is 180π cm³.

Gary thinks that the diameter of the cylinder is 36 cm.

a Explain the mistake that Gary has made.　　**b** What is the diameter of this cylinder?

a Volume of cylinder = 180π cm³

area of circle $\times 5 = 180\pi$

area of circle = 36π

Gary has mistaken the formula for the circumference of the circle (πd) for the area of the circle and used this to find the diameter.

b $\pi r^2 = 36\pi$

$r^2 = 36$, radius = 6 cm,

diameter = 12 cm

11 These are three cake tins.

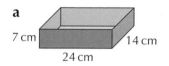

a 7 cm, 14 cm, 24 cm

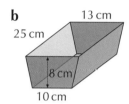

b 13 cm, 25 cm, 8 cm, 10 cm

c 20 cm, 7·5 cm

Which tin has the greatest volume?

12 A circular pool has a radius of 1.5 m and a depth of 60 cm.

a What is the capacity of the pool to the nearest 100 litres?

b How much water does it contain when two-thirds full?

13 A cylinder has a volume of 340 cm³ and a radius of 3 cm.

a What is its height to the nearest centimetre?

b Another cylinder has a volume of 37 m³ and it is 6 m high.
What is its radius to one decimal place?

c A third cylinder has a capacity of 2 litres, with a diameter of 8 cm.
How high is it to the nearest centimetre?

14 Are these statements always, sometimes or never true? Explain your reasoning.

a If you double the height of a cylinder, the volume also doubles.

b If you double the radius of a cylinder, the volume also doubles.

c If you double the diameter of a cylinder, the volume doesn't change.

Solve problems

A drinks manufacturer wants a can to hold 250 ml of a fizzy drink. The diameter of the can is 6 cm. How tall does the can need to be?

$250\,ml = 250\,cm^3$

Radius $= 6 \div 2 = 3\,cm$

$250 = \pi \times 3^2 \times height$

Height $= 250 \div 9\pi = 8.85\,cm\ (2\ dp)$

15 A water pipe has an internal diameter of 40 cm.

    a Calculate:

        i the radius **ii** the cross-sectional area **iii** the volume per metre of the pipe

        iv the quantity of water, to the nearest litre, in every metre of the pipe.

    b If the water in the pipe moves at a speed of 0.8 metres per second, how many litres, to the nearest 1000, will pass through it in one hour?

16 A component is manufactured by cutting a circular hole of radius 6 mm through a metal block.

    a Find the volume of the block after the hole has been drilled.

    b What percentage of the block has been removed? What is this to the nearest unit fraction?

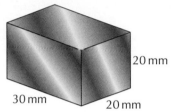 

20 mm

30 mm   20 mm

    c The original block has a mass of 252 g. What is its mass after the hole is removed?

17 An urn contains hot water and is a vertical cylinder. It has an internal diameter of 24 cm and is 40 cm high. The serving tap is located 5 cm above the base, and once the water reaches this level, no more will come out.

    a Calculate the total capacity of the urn to the nearest millilitre.

    b Calculate the usable amount of water in the urn to the nearest millilitre.

    c What percentage of the total capacity is usable?

    d In the canteen where it is used, tea mugs hold about 25 cl, but coffee is served in special 20 cl cups.

        If an equal number of both drinks are served, what is the maximum number of cups that can be filled?

18 A paint tin has a capacity of 10 litres and a height of 40 cm.

    a Write down the volume of the tin, in cubic centimetres.

    b Calculate the area of the base of the tin.

    c Calculate the diameter of the base. Give your answer correct to one decimal place.

36.5 Surface area of a cylinder

- I can calculate the curved surface area of a cylinder
- I can calculate the total surface area of a cylinder

A cylinder without a top and bottom is called an **open cylinder**.

When an open cylinder is cut and opened out, it forms a rectangle with the same length as the circumference of the base of the cylinder.

The **curved surface area** of the cylinder is the same as the area of the rectangle.

The area of the rectangle is $2\pi rh$. The formula for the curved surface of a cylinder is:

$A = 2\pi rh$

The total surface area of the cylinder is the curved surface area plus the area of the circles at each end. The formula for the total surface area of a cylinder is:

$A = 2\pi rh + 2\pi r^2$

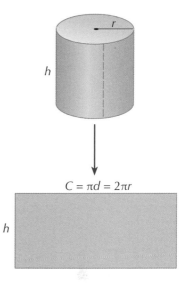

$C = \pi d = 2\pi r$

Develop fluency

A cylinder has a height of 1.2 m and a radius of 25 cm.

Calculate the total surface area of the cylinder.

Give your answer in square metres, correct to one decimal place.

$r = 25\,cm = 0.25\,m$

The total surface area is given by:

$A = 2\pi rh + 2\pi r^2$

So, $A = 2 \times \pi \times 0.25 \times 1.2 + 2 \times \pi \times 0.25^2 = 2.3\,m^2$ (1 dp)

In this exercise take $\pi = 3.14$ or use the pi key on your calculator.

1 Find the surface areas of these cylinders.
   a radius 4 cm, length 10 cm
   b radius 2.5 m, height 6 m
   c diameter 12 cm, length 15 cm
   d diameter 8.4 m, height 7.5 m

2 Calculate the total surface area of each cylinder. Give your answers correct to one decimal place.

a
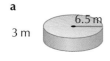
3 m   6.5 m

b
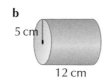
5 cm   12 cm

c

6 cm   9 cm

d
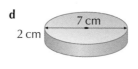
2 cm   7 cm

e

3 m   10 m

3 Calculate the curved surface area of each cylinder.
Give your answers correct to one decimal place.

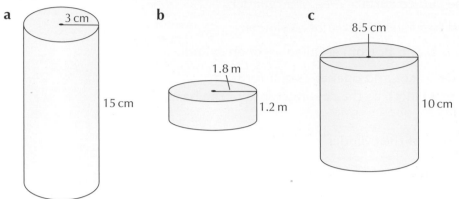

a   3 cm   15 cm

b   1.8 m   1.2 m

c   8.5 cm   10 cm

4 Calculate the total surface area of a cylinder that has a radius of 4.5 cm and a height of 7.2 cm.
Give your answer correct to one decimal place.

5 Calculate the curved surface area of a cylinder that has a diameter of 1.2 m and a length of 3.5 m.

6 Calculate the surface area of each cylinder. Give your answer in terms of π.
   a radius 4 m, height 12.5 m   **b** diameter 10 m, height 3 m   **c** diameter 12 m, height 5 m

7 Which of these cylinders, A or B, has the larger surface area?
A: radius 4 cm and height 10 cm   B: diameter 4 cm and height 20 cm

8 Calculate the surface area of each cylinder. Give your answer in terms of π.
   a radius 4 cm, height 3 cm   **b** diameter 4 cm, height 6 cm
   c diameter 10 cm, height 0.6 cm

9 Calculate the surface area of each open cylinder. Give your answer in terms of π.
   a radius 5 cm, height 10 cm   **b** diameter 10 m, height 5 m
   c diameter 1 m, height 50 cm   **d** radius 250 cm, height 10 m

Reason mathematically

Lily's pencil pot is in the shape of a cylinder.

The diameter of the cylinder is 16 cm and the height is 20 cm.

She paints the surface area of the pot including the base. She uses 1 tube of paint for every 128 cm². How many tubes of paint will Lily need?

Area of the base $= \pi \times 8^2 = 64\pi$ cm²

Curved surface area $= 2 \times \pi \times 8 \times 20 = 320\pi$ cm²

Total surface area $= 384\pi$ cm² $= 1206.37$ cm²

Number of tubes of paint $= 1206.37 \div 128 = 9.42$ tubes

Lily will need 10 tubes of paint.

10 A tin of salmon has a diameter of 8.5 cm.
    A label, with a height of 4 cm, goes around the curved surface of
    the tin.
    Calculate the area of the label, if there is an overlap of 1 cm to
    glue the ends together.
    Give your answer correct to the nearest square centimetre.

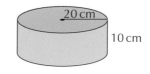

11 Tom calculates the area of the cylinder. He writes:
    Area of the circle $= \pi \times 10^2 = 100\pi$ cm²
    Curved surface area $= \pi \times 20 \times 10 = 200\pi$ cm²
    Total surface area $= 300\pi$ cm²
    Do you agree with Tom? Explain your answer.

12 A trampoline with a diameter of 12 m has protective netting all the way round to a height
    of 2 m. What area of netting is required?

13 a Which of these cylinders has the largest surface area? Explain why.
        Cylinder A: radius R, height 10 cm   Cylinder B: radius $2R$, 5 cm
    b Substitute R for 5 cm and check whether your answer to part **a** is still correct.

Solve problems

The curved surface area of a cylinder is 300π cm². Its height is 5 cm.

Work out its radius.

Let the radius of the cylinder $= x$ cm.

Curved surface area $= 2 \times \pi \times x \times 5 = 10\pi x$

$10\pi x = 300\pi$ Set up an equation and solve it.

$\quad 10x = 300$

$\qquad x = 30$

The diameter of the cylinder is 60 cm.

14 A circular sandpit is to be built in a children's play area, kept in place by a wooden fence
    which needs to be wood stained on both sides before the sand goes in.
    a The sandpit has a radius of 2.4 m and the fence is 40 cm high.
       What is the area to be stained?
    b The fence needs two coats of wood stain. A 2.5 litre tin claims to cover around 30 m².
       What proportion of the tin will not be used?

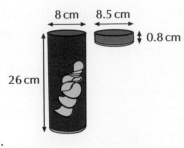

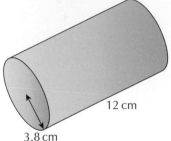

15 A cylindrical tube has a diameter of 8 cm and
a height of 26 cm.

  a Calculate the outside surface area of the tube.

  b The tube has a lid with a diameter of 8.5 cm.
Calculate the outside surface area of the lid, which is
0.8 cm deep.

  c Calculate the total surface area of the tube with the lid on.
Give all of your answers correct to one decimal place.

16 A firm sells butter in the form of a cylinder with radius
3.8 cm and length 12 cm. In order to wrap it fully they
allow an extra 10% of the surface area for packaging.
What area of packaging is used? Give you answer to the
nearest whole number?

17 Joanne makes a pencil pot in the shape of a cylinder using a sheet of thin cardboard with
dimensions 20 cm by 40 cm. She uses a circular piece of wood for its base.

  a Find the two possible values that the radius of the base can be.

  b Calculate the surface area of the pencil pot.

Now I can...

convert from one metric unit to another	calculate the volume of a cylinder	calculate the total surface area of a cylinder
calculate the surface area of a prism	calculate the curved surface area of a cylinder	
calculate the volume of a prism		

CHAPTER 37 Compound units

37.1 Speed

● I can understand and use measures of speed

Jasmine is taking part in a race. She is running at a constant **speed**. She runs 100 metres in 20 seconds.

A formula for working out speed is:

$$\text{speed} = \frac{\text{distance travelled}}{\text{time taken}} \text{ or } \frac{\text{distance}}{\text{time}}$$

Jasmine's speed is $\frac{100}{20} = 5$ metres per second.

The units are metres per second (m/s) because the distance is in metres and the time is in seconds. This is an example of a **compound unit** that involves other units: in this case, metres and seconds.

The units of speed depend on the units used to measure the distance and the time.

Develop fluency

a A car travels 45 km in 30 minutes.

Work out its speed. Include units in your answer.

b A car is travelling at a constant speed of 30 m/s.

How far does the car travel in one minute?

a $Speed = \frac{distance}{time} = \frac{45}{30}$

  $= 1.5\,km/minute$ The units are km/minute because the time is in minutes.

An alternative answer is $\frac{45}{0.5} = 90\,km/h$ since 30 minutes = 0.5 hours.

b $speed = \frac{distance}{time}$ In this case the speed is 30 m/s and the time is 60 s.

$30 = \frac{d}{60}$ The time must be in seconds. Use d for distance.

$30 \times 60 = d$ Multiply by 60 to solve the equation.

$d = 1800$

The distance is 1800 m or 1.8 km.   You must include the units.

1 A marathon runner runs 40 km in $2\frac{1}{2}$ hours.

Work out his speed, in kilometres per hour (km/h).

2 A train is travelling at a constant speed.
It takes 30 minutes to travel 45 km.

  a Work out the speed, in kilometres per minute (km/minute).

  b Work out the speed in kilometres per hour (km/h).

3 Calculate the speed in each case. Put units in each answer.

   a Peter runs 320 metres in 50 seconds.

   b A car travels 15 km in 10 minutes.

   c An aeroplane flies 400 km in half an hour.

   d A cyclist travels 1500 m in 4 minutes.

4 Matthew is cycling at 18 km/h.
Calculate how far he travels in:

   a 1 hour   **b** 4 hours   **c** 1.5 hours   **d** 2 hours and 30 minutes.

5 Paul can swim at a constant speed of 3 km/h.
Calculate how far he can swim in:

   a 1.5 hours   **b** $\frac{1}{2}$ hour   **c** $\frac{1}{4}$ hour   **d** 1 minute.

6 Calculate the distance travelled in each case.

   a Sharon walks at 3 m/s for two minutes.

   b Nathan drives at 80 km/h for 15 minutes.

   c A plane flies at 700 km/h for 4.5 hours.

   d A snail moves at 0.2 m/minute for 150 seconds.

7 The top speed of a sprinter is 8 m/s.
Calculate the time it takes at that speed to sprint:

   a 40 m   **b** 80 m   **c** 50 m   **d** 200 m.

8 Calculate the time taken to travel:

   a 40 km at 120 km/h   **b** 22 km at 4 km/h

   c 200 m at 5 m/s   **d** 5 km at 4 m/s.

9 Anita is walking in the countryside. She is travelling at 6 km/h.

   a Copy and complete this table.

Time (*t* hours)	0.5	1	1.5	2	2.5
Distance (*d* km)					

   b Work out how long it takes Anita to travel 8 km.

10 An aeroplane is flying at 850 km/h.

   a Calculate how far the aeroplane flies in $3\frac{1}{2}$ hours.

   b Calculate the time the plane takes to fly 5000 km.

Reason mathematically

The graph shows the journey that Kate and Charlie made from their homes to the park.

Kate forgets her phone and has to go back home to pick it up. Charlie meets a friend on the way to the park.

a Which line represents Kate's journey? Which line represents Charlie's journey?

b Who lives the closest to the park? How much closer?

c Who walks at a faster pace? Give a reason for your answer.

d Charlie says his average speed is about 1.87 km/h. Is he correct?

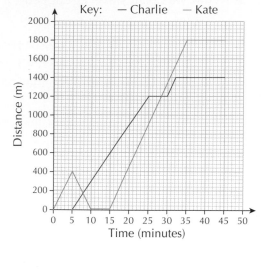

Key: — Charlie   — Kate

a The green line represents Kate's journey and the brown line represents Charlie's journey.

b Charlie lives 1800 – 1400 = 400 m closer to the park.

c Kate walks at a faster pace. Her graph is steeper. She walks 400 m in 5 minutes. Charlie walks 600 m in 10 minutes or 300 m in 5 minutes.

d Charlie travels 1.4 km in 40 minutes = $\frac{2}{3}$ hours.

Average speed = $1.4 \div \frac{2}{3} = 2.1$ km/h. Charlie is not correct.

11 This graph shows the journey of a car.

   a How far does the car travel in 20 minutes?

   b Explain how you know the car is travelling at a constant speed.

   c Work out the speed of the car, in kilometres/minute.

   d How long does it take the car to travel 50 km?

12 A plant grows 3.6 cm in 2 days.
What is the rate of growth in millimetres per hour (mm/h)?

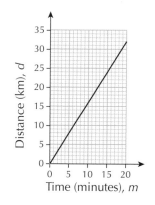

13   Lewis and Neil ran a 2000-metre race.
The distance–time graph below shows the race.

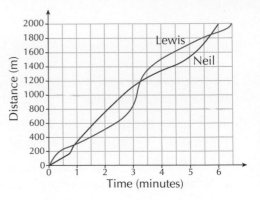

a   What was Neil's average speed for the race in

 i   m/s   ii   km/h?

b   After 800 m Lewis managed to sprint for about
15 seconds in order to catch up with Neil.
What was Lewis' speed for that part of the race
in m/s and km/h?

c   Use the graph to help you fill in the gaps in the report of this race:
'Lewis started well but soon tired and Neil took the lead after about_____ seconds.
Lewis made an effort to draw level at _____ metres, but could only hold the lead
for _____ minutes. Neil won the race in a time of _____ minutes, and Lewis finished
_____ seconds later.'

14   Michael draws a graph to show the journey from his home to his grandparents' house.
Explain which of these graphs can't be his graph.

a
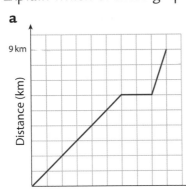

b
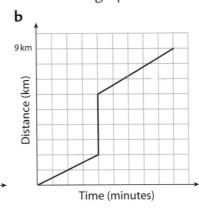

c
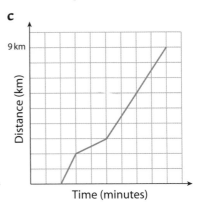

Solve problems

Matilda took 10 minutes to cycle one-third of the distance to her home. She cycled at
an average speed of 9 km/h. She walked the rest of the distance at an average speed
of 4.5 km/h. What is Matilda's average speed for the journey?

10 minutes = $\frac{10}{60}$ of an hour = $\frac{1}{6}$ hour

Distance cycled = $\frac{1}{6}$ hour × 9 km/h = 1.5 km

Second part of the journey = 1.5 km × 2 = 3 km

Time taken for the second part = $\frac{3 \text{ km}}{4.5 \text{ km/h}}$ = $\frac{3}{4.5}$ = $\frac{30}{45}$ = $\frac{2}{3}$ hour = 40 minutes

Total distance = 1.5 km × 3 = 4.5 km

Total time = 50 minutes = $\frac{50}{60}$ of an hour = $\frac{5}{6}$ hour

Average speed = $\frac{\text{distance}}{\text{time}}$ = $\frac{4.5 \text{ km}}{\frac{5}{6} \text{ hour}}$ = 5.4 km/h

15 Some friends took part in a K1 kayak slalom competition. Over the 1 km course, there were 12 down gates and 6 up gates. It is reckoned that every down gate adds 1 second to a competitor's time and each up gate adds 3 seconds. Any missed gate adds a 15 second time penalty.

    a Lucy gets through every gate successfully and finishes in 2 minutes 30 seconds. What was her average speed?

    b Gemma's usual average speed without any gates is 33 km/h. Adding extra time for the gates, how long did she take if she got through every gate successfully?

    c Sarah paddles at an average 36 km/h but misses one gate. In what order did these three girls finish?

16 The speed of sound is 340 m/s. There is an explosion 2 km away from Sam. Calculate how many seconds will pass before Sam hears the explosion.

17 On a journey lasting 23 minutes, this bicycle wheel rotated at a rate of 95 revolutions per minute.

42 cm

    a Calculate the circumference of the wheel.

    b Calculate the length of the journey. Give your answer in km, correct to the nearest 100 m.

18 Amy takes 2 hours to travel a third of her journey at a speed of 70 km/h. At what speed must she travel the rest of the journey so that she can complete the whole journey in 7 hours?

37.2 More about proportion

● I can understand and use density and other compound units

Another example of compound units is the **rate** of flow, which is a measure of how quickly a liquid is flowing.

Another example of a compound unit is **density**. This is calculated as 'mass per unit volume':

$$\text{density} = \frac{\text{mass}}{\text{volume}}$$

Examples of possible units are:

· grams per cubic centimetre (g/cm³)

· grams per litre (g/litre).

If you compare equal volumes of different substances, the denser one will be heavier.

Develop fluency

Water is flowing out of a tap. In 5 minutes, 24 litres flow out of the tap.

a Work out the rate of flow, in litres per minute.

b How long does it take to fill a 7.5 litre bucket?

a The rate of flow is measured as $\dfrac{\text{litres}}{\text{minutes}}$.

This is the number of litres ÷ number of minutes and the unit is litres/minute.

Rate of flow $= \dfrac{24}{5} = 4.8$ litres/minute.

b $4.8 = \dfrac{7.5}{m}$ Call the number of minutes m.

$4.8m = 7.5$ Multiply by m.

$m = \dfrac{7.5}{4.8} = 1.5625$ Divide by 4.8.

It takes 1.56 minutes.

Round the answer. It is just over $1\dfrac{1}{2}$ minutes.

A piece of iron has a volume of $20\,\text{cm}^3$ and a mass of $158\,\text{g}$.

a Calculate the density of iron. b Calculate the mass of $36\,\text{cm}^3$ of iron.

a Density $= \dfrac{\text{mass}}{\text{volume}} = \dfrac{158}{20}$

$= 7.9\,\text{g/cm}^3$ The units must involve grams and cm^3.

b Density $= \dfrac{\text{mass}}{\text{volume}}$

$7.9 = \dfrac{m}{36}$ Use the answer from part a. Use m for the mass.

$m = 7.9 \times 36 = 284.4$ Multiply by 36.

The mass is $284\,\text{g}$ to three significant figures.

1 A driver takes 20 seconds to put 56 litres of petrol in the fuel tank of her car.
   What is the rate of flow of the petrol? Give units in your answer.

2 A tap is dripping. In 20 minutes, 0.3 litres drip from the tap.
   What is the rate of flow, in litres per hour (litres/h)?

3 Water flows down a stream at a rate of 6 litres/s.
   a Calculate how much water flows in one minute.
   b Calculate how much water flows in half an hour.
   c Work out how long it takes for 1000 litres to flow past a particular point.

4 A shower has a rate of flow of 12 litres/minute.
   a Work out how much water is used when Sam has a shower that lasts four minutes.
   b Work out how long it takes to use 1 litre of water.

5 Balsa wood is used to make model aeroplanes because it has a low density of $0.16\,g/cm^3$.
  a A piece of balsa wood has a volume of $25\,cm^3$. Calculate its mass.
  b Calculate the volume of $1\,g$ of balsa wood.

6 Oxygen has a density of $1.43\,g/litre$.
Work out the mass of 200 litres of oxygen.

7 Graphite is used to make pencils. It has a density of $2.3\,g/cm^3$.
  a Work out the volume of $50\,g$ of graphite.
  b A pencil contains $1.2\,cm^3$ of graphite. Work out the mass of graphite in 10 pencils.
  c Work out the mass of $100\,cm^3$ of graphite.

8 A piece of copper has a mass of $45\,g$ and a volume of $5\,cm^3$.
  a Work out the density of copper.
  b Work out the volume of $120\,g$ of copper.
  c Work out the mass of a cube of copper with a side of $5\,cm$.

Reason mathematically

The graph shows the relationship between the mass ($m\,kg$) of metal A and its volume ($V\,cm^3$). Georgina says that she used $300\,g$ of metal A to make a cube with a side length $3\,cm$. Explain why Georgina is incorrect.

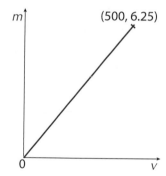

Density of metal A $= \dfrac{6.25}{500} = 0.0125\,kg/cm^3$

Volume of cube A $= 3^3 = 27\,cm^3$

Mass of cube A $= 0.0125 \times 27 = 0.3375\,kg$

$1\,kg = 1000\,g$

Mass of cube A $= 337.5\,g$. It is not $300\,g$.

9 This graph shows the connection between mass and volume for a type of steel.
  a Use the graph to find the mass of $2\,m^3$ of steel.
  b Work out the density of the steel. Give your answer in kg/m^3.
  c The mass of a particular type of steel beam is $400\,kg$. Show that the volume of the beam is $0.05\,m^3$.
  d Explain how you know that the mass of the beam is proportional to its volume.
    Hint: $1\,m^3 = 1\,000\,000\,cm^3$

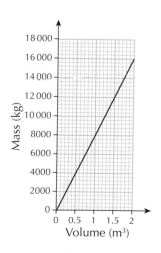

10 A hosepipe is able to fill this can at a rate of 15 litres/minute.

   a Calculate:

      i the volume of the can in cubic centimetres

      ii the capacity of the can in litres.

   b How long does it take to fill the can?
      Give your answer to the nearest second.

11 The ratio between the mass (m, g) and volume (V, cm^3) of substance P, is $3 : 10$.

   a Write a formula connecting the mass and volume of substance P.

   b What is the density of substance P?

   c Which of the graphs below can represent substance P? Explain why.

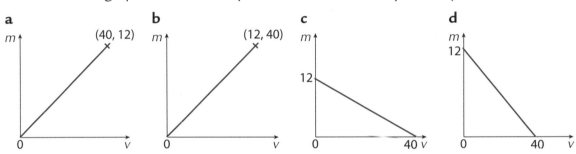

Solve problems

The density of copper is 8.96 g/cm^3. The mass of 50 cm^3 of zinc is 356.50 g.

1344 g of copper and 356.50 g of zinc were mixed together. Work out the density of the mixture. Give your answer to two decimal places.

density $= \dfrac{mass}{volume}$

Mass of mixture = mass of copper + mass of zinc = 1344 g + 356.50 g = 1700.50 g

Volume of copper = 1344 ÷ 8.96 = 150 cm^3

Volume of mixture = volume of copper + volume of zinc = 150 + 50 = 200 cm^3

Density of mixture = 1700.50 ÷ 200 = 8.50 g/cm^3

12 a A hosepipe is accidentally left on for 15 minutes. The water flows at 9 litres per minute. How much water is wasted?

     b The charge for water at this house is 80p per cubic metre. How much has this wastage cost?

13 **a** Car A takes 35 seconds to fill its fuel tank with 70 litres of diesel. What is the rate of flow of the fuel at this pump?

b At the same pump, Car B takes 22 seconds to fill up. How much fuel does it hold?

c Mary's car holds 16 gallons of petrol. How long does it take to fill up, to the nearest second?

d If diesel costs £1.39 per litre and petrol costs £1.32, which of the three cars costs most to fill up?

14 The density of gold is 19.3 g/cm³.

a Work out the mass of: **i** 3.5 cm³ of gold **ii** 3.5 g of gold..

b The largest gold bar in the world has a mass of 250 kg. Work out its volume.

15 An alloy is made from 3.2 kg of copper, 1.5 kg of lead and 600 g of tin. Calculate the density of the alloy.

Metal	Density (g/cm³)
copper	8.96
lead	11.40
tin	7.30

37.3 Unit costs

● I can understand and use unit pricing

Packets and containers of food and other items are sold in different sizes. If you want to compare the prices of the same item, in different sized containers, it is helpful to be able to work out a **unit price**. This is the price of one gram or one litre, or any other suitable unit.

Develop fluency

Compare the prices of the rice in these packets.

* You could find the cost/gram of rice from each packet.

  For the smaller packet, 89p ÷ 200 = 0.445 p/g (pence per gram).

  For the larger packet, 209p ÷ 500 = 0.418 p/g.

  The larger packet is better value, as it has the lower cost per gram.

* You could find how much you can buy for 1p.

  For the smaller packet, the number of grams/p is 200 ÷ 89 = 2.247... g/p.

  For the larger packet, it is 500 ÷ 209 = 2.392... g/p.

  The larger packet is better value because you can buy more for 1p.

1 1 kg of margarine costs £2.50.

a Calculate the cost: **i** per 100 g **ii** per gram.

b Calculate the number of grams bought for: **i** 1p **ii** £1.

2 A bag of pasta has a mass of 250 g and costs 87p.

   a Calculate the cost: **i** per 100 g **ii** per gram.

   b Calculate the number of grams bought for: **i** 1p **ii** £1.

3 A 160 g can of tuna costs £1.85. A pack of four cans costs £6.20.

   a Work out the cost per 100 g for one can.

   b Work out the cost per 100 g if you buy the pack of four cans.

   c Which is better value? Give a reason for your answer.

4 You can buy tomato puree in tubes or jars.
A shop sells 140 g tubes for 69p and 200 g jars for £1.09.

   a Work out the cost per 100 g in a tube of tomato puree.

   b Work out which is better value for money. Justify your answer.

5 Aziz buys 700 g of apples for £1.68. Calculate the cost per kilogram.

6 Miriam downloaded 928 kB of data from the Internet in 18 seconds.

   a How much data did she download each second, to the nearest kB?

   b How much data could she download at the same rate in:

      i 45 seconds   **ii** 7 minutes?

Reason mathematically

Tom and Ben travel to work. Tom's car uses 10.6 litres of petrol on a 160 km journey. The price of petrol is £1.26/litre. Ben travels 256 km by train. His train ticket costs £25.50. Which is better value, to drive or get the train? Explain why.

Cost of driving per km $= \dfrac{10.6 \times 1.26}{160} =$ £0.083475 per km

Cost of travelling by train per km $= \dfrac{25.50}{256} =$ £0.0996 per km

£0.083475 per km is less than £0.0996 per km. It is cheaper to drive.

7 A shop sells cans of pineapple in two sizes. The prices are shown in this table.

Mass	225 g	435 g
Price	83p	£1.45

Which is better value? Justify your answer.

8 A 600 g box of muesli costs £2.65. An 850 g box costs £3.95. Which is better value? Justify your answer.

9 Toothpaste comes in tubes of different sizes. A 125 ml tube costs £2.99 and a 75 ml tube of the same brand costs £1.89. Which is better value for money? Justify your answer.

10 A pack of four 120 g pots of yogurt costs £2.00. A 450 g pot of the same yogurt costs £1.79. Show that the large pot is better value.

Solve problems

Milk is sold in litres or pints. One pint is 568 ml.

A supermarket sells 1 litre of milk for 95p or 1 pint for 49p. Which is better value?

Work out the cost per litre.

1 pint = 568 ml = 0.568 litre

Cost per litre = 49p ÷ 0.568 = 86.3p per litre

The pint of milk is better value, because it is cheaper per litre.

11 a Eighteen eggs cost a shopkeeper £3.60. How much does each egg cost him?

   b He puts them in boxes of six and charges £1.80 per box. How much profit does he make on each box? Express this as a percentage of the cost.

   c He does an offer of three boxes for five pounds. What is his percentage profit here?

12 A 400 g packet of beef costs £4.60. What price would you expect to pay for a 750 g pack? Justify your answer.

13 The diagram shows how two businesses charge for the hire of a canoe.

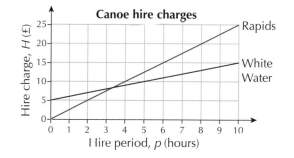

   a For which business is the charge directly proportional to the hire period? Give a reason for your answer.

   b For this business, what is the hire charge for one hour?

   c For this business write an equation connecting the hire charge, H, with the hire period, p.

   d i What would this company charge to hire the canoe for 4.5 hours?

   ii What is the hire period corresponding to a hire charge of £18 with this company? Give your answer in hours and minutes.

Now I can...

understand measures of speed	use measures of speed	use unit pricing
understand density and other compound units	use density and other compound units	
understand unit pricing		

CHAPTER 38 Solving equations graphically

38.1 Graphs from equations of the form $ay \pm bx = c$

- I can draw any linear graph from any linear equation
- I can solve a linear equation from a graph

You have already met the **linear equation** $y = mx + c$, which generates a straight-line graph. It is this equation that is seen here as $ay \pm bx = c$. When its graph is plotted it will, of course, still produce a straight line.

To construct the graph of $ay \pm bx = c$, follow these steps:

- Substitute $x = 0$ into the equation so that it becomes $ay = c$.

- Solve for y, i.e. $y = \dfrac{c}{a}$ so that one point on the graph is $\left(0, \dfrac{c}{a}\right)$.

- Substitute $y = 0$ into the equation so that it becomes $bx = c$.

- Solve for x, i.e. $x = \dfrac{c}{b}$, so that one point on the graph is $\left(\dfrac{c}{b}, 0\right)$.

- Plot the two points on the axes and join them up.

Develop fluency

Draw the graph of $4y - 5x = 20$.

Substitute $x = 0$:　$4y = 20$

$y = 5$

The graph passes through $(0, 5)$.

Substitute $y = 0$:　$-5x = 20$

$x = -4$

The graph passes through $(-4, 0)$.

Plot the points and join them up.

Note that this method is sometimes called the 'cover-up' method as all you have to do to solve x or y is cover up the other term.

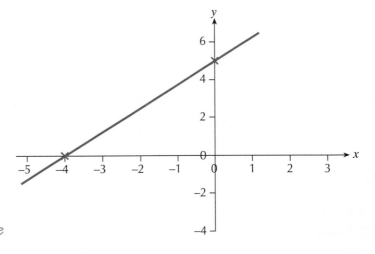

1. Draw the graph for each equation. Use a grid that is numbered from −10 to +10 on both the x-axis and the y-axis.

   a $y = x + 3$　　　**b** $y = x - 4$　　　**c** $y = 2x$　　　**d** $y = 3x + 1$

2. Draw the graph for each equation. Use a grid that is numbered from 0 to +10 on both the x-axis and the y-axis.

   a $2y + 3x = 6$　　**b** $4y + 3x = 12$　　**c** $y + 2x = 8$　　**d** $3y + 2x = 6$　　**e** $5y + 2x = 10$

3 Draw the graph for each equation. Use a grid that is numbered from –10 to +10 on both the x-axis and the y-axis.

 a $y - 5x = 10$ **b** $2y - 3x = 12$ **c** $2y - 5x = 10$ **d** $3x - 4y = 12$ **e** $5y - 2x = 10$

4 For each graph, find the coordinates of the two points where the graph intersects the x-axis and the y-axis.

 a $3x + 2y = 18$ **b** $5y - x = 15$ **c** $4y - 7x = -28$

5 Draw the graph for each equation. Use a grid that is numbered from –10 to +10 on both the x-axis and the y-axis.

 a $3y + 2x = 12$ **b** $4x + 5y = 40$ **c** $3y + 7x = 21$

 d $6x - y = 6$ **e** $3y - 4x = 24$

6 **a** Using a grid with axes numbered from –2 to 10, draw the graph of $y = 4x + 3$.

 b Use the graph to solve these equations.

    i $4x + 3 = 7$ **ii** $4x + 3 = 9$ **iii** $4x + 3 = 5$

7 **a** Using a grid with axes numbered from –5 to 8, draw the graph of $y = 3x - 1$.

 b Use the graph to find the value of y when

    i $x = -1$ **ii** $x = 0$ **iii** $x = 3$.

 c Use the graph to find the value of x when

    i $y = -1$ **ii** $y = 1$ **iii** $y = 5$.

8 **a** Using a grid with axes numbered from –2 to 10, draw the graph of $y = 3 - 2x$.

 b Write down the coordinates of the point where your graph crosses the

    i x-axis   **ii** y-axis.

 c Use your graph to find the value of x when $y = 6$.

9 **a** Using a grid with axes numbered from –3 to 4, draw the graph of $y = 3 - x$.

 b The point with coordinates $(p, 4.5)$ lies on the graph of $y = 3 - 3x$. Use your graph to find the value of p.

Reason mathematically

Vicky says that the graph shown has the equation $3x + 4y = 12$.

Is she correct? Explain your answer.

Find the x and y intercept of $3x + 4y = 12$.

When $y = 0$, $3x = 12$ so $x = 4$.

When $x = 0$, $4y = 12$ so $y = 3$.

Vicky is incorrect as the line intercepts the axes at $x = 3$ and $y = 4$.

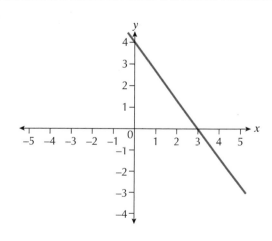

10 a Draw the graphs of all these equations on the same grid. Use a grid that is numbered from −10 to +10 on both the x-axis and the y-axis.

    i $y = \frac{1}{2}x + 4$ **ii** $y = \frac{1}{2}x - 2$ **iii** $y = \frac{1}{2}x$

  b What do you notice about all these graphs?

  c Explain how you could now draw the graph of $y = \frac{1}{2}x + 6$.

11 Draw the graphs for all these equations on the same grid. Use a grid that is numbered from −10 to +10 on the x-axis and from −2 to +10 on the y-axis.

  a i $3y + x = 6$ **ii** $4x - 5y = -10$ **iii** $x + y = 2$ **iv** $2x - 9y = -18$

  b What do you notice about all these graphs?

12 a Draw the graphs of all these equations on the same grid.
    Number both axes from −2 to 10.

    i $x + y = 6$ **ii** $x + y = 8$ **iii** $x + y = 10$ **iv** $x + y = 2$ **v** $x + y = 1$ **vi** $x + y = 7$

  b What do you notice about all these graphs?

  c Explain how you could now draw the graph of $x + y = -5.3$.

13 a Copy and complete the table for the graph $y = \frac{6}{x}$.

  b Why can you not find the value of y when $x = 0$?

  c Using a grid with axes numbered from 0 to 6, draw the graph of $y = \frac{6}{x}$.

x	1	2	3	4	5	6
y						

Solve problems

At 0 hours, there are 42 litres of fuel in the fuel tank of a car. After 7 hours, there is no fuel left in the fuel tank. Assume that the fuel is used at a constant rate.

  a Draw a graph to show the amount of fuel left in the tank against time.

  b Use your graph to estimate:

    i the amount of fuel used per hour of driving

    ii the volume of fuel in the tank after 3 hours

    iii the number of hours it takes for the tank to be half-full.

a
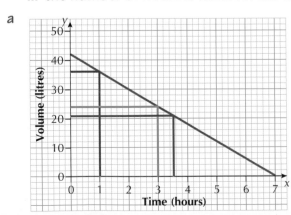

b i After 1 hour, the volume of the tank is 36 litres.

    The amount of fuel used per hour is 6 litres.

  ii When $t = 3$ hours, $V = 24$ litres.

  iii When $V = 21$ litres, $t = 3\frac{1}{2}$ hours

14 Angela works on Saturdays at the local cafe.
One day, Angela receives £64 for 8 hours work.

  a Draw a graph to show the monthly pay, P, that Angela receives based on the time, T, that she works.

  b What is Angela's hourly rate?

  c Use the graph to calculate the number of hours Angela needs to work to earn £100.

  d Why is it not realistic to find a value for P when $T = 100$?

15 A car is slowing down. The speed of the car, s metres per second, after time t seconds is given by the formula $s = 30 - 4t$

  a Calculate the initial speed of the car.

  b Sketch the graph showing the relationship between speed (s) and time (t).

  c Use the graph to find the time that it will take for the car to stop.

16 A bamboo shoot is 10 cm high. It then grows 4 cm every day.

  a Draw a graph to show the relationship between the height of the bamboo and the number of days.

  b Use your graph to calculate the number of days it takes for the bamboo to reach the height of 0.9 m.

  c Ben thinks that he can't use the same graph to find the height of the bamboo after two years. Explain why.

17 For each graph sketch, think of a real-life representation it could represent. State the quantities that x and y can represent and create a problem based on the graph.

  a
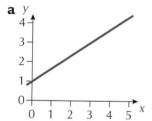
  b
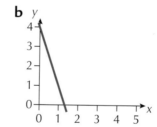

38.2 Graphs from quadratic equations

● I can draw graphs from quadratic equations

A **quadratic** equation involves two **variables** where the highest power of one of the variables is a square.

Some examples of quadratic equations are $y = x^2$, $y = x^2 + 3x$, $y = 2x^2 + x - 1$, $y = (x + 2)(x + 1)$.

Develop fluency

Draw the graph of the equation $y = x^2 + 3x$.

First, draw up a table of values for x, then substitute each value of x into x^2 and $3x$ to determine the y-value.

x	−4	−3	−2	−1	0	1	2
x^2	16	9	4	1	0	1	4
$3x$	−12	−9	−6	−3	0	3	6
$y = x^2 + 3x$	4	0	−2	−2	0	4	10

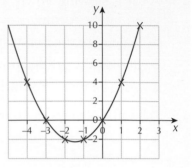

Now take the pairs of (x, y) coordinates from the table, plot each point on a grid, and join up all the points.

Note that the shape is a smooth curve. It is important always to try to draw a quadratic graph as smoothly as possible, especially at the bottom of the graph where it needs to be a smooth curve.

1 a Copy and complete this table of values for $y = x^2 + 2x$.

x	−3	−2	−1	0	1	2
x^2	9	4	1	0	1	4
$2x$			−2	0		
$y = x^2 + 2x$				0		

   b Draw a grid with the x-axis numbered from −3 to 2 and the y-axis from −2 to 10.
   c Use the table to help you draw, on the grid, the graph of $y = x^2 + 2x$.

2 a Copy and complete this table of values for $y = x^2 + 4x$.

x	−5	−4	−3	−2	−1	0	1
x^2	25						
$4x$	−20						
$y = x^2 + 4x$	5						

   b Draw a grid with the x-axis numbered from −5 to 1 and the y-axis from −5 to 6.
   c Use your table to help draw, on the grid, the graph of $y = x^2 + 4x$.

3 a Copy and complete this table of values for $y = x^2 + 3x + 2$.

x	–4	–3	–2	–1	0	1
x^2	9				0	
$3x$	–9				0	
2	2	2	2	2	2	2
$y = x^2 + 3x + 2$		2			2	

b Draw a grid with the x-axis numbered from –4 to 2 and the y-axis from –1 to 8.

c Use your table to help you draw, on the grid, the graph of $y = x^2 + 3x - 2$.

4 a Copy and complete this table of values for $y = x^2 + 2x - 3$.

x	–4	–3	–2	–1	0	1	2
x^2							
$2x$							
–3							
$y = x^2 + 2x - 3$							

b Draw a grid with the x-axis numbered from –4 to 2 and the y-axis from –5 to 6.

c Use your table to help you draw, on the grid, the graph of $y = x^2 + 2x - 3$.

5 Copy and complete this table of values for $y = x^2 + x$.

x	–1	0	1	2	3
x^2					
x					
y					

6 a Copy and complete this table of values for $y = x^2 + 6x$.

x	–7	–6	–5	–4	–3	–2	–1	0	1
x^2									
$6x$									
y									

b Draw a grid with the x-axis numbered from –7 to 1 and the y-axis from –10 to 10.

c Use the table to help you draw, on the grid, the graph of $y = x^2 + 6x$.

7 a Copy and complete this table of values for
$y = 2x^2 + 1$.

x	−2	−1	0	1	2
x^2					
$2x^2$					
y					

b Draw a grid with the x-axis numbered from −2 to 2 and the y-axis from 0 to 10.

c Use the table to help you draw, on the grid, the graph of $y = 2x^2 + 1$.

8 Construct a table of values for each equation. Using suitable scales, draw the graphs.

a $y = x^2 + 5x$ **b** $y = x^2 + 3x + 1$ **c** $y = x^2 + 4x - 3$

9 a Copy and complete the table of values for $y = 3x^2 - 5$.

x	−2	−1	0	1	2
x^2	4		0		
$3x^2$	12		0		
$y = 3x^2 - 5$	7		−5		

b Draw a grid with the x-axis numbered from −2 to 2 and the y-axis numbered from −7 to 10.

c Use the table to help you draw, on the grid, the graph of $y = 3x^2 - 5$.

Reason mathematically

a Construct a table of values for each equation.

   i $y = x^2$ **ii** $y = x^2 + 1$ **iii** $y = x^2 + 4$

b Plot all the graphs on the same pair of axes.
Number the x-axis from −3 to 3 and the y-axis from 0 to 13.

c Comment on your graphs.

d Sketch onto your diagram the graph with the equation $y = x^2 + 3$.

a

x	−3	−2	−1	0	1	2	3
$y = x^2$	9	4	1	0	1	4	9
$y = x^2 + 1$	10	5	2	1	2	5	10
$y = x^2 + 4$	13	8	5	4	5	8	13

b See the graph.

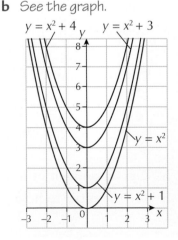

c The graphs are the same shape, however they have been translated in the positive y direction.

$y = x^2$ translated by 1 unit in the y-direction is $y = x^2 + 1$.

$y = x^2 + 1$ translated by 3 units in the positive y direction is $y = x^2 + 4$.

10 a Construct a table of values for each equation, then plot all their graphs on the same pair of axes. Number the x-axis from –4 to 4 and the y-axis from –10 to 55.

    i $y = x^2 - 10$ **ii** $y = 2x^2 - 10$ **iii** $y = 3x^2 - 10$ **iv** $y = 4x^2 - 10$

  b Comment on your graphs.

  c Sketch onto your diagram the graphs with these equations.

    i $y = \frac{1}{2}x^2 - 10$ **ii** $y = 2\frac{1}{2}x^2 - 10$ **iii** $y = 5x^2 - 10$

11 a Construct a table of values for each equation. Then plot all their graphs on the same pair of axes. Number the x-axis from –2 to 2 and the y-axis from –3 to 16.

    i $y = 3x^2 - 2$ **ii** $y = 3x^2$ **iii** $y = 3x^2 + 1$ **iv** $y = 3x^2 + 3$

  b Comment on your graphs.

  c Sketch onto your diagram the graph with the equation $y = 3x^2 + 2$.

12 a Copy and complete this table of values for $y = 8 - 2x - x^2$.

x	−5	−4	−3	−2	−1	0	1	2	3
8									
−2x									
− x²									
y									

  b Draw a grid with the x-axis numbered from −5 to 3 and the y-axis from −8 to 10.

  c Use the table to help you draw, on the grid, the graph of $y = 8 - 2x - x^2$.

  d Comment on the shape of the graph.

13 a Copy and complete this table for $y = (x + 1)^2$.

x	−5	−4	−3	−2	−1	0	1	2	3
x + 1									
(x + 1)²									

  b Using the method in part **a**, construct a table for each equation. Plot all the graphs on the same pair of axes. Number the x-axis from −5 to 3 and the y-axis from 0 to 50.

    i $y = x^2$ **ii** $y = (x - 1)^2$ **iii** $y = (x - 2)^2$

  c Comment on your graphs. What do you notice?

  d Sketch onto your diagram the graph with the equation $y = (x - 4)^2$.

Solve problems

Bob sows grass seeds in the garden.
This chart shows the height of the grass over the next 30 days.

Days	4	5	10	15	20	25	30
Height (mm)	0	0.5	2.0	6.0	7.5	9.0	10.5

  a Use this information to draw a graph.

  b How many days is it before the grass shows above the ground?

  c When is the grass growing at its slowest pace?

  d What is the height of the grass after 12 days?

  e How many days is it before the grass is 10 mm long?

a

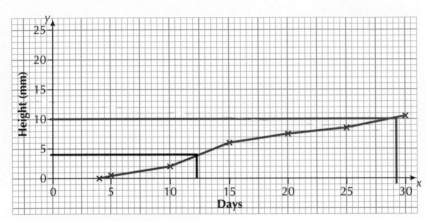

  b 5 days

  c From day 4 to day 5

  d 4 cm

  e 29 days

14 Andrew makes gold rings.
    This chart shows his charges for rings of different diameters.

Diameter (mm)	5	10	15	20	25
Cost (£)	45	61	87	124	170

  a Use this information to draw a graph.

  b Use your graph to estimate the cost of a ring with diameter:
    i 12 mm   **ii** 24 mm.

  c Brian bought a gold ring from Andrew for £100.
    What was the diameter of the ring?

15 The length (L) of a pendulum is related to the period of its swing cycle (T).
The table shows the lengths of five pendulums and their periods.

T (seconds)	2	4	6	8	10
L (metres)	1.0	4.0	8.9	15.9	24.8

a Draw a graph with T on the x-axis and L on the y-axis.

b Use your graph to estimate the length of a pendulum with a period of 7.5 seconds.

c Use your graph to estimate the period of a pendulum with a length of 5 m.

16 The time (T) it takes to complete a 5 km run is related to the speed (S) that a person is running.
The table shows the time it took to complete a 5 km run.
The measurements were taken every kilometre.

Length, L (km)	1	2	3	4	5
Time, T (minutes)	5.8	11.4	18.1	23	29.2

a Draw a graph with L on the x-axis and T on the y-axis.

b Use your graph to estimate the time it took to run 2.5 km.

c Use your graph to estimate how far can a person run in 20 minutes.

d Ashley uses the graph to estimate the time she will take to run 10 km.
Explain why this may not be accurate.

17 A ball is kicked from the ground up in the air.
The table shows the height of the ball 10 seconds after it was kicked in the air.

Time, T (seconds)	0	1	2	3	4	5	6	7	8	9	10
Height, H (m)	0	9	16	21	24	25	24	21	16	9	0

a Draw a graph with time, T, on the x-axis and height, H, on the y-axis.

b Use your graph to estimate the length of the ball's journey.

c Use your graph to estimate the maximum height the ball reached.

d Use your graph estimate the time the ball was 20 m or higher above the ground.

38.3 Solving quadratic equations by drawing graphs

● I can solve a quadratic equation by drawing a graph

You can find out a lot of information from a quadratic graph, when you know how to do it.

Develop fluency

a Draw the graph of $y = x^2 + 4x - 5$.

b Use your graph to find the value of y when $x = -3.5$.

c Find the solution to the equation $x^2 + 4x - 5 = 3.2$.

d What are the solutions to the equation $x^2 + 4x - 5 = 0$?

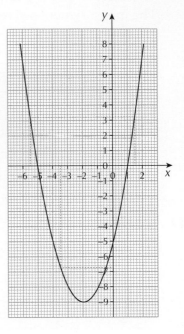

a *Draw the graph, as shown.*

b *Draw a dotted line from $x = -3.5$ to the graph.*

Following this across to the y-axis, you can see that when $x = -3.5$, $y = -6.75$.

c *To find the solution of the equation $x^2 + 4x - 5 = 3.2$, draw a dotted line across the graph of $y = x^2 + 4x - 5$ where $y = 3.2$.*

The dotted line for $y = 3.2$ cuts the graph in two places.

So there are two solutions to this equation.

Drawing dotted lines down from the graph to the x-axis, you can see that the solutions are $x = -5.5$ and $x = 1.5$.

d *You can find the solution of the equation $x^2 + 4x - 5 = 0$ on the graph where $y = 0$.*

This is where the graph cuts the x-axis.

You can see that this will be where $x = -5$ and $x = 1$.

Note that the solution of a quadratic equation will often give two answers, but not always!

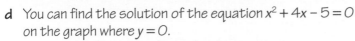

For all graphs in this exercise, use a scale of 2 cm to 1 unit on each axis.

1 a Draw the graph of $y = x^2$ from $x = -3$ to 3.

  b Write down the value of y when $x = 2.1$.

  c Use the graph to find the solutions to these equations.
   i $x^2 = 3$ **ii** $x^2 = 6$ **iii** $x^2 = 7.5$

2 a Draw the graph of $y = x^2 + 2x$ from $x = -3$ to 3.

  b Write down the value of y when $x = 0.7$.

  c Use the graph to find the solutions to these equations.
   i $x^2 + 2x = 2$ **ii** $x^2 + 2x = 1$ **iii** $x^2 + 2x = 0$

3 a Draw the graph of $y = x^2 - x$ from $x = -2$ to 3.
   b Write down the value of y when $x = -0.9$.
   c Use the graph to find the solutions to these equations.
      i $x^2 - x = 3$ ii $x^2 - x = 1.5$ iii $x^2 - x = 0.5$

4 a Draw the graph of $y = x^2 + 3x - 2$ from $x = -4$ to 2.
   b Write down the value of y when $x = 1.6$.
   c Use the graph to find the solutions to the following equations.
      i $x^2 + 3x - 2 = 0$ ii $x^2 + 3x - 2 = -1$ iii $x^2 + 3x = 3$

5 a Draw the graph of $y = x^2 + 2x - 3$ from $x = -4$ to 4.
   b Write down the value of y when $x = -1.7$.
   c Use the graph to find the solutions to the following equations.
      i $x^2 + 2x - 3 = 1$ ii $x^2 + 2x - 3 = 0$ iii $x^2 + 2x - 3 = -1$

6 Draw a graph to find the solutions of $x^2 + x - 7 = 0$.

7 Draw a graph to find the solutions of $x^2 - 4x + 1 = 0$.

8 a Draw the graph of $y = x^2 - 2x - 4$ from $x = -3$ to 5.
   b Use the graph to find solutions to these equations.
      i $x^2 - 2x - 4 = 0$ ii $x^2 - 2x = 1$ iii $x^2 - 2x = 10$

9 Draw graphs to find the solutions of:
   a $x^2 - 3x - 1 = 0$ b $x^2 + 4x - 6 = 2$ c $x^2 - x - 4 = 5$.

10 Draw a graph to find the solutions of $2x^2 - 3x - 1 = 0$.

Reason mathematically

a Draw the graph of $y = x^2 + 2x + 4$ from $x = -4$ to 2.
b Use the graph to explain why $x^2 + 2x + 4 = 0$ has no solution.

a
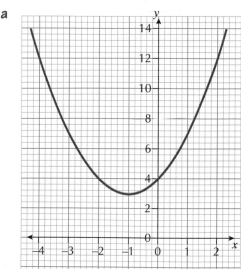

b The graph does not intersect the x-axis, so there are no solutions.

11 Look at the graph of $y = x^2 - 2x - 4$ from question 8. Explain why there is no solution for values of y that are less than -5.

12 Look at the graph of $y = x^2 - 3x - 1$ from question 9a.

    a Explain why there is only one solution for $y = -\frac{13}{4}$.

    b Explain why there are no solutions for $y < -\frac{13}{4}$.

13 Look at the graph of $y = 2x^2 - 3x - 1$ from question 10. Explain whether each statement is true or false.

    a The minimum point of $y = 2x^2 - 3x - 1$ is $(\frac{3}{4}, -3)$.

    b When $x < \frac{3}{4}$, $2x^2 - 3x - 1 = 0$ has no solutions.

    c $2x^2 - 3x - 1 = -3$ has no solutions.

Solve problems

A rectangle is $(x + 3.5)$ cm long and x cm high.

    a Show that the area of the rectangle is $x^2 + 3.5x$.

    b Sketch a graph for the area for $0 < x < 5$.

    c The area of the rectangle is 30 cm^2. Use your graph to find the length and height of the rectangle.

a Area of rectangle $= x \times (x + 3.5)$

$= x^2 + 3.5x$

b
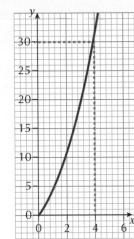

c $x = 4$ cm, so height is 4 cm and length is 7.5 cm.

14 **a** Draw the graph of $y = 2x^2 + x - 3$ from $x = -3$ to 3.

    b Write down the value of y when $x = -1.7$

    c Use the graph to find the solutions to the equations:

      i $2x^2 + x - 3 = 0$ **ii** $2x^2 + x = 7$ **iii** $2x^2 + x - 2 = 0$

    d Draw a straight line on your graph to solve the equations:

      i $2x^2 + x - 3 = x + 5$ **ii** $2x^2 + 2x - 7 = 0$

15 The sides of a rectangle are x cm and $(2x - 2)$ cm.

    a Show that the area of the rectangle is $A = 2x^2 - 2x$.

    b Sketch a graph for the area, A, for $0 < x < 6$.

    c The area of the rectangle is 31.5 cm². Use the graph to find a value for the width and the length of the rectangle.

16 The base of a triangle is $(3x + 5.5)$ cm long. The height of the triangle is $2x$ cm.

    a Show that the area of the triangle is $A = 3x^2 + 5.5x$.

    b Sketch a graph for the area.

    c The area of the triangle is 15 cm². Use the graph to find the value of x when $A = 15$.

    d Find the length of the base and the height of the triangle for this value of x.

17 A square has side length of $(x + 1)$ cm.

    The area of the square is 30 cm².

    a Show that $x^2 + 2x - 29 = 0$.

    b By drawing a suitable graph, find the solution of $x^2 + 2x - 29 = 0$.

    c What is the side length of the square?

Now I can...

draw any linear graph from any linear equation	solve a linear equation from a graph	solve a quadratic equation by drawing a graph
draw graphs from quadratic equations		

CHAPTER 39 Pythagoras' theorem

39.1 Calculating the length of the hypotenuse

● I can calculate the length of the hypotenuse in a right-angled triangle

Pythagoras was a Greek philosopher and mathematician, who was born in about 581 BC, on the island of Samos, just off the coast of Turkey. A very famous theorem about right-angled triangles is attributed to him.

In a right-angled triangle, the longest side, opposite the right angle, is called the **hypotenuse**.

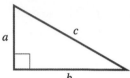

In any right-angled triangle, the square of the hypotenuse is equal to the sum of the squares of the other two sides.

Pythagoras' theorem is usually written as:

$c^2 = a^2 + b^2$

Develop fluency

Calculate the value of x in this triangle.

Using Pythagoras' theorem:

$x^2 = 6^2 + 5^2 = 36 + 25$

$= 61$

So $x = \sqrt{61} = 7.8\,cm$ (1 dp).

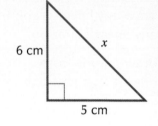

You should be able to work this out on a scientific calculator.

Try this sequence of keystrokes.

$$\boxed{6} \; \boxed{x^2} \; \boxed{+} \; \boxed{5} \; \boxed{x^2} \; \boxed{=} \; \boxed{\sqrt{}} \; \boxed{=}$$

This may not work on every calculator. You may need to ask your teacher to help you.

Calculate the length of the side labelled x in this triangle.

Using Pythagoras' theorem:

$x^2 = 2^2 + 4^2$

$= 4 + 16 = 20$

So $x = \sqrt{20} = 4.5\,cm$ (1 dp).

1 Calculate the length of the hypotenuse correct to one decimal place.

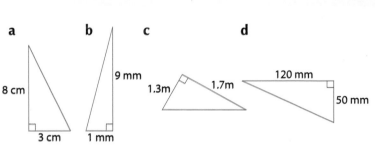

2 Calculate the length of the hypotenuse in each right-angled triangle.
Give your answers correct to one decimal place.

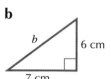 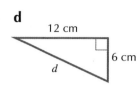 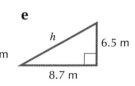

a 3 cm, a, 5 cm   **b** b, 6 cm, 7 cm   **c** 9 cm, 8 cm, c **d** 12 cm, 6 cm, d **e** h, 6.5 m, 8.7 m

3 Copy and complete, filling in the missing numbers.

a $2^2 + 3^2 = c^2$
$4 + 9 = c^2$
$c^2 = ...$
$c = \sqrt{...}$

b $10^2 + 9^2 = c^2$
$... + 81 = c^2$
$... = c^2$
$c = \sqrt{...}$

c $8^2 + b^2 = 17^2$
$64 + b^2 = 289^2$
$b^2 = ...$
$b = ...$

d $...^2 + ...^2 = c^2$
$144 + 324 = c^2$
$... = c^2$
$c = ...$

Now copy and complete the table for $a^2 + b^2 = c^2$.

a	b	c	a^2	b^2	$a^2 + b^2$	c^2
2	3					
10	9					
8		17				
			144	324		

4 Which of these sets of numbers are Pythagorean triples?

a 1, 2, 3   **b** 15, 20, 25   **c** 6, 9, 12   **d** 24, 32, 40   **e** 90, 120, 150

5 Calculate the length of the diagonal AC in the rectangle ABCD.
Give your answer correct to one decimal place.

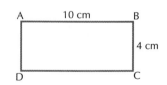

6 Calculate the length of the diagonal of a square with side length
5 cm. Give your answer correct to one decimal place.

7 The triangle PQR is drawn on a coordinate grid, as
shown in the diagram.

a The length of PQ is 3 units. Write down the length
of QR.

b Calculate the length of PR.

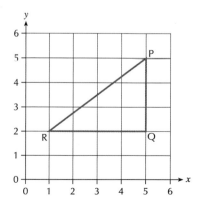

8 Calculate the sides of each of the triangles to the nearest 10 m.
  Each square represents 100 m.

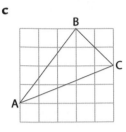

a

b

c

9 In right-angled triangle XYZ, the lengths of the perpendicular sides are 3 cm and 4 cm.
  What is the length of the hypotenuse?

10 In right-angled triangle ABC, the lengths of the perpendicular sides are 10 cm and 12 cm.
   What is the length of the hypotenuse?

Reason mathematically

Is triangle ABC with sides of lengths 12 cm, 9 cm and 15 cm, right-angled?
Explain how you know.

If the triangle is right-angled then $15^2 = 12^2 + 9^2$

$225 = 225$ so yes, it is right-angled.

11 Decide whether these triangles could exist with the given dimensions.

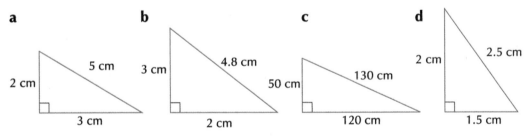

a

b

c

d

12 Here are some more Pythagorean triples that follow different patterns.

  5, 12, 13   7, 24, __   9, __, 41   __, 60, 61

  Find the missing numbers.

13 In triangle XYZ, XY = 7 cm and YZ = 10 cm. There is a right angle at Y.
   Calculate the length of XZ. Give your answer correct to one decimal place.

14 Calculate the length between the coordinate points in each pair.
   a A(2, 3) and B(4, 7)   b C(1, 5) and D(4, 3)
   c E(–1, 0) and F(2, –3)   d G(–5, 1) and H(4, –3)

Solve problems

How long is the diagonal of a square with sides 10 cm?

diagonal^2 = 10^2 + 10^2 = 200

$\sqrt{200}$ = 14.1

The diagonal is 14.1 cm long.

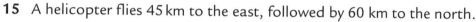

15 A helicopter flies 45 km to the east, followed by 60 km to the north.

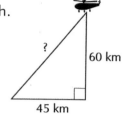

a How far is it back to its base in a straight line?

b The aircraft only has enough fuel to cover 200 km.
   What fraction of fuel is left after this journey?

16 Television screens are sized according to the length of their diagonals, to the nearest whole number.

a A television measures 32 inches across and is 18 inches high.
   What size would it be classed as?
   Give your answer to the nearest whole number.

b A television is advertised as 42 inches, and measures 37 inches across.
   How high is the screen?

c The smaller of these two costs £555.
   How much should the larger one cost, to give the same value for money?

17 Kath and Mark are on a hike. The footpath goes around the edge of a farmer's field, but they decide to cut across the diagonal to save time.

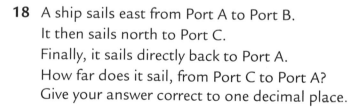

a How much shorter is it to cut directly across the diagonal?

b How much time do they save by taking the shortcut, if they are walking at 5 km/h?

18 A ship sails east from Port A to Port B.
It then sails north to Port C.
Finally, it sails directly back to Port A.
How far does it sail, from Port C to Port A?
Give your answer correct to one decimal place.

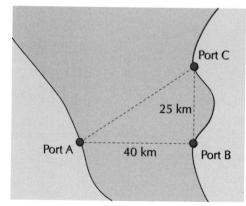

39.2 Calculating the length of a shorter side

- I can calculate the length of a shorter side in a right-angled triangle
- I can show that a triangle is right-angled

You can use Pythagoras' theorem to calculate the length of a shorter side in a right-angled triangle, but you will need to use subtraction. Pythagoras' theorem states that:

$c^2 = a^2 + b^2$

where c is the hypotenuse and a and b are the shorter sides.

You can rearrange the formula to obtain the form you need.

$a^2 = c^2 - b^2$ or $b^2 = c^2 - a^2$

Then you can use this version of Pythagoras' theorem to calculate the length of a shorter side, when you know the hypotenuse and another side.

Develop fluency

Calculate the length labelled x in this triangle.

The side labelled x is a shorter side.

Using Pythagoras' theorem:

$x^2 = 9^2 - 7^2$

$\quad = 81 - 49$

$\quad = 32$

So, $x = \sqrt{32} = 5.7\,cm$ (1 dp).

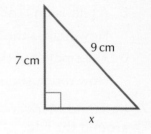

Try this sequence of keystrokes on your calculator.

Show that triangle ABC is a right-angled triangle.

If the triangle is right-angled the square of the hypotenuse AC should equal the sum of the squares of the other two sides AB and BC or
$AC^2 = AB^2 + BC^2$.

$AC^2 = 4^2 = 16\,cm$

$AB^2 + BC^2 = ?$

$2.4^2 + 3.2^2 = 5.76 + 10.24 = 16\,cm$

So, the triangle has a right angle at B.

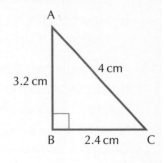

1   Calculate the length of the unknown shorter side in each right-angled triangle.
    Give your answers correct to one decimal place.

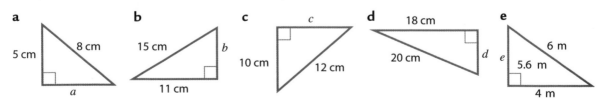

2 Find the lengths of the unknown sides.

a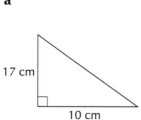

17 cm

10 cm

b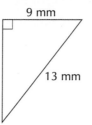

9 mm

13 mm

c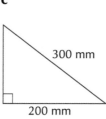

300 mm

200 mm

d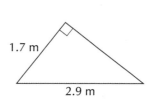

1.7 m

2.9 m

3 In triangle ABC, AC = 15 cm and AB = 10 cm. The angle at B is a right angle. Calculate the length of BC.
Give your answer correct to one decimal place.

4 Show that these are all right-angled triangles.

a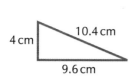

4 cm

10.4 cm

9.6 cm

b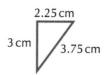

2.25 cm

3 cm

3.75 cm

c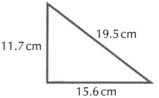

19.5 cm

11.7 cm

15.6 cm

5 To rescue a cat, its owner places a ladder of length 12 m against a tree, with the foot of the ladder 3 m away from the tree. How high up the tree is the cat?

6 A triangle has sides of lengths of 10 metres, 16 metres and 20 metres.
Is it a right-angled triangle? Explain your reasoning.

7 To wash a window that is 10 metres off the ground, Lucas leans a 11-metre ladder against the side of the building. How far from the building should Lucas place the base of the ladder so that he can reach the window?

8 One side of a right-angled triangle is 8 cm long. The other two are both of length x.
Calculate x to two decimal places.

9 One side of a right-angled triangle is 14 cm long. The other two are both of length x.
Calculate x to two decimal places.

10 a One side of a right-angled triangle is 10 cm long. The other two are both of length x.
Calculate x to two decimal places.

b Find the perimeter of the triangle in part **a**.

Reason mathematically

My favourite pen is 15 cm long.
How wide does my tray need to be for it to fit in, if its length is 14 cm?

$15^2 - 14^2 = \text{width}^2$

$\text{Width} = 5.4 \text{ cm}$

11 This shed is symmetrical. It is 4 m wide, and the sloping side of the roof is 2.8 m long. How high is it in total?

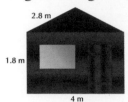

2.8 m

1.8 m

4 m

12 A radio mast 5.25 m tall is anchored to the ground by a cable that is 8.75 m long. The cable is anchored to a point 7 m from the base of the mast.

Is the mast vertical? Explain your answer.

13 What is the perimeter of this isosceles triangle?

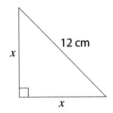

12 cm

x

x

14 This rectangle is twice as long as it is wide and has the same perimeter as the triangle. What is the area of the rectangle?

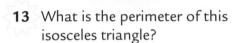

a

$2a$

10 cm

8 cm

Solve problems

A picture in a frame measuring 90 mm by 305 mm is placed diagonally in a box.

The box measures 90 mm by 300 mm. How deep does it need to be for the picture to fit?

$depth^2 = 305^2 - 300^2$

$\quad\quad\quad = 93\,025 - 90\,000$

$\quad\quad\quad = 3025$

$depth = \sqrt{3025} = 55\,mm$

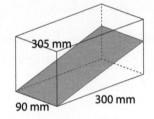

305 mm

90 mm

300 mm

15 Calculate the length of CD on this diagram.

Give your answer correct to one decimal place.

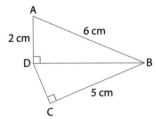

A

2 cm

6 cm

D

B

5 cm

C

16 An equilateral triangle has sides of length 10 cm.

a Calculate the perpendicular height, h, of the triangle.

b Calculate the area of the triangle.

Give your answers correct to one decimal place.

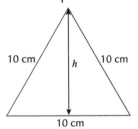

10 cm

h

10 cm

10 cm

17 Calculate the area of triangle ABD.
    Give your answer correct to one decimal place.

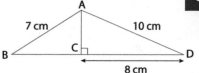

18 The maximum allowable gradient for wheelchair ramps is $1 : 12$, as shown.

1 m

12 m

   a A community centre's main entrance is 4.5 m above the road level.
          How long would the ramp have to be, to conform to the legislation?
          Give your answer to one decimal place.

   b The centre is only 15 m wide and the ramp cannot be wider than this, so the ramp will
          have to be in sections. How many sections will the ramp need to be split into?

39.3 Using Pythagoras' theorem to solve problems

● I can use Pythagoras' theorem to solve problems

You can use Pythagoras' theorem to solve various practical problems in 2D.

● Draw a diagram for the problem, clearly showing the right angle.

● Decide whether you need to find the hypotenuse or one of the shorter sides.

● Label the unknown side x.

● Use Pythagoras' theorem to calculate the value of x.

● Round your answer to a suitable degree of accuracy.

Develop fluency

A ship sails 4 km to the east, then a further 5 km to the south. Calculate the distance the
ship would have travelled, if it sailed directly from its starting point to its finishing point.

First, draw a diagram to show the distances sailed by the ship.

Then label the direct distance, x.

Now use Pythagoras' theorem to calculate the length of the hypotenuse.

$x^2 = 4^2 + 5^2$

$\quad = 16 + 25$

$\quad = 41$

So, $x = \sqrt{41} = 6.4$ km (1 dp).

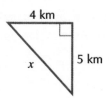

4 km

5 km

x

1 Calculate the length, d, of rope
   connecting the sailboarder to the kite.

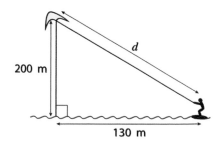

d

200 m

130 m

2 Calculate the marked lengths.

a

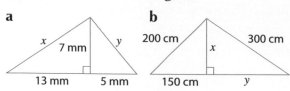

b

3 Calculate the width, x, of the cellar opening.

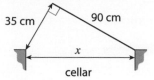

4 Calculate the marked length in each picture.

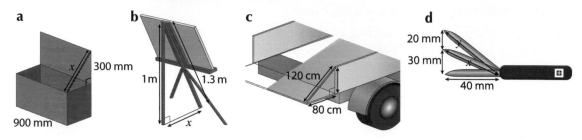

a **b** **c** **d**

5 An aircraft flies 80 km to the north. Then it flies 72 km to the west.
Calculate how far the aircraft would have travelled if it had travelled directly from its starting point to its destination.

6 A flagpole is 10 m high. It is held in position by four ropes that are each fixed to the ground, 4 m away from the foot of the flagpole, to make a square. The diagram shows the flagpole and two ropes that are opposite each other.
Calculate the length of each rope.

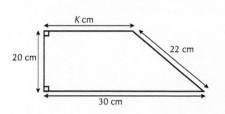

7 A hockey pitch measures 90 m by 55 m.
Calculate the length of a diagonal of the pitch.

Reason mathematically

Use Pythagoras' theorem to find the value of k in this trapezium. Give your answer to one decimal place.

$(30 - K)^2 + 20^2 = 22^2$

$22^2 - 20^2 = 84$

$\sqrt{84} = 9.17$

$K = 30 - 9.17 = 20.83 \, cm$

8 The lengths of the sides of an equilateral triangle are 10 cm.
Calculate the perpendicular height of the triangle.

9 a Calculate the unknown lengths to one decimal place.

b Explain why triangle ABC is similar to triangle PQR.

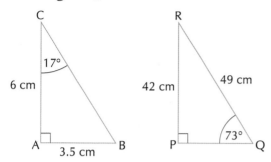

10 The diagram shows the sail of a yacht.

a Explain why triangles PST and PQR are similar.

b Calculate the length of the spar ST.

c How high up the sail is the spar positioned?

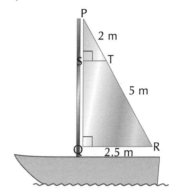

11 The lengths of two sides of a right-angled triangle are 20 cm and 30 cm. Calculate the length of the third side, if it is:

a the hypotenuse

b not the hypotenuse.

Solve problems

An 8 m ladder is placed against a wall so that the foot of the ladder is 2 m away from the bottom of the wall.

Calculate how far the ladder reaches up the wall.

$8^2 - 2^2 = 60$

$\sqrt{60} = 7.7$

The ladder reaches 7.7 m up the wall.

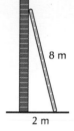

12 The diagram shows the side wall of a shed.

Calculate the length of the sloping roof.

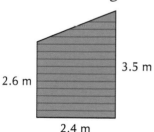

13 ABC is an isosceles triangle.

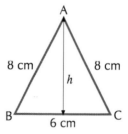

a Calculate the perpendicular height, h.

b Hence calculate the area of the triangle.

14 The diagram shows the positions of three towns X, Y and Z connected by straight roads.
Calculate the shortest distance from X to the road connecting Y and Z.

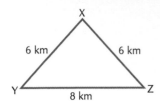

15 In rectangle ABCD, AB = 20 cm and CD = 16 cm.

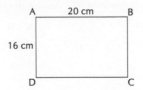

E is a point on AB and AE : EB = 2 : 3.
F is a point on BC and BF : FC = 3 : 1.
Calculate the length of EF.

Now I can...

calculate the length of the hypotenuse in a right-angled triangle	show that a triangle is right-angled	use Pythagoras' theorem to solve problems
calculate the length of a shorter side in a right-angled triangle		

CHAPTER 40 Working with decimals

40.1 Negative powers of 10

● I can understand and work with positive and negative powers of ten

This table shows you some powers of 10, including **negative powers** of ten.

Power	10^4	10^3	10^2	10^1	10^0	10^{-1}	10^{-2}	10^{-3}	10^{-4}
Value	10 000	1000	100	10	1	0.1	0.01	0.001	0.0001

Remember:

When you are multiplying by 10^n (positive n) you move all the digits n places to the left.

When you are dividing by 10^n (positive n) you move all the digits n places to the right.

When you are multiplying by 10^{-n} (negative n) you move all the digits n places to the right.

When you are dividing by 10^{-n} (negative n) you move all the digits n places to the left.

Develop fluency

Calculate the value of each expression.

a 0.752×10^2

b 15.08×10^{-3}

a $0.752 \times 10^2 = 75.2$

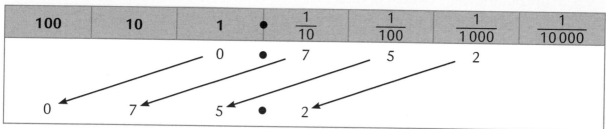

Move the digits 2 places to the *left*.

b $15.08 \times 10^{-3} = 0.01508$

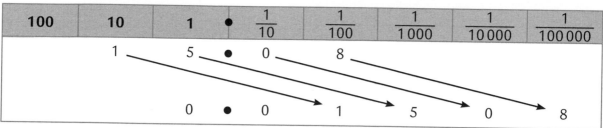

Move the digits 3 places to the *right*.

1 Calculate the value of each expression.

   a 6.34×100 **b** $6.34 \div 100$ **c** 6.34×1000 **d** $6.34 \div 1000$

2 Multiply each number by 10^3.

   a 8.7   **b** 87   **c** 873.1   **d** 0.0871   **e** 87.31

3 Multiply each number by 10^5.

   a 6.37   **b** 0.637   **c** 637   **d** 63.07   **e** 0.06307

4 Multiply each number by 10^{-1}.

   a 12.4   **b** 0.124   **c** 124.5   **d** 0.00124   **e** 12450

5 Multiply each number by 10^{-3}.

   a 7.85   **b** 0.785   **c** 78.5   **d** 7850   **e** 785.01

6 Multiply each number by 0.01.

   a 8.3   **b** 0.83   **c** 83   **d** 830   **e** 0.0083

7 Multiply each number by 0.001.

   a 12.7   **b** 0.127   **c** 127.01   **d** 1.27   **e** 1207.1

8 Calculate the value of each expression.

   a 3.76×10^2 **b** $2.3 \div 10^3$ **c** $3.09 \div 10^3$ **d** 2.35×10^2 **e** $0.01 \div 10^4$

9 **a** Write each number as a decimal.

     i 10^{-1} **ii** 10^{-2} **iii** 10^{-3} **iv** 10^{-4}

   b Work out the value of each of these.

     i 9.2×10^{-1} **ii** 0.71×10^{-3} **iii** 4.2×10^{-1} **iv** 0.98×10^{-2} **v** 2.14×10^{-3}

10 Multiply each of the numbers in **a–e** by: **i** 10^3 **ii** 10^7 **iii** 10^{-1} **iv** 10^{-3}

   a 600   **b** 5   **c** 23 000   **d** 0.6   **e** 25.2

Reason mathematically

Fill in the boxes to make each equation true.

   a $1234 \times 10^{\square} = 12.34$ **b** $1234 \div 10^{\square} = 12.34$

   a multiplying by 10^{-2} moves the digits two places to the right.

   $1234 \times 10^{-2} = 12.34$

                                      b dividing by 10^2 moves the digits two places to the left.

                                        $1234 \div 10^2 = 12.34$

11 Fill in the boxes to make each equation true.

   a $4798 \times 10^{\square} = 479\,800$ **b** $4798 \div 10^{\square} = 479\,800$

   c $\square \times 10^{-3} = 0.4798$ **d** $\square \div 10^{-5} = 0.004798$

12 Explain why each of these calculations gives the same answer.

   35×0.01 35×10^{-2} $35 \div 100$ $35 \div 10^2$

13 Write down four calculations using different positive and negative powers of ten and the number cards below that all give the same answer.

14 Metric units use prefixes to show how much of the unit they are.

For example, the prefix kilo means 1000, as in 1 kilogram is 1000 grams.

The prefix centi- means one hundredth.

The prefix milli- means one thousandth.

This table gives the main prefixes and their equivalent multiples, written as powers of 10.

giga	mega	kilo	centi	milli	micro	nano	pico
10^9	10^6	10^3	10^{-2}	10^{-3}	10^{-6}	10^{-9}	10^{-12}

For example, 0.000 007 grams could be written as 7 micrograms.

Use suitable prefixes to write each quantity in a simpler form.

a 0.004 grams **b** 8 000 000 watts **c** 0.000 000 007 5 metres

Solve problems

A stack of 1000 pieces of A4 paper is 2.5 cm high. How many mm thick is a single piece of A4 paper?

Show your working, using powers of 10, and give your answer in cm.

Divide the height of the stack by the number of pieces of paper.

Written as a power of 10, $1000 = 10^3$

$2.5 \div 10^3 = 0.0025$ cm   Move the digits 3 places to the right.

Convert cm to mm. There are 10 mm in 1 cm.

$0.0025 \times 10 = 0.025$ mm   Move the digits 1 place to the left.

15 A stack of 1 000 000 £5 notes weighs 70 kg. How much does one £5 note weigh? Show your working, using powers of 10, and give your answer in g.

16 A single plastic brick is 9.6 mm high. How high would a stack of 100 000 plastic bricks be? Give your answer in metres.

17 The population of Canada is approximately 37 060 000 people. Canada covers an area of approximately ten million square kilometres. What is the average number of people per square kilometre in Canada? Give your answer correct to two significant figures.

18 China covers an area of approximately ten million square kilometres and has a population of 1.39 billion. What is the average number of people per square kilometre in China? Give your answer correct to two significant figures.

40.2 Standard form

● I can understand and work with standard form, using positive and negative powers of ten

You have already used **standard form** to express large numbers.

For example, you can write 73 000 000 000 000 as 7.3×10^{13}.

Standard form can also be used with negative powers of ten. These are used for very small numbers, such as 0.000 000 067 which can be expressed as 6.7×10^{-8} in standard form.

There are two things to remember about numbers expressed in standard form.

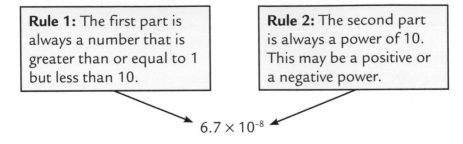

Rule 1: The first part is always a number that is greater than or equal to 1 but less than 10.

Rule 2: The second part is always a power of 10. This may be a positive or a negative power.

6.7×10^{-8}

The power tells you two things:

- The number tells you how many places the digits have moved.
- The sign of the power tells you how the digits have moved.
- A positive power tells you that the digits have moved to the *right*.
- A negative power tells you that the digits have moved to the *left*.

For example, 6.7×10^{-8} tells you that the digits 6 and 7 in the ordinary number have moved eight places to the left to be written in standard form.

Develop fluency

a Write 0.005 42 in standard form.

b Write 5.43×10^{-4} as an ordinary number.

a *To convert an ordinary number to standard form, decide which direction the digits will need to move and count how many places they will need to move.*

In 0.005 42 that will be three places to the left.

$0.005\,42 = 5.42 \times 10^{-3}$

The sign of the power is negative because the digits move left.

b *To convert a number in standard form to an ordinary number, use the sign of the power to decide which direction the digits have moved and use the number in the power to move the digits that many places in the opposite direction.*

In 5.43×10^{-4} that will be 4 places to the right.

$5.43 \times 10^{-4} = 0.000\,543$

Note that the power of 10 in the number written in standard form is the same as the number of zeros before the number in ordinary form *including* the zero before the decimal point.

a Add $1.32 \times 10^{-8} + 3.47 \times 10^{-8}$ and write your answer in standard form correct to three significant figures.

b Add $2.86 \times 10^{-2} + 8.91 \times 10^{-3}$ and write your answer in standard form correct to three significant figures.

a $1.32 \times 10^{-8} + 3.47 \times 10^{-8} = 4.79 \times 10^{-8}$

b Note that the powers of 10 are not the same.
Convert both numbers to ordinary numbers.

$2.86 \times 10^{-2} = 0.0286$ and $8.91 \times 10^{-3} = 0.00891$

Then $0.0286 + 0.00891 = 0.03751$

$0.03751 = 3.75 \times 10^{-2}$ in standard form to three significant figures.

1 Write each number in standard form.

   a 0.85   **b** 0.0127   **c** 0.432   **d** 0.005 12

2 Write each standard-form number as an ordinary number.

   a 6.41×10^{-3} **b** 9.03×10^{-4} **c** 8.0×10^{-2} **d** 7.1×10^{-4}

3 Write each standard-form number as an ordinary number and each ordinary number in standard form.

   a 3.142×10^{-3} **b** 5.01×10^{-4} **c** 9.852×10^{-7} **d** 0.0038   **e** 0.000709

4 Find the square of each number, giving your answer in standard form.

   a 0.08   **b** 0.0015   **c** 0.0012   **d** 0.000 04

5 Find the square of each number, giving your answer in standard form.

   a 30   **b** 400   **c** 3000   **d** 40

6 Write each answer in standard form, correct to two significant figures.

   a $0.001358 + 0.012$ **b** $0.038 + 0.67$ **c** $0.00238 + 0.0863$

7 Write each answer in standard form, correct to two significant figures.

   a $3.45 \times 10^{7} + 2.7 \times 10^{7}$ **b** $9.25 \times 10^{-1} + 1.7 \times 10^{-1}$

   c $2.26 \times 10^{4} + 4.6 \times 10^{5}$

8 Write each answer in standard form, correct to two significant figures.

   a $1.88 \times 10^{-2} + 5.4 \times 10^{-3}$ **b** $6.49 \times 10^{4} - 2.7 \times 10^{3}$ **c** $3.45 \times 10^{-2} - 5.7 \times 10^{-3}$

9 Write each answer in standard form, correct to two significant figures.

   a $6.5 \times 10^{10} + 3.4 \times 10^{10}$ **b** $6.4 \times 10^{4} + 2.8 \times 10^{3}$ **c** $2.525 \times 10^{9} + 3.131 \times 10^{8}$

10 Write each answer in standard form, correct to two significant figures.

   a $8.3 \times 10^{7} - 4.6 \times 10^{6}$ **b** $4.3 \times 10^{-1} + 2.8 \times 10^{-1}$ **c** $3.7 \times 10^{-2} - 7.3 \times 10^{-3}$

Reason mathematically

Explain why 13.8×10^{-3} is not in standard form and write it in standard form.

It is not standard form because the first part, 13.8, is not a number between 1 and 10.

To write it in standard form, it must be 1.38 multiplied by some power of 10.

13.8×10^{-3} means that the digits in 13.8 have moved three places to the left.

$13.8 \times 10^{-3} = 0.0138$ in ordinary form, which is 1.38×10^{-2} in standard form.

$13.8 \times 10^{-3} = 1.38 \times 10^{-2}$ in standard form.

11 Which of these numbers are written in standard form?
    If the number is not in standard form, write it in standard form.
    a 40×10^3　　**b** 2.3×10^7　　**c** 4×10^{-3}　　**d** 10^9　　**e** 0.8×10^5

12 Order these numbers by size, smallest to largest.
    0.0387×10^6　　　0.00387　　　387×10^{-6}　　　38.7×10^{-3}

13 Explain why $3 \times 10^8 + 2 \times 10^7$ does not equal 5×10^{15}.

14 Explain why, when you multiply a number by 10^{10} you move the digits ten places to the left, but when you multiply by 10^{-10} you move the digits ten places to the right.

Solve problems

A mole is a scientific unit, $602\,214\,076\,000\,000\,000\,000\,000$

The mass of one mole of atoms of argon is 39.948 g.

  a Write one mole in standard form correct to four significant figures.

  b What is the mass of one atom of argon?
     Write your answer in standard form correct to four significant figures.

  a *The digits in 602 214 076 000 000 000 000 000 must move 23 places to the right to become 6.022 multiplied by a power of 10, so one mole is 6.022×10^{23} in standard form.*

  b *Divide the mass of one mole of argon by the number of atoms in one mole.*
     $39.948 \div (6.022 \times 10^{23}) = 6.634 \times 10^{-23}$ g

15 The mass of one electron is $0.000\,000\,000\,000\,000\,000\,000\,000\,000\,92$ grams.
    a Write this number in standard form.
    b What is the mass of 8 electrons?
    c What is the mass of three million electrons?

16 In biology, the diameters of cells are usually measured in microns. (1 micron = 10^{-3} mm)
    Yeast cells are little balls about 2 microns in diameter.
    Most human cells are about 20 microns in diameter.
    a How long would a string of one million yeast cells be?
    b How long would one thousand human cells be, if they were all in one long string?

17 A unit of length used by scientists is the angstrom, Å. (1 angstrom = 10^{-10} m)
   A hydrogen atom is about 1 Å in diameter, a carbon atom is about 2 Å in diameter.
   a What would be the length of a million hydrogen atoms, laid in a row?
   b What would be the length of a string of a billion carbon atoms?

18 A dust mite is 1.25×10^{-2} cm wide and long.
   a What is the width of four dust mites, side by side?
   b What is the width of 400 dust mites, side by side?
   c A mat measuring 50 cm by 80 cm is full of dust mites.
   How many dust mites will there be on the mat?

40.3 Rounding appropriately

● I can round numbers, where necessary, to an appropriate or suitable degree of accuracy

There are two main reasons for rounding, both of which you have met before.

·   Rounding enables you to make an **estimate** of the answer to a problem.

·   Rounding gives an answer to an **appropriate degree of accuracy**.

Develop fluency

Estimate the answer to each calculation.
   a 21% of £598
   b $3.9^2 \div 0.0378$

Round the numbers to one significant figure each time.
   a 21% of £598 ≈ 20% of £600 = £120
   b $3.9^2 \div 0.0378 \approx 4^2 \div 0.04 = 16 \div 0.04 = 160 \div 0.4 = 1600 \div 4 = 400$

The distance from my house to the local post office has been rounded to:

   i 721.4 m   **ii** 721 m   **iii** 700 m

Which measurement has been rounded to the most appropriate degree of accuracy?

Distances are usually rounded to one or two significant figures.

Hence, 700 m is the most appropriate answer.

1 Round each number to the nearest whole number and use these answers to estimate the
   value of each calculation.
   a 6.2 + 8.7   **b** 9.4 − 2.9   **c** 4.2 × 8.9   **d** 9.6 ÷ 2.1

2 Round each number to the nearest 10 and use these answers to estimate the value of
   each calculation.
   a 43 + 91   **b** 62 − 28   **c** 51 × 78   **d** 76 ÷ 18

3 Estimate the answer to each calculation.
   a 19% of £278   **b** 23.2 ÷ 0.018   **c** 12.32 × 0.058   **d** $\dfrac{23.1 + 57.3}{16.5 + 7.3}$

4 Estimate the answer to each calculation.

a $\dfrac{0.245 \times 0.03}{1.89 \times 3.14}$ b $\dfrac{45.9 \times 83.2}{26.7 - 9.8}$ c 14% of 450 kg d 59.5 ÷ 0.132

5 Estimate the answer to each calculation.

a $(3.95 \times 0.68)^2$ b 28% of 621 km c 4% of £812 d 0.068×0.032

6 Round each quantity to a sensible degree of accuracy.

a Average speed of a journey: 63.7 mph

b Size of an angle in a right-angled triangle: 23.478°

c Mass of a sack of potatoes: 46.89 kg

7 Round each quantity to a sensible degree of accuracy.

a Time to run a marathon: 2 hours 32 minutes and 44 seconds

b World record for running 100 m: 9.58 seconds

c Kailash held his breath for 82.71 seconds.

8 Round these quantities to an appropriate degree of accuracy.

a Jeremy is 1.8256 m tall.

b A computer disk holds 688 332 800 bytes of information.

c An English dictionary contains 59 238 722 words.

9 Use a calculator to work out each answer, then round each result to an appropriate degree of accuracy.

a $\dfrac{56.2 + 48.9}{17.8 - 12.5}$ b $\dfrac{12.7 \times 13.9}{8.9 \times 4.3}$ c 1 ÷ 32

10 Use a calculator to work out each answer, then round each result to an appropriate degree of accuracy.

a 0.58^2 b 1 ÷ 45 c 23.478 ÷ 0.123

Reason mathematically

A table measures 2.38 m by 1.24 m. Find the area of the table and round the answer to an appropriate degree of accuracy.

2.38 × 1.24 = 2.9512

However, since the measurements were only given to the nearest cm, the area should be given to the nearest cm².

The area is 2.95 m².

11 A large rectangular park measures 1.02 km by 1.59 km. Find the area of the park and round the answer to an appropriate degree of accuracy.

12 Billy rounded a number to 8. His brother Isaac rounded the same number to 7.8 but they were both correct. Explain how this could be.

13 Here are four calculator displays, each rounded to one decimal place, four ordinary numbers and four numbers in standard form, rounded to three significant figures. Match them up.

      7.2³

37842   7234   0.0003784   0.007234

7.23 × 10⁻³   3.78 × 10⁴   3.78 10⁻⁴   7.23 × 10³

14 Grace, Victoria and Zeenat all measured the length of a boat, each giving their answer to a different degree of accuracy.
Grace measured the length as 20 m.
Victoria measured the length as 18 m.
Zeenat measured the length as 18.5 m.
   a If all three were correct, write down a possible length for the boat.
   b Write down what rounding each person did.

Solve problems

Zeke is hosting a large party with three friends. They are inviting 165 people at a cost of £7.95 each. How much must each host pay towards the cost of the party?

$165 \times 7.95 = 1311.75$

$1311.75 \div 4 = 327.9375$

Since the question is asking about money, it is appropriate to round the answer to the nearest penny.

Each host must pay £327.94.

15 Charlie and five friends go to dinner. Their bill comes to £173.87 including tip. They split the bill equally amongst them. How much must each person pay?

16 A shop sells wire on rolls that hold 46 metres of wire.
Helen needs to cut lengths of wire that are 9.1 cm long for some artwork with her class.
Estimate how many 9.1 cm lengths she could cut from the 46 m roll.

17 This table shows the prices of some metals.
Estimate the cost of buying:
   a 2474 kg of lead
   b 0.2856 g of silver
   c 0.036 kg of gold
   d 19.08 g of platinum.
   Show your working.

Metal	Price
Lead	£0.2412 per kg
Silver	£0.085 per g
Gold	£2575 per kg
Platinum	£11.63 per g

18 Use the table in question 17 to estimate your answers. Show your working.

   a How much gold could be bought for £64 250?

   b How much lead could be bought for £124.59?

   c How much platinum could be bought for £289 320?

   d How much silver could be bought for £0.94?

40.4 Mental calculations

● I can use some routines that can help in mental arithmetic

There are many methods of **mental calculation**. Here are a few examples.
The steps in each of these examples can be done in any order.

Doubling and halving is often very useful and can be used in combination.

Calculation	Method	Calculation	Method
× 4	Double and double	÷ 4	Halve and halve
× 5	Halve and × 10	÷ 5	Double and ÷10
× 20	Double and × 10	÷ 20	Halve and ÷10
× 25	÷ 4 and × 100	÷ 25	× 4 and ÷ 100
× 15	× 3, ÷ 2, × 10		

Note that these methods can often be extended to other calculations. For example, to divide by 8, halve, then halve, then halve again. To multiply by 30, multiply by 3 and multiply by 10. To multiply by 24, multiply by 20 and multiply by 4 then add the answers together.

Develop fluency

Work these out mentally.

a 88 × 25

  Divide by 4   88÷2÷2= 22

  Multiply by 100   22 × 100 = 2200

  Then, 88 × 25 = 2200

b 13.6 ÷ 5

  Double   13.6 × 2 = 27.2

  (13 × 2 = 26 and 0.6 × 2 =1.2)

  Divide by 10   27.2 ÷ 10 = 2.72

  13.6 ÷ 5 = 2.72

Do not use a calculator to answer any of these questions.

For questions 1–8, use mental methods to work out the answers.

1 a 250 × 4   **b** 250 × 8   **c** 540 × 4   **d** 540 × 8

   e 250 ÷ 4   **f** 250 ÷ 8   **g** 540 ÷ 4

2 a 120×5 **b** 36×5 **c** 28×5 **d** 184×5
   e $120 \div 5$ **f** $36 \div 5$ **g** $28 \div 5$

3 a 18×20 **b** 37×20 **c** 54×20 **d** 216×20
   e $18 \div 20$ **f** $37 \div 20$ **g** $54 \div 20$

4 a 400×25 **b** 72×25 **c** 108×25 **d** 8.4×25
   e $600 \div 25$ **f** $450 \div 25$ **g** $1200 \div 25$

5 a 28×15 **b** 82×15 **c** 104×15 **d** 36×15

6 a 230×4 **b** $230 \div 4$ **c** 230×5 **d** $230 \div 5$
   e 18×30 **f** 18×25 **g** $18 \div 25$

7 a $0.8 \div 0.16$ **b** 0.8×16 **c** 2.6×5 **d** $2.6 \div 5$
   e 3.2×2.5 **f** $3.2 \div 2.5$ **g** 1.6×15

8 a $12 \times £4.98$ **b** $£4.98 \div 4$ **c** $15 \times £4.98$ **d** $£4.98 \div 5$

9 Use mental methods and suitable rounding to estimate your answers to these.
   a $678 \div 19$ **b** 92×24 **c** $7.3 \div 0.24$ **d** $219 \div 16.3$

10 Use mental methods and suitable rounding to estimate your answers.
   a 1199×24 **b** $8641 \div 19.6$ **c** $6012 \div 16.03$ **d** $20\,202\,020 \times 14.7$

Reason mathematically

One brick weighs 1.48 kg. How much does a stack of 25 bricks weigh?

To multiply 1.48 × 25, divide by 4 and multiply by 100.

Divide by 4   1.48 ÷ 2 = 0.74, then 0.74 ÷ 2 = 0.37

Multiply by 100   0.37 × 100 = 37

A stack of 25 bricks weighs 37 kg.

11 Use mental methods to work out the answers.

   a A key cutter charges £2.40 to repair a pair of shoes. How much would she charge to repair 25 pairs of shoes?

   b The mass of a block of platinum is 8400 kg. What is the mass of $\frac{1}{16}$ of the block?

   c Valerie has 860 lottery tickets to sell for a charity event. If she only sells $\frac{1}{20}$ of them, how many has she sold?

12 Bethany bought eight DVDs. Each DVD cost £5.98.
    How much change should Bethany get from a £50 note?

13 Use mental methods to work these out and just write down the answers.

    a A bar of chocolate costs one pound and eighty pence.
       How much would twenty-five bars cost?

    b A school has twice as many girls as boys. There are 286 girls.
       How many pupils are there in the school?

14 A theatre makes £28 profit for each ticket sold for a musical.
    If 391 tickets are sold, find the approximate profit.

Solve problems

Henry is hosting a party. It costs £8.95 per guest. He will have 15 guests.

What is the approximate cost of the party?

Round 8.95 to 9

Multiply 9 × 15.

Multiply by 3	9 × 3 = 27
Divide by 2	27 ÷ 2 = 13.5
Multiply by 10	13.5 × 10 = 135

The party will cost approximately £135.

15 I buy 42.8 litres of petrol at £1.29 per litre.
    What is the approximate cost?

16 A rectangle measures 2.46 m by 0.27 m.
    What is the approximate area of the rectangle?

17 The *Daily Mail* reported that in 2013 it sold an average of 1 863 000 copies per working day.
    Approximately, how many copies did it sell in the year?

18 A rectangle has an area of 1203 cm².
    Its base is 24.7 cm.
    Find an estimate for the perimeter of the rectangle.

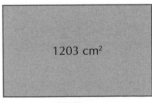

1203 cm²

24.7 cm

40.5 Solving problems

● I can solve real-life problems involving decimals

Being able to apply what you have learned to everyday life is an important skill. For example, decimal calculations can be an important tool in saving money because you can compare prices of different products and different quantities.

Reason mathematically

The total cost of five cups and three mugs is £23.50.

The total cost of five cups and four mugs is £26.50.

a Find the cost of a mug.

b Find the cost of a cup.

a Cost of 5 cups + 3 mugs = £23.50

Cost of 5 cups + 4 mugs = £26.50

So the cost of 1 mug must be £26.50 – £23.50 = £3.00

b Since one mug cost £3.00, then 3 mugs cost 3 × £3.00 = £9.00.

Subtract the cost of three mugs from the cost of five cups and three mugs to find the cost of five cups.

£23.50 – £9.00 = £14.50

So, one cup costs £14.50 ÷ 5 = £2.90.

1 Two families went to the cinema.

It cost the Ahmed family of one adult and two children £11.50.

It cost the Smith family of two adults and two children £16.

What is the cost of an adult's ticket and a child's ticket?

2 The value of Euler's number e is 2.718 282, correct to six decimal places.

a Write the value of e correct to four decimal places.

b Which value below is closest to the value of e?

$$\frac{100}{37} \qquad 2\frac{8}{11} \qquad \left(\frac{8}{6}\right)^2 \qquad \frac{590}{217}$$

3 In a right-angled triangle, the lengths of the two shorter sides are 3.15×10^5 mm and 1.87×10^5 mm.

What is the length of the hypotenuse? Give your answer in standard form and correct to two significant figures.

4 The TV programme Wacko Magic lasts 25 minutes and is shown every Tuesday and Thursday. Country Facts lasts 15 minutes and is shown every Wednesday. Karen records these programmes each week. She has recorded 7 hours and 35 minutes on a videotape.

a How many weeks did she record?

b How many of each programme did she record?

5 Rita and Norris work in a shop. Norris works three times as many hours as Rita. Altogether they work 42 hours. How many hours does Rita work?

6 Mr Marshall has two children, Daniel aged 8 and Kate aged 12. He divided £735 between them in the ratio of their ages. The children bought a computer between them for £417.96. Kate paid twice as much as Daniel for the computer.

How much money does Daniel have left after buying the computer?

7 Fibonacci sequences are formed by adding the previous two numbers to get the next number. For example: 1, 1, 2, 3, 5, 8, 13, ...

    a Write down the next three terms of the Fibonacci sequence that starts:
      2, 2, 4, 6, 10, 16, ...

    b Work out the missing values of this sequence: 0.2, 0.3, 0.5, ..., 1.3, ..., ..., 5.5, ...

    c Work out the first three terms of this sequence: ..., ..., ..., 1.25, 2, 3.25, 5.25 , ...

Solve problems

Jam is sold in jars holding 454 g for £0.89 and jars holding 2000 g for £4.00.

Which jar of jam offers the better value?

The smaller jar gives 454 ÷ 89 = 5.1 grams/pence

The larger jar gives 2000 ÷ 400 = 5 grams/pence

Do not forget to compare prices in either pounds or pence.

So, the smaller jar offers the better value.

8 Biscuits are sold in packs of 20 for 74p, packs of 35 for £1.19 and packs of 36 for £1.26. Which pack gives the best value? Explain your answer.

9 Tea bags are sold in boxes of 200, 300 and 500.
    A box of 200 costs £1.72.
    A box of 300 costs £2.61.
    A box of 500 costs £4.45.
    Which box is the best value for money? Explain your answer.

10 A school buys some books as prizes. The books cost £4.25 each.
    The school had £40 to spend on prizes. They buy as many books as possible.
    How much money is left?

11 a Tamsin converted £200 to euros and then bought some fruit juice.
      The exchange rate was £1 = €1.48. Fruit juice costs €3.50 per bottle.
      How many euros did she receive?

    b She bought as many bottles of fruit juice as possible.
      How much fruit juice did she buy?

    c i How much money did she have left over?
      ii Convert your answer to British pounds.

12 A cash-and-carry shop sells crisps in boxes.
    A 12-packet box costs £3.00.
    An 18-packet box costs £5.00.
    A 30-packet box costs £8.00.
    Which box gives the best value?

13 A supermarket sells crisps in different-sized packets.

An ordinary bag contains 30 g and costs 28p.

A large bag contains 100 g and costs 90p.

A jumbo bag contains 250 g and costs £2.30.

Which bag is the best value? You must show all your working.

14 Leisureways sell Kayenno trainers for £69.99 but give 10% discount for Jiggers Running Club members.

All Sports sell the same trainers for £75 but are having a sale in which everything is reduced by 15%.

In which shop are the trainers cheaper for Jiggers Running Club members?

15 A recipe for marmalade uses 65 g of oranges for every 100 g of marmalade.

Mary has 10 kg of oranges.

How many 454 g jars of marmalade can Mary make?

16 The charge for hiring a car is £25 plus £16.00 per day.

a How much will it cost to hire the car for five days?

b John pays £137.00 for car hire. How many days did he have the car?

Now I can...

understand and work with positive and negative powers of ten	understand and work with standard form, using positive and negative powers of ten	solve real-life problems involving decimals
round numbers, where necessary, to an appropriate or suitable degree of accuracy	use some routines that can help in mental arithmetic	

CHAPTER 41 Manipulating brackets

41.1 More about brackets

● I can expand a term with a variable or constant outside brackets

Look at this example.

$2x(3x - 5)$

x is the **variable**. Expanding gives two terms. The first is $2x \times 3x$ and the second is $2x \times (-5)$.

You write $2x \times 3x$ as $6x^2$ because $2 \times 3 = 6$ and $x \times x = x^2$.

You write $2x \times (-5)$ as $-10x$.

So $2x(3x - 5) = 6x^2 - 10x$.

It can be helpful to draw lines to remind you what to multiply by what.

$2x(3x - 5)$

Develop fluency

Expand the brackets.

  a $2x(x + 8)$ **b** $3a(a + 4b)$ **c** $3t(5 - 2t)$ **d** $2x(5x - 3)$

  a $2x(x + 8) = 2x^2 + 16x$ $2x \times x = 2x^2$ and $2x \times 8 = 16x$

  b $3a(a + 4b) = 3a^2 + 12ab$ $3a \times a = 3a^2$ and $3a \times 4b = 12ab$

  c $3t(5 - 2t) = 15t - 6t^2$ $3t \times 5 = 15t$ and $3t \times (-2t) = -6t^2$

  d $2x(5x - 3) = 10x^2 - 6x$ $2x \times 5x = 10x^2$ and $2x \times (-3) = -6x$

1 Expand the brackets.

  a $4(x + 1)$ **b** $3(y - 7)$ **c** $8(2 - x)$ **d** $4(5 - t)$

2 Expand the brackets.

  a $4(x + 5)$ **b** $x(x + 5)$ **c** $7(y + 10)$ **d** $y(10 - y)$

3 Expand the brackets.

  a $2(x + y)$ **b** $x(x + 2)$ **c** $x(x + y)$ **d** $m(m - 7)$

4 Expand the brackets.

  a $2(2x + 1)$ **b** $4(3a - b)$ **c** $3x(x - z)$ **d** $x(3x + 2y)$

5 Expand the brackets and simplify as much as possible.

  a $x(x + 4) + 3x$ **b** $x(x - 3) - x$ **c** $x(x + 7) - 5x$ **d** $x(4 + x) - 3x$

6 Simplify each expression.

   a $2x \times 3x$ **b** $2t \times 5t$ **c** $6x \times 6x$ **d** $(2x)^2$ **e** $(3x)^2$

7 Multiply out the brackets.

   a $2t(t + 5)$ **b** $2t(2t + 5)$ **c** $2t(3t + 5)$ **d** $2t(5x + 3)$

8 Expand the brackets.

   a $2x(4x - 1)$ **b** $3x(2x - 3)$ **c** $2x(3x + y)$ **d** $6t(6t - 1)$

9 Expand the brackets and simplify as much as possible.

   a $g(g - 3) + 7g$ **b** $g(4 + g) - 5g$ **c** $g(2g - 9) + g^2$ **d** $g(12 - g) - 8g$

10 Expand the brackets.

   a $3x(3x + 5)$ **b** $4x(2x^2 + 9)$ **c** $5x^2(11 - x)$ **d** $6x^2(10x - 7)$

Reason mathematically

A coffee costs c pence. A hot chocolate costs 20 pence more than a coffee.

Write an expression for 3 hot chocolates.

Write an expression for the hot chocolate first: $c + 20$

Next write an expression for three of them: $3(c + 20)$

Finally expand your brackets: $3c + 60$

11 An envelope costs 5 pence and a stamp costs s pence.

   a Write down an expression for the cost of:

      i a stamped envelope   **ii** 6 stamped envelopes   **iii** n stamped envelopes.

   b Expand the brackets for your answers to parts **ii** and **iii**.

12 Check these answers and correct any mistakes.

   a $7x(x + 3) = 7x^2 + 10x$ **b** $x(3x + 5y) = 3x^2 + 5y$ **c** $2t(t - 9) = 2t - 18t$

13 A box contains b grams of salt. The mass of the box is $10\,g$.

   a Write down an expression for the total mass of:

      i a box of salt   **ii** 3 boxes of salt   **iii** b boxes of salt.

   b Expand the brackets for your answers to parts **ii** and **iii**.

14 a Find an expression for the unknown length in terms of a and b.

   b Find two different expressions for the area of the shaded region.

   c Show that your expressions in part **b** are equivalent.

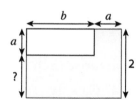

Solve problems

A rectangular garden has dimensions $2x$ by $3x + 5$. Draw a diagram to represent the garden. Find the total area of the garden.

The area is $2x(3x + 5)$.

Expand the brackets to get an area of $6x^2 + 10x$.

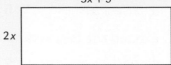

15 Write down an expression for the area of:

   a square A

   b rectangle B

   c the whole shape.

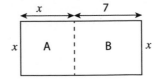

16 Write an expression for the area of each rectangle.
   Remember to use brackets. Then expand the brackets.

   a   **b** $y - 4$ **c**

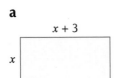

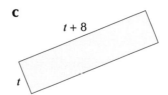

17 The length of one side of a rectangular field is r and area is $r^2 + 8r$. What was the length of the other side of the rectangle?

18 Expand this expression.
   $2x(3x + 2y + 4) + 3x(4x - y - 3)$

41.2 Factorising expressions containing powers

● I can take out a variable as a factor

When you **factorise** an expression you take any common factor and write it outside the brackets. It might happen that a variable is a common factor, or part of it. If so, it can be put outside the brackets.

Develop fluency

Factorise each expression as much as possible.

   a $6xy - 2x$ **b** $t^2 + 6t$ **c** $4x^2 - 10x$

   a $6xy - 2x = 2(3xy - x)$ 2 is a factor of 6 and 2. However, you can factorise further.

            $= 2x(3y - 1)$ x is also a factor of both terms. Take it outside the brackets.

   b $t^2 + 6t = t(t + 6)$ t is a common factor. Check this by expanding the brackets.

   c $4x^2 - 10x = 2x(2x - 5)$ Both 2 and x are common factors.

1 Factorise each expression.

    a $4x + 8$ **b** $12y - 15$ **c** $14 - 7x$ **d** $32y + 40$

2 Complete each factorisation.

    a $x^2 - 3x = x(...)$ **b** $t^2 + 5t = t(...)$ **c** $y^2 - 4y = y(...)$ **d** $6n + n^2 = n(...)$

3 Factorise each expression as much as possible.

    a $x^2 + 6x$ **b** $2n - n^2$ **c** $20n + n^2$ **d** $3x - x^2$

4 Factorise each expression as much as possible.

    a $x^2 + kx$ **b** $2cx + x^2$ **c** $3x^2 + x$ **d** $4n^2 - n$

5 Factorise each expression as much as possible.

    a $6x + 12$ **b** $12x - 8y$ **c** $9t^2 - 6$ **d** $6a + 9c$

6 Complete each factorisation.

    a $6x^2 + 12x = 6x(...)$ **b** $9y + 6y^2 = 3y(...)$ **c** $24x^2 + 16x = ...(3x + 2)$

7 Factorise each expression as much as possible.

    a $4x^2 + 4x$ **b** $6y^2 - 6y$ **c** $2t^2 + 10t$ **d** $6x^2 - 2$

8 Complete these factorisations.

    a $2x^2 + 4xy = 2x(... + ...)$ **b** $6a^2 - 9ab = ...(... - ...)$ **c** $12pq - 16q^2 = 4q(... - ...)$

9 Factorise each expression as much as possible.

    a $12x^2 + 12xy$ **b** $4a^2 - 8ab$ **c** $5xy + 5xz$ **d** $2ab + 12b$

10 Factorise each expression as much as possible.

    a $a^3 + 2a$ **b** $x^2 - 2x^3$ **c** $6n + 3n^3$ **d** $4x^3 - 2x^2$

Reason mathematically

The area of a game board is $4pn - 2p$. Given that the width of the board is $2p$, what is its length?

You need to think of an expression with two terms that can be multiplied by $2p$ by to give $4pn - 2p$.

$4pn - 2p = 2p(? - ?)$
$= 2p(2n - 1)$

11 Jenson factorises $8rn + 4r$ and gets the answer $2r(4n + 2r)$.
    What has he done wrong? Correct his mistake.

12 Some of these expressions can be factorised and some cannot.
    Factorise them if you can. Write 'Not possible' if they cannot be factorised.

    a $x^2 + 4$ **b** $x^2 - 6x$ **c** $4x - 10$ **d** $4x^2 + 1$

13 **a** Find the value of $4u^2 + u$ when $u = 5$.

    b Factorise $4u^2 + u$.

    c Show that you get the same value as part **a** when you substitute $u = 5$ into the factorisation from part **b**.

14 Some of these expressions can be factorised and some cannot.
Factorise them if you can. Write 'Not possible' if they cannot be factorised.

   a $5a + a^2$ **b** $3b - 10$ **c** $4c^2 - 5c$ **d** $6e^2 + 1$

   e $8g - g^2$ **f** $7h - 14$ **g** $i^2 + 10j$

Solve problems

Luke factorised $w^2 - 7w$ and got the answer $w(w + 7)$. What has he done wrong?

Hint: This is a common mistake when factorising – be careful with the signs.

He has written $w + 7$, where he should have written $w - 7$.

The correct answer is $w^2 - 7w = w(w - 7)$.

15 Factorise each expression.

   a $2x^2 + 8x$ **b** $3x^2 + 9x$ **c** $12x^2 + 36x$ **d** $8x^3 - 4x^2$

16 Factorise each expression.

   a $5xy + 10x$ **b** $6x^2y + 3xy$ **c** $4ab - 3a$ **d** $10a^2b + 15ab$

17 Explain why $8x + 9y$ cannot be factorised.

18 Sam correctly factorised an expression as $2(7y - 3)$. What expression did he factorise?

41.3 Expanding the product of two brackets

● I can multiply out two brackets

This expression has two brackets: $(x + 3)(x - 6)$

To multiply out the brackets you must multiply each term in the first set of brackets by each term in the second set of brackets.

● You multiply x by x and by -6. That gives x^2 and $-6x$.

● You multiply 3 by x and by -6. That gives $3x$ and -18.

So: $(x + 3)(x - 6) = x^2 - 6x + 3x - 18$

Then you can combine the middle two terms (they are **like terms**) to give:

$(x + 3)(x - 6) = x^2 - 3x - 18$

Be careful! Remember to do all four multiplications.

You can draw lines to remind you, like this.

They make a 'smiley face'.

$(x + 3)(x - 6)$

Or you can put the terms in a grid, like this.

Always collect like terms if you can.

	x	$+3$
x	x^2	$+3x$
-6	$-6x$	-18

Develop fluency

Expand the brackets. Write the result as simply as possible.

a $(x - 3)(x - 5)$

b $(a + b)(a - c)$

a Here is a grid for $(x - 3)(x - 5)$.

$(x - 3)(x - 5) = x^2 - 3x - 5x + 15$ Note that $-3 \times -5 = +15$.

$\qquad\qquad\quad = x^2 - 8x + 15$ Combine like terms: $-3x + -5x = -8x$.

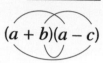

b Here is a 'smiley face' for $(a + b)(a - c)$.

$(a + b)(a - c) = a^2 - ac + ba - bc$

This cannot be simplified. There are no like terms.

$(a + b)(a - c)$

1 Expand the brackets.

a $(a + 1)(b + 1)$ **b** $(c + 1)(d + 3)$ **c** $(p + 4)(q + 2)$ **d** $(s + 5)(t + 7)$

2 Expand the brackets.

a $(a + b)(c + d)$ **b** $(a + b)(c - d)$ **c** $(a - b)(c + d)$ **d** $(a - b)(c - d)$

3 Expand the brackets.

a $x(x + 2)$ **b** $a(a - 3)$ **c** $p(p + 5)$ **d** $y(y - 5)$

4 Expand the brackets. Simplify the result by collecting like terms.

a $(x + 3)(x + 1)$ **b** $(x + 9)(x + 2)$ **c** $(x + 5)(x + 4)$ **d** $(x - 8)(x + 9)$

5 Expand the brackets. Simplify the result by collecting like terms.

a $(a + 2)(a + 1)$ **b** $(n + 4)(n + 3)$ **c** $(x - 3)(x - 4)$ **d** $(z - 10)(z - 4)$

6 Expand the brackets. Simplify each answer as much as possible.

a $(x + 3)(2 + x)$ **b** $(4 + t)(5 + t)$ **c** $(3 - n)(5 + n)$ **d** $(8 - y)(2 - y)$

7 Fiona's homework has been smudged. Fill in the gaps.

a $(a + 5)(a - 2) = a^2 + \ldots\ldots - 10$

b $(b + 7)(b + 8) = b^2 + 15b + \ldots\ldots$

c $(c - 3)(c - 7) = c^2 - \ldots\ldots + \ldots\ldots$

d $(4 + d)(9 + d) = d^2 + \ldots\ldots + 36$

e $(e + 11)(e + 5) = \ldots\ldots + 16e + \ldots\ldots$

f $(f - 9)(f + 1) = f^2 - \ldots\ldots - \ldots\ldots$

8 You can write $(x + 2)(x + 2)$ as $(x + 2)^2$.

Expand these brackets and simplify as much as possible.

a $(x + 2)^2$ **b** $(x + 1)^2$ **c** $(x + 4)^2$ **d** $(x + 7)^2$

9 Expand and simplify these expressions.

a $(x + 5)^2$ **b** $(x - 8)^2$ **c** $(x + 100)^2$ **d** $(x - 200)^2$

Reason mathematically

a Explain why the area of the blue rectangle can be written as $(s + t)(u - v)$.

b Use the diagram to explain why the area of the blue rectangle can also be written as $su + tu - sv - tv$

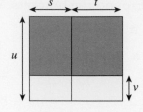

a The length of the top is $s + t$. The entire height is u and the height of the yellow part is v, so the height of the blue part is $u - v$.

b Link s and u and then link the link to su.
Link t and v and link the link to tv.
Link t and u and then link the link to tu..
Link s and v and then link the link to sv
$(s + t)(u - v) = su + tu - sv - tv$

10 a Find the value of 13×12.

b Expand and simplify $(x + 3)(x + 2)$.

c Substitute $x = 10$ into your answer to part **b**.

d Your answers to parts **a** and **c** should be the same. Explain why.

11 a Expand the brackets. Simplify the result by collecting like terms.

    i $(c + 5)(c - 5)$ **ii** $(f + 4)(f - 4)$ **iii** $(t + 9)(t - 9)$

b Explain why each answer has only two terms.

12 Explain why $(x + 3)^2 = x^2 + 6x + 9$.

13 a Explain why the area of this rectangle can be written as $(p + q)(r + s)$.

b Use the diagram to explain why the area of the rectangle can be written as $pr + ps + qr + qs$.

14 a Explain why the area of the blue rectangle can be written as $(w - x)(y + z)$.

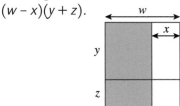

b Use the diagram to explain why the area of the blue rectangle can also be written as $wy + wz - xy - xz$.

Solve problems

Use this diagram to explain why $(x + 3)(x + 5)$ is equivalent to $x^2 + 8x + 15$.

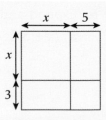

The area of the whole rectangle is given by $(x + 3)(x + 5)$.

Working out the area of each section in the diagram and adding them gives $x^2 + 5x + 3x + 15$.

Collecting like terms gives $x^2 + 8x + 15$.

So, $(x + 3)(x + 5) = x^2 + 8x + 15$.

15 Work out the number that is missing from each expansion.

    a $(x + 6)(x + 3) = x^2 + \ldots x + 18$ **b** $(x + 2)(x - 5) = x^2 - \ldots x - 10$

16 Use this diagram to explain why $(x + 6)(x - 2)$ is equivalent to $x^2 + 4x - 12$.

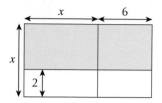

17 Spot the mistakes.

    a $(5y - 1)(y + 2) = 5y^2 + 10y - y + 2$ **b** $(y - 3)^2 = y^2 + 9$

                    $= 5y^2 + 9y + 2$

18 Expand and simplify $(x + 2)^2 + (x + 1)^2$.

41.4 Expanding expressions with more than two brackets

● I can multiply out three brackets

Look at this expression.

$(x + 2)(x - 1)(x + 3)$

It contains three bracketed terms. You can multiply out these brackets.

● First, choose two and expand them.

● Then multiply the result by the third one.

The answer to the last example is called a **cubic expression** because it includes an x^3 term.

Develop fluency

Expand the brackets in $(x + 2)(x - 1)(x + 3)$ and simplify the result as much as possible.

First expand $(x + 2)(x - 1)$.   You can choose any two brackets to start.

$(x + 2)(x - 1) = x^2 + 2x - x - 2$ You can use a grid to check this.

          $= x^2 + x - 2$ Collect like terms.

Now you have $(x^2 + x - 2)(x + 3)$.   Multiply by the third bracketed term.

$(x^2 + x - 2)(x + 3) = x^3 + x^2 - 2x + 3x^2 + 3x - 6$

                $= x^3 + 4x^2 + x - 6$ There are two pairs of like terms.

1 Look at these expressions.
   For each expression, expand the brackets and collect like terms.
   a $(x^2 + x - 2)(x + 2)$ **b** $(x^2 + 5x + 4)(x - 2)$

2 **a** Expand the expression $(x + 3)(x + 2)$.
   b Use your result to expand the expression $(x + 3)(x + 2)(x + 1)$.

3 **a** Expand $(x + 4)(x - 1)$.
   b Use your result to expand $(x + 4)(x - 1)(x + 2)$.

4 **a** Expand $(x - 3)(x - 1)$.
   b Use your result to expand $(x - 3)(x - 1)(x - 2)$.

5 Expand each expression and collect like terms.
   a $(x + 1)(x - 1)(x + 5)$ **b** $(x + 2)(x - 2)(x + 3)$ **c** $(x + 1)(x - 4)(x - 1)$

6 Expand and collect like terms.
   a $(x + 1)^2(x - 1)$ **b** $(x - 2)^2(x + 1)$ **c** $(x - 3)(x + 2)^2$

7 **a** Expand $(x + 1)(x + 2)^2$.
   b Use your result to expand $(x + 1)^2(x + 2)^2$.

8 Expand and collect like terms for $x(x + 2)^2$

Reason mathematically

Show that $(2x + 4)(x - 1)(4x - 3) = 8x^3 + 2x^2 - 22x + 12$.

$(2x + 4)(x - 1) = 2x^2 - 2x + 4x - 4$ First expand any two of the brackets.

$\qquad\qquad\qquad = 2x^2 + 2x - 4$ Simplify.

Now $(2x + 4)(x - 1)(4x - 3) = (2x^2 + 2x - 4)(4x - 3)$

$\qquad\qquad\qquad = 4x(2x^2 + 2x - 4) - 3(2x^2 + 2x - 4)$

Multiply your expansion by each term in the 3rd set of brackets.

$\qquad\qquad\qquad = 8x^3 + 8x^2 - 16x - 6x^2 - 6x + 12$

Remember the minus outside the 2nd bracket changes each sign inside the 2nd bracket.

$\qquad\qquad\qquad = 8x^3 + 2x^2 - 22x + 12$ Simplify.

9 Look at this expression.

$(x + 3)(x - 3)(x + 1)$

  a Expand the first two brackets and then multiply by the third.

  b Show that you get the same result when you expand the last two brackets and then multiply by the first.

10 a Expand $(x + 1)^2$.   **b** Use your result to expand $(x + 1)3$.

  c Expand $(x + 2)3$.   **d** Expand $(x - 1)3$.

  e Use your results so far to predict the expansion of $(x - 2)^3$. Check whether you are correct.

11 a Write down an expression for the volume of this cuboid.

  b Expand the brackets in your expression and collect like terms.

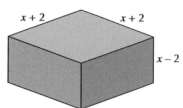

12 A cube has sides of length $2x + 1$. What is its volume?

Solve problems

$(x + 1)(x + 2)(x + c) = x^3 + 6x^2 + 11x + 6$

Find the value of c.

Expand the first two brackets to get $(x^2 + 3x + 2)(x + c)$.

The constant term in the final expansion is 6 and the constant term on the expansion of the first two brackets is 2.

The only way to get the 6 is to multiply 2 by 3, therefore $c = 3$.

13 Find the value of c when $(x + 3)(x + 7)(x + c) = x^3 + 14x^2 + 61x + 84$.

14 Expand and simplify $(3x + 2)(x + 1)(x + 5) + (x + 3)^3$.

15 Expand and simplify $(2x - 3)^3 - (x - 4)^3$.

16 The volume of a cuboid is $x^3 + x^2 - 22x - 40$. What is the missing side length, given that the other two are $(x + 4)$ and $(x + 2)$?

Now I can...

expand a term with a variable or constant outside brackets	multiply out two brackets	multiply out three brackets
take out a variable as a factor		

CHAPTER 42 Trigonometric ratios

42.1 Finding trigonometric ratios of angles

● I can understand what the trigonometric ratios sine, cosine and tangent are

In any right-angled triangle, given the angle *x*, you can label the sides as 'opposite' (opp), 'adjacent' (adj) and 'hypotenuse' (hyp). The opposite and adjacent sides are always labelled according to the angle you are looking at.

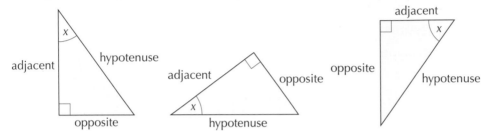

It doesn't matter which way round you see the right-angled triangle. For each angle there is only one adjacent side and one opposite side.

You need to learn and remember these relationships:

sine (sin) $= \dfrac{\text{opposite}}{\text{hypotenuse}}$ **cosine** (cos) $= \dfrac{\text{adjacent}}{\text{hypotenuse}}$ **tangent** (tan) $= \dfrac{\text{opposite}}{\text{adjacent}}$

One way to learn this is to remember a rhyme or a mnemonic, such as:

SOH CAH TOA

$S = \dfrac{O}{H}$ $C = \dfrac{A}{H}$ $T = \dfrac{O}{A}$ or

Tommy On A Ship Of His Caught A Herring

$T = \dfrac{O}{A}$ $S = \dfrac{O}{H}$ $C = \dfrac{A}{H}$

Make up some of your own mnemonics. Try to use family names or words you will easily remember.

You need to know these ratios by heart.

Develop fluency

For each triangle, identify and work out a trigonometric ratio for angle *x* from the lengths shown.

a    b    c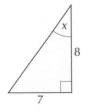

a Identify 3 as 'opposite', 8 as 'hypotenuse'.

So, use SOH.

$$\sin x = \frac{opp}{hyp} = \frac{3}{8} = 0.375$$

b Identify 5 as 'adjacent', 7 as 'hypotenuse'.

So, use CAH.

$$\cos x = \frac{adj}{hyp} = \frac{5}{7} = 0.714$$

c Identify 8 as 'adjacent', 7 as 'opposite'.

So, use TOA.

$$\tan x = \frac{opp}{adj} = \frac{7}{8} = 0.875$$

Sketch a triangle from each trigonometric ratio.

a $\tan x = \frac{5}{6}$

b $\cos x = \frac{2}{7}$

c $\sin x = \frac{5}{9}$

a $\tan x = \frac{5}{6} = \frac{opp}{adj}$

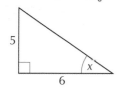

b $\cos x = \frac{2}{7} = \frac{adj}{hyp}$

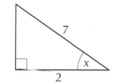

c $\sin x = \frac{5}{9} = \frac{opp}{hyp}$

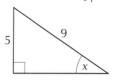

1 For each triangle, identify the opposite, the adjacent and the hypotenuse in relation to the angle labelled x.

a

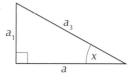

b

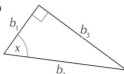

c

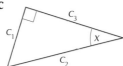

d

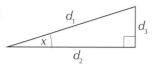

e

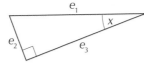

2 For each triangle, copy and complete the trigonometric ratios. The first one has been started for you.

a

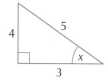

b

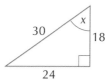

c

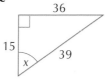

a $\sin x = \frac{opp}{hyp} = \frac{4}{5} = 0.8$

$\cos x = \frac{adj}{hyp} = \frac{3}{5} = \square$

$\tan x = \frac{opp}{adj} = \frac{4}{3} = \square$

b $\sin x = \frac{\square}{\square} = \frac{\square}{\square} = \square$

$\cos x = \frac{\square}{\square} = \frac{\square}{\square} = \square$

$\tan x = \frac{\square}{\square} = \frac{\square}{\square} = \square$

c $\sin x = \frac{\square}{\square} = \frac{\square}{\square} = \square$

$\cos x = \frac{\square}{\square} = \frac{\square}{\square} = \square$

$\tan x = \frac{\square}{\square} = \frac{\square}{\square} = \square$

3 For the angle labelled *x* in each triangle:
   i write down which side is 'opp', which is 'adj' and which is 'hyp'
   ii write down the fraction for sin, cos and tan.

   a   **b**   **c**   **d**

   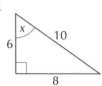 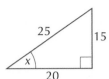 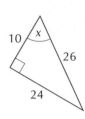 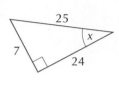

4 For the angle labelled *x* on each triangle:
   i write down which of the trigonometric ratios can be identified
   ii write down the fraction.

   a   **b**   **c**   **d**

   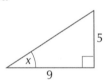 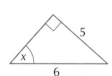

5 Look at this triangle.
   Write down the fraction for:

   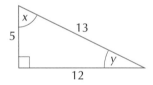

   a tan *x*   **b** sin *x*   **c** cos *x*.
   d tan *y*   **e** sin *y*   **f** cos *y*.

6 Calculate the value of each ratio as a decimal.

   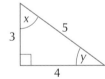

   a sin *x*   **b** sin *y*   **c** cos *x*
   d cos *y*   **e** tan *x*   **f** tan *y*

7 Write the value of each ratio as a decimal, where possible, correct to
   three decimal places.

   a sin 30°   **b** cos 30°   **c** tan 30°

8 What is the value of sin 45°? Compare this to the value of cos 45°. What do you notice?

9 What happens if you try to find the value of tan 90°?

10 Multiply these trigonometric expressions by the number shown to find their values, where
   possible, correct to three decimal places.
   a 200 sin 90°   **b** 200 sin 0°   **c** 500 cos 90°
   d 500 cos 0°   **e** 350 tan 45°   **f** 350 tan 0°

Reason mathematically

Sketch a right-angled triangle that includes an angle x, given that $\cos x = \frac{1}{2}$.

Remember that $\cos x = \dfrac{\text{adjacent}}{\text{hypotenuse}}$

Sketch the triangle in which the adjacent side is half the hypotenuse and label what you know.

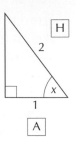

11 For each part, sketch a right-angled triangle. Label the sides, to illustrate the given facts about the triangle.

    a $\tan x = \dfrac{6}{7}$ **b** $\cos x = \dfrac{3}{5}$ **c** $\sin x = \dfrac{7}{12}$

    d $\tan x = \dfrac{5}{9}$ **e** $\cos x = \dfrac{9}{11}$ **f** $\sin x = \dfrac{7}{11}$

12 Here are some other trigonometric ratios. What do you notice in each set?

    a $\sin 0°$, $\sin 25°$, $\sin 45°$, $\sin 70°$, $\sin 90°$

    b $\cos 90°$, $\cos 65°$, $\cos 45°$, $\cos 20°$, $\cos 0°$

    c Compare your answers to parts **a** and **b** above. What can you say about the relationship between sines and cosines in the light of these results?

13 Make some observations about these trigonometric ratio sequences.

    a $\tan 0°$, $\tan 15°$, $\tan 30°$, $\tan 45°$, $\tan 60°$, $\tan 75°$

    b $\tan 88°$, $\tan 89°$, $\tan 89.5°$, $\tan 89.8°$, $\tan 89.9°$, $\tan 90°$

    c What happens to tangent ratios that does not happen with sine or cosine?

14 Look at your answers to question 5.

    a What do you notice about:

        i $\sin x$ and $\cos y$ **ii** $\sin y$ and $\cos x$?

    b What do you notice about $\tan x$ and $\tan y$?

Solve problems

You are told that $\cos 30° = \dfrac{\sqrt{3}}{2}$.

Sketch a triangle in which one angle is $30°$.

Sketch the triangle and label what you know.

15 You are told that sin $30° = \frac{1}{2}$.

From this information, sketch three different-sized triangles that each have angles of 90° and 30°.

Label the angles and the lengths of two appropriate sides.

16 You are told that tan $45° = \frac{1}{1}$.

From this information, sketch three different-sized triangles that each have angles of 90° and 45°.

Label the angles and the lengths of two appropriate sides.

17 You are told that cos $60° = \frac{1}{2}$.

From this information, sketch three different-sized triangles that each have angles of 90° and 60°.

Label the angles and the lengths of two appropriate sides.

18 The base of the triangle is 5.2 m. How would you find the area of this triangle?

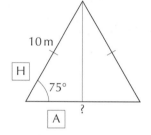

42.2 Using trigonometric ratios to find the sizes of angles

● I can find the size of an angle identified from a trigonometric ratio

You can use your calculator to find the angle from the trigonometric function. You can use the inverse trigonometric function buttons sin⁻¹ cos⁻¹ and tan⁻¹.

Not all calculators work in this way so make sure you know how to use these keys on your own calculator.

Develop fluency

A ladder 4 m long leans against a wall.

It just reaches a windowsill that is known to be 3.8 m above the ground.

What angle does the ladder make with the ground?

Draw a sketch.

Call the angle at the ground x.

Then you can name the opposite side and the hypotenuse.

Using SOH:

$\sin = \dfrac{opp}{hyp}$

$\sin x = \dfrac{opp}{hyp} = \dfrac{3.8}{4} = 0.95$

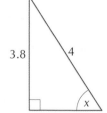

    = 71.8°

A ramp is built for wheelchair access.

It starts at a horizontal distance of 6 m from the step, which is 0.5 m high.

What angle of slope does the ramp have?

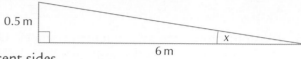

Draw a sketch and call the angle of the slope x.

Then you can name the opposite and the adjacent sides.

Using TOA:

$$\tan = \frac{opp}{adj}$$

$$\tan x = \frac{opp}{adj} = \frac{0.5}{6} = 0.083\,333\,3$$

 $\text{tan}^{-1} = 4.8°$

A builder knew that his ladders were 4.5 metres long when fully extended and that it was in the safest position when the foot of the ladder was 1 metre away from the wall.

What is the 'safe angle' at the floor?

Draw a sketch and call the safe angle at the floor x.

Then you know the adjacent side and the hypotenuse.

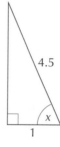

Using CAH:

$$\cos = \frac{adj}{hyp}$$

$$\cos x = \frac{adj}{hyp} = \frac{1}{4.5} = 0.222\,222\,2$$

$\boxed{0} \cdot \boxed{2}\,\boxed{2}\,\boxed{2}\,\boxed{2}\,\boxed{2}\,\boxed{2}\ \text{cos}^{-1} = 77.2°$

1 These numbers are all sines of angles. Use your calculator to find the size of each angle, correct to one decimal place.

    a 0.828   **b** 0.127   **c** 0.632   **d** 0.512

    e 0.395   **f** 0.505   **g** 0.67   **h** 0.99

2 These numbers are all tangents of angles. Use your calculator to find the size of each angle, correct to one decimal place.

    a 0.641   **b** 0.903   **c** 0.807   **d** 1.56

    e 3.14   **f** 0.845   **g** 5.01   **h** 0.752

3 These numbers are all cosines of angles. Use your calculator to find the size of each angle, correct to one decimal place.

    a 0.428   **b** 0.705   **c** 0.129   **d** 0.431

    e 0.137   **f** 0.104   **g** 0.811   **h** 0.905

4 These fractions are all sines of angles. Use your calculator to find the size of each angle, correct to one decimal place.

   a $\dfrac{4}{5}$ **b** $\dfrac{7}{8}$ **c** $\dfrac{1}{9}$ **d** $\dfrac{4}{11}$ **e** $\dfrac{1}{7}$

5 These fractions are all tangents of angles. Use your calculator to find the size of each angle, correct to one decimal place.

   a $\dfrac{6}{7}$ **b** $\dfrac{9}{5}$ **c** $\dfrac{8}{7}$ **d** $\dfrac{1}{6}$ **e** $\dfrac{3}{14}$

6 These fractions are all cosines of angles. Use your calculator to find the size of each angle, correct to one decimal place.

   a $\dfrac{4}{7}$ **b** $\dfrac{7}{15}$ **c** $\dfrac{1}{9}$ **d** $\dfrac{4}{13}$ **e** $\dfrac{13}{17}$

7 Calculate the size of the angle labelled x in each triangle.
   Give your answers correct to one decimal place.

   a 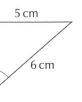 **b** 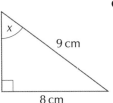 **c** 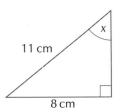 **d**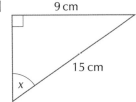

8 Calculate the size of the angle labelled x in each triangle.
   Give your answers correct to one decimal place.

   a 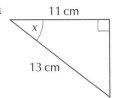 **b**   **c** 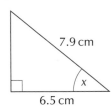 **d**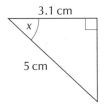

9 Calculate the size of the angle labelled x in each triangle.
   Give your answers correct to one decimal place.

   a 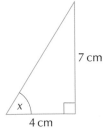 **b** 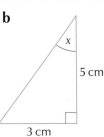 **c** 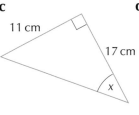 **d**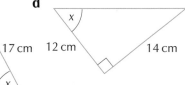

10 Calculate the size of the angle labelled *x* in each triangle.
Give your answers correct to one decimal place.

a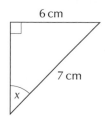
6 cm
7 cm
x

b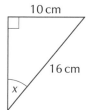
x
6 cm
10 cm

c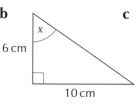
12 cm
x
14 cm

d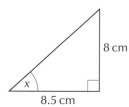
8 cm
x
8.5 cm

e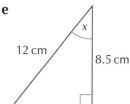
x
12 cm
8.5 cm

f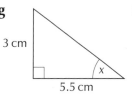
10 cm
16 cm
x

g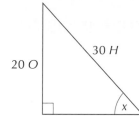
3 cm
x
5.5 cm

h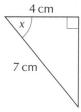
4 cm
x
7 cm

Reason mathematically

A cat is stuck at the top of a tree that is 20 m high. You have a ladder that is 30 m long, and you place the ladder so that the top is right by the cat.
What angle does the ladder make with the ground?

Start by drawing a diagram.

Label the sides H and O.

$$\sin x = \frac{O}{H}$$

$$x = \sin^{-1}\frac{20}{30}$$

$$= 41.8°$$

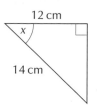
30 H
20 O
x

11 Find the height of this triangle (note that it is not right-angled).

6 m
30°
10 m

12 Find the area of this triangle.

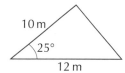
10 m
25°
12 m

13 Find the value of *x* in the two images.

a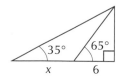
35° 65°
x 6

b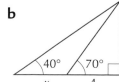
40° 70°
x 4

Solve problems

A windsurfer travelled 100 m in a straight line and then jumped 5 m vertically.
Calculate the angle made during the jump.

Always make a sketch and label what you know.

$\tan x = \dfrac{O}{A}$

$x = \tan^{-1} \dfrac{5}{100}$

$= 2.9°$

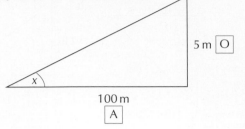

14 As a train travelled 2 km along a straight track it climbed 300 m vertically.
Calculate the angle the track makes with the horizontal.

15 When sliding down a 400 m slope, a skier dropped through a vertical height of 320 m.
At what angle is the slope to the horizontal?

16 A ladder 4.8 m long rests against a wall. It reaches 4 m up the wall.
At what angle is the ladder leaning against the wall?

17 Ewan is 1.5 m tall. He stood on level ground, looking up at the top of a tower that he
knew to be 80 m tall. He was 200 metres away from the bottom of the tower.
What angle from the horizontal did he need to look up, to see the top of the tower?

42.3 Using trigonometric ratios to find lengths

● I can find an unknown length in a right-angled triangle, given one side and another angle

When you are told an angle in a right-angled triangle, as well as an appropriate length, then
you can calculate the lengths of the other sides.

Develop fluency

Find the length of the side labelled x in the diagram.

You can see that the opposite (x), the hypotenuse (6) and the angle 62° are all labelled.

Remember that:

$\sin 62° = \dfrac{opp}{hyp} = \dfrac{x}{6}$

Rearranging, you can write:

$\dfrac{x}{6} = \sin 62°$

Then $x = 6 \times \sin 62°$ Multiply both sides by 6.

This is calculated on most calculators as:

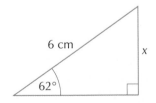

So $x = 5.3$ cm (1 dp).

Hint: Check exactly how to use trigonometric functions on your calculator.

Find the length of the side labelled x in the diagram.

The adjacent (x), the hypotenuse (15) and the angle 27° are all labelled.

Remember that:

$$\cos 27° = \frac{\text{adj}}{\text{hyp}} = \frac{x}{15}$$

So $\frac{x}{15} = \cos 27°$

Then $x = 15 \times \cos 27°$ Multiply both sides by 15.

This is calculated on most calculators as:

 13.4

So $x = 13.4$ cm (1 dp).

Find the length of the side labelled x in the diagram.

The opposite (x), the adjacent (7) and the angle 70° are all labelled.

Remember that:

$$\tan 70° = \frac{\text{opp}}{\text{adj}} = \frac{x}{7}$$

So $\frac{x}{7} = \tan 70°$

Then $x = 7 \times \tan 70°$ Multiply both sides by 7.

This is calculated on most calculators as:

 19.2

So $x = 19.2$ cm (1 dp).

1 Calculate the length labelled x in each right-angled triangle.
   Give your answers correct to one decimal place.

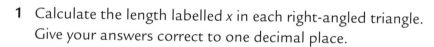

a 12 cm, x, 52°

b 71°, 4 cm, x, x

c x, 17 cm, 18°

d 44°, 9 cm, x

2 Calculate the length labelled x in each right-angled triangle.
   Give your answers correct to one decimal place.

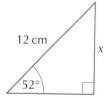 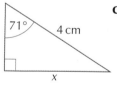  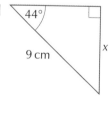

a x, 5 cm, 55°

b x, 43°, 7 cm

c 11 cm, 43°, x

d 69°, 18 cm, x

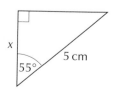 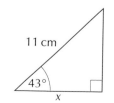 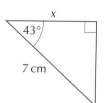 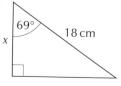

3 Calculate the length labelled x in each right-angled triangle.
Give your answers correct to one decimal place.

a

b

c

d

4 Find the values of these trigonometric expressions, giving your answers correct to three decimal places.

a sin 16°   **b** sin 13.4°   **c** cos 74°   **d** cos 2.5°

e tan 1°   **f** tan 89.3°

5 Calculate the value of each expression, giving your answers correct to three significant figures.

a 10 sin 67°   **b** 4 sin 28°   **c** 5 cos 50°

d 3.5 cos 88.3°   **e** 7 tan 12°   **f** 100 tan 87°

6 Calculate the length labelled x in each right-angled triangle.
Give your answers correct to one decimal place.

a 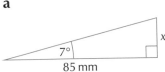   **b** 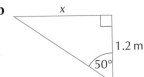   **c** 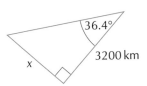   **d**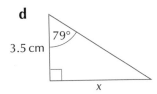

7 Calculate the value of x, correct to three significant figures.

a

b

c

8 In the triangle LMN, L = 33°, M = 90° and LM = 1.7 m.
Find the length of MN. Give your answer correct to three significant figures.

9 In the triangle XYZ, X = 90°, Y = 68° and YZ = 42 mm.
Find the length of XZ. Give your answer correct to three significant figures.

10 In the triangle ABC, A = 90°, B = 21° and AB = 6.3 m.
Find the length of AC. Give your answer correct to three significant figures.

Reason mathematically

Find the height of the trapezium.

As this is an isosceles trapezium it can be split it into two triangles and a rectangle.

$$\tan x = \frac{height}{a}$$

$$\tan 60° = \frac{height}{2.5}$$

height = tan 60° × 2.5 = 4.3 cm

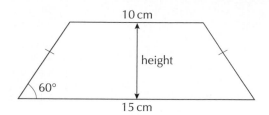

11 A child is on a swing. The highest position that he reaches is shown on the diagram. How high is the swing seat off the ground?

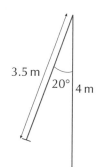

3.5 m
20°
4 m

12 Explain why sin 45° = cos 45°.

13 Find the shadow cast by a 6 m lamp post when the angle of elevation of the Sun is 58°. Find the length to the nearest decimal point.

14 Tony is flying a kite. How high up is the kite when the string is 12.5 m long and Tony is holding it at an angle of 60°?

Solve problems

An equilateral triangle has sides of length 12 cm. Use trigonometry to calculate the shortest distance from a vertex to its opposite side.

$$\tan x = \frac{height}{a}$$

$$\tan 60° = \frac{height}{6}$$

height = tan 60° × 6 = 10.39 cm

15 A right-angled triangle has another angle of 50° and the length of its hypotenuse is 5 cm. Calculate the area of the triangle.

16 A plank 8 metres long is leaning against a wall at an angle of 30° with the horizontal. How far up the wall does the plank reach?

17 A ship sails at 75⁰ from North for 150 km.

   a Draw a sketch of the ships journey, showing a north line and the east direction.

   b Use your diagram to help you find how far east the ship has sailed.

18 You fly for 300 km on at 150°.

   How far: a east have you flown   b south have you flown?

Now I can...

understand what the trigonometric ratios sine, cosine and tangent are	find the size of an angle identified from a trigonometric ratio	find an unknown length of a right-angled triangle given one side and another angle

ANSWERS

1.1 Factors and highest common factors

1. **a** 1, 3, 5, 15 **b** 1, 2, 4, 5, 10, 20
   c 1, 2, 4, 8, 16, 32 **d** 1, 5, 7, 35
   e 1, 2, 3, 4, 5, 6, 10, 12, 15, 20, 30, 60

2. **a** 5 **b** 15 **c** 20 **d** 4

3. **a** 1, 2, 4 **b** 1, 2, 3, 4, 6, 12 **c** 1, 5
   d 1, 3, 5, 15

4. **a** 3 **b** 2 **c** 9 **d** 1

5. **a** 9 **b** 4 **c** 28 **d** 50

6. **a** $\frac{2}{3}$ **b** $\frac{2}{3}$ **c** $\frac{3}{5}$ **d** $\frac{2}{5}$ **e** $\frac{3}{8}$ **f** $\frac{4}{5}$
   g $\frac{9}{20}$ **h** $\frac{8}{9}$

7. **a** It is even
   b Digits sum to a multiple of 3
   c It is even and digits sum to a multiple of 3
   d Units digit is 5 or 0

8. **a** $\frac{24}{36}$ and **b** $\frac{120}{132}$

9. **a** $\frac{25}{40}$ **b** $\frac{35}{45}$ and **d** $\frac{90}{95}$

10. **a** 7 is prime and 24 is not a multiple of 7
    b 23 is prime
    c Only 1 difference between numerator and denominator
    d eg 35 is 5 × 7 and 39 is not a multiple of either 5 or 7

11. $\frac{15}{20} = \frac{3}{4}$ and $\frac{18}{24} = \frac{3}{4}$

12. 12 and 24

13. 32 and 36

14. 24 and 42 or 30 and 48

15. 6, 8, 12, 24, 16

16. 1, 2, 3, 4, 6, 9, 12, 18, 36

17. The largest group size will be 9
    Class A have 4 groups
    Class B have 3 groups

18. **a** 13 **b** 5 boys and 4 girls

1.2 Multiples and common multiples

1. **a** 10, 4, 18, 8, 72, 100
   b 18, 69, 81, 33, 72
   c 10, 65, 100
   d 18, 81, 72

2. **a** 4, 8, 12, 16, 20, 24, 28, 32, 36, 40
   b 5, 10, 15, 20, 25, 30, 35, 40, 45, 50
   c 8, 16, 24, 32, 40, 48, 56, 64, 72, 80
   d 15, 30, 45, 60, 75, 90, 105, 120, 135, 150
   e 20, 40, 60, 80, 100, 120, 140, 160, 180, 200

3. **a** 40 **b** 20 **c** 30 **d** 40

4. **a** 45 **b** 25 **c** 63 **d** 77

5. **a** 30 **b** 90 **c** 240 **d** 231

6. **b** 3 and 8 and **c** 3, 4 and 8

7. **a** 6 and 14 **b** 2, 3 and 7 and **d** 2, 6 and 7

8. Because all multiples of 2 are even

9. Because even numbers can always be divided again, by 2.

10. Because 2 is a common factor

11. Multiples of 10: 10, 20, 30, 40, 50, 60, ...
    Multiples of 12: 12, 24, 26, 48, 60, ...
    LCM of 10 and 12 is 60, so after 60 seconds

12. **a** 63 **b** Ali, fewer numbers to sing

13. 165 = 3 × 5 × 11, so 3 × 5 = 15 chairs is most per row.

14. Numbers are 4 and 10, 4 + 10 = 14

15. 480 pupils

16. 112 seconds

17. 630 cm

18. 4 days in a year

1.3 Prime factors

1. **a** 12 **b** 90 **c** 270

2. **a** 8 = 2 × 2 × 2 **b** 28 = 2 × 2 × 7
   c 35 = 7 × 5 **d** 52 = 2 × 2 × 13
   e 180 = 2 × 2 × 3 × 3 × 5

3. **a** 42 = 2 × 3 × 7 **b** 75 = 3 × 5 × 5
   c 140 = 2 × 2 × 5 × 7
   d 250 = 2 × 5 × 5 × 5
   e 480 = 2 × 2 × 2 × 2 × 2 × 3 × 5

4. 2 = 2, 3 = 3, 4 = 2 × 2, 5 = 5, 6 = 2 × 3, 7 = 7, 8 = 2 × 2 × 2, 9 = 3 × 3, 10 = 2 × 5, 11 = 11, 12 = 2 × 2 × 3, 13 = 13, 14 = 2 × 7, 15 = 3 × 5, 16 = 2 × 2 × 2 × 2, 17 = 17, 18 = 2 × 3 × 3, 19 = 19, 20 = 2 × 2 × 5

5. **a** 2, 3, 5, 7, 11, 13, 17, 19
   b Prime numbers

6. **a i** $200 = 2^3 \times 5^2$ **ii** 2×5^2
   iii $2^3 \times 5^3$ **b** $2^6 \times 5^6$

7. **a** $32 = 2^5$ **b** $64 = 2^6$ **c** $128 = 2^7$
   d $1024 = 2^{10}$

8. 420

9. 2 has to be a prime factor

10. It is an odd number

11. $18 = 2 \times 3^2, 72 = 2^3 \times 3^2$

12. $\frac{216}{600} = \frac{2^3 \times 3^3}{2^3 \times 3 \times 5^2} = \frac{3^2}{5^2} = \frac{9}{25}$

13. **a** 10 = 2 × 5 **b** 30 = 2 × 3 × 5

14. **a i** $60 = 2^2 \times 3 \times 5$ **ii** $100 = 2^2 \times 5^2$
    b 300
    c HCF of 60 and 100 is $2^2 \times 5$
    LCM is $2^2 \times 5 \times 3 \times 5 = 2^2 \times 5^2 \times 3 = 300$

15. 3, 5 and 7 (any order)

16. $x = 11$ and $y = 7$

17. **a** $625 = 5^4$ **b** $\sqrt{625} = 5^2$

18. **a** $324 = 2^2 \times 3^4$ **b** $\sqrt{324} = 2 \times 3^2$

2.1 Sequences and rules

1. **a** 1, 4, 7, 10, 13
   b 1, 3, 9, 27, 81
   c 1, 6, 11, 16, 21
   d 1, 10, 100, 1000, 100000
   e 1, 12, 23, 34, 45
   f 1, 4, 16, 64, 256
   g 1, 9, 17, 25, 33
   h 1, 106, 211, 316, 421

2. **a** 5, 8, 11, 14, 17
   b 5, 15, 45, 135, 405
   c 5, 10, 15, 20, 25
   d 5, 50, 500, 5000, 50000
   e 5, 16, 27, 38, 49
   f 5, 20, 80, 320, 1280
   g 5, 13, 21, 29, 37
   h 5, 110, 215, 320, 425

3. **a** 8, 10 rule: add 2
   b 1000, 10000 rule: multiply by 10
   c 250, 1250 rule: multiply by 5
   d 21, 28 rule: add 7
   e 19, 24 rule: add 5
   f 16, 20 rule: add 4
   g 48, 60 rule: add 12
   h 54, 162 rule: multiply by 3

4. **a** 25, 20 subtract 5
   b 20, 17 subtract 3
   c −11, −15 subtract 4
   d −18.5, −23.5 subtract 5

5. **a** 3, 16, 55, 172 **b** 32, 12, 2, −3

6. **a** 3, 2, −1, −10 **b** 5, 28, 120, 488

7. **a** 5, 21, 69, 213 **b** 32, 12, 7, 5.75

8. **a** 4, 18, 60, 186 **b** 0, 4, 5, 5.25

9. $\frac{7}{16}, \frac{9}{32}$ 10. $\frac{12}{25}, \frac{24}{35}$

11. **a** 3, 5 add 2 **b** 6, 9 add 3
    c 8, 11 add 3 **d** 14, 26 add 12

12. **a** 5, 9 add 4 **b** 4.75 add 0.75
    c 75, 70, 65, 60 subtract 5
    d 1, 0 subtract 1

13. **a** 8 multiply by 2 **b** 6 multiply by 2
    c 10 multiply by 2
    d 20 multiply by 5

14. **a** Add 1 **b** Divide by 2
    c Multiply by 2

15. **a** 1, 4, 7, 10 add 3 1, 4, 16, 64 multiply by 4
    b 3, 9, 27, 81 multiply by 3 3, 9, 15, 21 add 6
    c (d in the book) 3, 6, 9, 12 add 3 3, 6, 12, 24 multiply by 2
    d 5, 15, 25, 35 add 10 5, 15, 45, 135 multiply by 3

16. **a** 81, 243 multiply by 3
    b 5, 0 subtract 5
    c 0.1, 0.01 divide by 10
    d 4, −4 subtract 8
    e 80, 160 multiply by 2
    f 1.25, 0.625 divide by 2

2.2 Working out missing terms

1. **a** Students' own diagrams **b** 31

2. **a** Students' own diagrams **b** 124

3. **a** Students' own diagrams **b** 50

4. **a** Students' own diagrams **b** 76

5. **a** 12 and 102 **b** 21 and 246
   c 31 and 346 **d** 17 and 152
   e 17 and 197 **f** 34 and 394
   g 60 and 510 **h** 46 and 451
   i 27 and 297

6. **a** −10 and −100 **b** −22 and −247
   c −32 and −347 **d** −4 and −139
   e −21 and −246 **f** −44 and −404
   g −15 and −285 **h** −1 and 404
   i −9 and 126

7. 43, 83 and 123

8. 37, 67 and 157

9. 14, −26 and −66

10. −30, −70 and −190

11 **a** 22 and 220 **b** 9 and 504
c 67 and –230 **d** 29 and –70

12 **a** 7 and 55 **b** 3 and 123
c 7 and 103 **d** 5 and 221

13 **a** 4 **b** 13
c Students' own diagrams

14 **a** The rule is "add 2", The first term
is 0, The fourth term is 6.
b The rule is "multiply by 2", The
first term is 1, The fourth term is 8.

15 **a** 21 **b** 9 **c** 30

2.3 Other sequences

1 $1 \times 1 = 1, 2 \times 2 = 4, 3 \times 3 = 9,$
$4 \times 4 = 16, 5 \times 5 = 25, 6 \times 6 = 36,$
$7 \times 7 = 49, 8 \times 8 = 64, 9 \times 9 = 81,$
$10 \times 10 = 100, 11 \times 11 = 121,$
$12 \times 12 = 144, 13 \times 13 = 169,$
$14 \times 14 = 196, 15 \times 15 = 225$

2 **c** $13 = 4 + 9$ **d** $17 = 1 + 16$
e $20 = 4 + 16$ **f** $25 = 9 + 16$
g $26 = 1 + 25$ **h** $29 = 4 + 25$
i $34 = 9 + 25$ **j** $45 = 9 + 36$
k $50 = 1 + 49$ **l** $52 = 16 + 36$

3 1, 3, 6, 10, 15, 21, 28, 36, 45, 55,
66, 78, 91, 105, 120

4 **c** $9 = 3 + 6$ **d** $11 = 1 + 10$
e $13 = 3 + 10$ **f** $16 = 6 + 10$
g $24 = 3 + 21$ **h** $25 = 10 + 15$
i $27 = 6 + 21$ **j** $29 = 1 + 28$
k $31 = 3 + 28$ or $10 + 21$
l $34 = 6 + 28$

5 1 and 36   **6** a, b and d

7 **a** 4, 9, 16, 25, 36, 49, 64, 81,
100, 121
b Square numbers

8 **a** 3, 5, 7, 9, 11, 13, 15, 17, 19, 21
b Odd numbers

9 **a** 2,3, 4, 5, 6, 7, 8, 9, 10, 11
b Integers

10 **a** 1, 3, 6, 10, 15
b Triangular numbers

11 The number of games is $5 \times 6 \div 2 = 15$, which is a triangular number

12 **a** 5, 13, 25, 41, 61
b Not possible as always odd + even = odd

13 2, 3, 5, 8, 12, … Add 1 then 2 then 3 and so on.
2, 3, 5, 7, 11, 13, …
Prime numbers
2, 3, 5, 8, 13, 21, 34, … Sum of the previous two terms

14 **a** $1+3+5+7+9+11 = 36 = 6^2$
$1+3+5+7+9+11+13 = 49 = 7^2$
b Odd numbers **c i** $100 = 10^2$
ii. $225 = 15^2$

15 **a** $1+2+3+4+5+6 = 21$
$1+2+3+4+5+6+7 = 28$
b Integers **c** Triangular numbers
d i. 55 **ii.** 120

16 **a**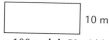

---

b 0, 3, 9, 18 **c** 0, 1, 3, 6 **d** 10, 15

2.4 The nth term of a sequence

1 **a i.** 7, 9, 11 **ii.** 205 **b i.** 2, 7, 12
ii. 497 **c i.** 9, 13, 17 **ii.** 405
d i. 11, 21, 31 **ii.** 1001
e i. 6, 13, 20 **ii.** 699 **f i.** 9, 8, 7
ii. –90 **g i.** 18, 16, 14 **ii.** –180
h i. 4, 1, –2 **ii.** –293

2 **a i.** 3, 5, 7, 9 **ii.** 3 **iii.** 2
b i. 4, 6, 8, 10 **ii.** 4 **iii.** 2
c i. 5, 7, 9, 11 **ii.** 5 **iii.** 2
d i. 6, 8, 10, 12 **ii.** 6 **iii.** 2

3 **a i.** 4, 9, 14, 19 **ii.** 4 **iii.** 5
b i. 7, 12, 17, 22 **ii.** 7 **iii.** 5
c i. 1, 6, 11, 16 **ii.** 1 **iii.** 5
d i. 8, 13, 18, 23 **ii.** 8 **iii.** 5

4 **a i.** 2, 5, 8, 11 **ii.** 3
b i. 6, 10, 14, 18 **ii.** 4
c i. 2, 8, 14, 20 **ii.** 6
d i. 13, 23, 33, 43 **ii.** 10

5 **a** $a = 4$, $d = 5$ **b** $a = 1$, $d = 2$
c $a = 3$, $d = 6$ **d** $a = 5$, $d = -2$

6 **a i.** 4, 9, 14, 19 **ii.** 5 **b i.** 10, 18,
26, 34 **ii.** 8 **c i.** –3, 3, 9, 15 **ii.** 6
d i. 13, 23, 33, 43 **ii.** 10

7 **a** $a = 1$, $d = 7$ **b** $a = 2$, $d = 2$
c $a = 3$, $d = 9$ **d** $a = 6$, $d = -3$

8 **a** 1, 9, 17, 25, 33, 41
b 5, 12, 19, 26, 33, 40
c 4, 2, 0, –2, –4, –6
d 1.5, 2, 2.5, 3, 3.5, 4
e 10, 7, 4, 1, –2, –5
f 2, 1.5, 1, 0.5, 0, –0.5

9 **a** 1, 4, 9, 16 **b** Square numbers

10 **a** 1, 3, 6, 10 **b** Triangular numbers

11 **a** $3 \times 1 + 4 = 7$, $3 \times 2 + 4 = 10$,
$3 \times 3 + 4 = 13$
b When n = 12, 3n + 4 = 40 so it is in the sequence (12th term)

12 **a** $1^2 + 1 = 2$, $2^2 + 1 = 5$,
$3^2 + 1 = 10$
b 34 is not a square number so 35 is not in the sequence

13 d equals the coefficient of n

14 **i. a** $a = 3$, $d = 4$ **b** $a = 6$, $d = 4$
c $a = 0$, $d = 4$ **d** $a = 9$, $d = 4$
ii. Same as coefficient of n

15 **i. a** $a = 4$, $d = 5$ **b** $a = 7$, $d = 5$
c $a = 2$, $d = 5$ **d** $a = 9$, $d = 5$
ii. d = coefficient of n

16 11   **17** –6   **18** 36

2.5 Finding the nth term

1 **a** $6n - 2$ **b** $3n + 6$ **c** $6n + 3$
d $3n - 1$ **e** $7n - 5$ **f** $2n + 6$ **g** $4n + 6$
h $8n - 5$ **i** $10n - 1$ **j** $9n - 5$

2 **a** 95 – 5n **b** 50 – 7n **c** 31 – 3n
d 52 – 8n

3 **a** $6n - 2$ **b** $3n + 5$ **c** $6n + 3$
d $3n + 1$ **e** $7n + 6$

4 **a** 0.5n + 2 **b** 2.5n + 8 **c** 0.1n + 3
d 8.2 – 0.2n

---

5 **a** 4n , 160 **b** 7n + 1 , 281

6 **a** $4n - 3$ **b** $2n + 8$ **c** $6n$ **d** $2n + 2$

7 **a** $4n$ **b** $5n - 1$ **c** $3n + 1$ **d** $n + 3$

8 **a** $n^2 + 1$ **b** $3n^2$

9 **a** $n^2 + 4$ **b** $\frac{1}{2}n^2$

10 **a** $3n$ **b** $n^2 + 3n$

11 n^2 gives the sequence 1, 4, 9, 16, 25,…
n gives the sequence 1, 2, 3, 4,5,…
So, $n^2 + n$ = 2, 6, 12, 20, 30, …

12 **a** £11 **b** £(n + 1)

13 **a** 20p **b** 2n pence **c** 2p

3.1 Perimetre and area of a rectangle

1 **a** 20 cm, 25 cm² **b** 46 cm, 120 cm²
c 30 m, 56 m²
d 108 mm, 720 mm²

2 2.8 m

3 **a i.** 16 cm² **ii.** 16 cm **b i.** 84 cm²
ii. 38 cm **c i.** 60 m² **ii.** 32 m
d i. 400 mm² **ii.** 82 mm
e i. 150 cm² **ii.** 50 cm **f i.** 160 cm²
ii. 56 cm **g i.** 54 cm² **ii.** 30 cm
h i. 128 cm² **ii.** 48 cm

4 Length = 3 m, Width = 2 m

5 **a** 4 cm **b** 10 cm **c** 6 m **d** 8 cm

6 20 cm

7

	Length	Width	Perimetre	Area
a	8 cm	6 cm	28 cm	48 cm²
b	20 cm	15 cm	70 cm	300 cm²
c	10 cm	5 cm	30 cm	50 cm²
d	6 m	5 m	22 m	30 m²
e	7 m	6 m	26 m	42 m²
f	25 mm	10 mm	70 mm	250 mm²

8 **a** 30 m **b** 10 lengths

9 40 m

┌─────────────────┐
│   │ 10 m
└─────────────────┘

a 100 m **b i.** 50 widths
ii. 12.5 lengths

10 Yes; e.g. when the length of the square side is 4 units

11 length = 30 m and width = 15 m

3.2 Compound shapes

1 **a i.** 32 cm **ii.** 48 cm² **b i.** 40 cm
ii. 36 cm² **c i.** 60 cm **ii.** 88 cm²

2 **a i.** 13 cm² **ii.** 20 cm **b i.** 36 cm²
ii. 28 cm **c i.** 38 m² **ii.** 34 m
d i. 51 cm² **ii.** 36 cm **e i.** 75 cm²
ii. 40 cm **f i.** 24 cm² **ii.** 22 cm
g i. 50 cm² **ii.** 30 cm **h i.** 88 cm²
ii. 40 cm

3 **a** He didn't calculate the dimensions of the rectangles correctly after splitting the shape.
b Area = 10 × 4 + 4 × 5
= 40 + 20
= 60 cm²

Answers

4 **a** 336 cm² **b** 600 cm² **c** 264 cm²
5 9 cm
6 5 m, 6 m and 3 m
7 142 m²   **8** 11.4 m²
9 84 cm   **10** 300 cm²

3.3 Area of a triangle

1 **a** 35 cm² **b** 108 cm² **c** 24.5 cm²
  d 6 m² **e** 28 m²
2 **a** 15 cm² **b** 270 mm² **c** 1750 mm²
  d 120 mm²

3

Triangle	Base	Height	Area
a	6 cm	5 cm	15 cm²
b	8 cm	7 cm	28 cm²
c	11 m	5 m	27.5 m²

4 **a** Area 6

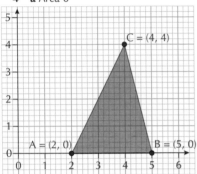

b Area 10

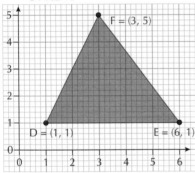

c Area 6

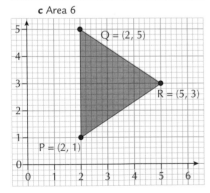

d Area 12

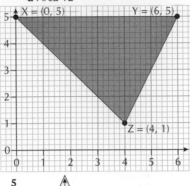

5

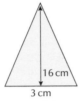 

6 **a** 21 cm² **b** 9.46 cm² **c** 12 cm²
  d 169 m² **e** 31.2 cm² **f** 12.4 cm²
7 **a** 6 m² **b** 45 cm² **c** 12 m²
8 e.g. 6 and 12, 18 and 4, 36 and 2
9 She is wrong; area = $\frac{1}{2} \times 5 \times 12$ = 30 cm²
10 **a** 135 cm² **b** 100 cm²
11 b = 12 cm   **12** 72 m²

13

Triangle	Base	Height	Area
a	8 cm	8 cm	32 cm²
b	6 cm	9 cm	27 cm²
c	11 cm	12 cm	66 cm²

14 10 cm²

3.4 Area of a parallelogram

1 **a** 36 cm² **b** 150 cm² **c** 49 m²
  d 80 cm² **e** 8820 mm²
  f 3000 mm²
2 **a** A = 15 cm² P = 22 cm
  b A = 16 m² P = 26 m
  c A = 5 m² P = 10 m
  d A = 57 cm² P = 34 cm
  e A = 400 mm² P = 90 mm
  f A = 16.5 m² P = 19 m

3

Parallelogram	Base	Height	Area
a	8 cm	4 cm	32 cm²
b	17 cm	12 cm	204 cm²
c	8 m	5 m	40 m²

4 **a** Area 20

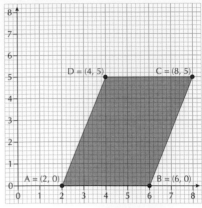

b Area 15

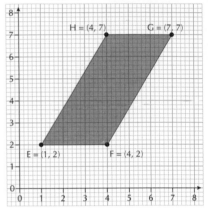

c Area 24

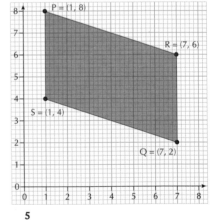

5

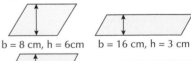

b = 8 cm, h = 6cm   b = 16 cm, h = 3 cm

b = 12 cm, h = 4 cm   b =24 cm, h = 2 cm

6 4.5 cm   **7** 6.25 cm
8 32 m   **9** 6 cm
10 5 cm
11 **a** 7 **b** 6 **c** 10 **d** 6.5 **e** 6 **f** 4 **g** 5
  h 5.5 **i** 7 **j** 4 **k** 3

12 204 cm²   **13** x = 14 cm

3.5 Area of a trapezium

1   **a** 72 cm² **b** 25 cm² **c** 22 m²
    d 325 mm² **e** 64 cm² **f** 26 cm²
    g 69 m² **h** 23.8 cm²

2   **a** A = 52.5 mm² P = 33 mm
    b A = 18 cm² P = 22 cm
    c A = 17.75 cm² P = 19.6 cm

3

Trapezium	Length, a	Length, b
a	4 cm	6 cm
b	10 cm	12 cm
c	9 m	3 m

Trapezium	Height, h	Area, A
a	3 cm	15 cm²
b	6 cm	66 cm²
c	5 m	30 m²

4   30 m²   **5**   A = 56

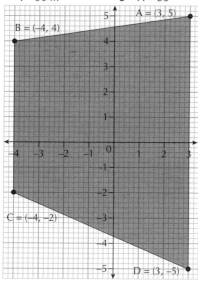

A = (3, 5)
B = (–4, 4)
C = (–4, –2)
D = (3, –5)

6   h = 8 cm

7   a = 6 cm b = 14 cm

8   2.1 m

9   **i.** h = 2, a = 1, b = 8
    ii. h = 2, a = 4, b = 5
    iii. h = 3, a = 2, b = 4

10   44 cm²   **11**   75 cm²

12   72 cm²

13   **i.** h = 2, a = 1, b = 5
    ii. h = 2, a = 2, b = 4
    iii. h = 3, a = 1, b = 3

4.1 The number line

1   **a** T **b** T **c** F **d** T

2   **a** < **b** > **c** < **d** <

3   **a** –8 °C **b** –9 °C **c** –1 °C **d** –3 °C

4   **a** –1, –0.5, 0 **b** –5, –1, 2
    c –1.9, –1.6, –1.1

5   **a** –1 **b** –5.6 **c** –1.2

6   **a** –2, –3, –5 **b** 1, 0, –1
    c –1.3, –3, –3.3

7   **a** –1.2, –0.8, 0.5 **b** –5.9, –1.5, 2.3
    c –3.8, –3.3, –2.7

8   **a** –108 **b** –29 **c** –99 **d** –61

9   **a** –5.3 **b** –7.5 **c** –4.1

10   Because the temperature outside could be 30 °C, and when it drops down by 5 °C it remains warm at 25 °C.

11   Up
    40 metres below sea level = –40
    20 metres below sea level = –20
    To move from –40 to –20 on the number line you move to the right direction, which means up with respect to the submarine.

12   On the number line, –5.9 is further to the left than –5.1, and the further left number is on the line, the lower the number is.

13   Because he paid for the bills more than he has in his bank balance.

14   Third floor   **15**   £–8

4.2 Arithmetic with negative numbers

1   **a** –4 **b** 6 **c** 2 **d** –4 **e** –5

2   **a** –2 **b** –7 **c** –1 **d** –6 **e** –5

3   **a** –2 **b** –5 **c** –2 **d** –2

4   **a** –6 **b** –4 **c** –7 **d** –5

5   **a** 0 **b** –2 **c** –7 **d** –2

6

–3	2	–7
–12	0	4
7	–10	–5

No, because 0 is not the same as –8, the diagonals and rows do not add up to the same number, so it is not a magic square.

7   £–11, since 124 – 135 = –11

8   **a** –10 **b** 15 **c** 7 questions
    d 5 questions
    e To score 49, you must answer 10 questions correctly and answer one question incorrectly. The total number of questions becomes 11, while there are only 10 questions in the competition.

9

–27	–6	0	15
6	9	–21	–12
–9	–24	18	–3
12	3	–15	–18

10   Floor –1

4.3 Subtraction with negative numbers

1   **a** 7 **b** 5 **c** 11 **d** 14

2   **a** 1 **b** –2 **c** –1 **d** –4

3   **a** 7 **b** 5 **c** 11 **d** 13

4   **a** 5 **b** 1 **c** 3 **d** 1

5   **a** 10 **b** 0 **c** 7

6   **a** 7 **b** 5 **c** –1

7   **a** 15 **b** 9 **c** 17

8   **a** 5 **b** –5 **c** –1

9   **a** 2 **b** 6 **c** 10

10   **a** 3 **b** 6 **c** –1

11   17 + 12 = 29 m

12   **a** 25 **b** Alice **c** 5 rolls

13   Biggest 38, Smallest –38

14   (8 – 10) – (4 – 9) = –2 – –5 = –2 + 5 = 3

4.4 Multiplication with negative numbers

1   **a** –2 **b** –12 **c** –42 **d** –7

2   **a** –6 **b** –15 **c** –30 **d** –28

3   **a** –3 **b** –21 **c** 56 **d** –24

4   **a** –5 **b** –6 **c** –4 **d** –10.5

5   **a** –7.5 **b** –12.4 **c** –18.6 **d** –17.2

6   **a** –5.6 **b** –12.8 **c** 63.9 **d** –21.5

7   **a** –12.9 **b** –9.4 **c** 45.5 **d** –16.2

8   **a** 6 **b** 6 **c** –30 **d** 16

9   **a** 16, –32, 64 **b** –81, –243, –729
    c –625, 3125, –15625
    d 256, –1024, 4096

10   (7 – 11) × (3 – 8) = –4 × –5 = 20

11   –4 and 6; 4 and –6; 3 and –8;
    –3 and 8; 2 and –12; 1 and –24

12   **a i**

    ii

    b No – if the negative is at an end of the bottom row you get a negative times a positive which is negative.

4.5 Division with negative numbers

1   **a** –3 **b** –2 **c** –4 **d** –3

2   **a** –4 **b** –7 **c** –12 **d** –4

3   **a** 3 **b** 3 **c** 2 **d** 5

4   **a** –5 **b** –5 **c** 3 **d** –7

5   **a** –1.3 **b** –1.1 **c** –4.2 **d** –0.5

6   **a** –1.5 **b** –3.2 **c** –2.3 **d** –1.2

7   **a** 0.5 **b** 3.1 **c** 0.9 **d** 0.3

8   **a** –2.1 **b** –1.6 **c** 0.9 **d** –1.3

9   **a** 3 **b** 7 **c** 5 **d** –4

10   **a** 4 **b** 15 **c** –12 **d** 21

11   (8 – 52) ÷ (–7 + 3) = –44 ÷ –4 = 11

Answers

12 No, because the product of two negative numbers is always positive.

13 $10 \div -5 = -2$, $16 \div -4 = -4$, $24 \div -3 = -8$, $72 \div -9 = -8$

14 **a** Yes, because division of positive number by negative one gives a negative number.
b $-1443 \div 13 = -111$

15 $-3 \div 1 = -3$, $3 \div -1 = -3$
$-6 \div 2 = -3$, $6 \div -2 = -3$
$-9 \div 3 = -3$, $9 \div -3 = -3$

16 $-5 \div 1 = -5$ $20 \div -4 = -5$
$-15 \div 3 = -5$ $10 \div -2 = -5$
$-30 \div 6 = -5$

17 30

5.1 Mode, median and range

1 **a** Red **b** Sun **c** E **d** ♥
2 **a** 5 **b** 34 **c** 13 **d** 101 **e** 5
3 **a** 19 **b** 6 **c** 27 **d** 14
4 **a** mode = £ 1.80, range = £ 2.4
b mode = no mode, range = 14
c mode = 132, range= 19
d mode = 32, range = 6
5 **a** £ 2.20 **b** 1.6
6 **a** no mode **b** 13
7 **a** 33 **b** 148 **c** 148.5
8 3, 2
9 The median = 28, the range = 34
10 mode = 11 median = 15
range = 12
11 **a** 5 years and 10 years or 1 year and 6 years **b** The range = 0
12 The missing value = 7
13 Age of the second child = 7 years
Age of the third child = 11 years
14 6, 13
15 **a** 2, 7, 8, 10, 12, 12, 12
b 5, 6, 7, 8, 12, 12, 14
c 4, 5, 6,10, 12, 12, 12
16 **a** Kathy **b i** Kathy **ii** Tim
17 1, 7, 8
18 **a** Ford. **b** Toyota, Fiat, Honda.
c Ford, in (b) the car that arrived is Honda.
d Because this data is non-numerical.
e Clothing brands, animals, students names.

5.2 The mean

1 **a** $\dfrac{18}{3} = 6$ **b** $\dfrac{20}{4} = 5$

2 **a** $\dfrac{21}{3} = 7$ **b** $\dfrac{12}{4} = 3$

3 **a** $\dfrac{1+5+6}{3} = \dfrac{12}{3} = 4$

b $\dfrac{1+4+7+8}{4} = \dfrac{20}{4} = 5$

4 **a** $\dfrac{2+4=11+13+15}{5} = \dfrac{45}{5} = 9$

b $\dfrac{1+3+8+8}{4} = \dfrac{20}{4} = 5$

5 **a** 7 **b** 31 **c** 15 **d** 2.8
6 **a** 4.7 **b** 14.7 **c** 74.8 **d** 9.4
7 8 Kg   **8** 26 s
9 92.7p   **10** 31 seconds
11 **a** 141 **b** 140 **c** 132
d The best average to use would be the mean as it takes into account all the heights.
12 **a** 1 **b** 2 **c** 2
d The best average to use would be the mean as it takes into account all the numbers of children.
13 **a** 3 **b** 4 **c** 4
d The best average to use would be the mean as it takes into account all the sizes.
14 **a** 17.5 **b** 17 **c** 20
d The mode, because it is not representative to the data
e Because the mean is higher than the median.
15 **a** 6.5 **b** 9 eggs.
16 **a** 2.52 **b** 0.42 kg
17 **a** Increase **b** 195
18 Higher, because the mean price for this year is 61.

5.3 Statistical diagrams

1 **a** 10 **b** By bus. **c** 30
2 **a** £ 60 **b** £ 70 **c** year 7 **d** £ 380
3 **a** Ceri **b** 16 **c** 18 **d** 3 **e** 69
4 **a** 15 **b** 17 **c** 9
d It shows trends and patterns in the data, generally over time.
5 **a** Blue **b** 10%
c Green and purple.
d Because we can use it to compare between data and find the percentage of the data easily.
6 **a** ITV1 **b** Channel 5
7 **a** 3 students **b** 7 **c** 48 students

d

Size	Number of students
2	◿
3	◖
4	⊕ ◿
5	⊕ ◖
6	⊕ ⊕ ◿
7	⊕ ⊕ ⊕
8	⊕ ◖
9	◖
10	
11	◿

⊕ = 4 students

8 **a** Cinema **b** 430 **c** 290

9 **a**

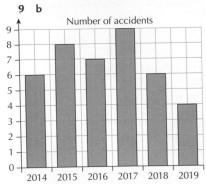

Year	Number of accidents
2014	⊕ ◠
2015	⊕ ⊕
2016	⊕ ◔
2017	⊕ ⊕ ◹
2018	⊕ ◠
2019	⊕

⊕ = 4 accidents

9 b

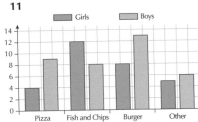

Number of accidents

c A bar chart, to show the change in the number of accidents over the last six years.

10 **a** 160 **b** 60

11

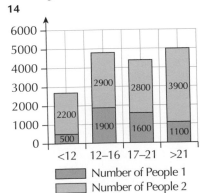

12 **a** 32 students **b** 15 students
c Smallest = 61, greatest = 99
13 **a** Wednesday. **b** 17
c * Add two columns to the chart on Friday, a green one for boys with frequency (1) and a pink one for girls with frequency (9)*

14

Number of People 1
Number of People 2

15 **a** 1250 m **b** 2500 m
c 2500 − 750 = 1750 m

6.1 Equivalent fractions

1 **a** 24 **b** 9 **c** 48 **d** 24

2 **a** 7 **b** 5 **c** 6 **d** 4

3 **a** $\frac{2}{3}$ **b** $\frac{2}{3}$ **c** $\frac{3}{5}$ **d** $\frac{1}{5}$ **e** 6 **f** $\frac{8}{15}$

4 **a** $\frac{2}{6}, \frac{3}{9}, \frac{4}{12}$ **b** $\frac{2}{8}, \frac{3}{12}, \frac{4}{16}$

   c $\frac{6}{8}, \frac{9}{12}, \frac{12}{16}$ **d** $\frac{50}{70}, \frac{75}{105}, \frac{100}{140}$

   e $\frac{30}{100}, \frac{3}{10}, \frac{150}{500}$ **f** $\frac{15}{20}, \frac{3}{4}, \frac{6}{8}$

5 **a** $\frac{5}{15}, \frac{14}{42}, \frac{19}{57}$ **b** $\frac{8}{12}, \frac{30}{45}, \frac{36}{54}$

   c $\frac{45}{60}, \frac{18}{24}, \frac{75}{100}$ **d** $\frac{14}{20}, \frac{49}{70}, \frac{42}{60}$

6 **a** 18 **b** 9 **c** 50 **d** 10

7 **a** $\frac{3}{4} = \frac{6}{8}, \frac{2}{3} = \frac{6}{9}$

   b $\frac{3}{4} = \frac{9}{12}, \frac{2}{3} = \frac{8}{12}$

8 **a i** $\frac{14}{18}, \frac{21}{27}, \frac{28}{36}, \frac{35}{34}, \frac{42}{54}$

   ii $\frac{16}{28}, \frac{24}{27}, \frac{32}{36}, \frac{40}{45}, \frac{48}{54}$

   iii $\frac{14}{16}, \frac{21}{24}, \frac{28}{32}, \frac{35}{40}, \frac{42}{48}$

   iv $\frac{16}{14}, \frac{24}{21}, \frac{32}{28}, \frac{40}{35}, \frac{48}{42}$

   b $\frac{18}{28}, \frac{27}{42}$

9 **a** $\frac{11}{20}$ **b** $\frac{11}{20}$ **c** $\frac{3}{40}$

10 8

11 **a** $\frac{1}{2}$ **b** $\frac{5}{8}$ **c** $\frac{1}{2}$ **d** $\frac{1}{2}$

12 **a i** $\frac{17}{20}$ **ii** $\frac{1}{4}$ **iii** $\frac{2}{5}$

   b i $\frac{9}{20}$ **ii** $\frac{9}{200}$ **iii** $\frac{39}{40}$

   c i $\frac{7}{12}$ **ii** $\frac{3}{4}$ **iii** $\frac{4}{5}$

13 $\frac{4}{7}$

6.2 Adding and subtracting fractions

1 **a** $\frac{1}{2}$ **b** $\frac{3}{5}$ **c** 1 **d** $\frac{1}{2}$

2 **a** $\frac{4}{5}$ **b** $\frac{3}{4}$ **c** $\frac{5}{6}$ **d** $\frac{11}{16}$

3 **a** $\frac{3}{8}$ **b** $\frac{1}{2}$ **c** $\frac{3}{8}$ **d** $\frac{5}{8}$

4 **a** $\frac{11}{24}$ **b** $\frac{26}{45}$ **c** $\frac{17}{56}$ **d** $\frac{1}{10}$

5 **a** $\frac{5}{6}$ **b** $\frac{1}{6}$ **c** $\frac{7}{12}$ **d** $\frac{11}{12}$

6 **a** $\frac{5}{12}$ **b** $\frac{7}{8}$ **c** $\frac{47}{40}$ **d** $\frac{19}{24}$

7 $\frac{1}{4}$ 8 $\frac{3}{20}$ 9 $\frac{1}{15}$ 10 $\frac{5}{72}$ 11 $\frac{11}{24}$

12 $\frac{11}{24}$

13 **a** $\frac{11}{12}$ of a Litre **b** $\frac{5}{24}$ of a Litre

14 $\frac{21}{40}$ Km

6.3 Mixed numbers and improper fractions

1 **a** $\frac{5}{2}$ **b** $\frac{7}{2}$ **c** $\frac{11}{2}$ **d** $\frac{23}{2}$ **e** $\frac{37}{9}$ **f** $\frac{40}{9}$

2 **a** $\frac{23}{12}$ **b** $\frac{41}{12}$ **c** $\frac{26}{11}$ **d** $\frac{37}{11}$ **e** $\frac{34}{9}$

   f $\frac{21}{9}$

3 **a** $4\frac{5}{6}$ **b** $5\frac{5}{6}$ **c** $5\frac{5}{8}$ **d** $3\frac{3}{8}$ **e** $2\frac{1}{5}$

   f $8\frac{1}{5}$

4 **a** $6\frac{1}{4}$ **b** $6\frac{3}{4}$ **c** 9 **d** $8\frac{1}{3}$ **e** $7\frac{2}{5}$

   f $3\frac{7}{10}$

5 **a** $\frac{25}{21}$ **b** $\frac{69}{56}$ **c** $\frac{26}{24}$ **d** $\frac{73}{24}$ **e** $\frac{9}{8}$

   f $\frac{13}{12}$

6 **a** $\frac{116}{40}$ **b** $\frac{140}{40}$ **c** $\frac{18}{8}$ **d** $\frac{15}{8}$ **e** $\frac{9}{8}$

   f $\frac{29}{24}$

7 $\frac{3}{8}, \frac{7}{12}, 1\frac{1}{3}, 2\frac{1}{4}, \frac{8}{3}, \frac{27}{8}, 4$

8 **i a** decreasing **b** increasing
   ii When the numerator is fixed, the higher the value of the denominator, the greater the total number of parts of the whole one, which means the value decreases, and vice versa.

9 No, to compare two fractions they should have the same denominator. $\frac{24}{5} = \frac{48}{10}$
   So, $\frac{27}{10}$ is smaller than $\frac{24}{5} = \frac{48}{10}$

10 Yes he is correct, because denominators are the same.

11 Yes $\frac{3}{4} + \frac{2}{3} + \frac{5}{8} = \frac{18}{24} + \frac{16}{24} + \frac{15}{24} = \frac{49}{24} = 2\frac{1}{24}$

12 $\frac{7}{8}$, because $1 - \frac{7}{8} = \frac{8}{8} - \frac{7}{8} = \frac{1}{8} = \frac{7}{56}$
    $\frac{8}{7} - 1 = \frac{8}{7} - \frac{7}{7} = \frac{1}{7} = \frac{8}{56}$
    So, $\frac{1}{8}$ is smaller than $\frac{1}{7}$.
    Then, $\frac{7}{8}$ is nearer to 1

13 $\frac{26}{11}$

14 **a i** decreasing **ii** comparing the first and last terms of the sequence gives that $\frac{17}{10} < \frac{21}{14}$, so the sequence is decreasing.
    b i increasing **ii** comparing the first and last terms of the sequence gives that $\frac{17}{10} < \frac{13}{6}$, so the sequence is increasing.

6.4 Adding and subtracting mixed numbers

1 **a** 4 **b** 3 **c** $6\frac{2}{5}$ **d** $8\frac{1}{3}$ **e** 2

2 **a** $5\frac{1}{6}$ **b** $6\frac{1}{6}$ **c** $5\frac{1}{5}$ **d** $4\frac{3}{10}$ **e** $3\frac{1}{4}$

3 **a** $2\frac{3}{7}$ **b** $2\frac{2}{3}$ **c** $3\frac{3}{5}$ **d** $3\frac{2}{3}$

4 **a** $1\frac{3}{8}$ **b** $1\frac{1}{12}$ **c** $2\frac{1}{8}$ **d** $1\frac{1}{8}$ **e** $1\frac{1}{8}$

5 **a** $6\frac{3}{4}$ **b** $8\frac{3}{8}$ **c** $5\frac{1}{2}$ **d** $4\frac{1}{2}$ **e** $2\frac{5}{6}$

6 $4\frac{1}{4}$ hours   7 $5\frac{1}{2}$ hours

8 **a** $5\frac{1}{2}$ **b** $5\frac{1}{2}$ 9 $5\frac{1}{2}$

10 $5\frac{1}{2}$ 11 $5\frac{1}{2}$

12 $5\frac{1}{2}$

7.1 Order of operations

1 **a** $2 + 3 \times 6 = 2 + 18 = 20$
   b $12 - 6 \div 3 = 12 - 2 = 10$
   c $5 \times 5 + 2 = 25 + 2 = 27$
   d $12 \div 4 - 2 = 3 - 2 = 1$
   e $(2 + 3) \times 6 = 5 \times 6 = 30$
   f $(12 - 3) \div 3 = 9 \div 3 = 3$
   g $5 \times (5 + 2) = 5 \times 7 = 35$
   h $12 \div (4 - 2) = 12 \div 2 = 6$

2 **a** $2 \times 3 + 4 = 6 + 4 = 10$
   b $2 \times (3 + 4) = 2 \times 7 = 14$
   c $2 + 3 \times 4 = 2 + 12 = 14$
   d $(2 + 3) \times 4 = 5 \times 4 = 20$
   e $4 \times 4 - 4 = 16 - 4 = 12$
   f $5 + 3^2 + 6 = 5 + 9 + 6 = 14 + 6 = 20$
   g $5 \times (3^2 + 6) = 5 \times (9 + 6) = 5 \times 15 = 75$
   h $3^2 - (5 - 2) = 9 - 3 = 6$

3 **a** 1 **b** 2 **c** 3 **d** 4 **e** 5 **f** 6 **g** 7 **h** 8

4 **a** 31 **b** 13 **c** 33 **d** 3

5 **a** 34 **b** 5 **c** 28 **d** 33

6 **a** 27 **b** 10 **c** 64 **d** 40

7 **a** 2 **b** 3 **c** 20 **d** – 2

8 **a** zero **b** 12 **c** 35 **d** 38

9 **a** $2 \times (5 + 4) = 18$
   b $(2 + 6) \times 3 = 24$
   c $(2 + 3) \times (1 + 6) = 35$
   d $5 + 2^2 \times 1 = 9$
   e $(3 + 2)^2 = 25$
   f $3 \times (4 + 3) + 7 = 28$
   g $(9 - 5) \times 2 = 8$
   h $(4 + 4 + 4) \div 2 = 6$
   i $(1 + 4)^2 - (9 - 2) = 18$

10 The wrong calculation is $2 \times 3^2 = 36$
    The correct calculation is $(2 \times 3)^2 = 36$

11 $6 \times 1.99 + 2.50 = 14.44$

12 **a** Yes, he has **b** $2.70 + 4.20 + 2.70$ **c** 9.6

13 $3 \times 10 + 4 \times 20 = 30 + 80 = 110$

14 $(3 \times 5) + (2 \times 4) = 23$

Answers

15 **a** $4 \div 2 \times 3 = 6$ **b** $(2 + 3) \times 4 = 20$

16 $(4 \div 4 + 4) \times 4 = 20$

7.2 Expressions and substitution

1 **a** $4 + m$ **b** $8t$ **c** $9 - y$ **d** $m \times m = m^2$
   e $n \div 5$ **f** $7 - t$ **g** $3n + 5$ **h** $6t$
   i $5m - 5$ **j** $x \times x = x^2$

2 **a i.** 8 **ii.** 20 **iii.** 44 **b i.** 3 **ii.** 7 **iii.** 4
   c i. 9 **ii.** 36 **iii.** 49

3 **a i.** 21 **ii.** 15 **iii.** 27 **b i.** 2 **ii.** 1
   iii. 4 **c i.** 12 **ii.** 18 **iii.** 30

4 **a i.** 10 **ii.** 11 **iii.** 1 **b i.** 3 **ii.** 9
   iii. –3 **c i.** 6 **ii.** 3 **iii.** 13

5 **a i.** 7 **ii.** 13 **iii.** 3 **b i.** 10 **ii.** 50
   iii. zero **c i.** 18 **ii.** 26 **iii.** 40

6 **a i.** 3 **ii.** 8 **iii.** zero **b i.** 26 **ii.** 37
   iii. 101 **c i.** 69 **ii.** 86 **iii.** 5

7 **a i.** 2 **ii.** 5 **iii.** 10 **b i.** 21 **ii.** 17
   iii. 85 **c i.** 7 **ii.** 63 **iii.** zero

8 **a i.** 3 **ii.** 12 **iii.** –3 **b i.** 12 **ii.** 28
   iii. 8 **c i.** zero **ii.** 25 **iii.** 60

9 11   **10** 50

11 **a** $P = a + 12 + 10 = a + 22$ **b** 30.5

12 **a** *Perimetre* $= 6 + p + q$ **b** 29 cm

13 **a** *Perimetre* $= d + 14 + d + 14$
   Perimetre $= d + d + 14 + 14$
   Perimetre $= 2d + 2 \times 14$
   Perimetre $= 2(d + 14)$ **b** 40 cm

14 **a** *Perimetre* $= 5s$ **b** 125 cm

15 **a** $x + 2$ **b** $x - 1.5$
   c *Perimetre* $= x + 2 + x - 1.5 + x$
   d 60.5 cm

16 **a** $3 \times 6 = 18$ **b** $2a$ **c** $2a + 18$
   d The length of the missing side in
   the horizontal rectangle equals 3
   cm. The length of the missing side
   in the vertical rectangle equals a
   cm. The length of the missing side
   in the shape equals $(a + 3)$ cm.
   e $18 + 2a$

17 **a** $A + 2B$ cm² **b** $A + B$ **c** $4A$
   d $2A + B$ **e** $2A + 2B$ **f** $4B$ **g** $\frac{A}{2}$
   h $A - B$ **i** $2A - 2B$

18 **a** 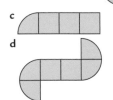 **b**
   c
   d

7.3 Simplifying expressions

1 **a** $2a$ **b** $4b$ **c** $8c$ **d** $5d$ **e** $6c$ **f** $10d$
   g $2p$ **h** $11x$

2 **a** $2x$ **b** $7y$ **c** z **d** $4q$

3 **a** $5t$ **b** $4m$ **c** $4q$ **d** $2g$

4 **a** $5x + 8y$ **b** $2w + 4t$ **c** $7m + 6n$
   d $x + 6y$ **e** $6x + 5$ **f** $5p + 5$

5 **a** $y + 5x - 3$ **b** $5d + 4c + 7$
   c $f + 3d + 1$

6 **a** $7x - 3y$ **b** $5x - 4y$ **c** $-a + 6b$
   d $-2x - 13y$ **e** $-3 + 12w$ **f** $-x - 4$

7 **a** $-7x - 5$ **b** $16x - 9y$ **c** $t - 14$

8 **a** $2x^2$ **b** $4x^2$ **c** $4x^2$ **d** $3x^2$ **e** $6x^2$ **f** $6x^2$
   g $4x^2$ **h** $9x^2$

9 **a** $3x^2 + 3$ **b** $7x^2 + 4$ **c** $2x^2 - 4$
   d $7x^2 - 7$ **e** $4x^2 + 8$ **f** $5x^2 - 2$
   g $3x^2 - 3$ **h** $9x^2 + 16$

10 **a** $4x^2 + y$ **b** $6x^2 + y$ **c** $x^2 - 2y$
   d $8x^2 + 7y^2$ **e** $4x^2 + 5x + 7$
   f $7x^2 - 3x + 5$ **g** $4y^2 - 12y + 3$
   h $6x^2 + 7x + 11$

11 **a** $2x + 8y + 5x + 6y = 7x + 14y$
   b $5w - 5t - 4w - 4t = w - 9t$
   c $8p + 5x - 7p + 2x = p + 7x$
   d $8p + 9q - 3p - 10q = 5p - q$

12 **a** $3x + 5y + 4x + 2y = 7x + 7y$
   b $8w - 4z - 7w - 5z = w - 9z$
   c $12y + 3x - 11y + 2x = y + 5x$
   d $9p + 5q - 6p - 6q = 3p - q$

13 **a** $4x + 8$ **b** $6x + 18$ **c** $3y + 12$

14 Bethan, because the variables are
   different, so we cannot make them
   any simpler.

15 **a** 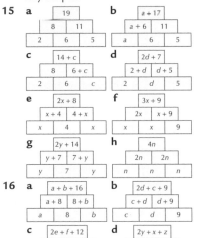 **b**
   c **d**
   e **f**
   g **h**

16 **a** **b**
   c  **d**

17 $2(4x + 6)$

18 **a** **b**

7.4 Using formulae

1 **a** 45 **b** 29 **c** 69

2 **a** 45 **b** 95 **c** 195

3 **a** 18 **b** 28 **c** 48

4 **a** 50 **b** 40 **c** 40

5 **a** 250 **b** 100 **c** 210 **d** 235

6 **a** 28 **b** 52   **7** **a** 3 **b** 10

8 **a** 17 **b** 35

9 **a** 68°F **b** 14°F **10** **a** 16 **b** 33

11 **a** $180 - \dfrac{360}{5} = 180 - 72 = 108°$

   b 120° **c** 135° **d** 144°

12 **a** $F = 12 - 8 + 2 = 6$
   b $F = 8 - 5 + 2 = 5$

13 **a** 37.5 **b** 41.7

14 **a** $R = 3$ **b** $R = 2$

15 **a** $A = L \times w$ $p = 2L + 2w$
   b When $L = 6$, $w = 5$ or when
   $L = 5, w = 6$

16 **a** $p = 2y + x$ **b** $y = 9$

17 **a** £32 **b** $A = 22$, $B = 32$

18 **a** $p = 12x + 12$ **b** $3x + 3$
   c $A = (3x + 3)^2$ **d** $A = 81$ cm²

7.5 Writing formulae

1 $p = n(n + 1) = n^2 + n$

2 $p = a \times b$

3 $s = 2n + 1$ **4** $y = 52w$

5 $d = b - e$ **6** $M = x + y$

7 $A = 2n + 3$ **8** $A = 3n - 5$

9 $C = 0.25p$

10 **a** $r = p + 2q + 6$

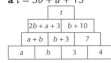

   b $r = 41$ **c** $r = 12$

11 **a** $d = m - b$ **b** $s = m + b$ **c** $s = 50$
   d $b + 6$ **e** $m + 6$ **f** $t = b + m + 12$

12 **a** $t = 2d + c + 13$ **b** $t = 53$ **c** $t = 48$

13 **a** $t = 3b + a + 13$

   b $t = 27$ **c** $t = 39$ **d** $t = 33$

14 **a** $c = 6h$ **b** $a = 1.40b$ **c** $w = 7d$
   d $m = 5q$ **e** $t = 20x + 30y$
   f $m = 30 + 20x$

15 **a** 3 cm, 8 cm **b** $p = 62$
   c $p = 2(2n + 5)$ **d** $p = 130$

16 **a** 15.5 cm, 17.5 cm, 13.5 cm,
   12.5 cm
   b $p = 4n + 5$ **c** $p = 205$

17 **a** $r = 4x + y$ **b** $r = 37$ **c** $r = 37$
   d $r = 58$

8.1 Calculating angles

1 **a** $a = 28°$ **b** $b = 47°$ **c** $c = 31°$
   d 45°

2 **a** 130° **b** 160° **c** 85° **d** 221°

3 **a** 50° **b** 130° **c** 80° **d** 41° **e** 50°
   f 25° **g** $d = 376°$ **e** $= 143°$ **h** 51°
   i 129° **j** 116° **k** $a = 20°$ **l** $c = 45°$
   m $a = 20°$ **n** $b = 25°$ **o** $c = 45°$

4 **a** 5° **b** 145° **c** $c = 36°$
   d $d = 100°$, $x = 65°$

5 **a** $x = 30°, y = 60°$
   b $x = 26°, y = 13°$

6 **a** $x = 20°$ **b** $x = 15°$

7 $360 = 100 + 20 + 2p + p$
$360 = 120 + 3p$
$240 = 3p$
$p = 80°$
On line xy this gives $80 + 100 = 180°$ so yes, xy is a straight line.

8 $85° + 60° + 35° = 180°$

9 **a** $m = 15°$ **b** $m = 30°$, $n = 150°$
c $m = 45°$, $n = 135°$ **d** $m = 60°$,
$n = 120°$ **e** Time table of 15

10 **a** $a + a + a + a + a + a = 360$
$6a = 360$
$a = 60°$
b $b + 80 + b = 180$
$2b + 80 = 180$
$2b = 100$
$b = 50°$

11 e

12 $85°, 95°$

13 $a = 24°$

14 Continuing the series:
$180° = 10x + 9x + 8x + 7x + 6x + 5x$
This gives:
$180° = 45x$ and then $x = 4$
So, Alice is wrong, because 4 is an integer.

8.2 Angles in a triangle

1 **a** $85°$ **b** $100°$ **c** $55°$ **d** $95°$

2 **a** $55°$ **b** $42°$ **c** $38°$ **d** $42°$ **e** $95°$
f $30°$

3 **a** $65°$ **b** $35°$ **c** $g = 48°$, $h = 48°$

4 **a** $a = 35°$, $b = 145°$
b $c = 62°$, $d = 118°$
c $e = 42°$, $f = 138°$

5 **a** $a = 45°$ **b** $b = 30°$ **c** $c = 72°$
d $d = 21°$

6 **a** $a = 35°$ **b** $b = 25°$ **c** $c = 112°$
d $d = 135°$

7 **a** $x = 50°$, $y = 25°$
b $x = 109°$, $y = 31°$

8 **a** $a = 35°$ **b** $b = 49°$ **c** $c = 10°$

9 **a** $x = 136°$ **b** $x = 109°$

10 **a** $x = 290°$, $y = 140°$
b $x = 23°$, $y = 40°$

11 **a** $a = 40°$, $y = 140°$

12

```
      /\                    /\
     /42°\                 /96°\
    /    \               /     \
   /      \             /       \
  /_____\           /_____\
 /69°    69°\         /42°      42°\
```

13 $a = 50°$, $b = 60°$, $c = 70°$

14 $a = 80°$, $b = 55°$, $c = 71°$, $d = 61°$

15 **a** $c = 110°$, $d = 30°$
b $f = 80°$, $e = 50°$, $g = 30°$

16 **a** $b = 305°$ **b** $c = 140°$ **c** $d = 23.5°$

17 **a** $x = 15°$, $y = 135°$ **b** $50°$, $80°$

18 **a** $y = 59°$, $z = 121°$ ($x - 10 = 62$,
$8x + 10 = 298$)
b $x = 24$ so, the three angles are
$50°$, $50°$ and $80°$.

Yes, the triangle is isosceles because 2 angles are the same size.

8.3 Angles in a quadrilateral

1 **a** $a = 109°$ **b** $b = 108°$ **c** $f = 145°$

2 **a** $71°$ **b** $135°$ **c** $105°$ **d** $61°$

3 **a** $125°$ **b** $73°$ **c** $122°$

4 **a** $92°$ **b** $63°$

5 a, c

6 **a** $86°$ **b** $71°$ **c** $116°$

7 **a** $131°$ **b** $113°$ **c** $103°$

8 **a** $a = 97°$ **b** $e = 67°$, $2b = 134°$
c $c = 18°$, $d = 90°$

9 **a** $a = 80°$, $a + 22 = 80 + 22 = 102°$
b $b = 100°$, $b + 4 = 104°$,
$b + 23 = 123°$
c $c = 29.5°$, $3c + 6 = 94.5°$
$4c + 10 = 128°$
$2c - 4 = 55°$

10 $x = 18°$ so, angles are $54°, 72°$,
$108°$ and $126°$.
$y = 108°$ so yes, the two shapes are similar.
The have angles that match.

11 Interior angles are only those angles contained by the outer edge of the shape, the perimetre in this case 4 angles of $90°$. Paula has included angles that are part of a different, interior and pattern.

12 **a** $a = 58°$, $b = 122°$
b $c + 108 = 180$
$c = 180 - 108$
$c = 72°$
$110 + 80 + 72 + d = 360$
$262 + d = 360$
$d = 360 - 262$
$b = 98°$

13 **a**

```
    /\
   /  \
  /    \
 <      >
  \    /
   \  /
    \/
```

b The angles p and q are equal.
c $p = q = 115°$

14 No, three right angles = $3 \times 90 = 270°$, remains $360 - 270 = 90°$ which is also a right angle.

15 **a** rhombus **b** square
c rectangle, kite or parallelogram
d parallelogram or kite

16 **a** 3 acute angles, 1 obtuse and 1 reflex.
b Shade the right part of the shape along the line of the window.
c $40°$.

17 **a** The total angle of a quadrilateral = $360°$ but the measurements add up to $378°$.
b $97°$ is incorrect, it should be $79°$.

---

18 **a** $36°, 72°$, $108°$, $144°$ **b** The angle at the centre has to be a factor of $360°$. Both $36°$ and $72°$ are factors of $360°$. Using $36°$ at the centre requires 10 diamonds, but $72°$ requires only 5 diamonds.

8.4 Angles within parallel lines

1 **a** allied. **b** corresponding.
c corresponding. **d** alternate.
e allied. **f** alternate.

2 **a** e **b** f **c** g **d** h **e** d **f** c

3 **a** $70°$ **b** $125°$ **c** $160°$ **d** $121°$
e $34°$

4 **a** $x = y = 91°$ **b** $y = 63°$
c $x = y = 108°$

5 **a** $a = 108°$ **b** $b = 64°$ **c** $99°$

6 **a** $a = 50°$ **b** $b = 62°$ **c** $f = 108°$

7 **a** $x = y = 105°$ **b** $b = 81°$
c $x = y = 140°$

8 **a** $x = a = 67°$, corresponding
b $x = 117°$, $b = 63°$, allied
c $x = c = 120°$, alternate

9 **a** $a = 60°$ **b** $b = 36°$

10 **a** $a = 122°$, $b = 58°$, $c = 58°$
b $d = 60°$, $e = 60°$, $f = 120°$

11 i, because all the shapes have equal angles except this one.

12 $F_1 = 20°$ (alternate angle)
$F_2 = 55°$ (alternate angle)
$F = F_1 + F_2 = 20 + 55 = 75°$

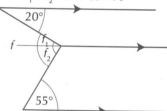

13 **a** $a = 35°$ **b** $b = 68°$

14 **a** $x = 12°$, $62°$ **b** $x = 20°$, $105°$

15 **a** $a = 52°$ **b** $b = 90°$

16 $7x - 10$ is incorrect; it should be $7x - 17$.
$x = 19$, so the four angles sizes:
$116°$, $64°$, $116°$ and $64°$

17 $\angle ABD = \angle BDC = x$ (alternate angles are equal)
$\angle DAC = \angle ACB = y$ (alternate angles are equal)
$\angle DAB = x + y$ and $\angle BCD = x + y$
Therefore the opposite angles are equal

8.5 Constructions

1

```
           \  /
            \/C
            /\
           /  \
         __|___
    A ___|___|___ B
       4 cm X 4 cm
           \  /
            \/
            /\D
           /  \
```

Answers

2 The same as Q1

3

4

5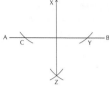

6 The same as Q4

7 Student's answer

8

9

10 Draw a base of 5 cm. Set compasses to 5 cm and draw an arc above the line from each end. Join between the point where the two arcs intersect with the ends of the base.

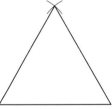

11 Construct an angle of 60° (using method from Q10) and then bisect it to create 30°

12 a The same as Q1. **b** Make the length of the perpendicular bisector 5 cm, then join the 4 ends of xy and the perpendicular bisector to get a rhombus.

13 a

---

b 9 m **c** 72°

14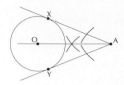

The bisector will pass through the centre of the circle and form a diametre.

15 Create ∠ACB, and then bisect it. The bisecting line is the path for the ball.

16 The shortest distance between a point and a line is the perpendicular path from the point towards that line, so we have to construct a perpendicular from the point C to XY.

17 Form two chords connecting the three points you have chosen, construct a perpendicular bisect at each chord and where the bisectors meet is the centre of the circle.

18 Construct the triangle according to the scale. The zip wire meets the tree 50m (2.5 × 20) up the tree. The zip wire will be 210m (19.5 × 20) long.

9.1 Rounding numbers

1 a 20 **b** 20 **c** 40 **d** 80 **e** 160 **f** 270 **g** 150

2 a 600 **b** 400 **c** 500 **d** 700 **e** 900 **f** 1000 **g** 100

3 a 6000 **b** 4000 **c** 6000 **d** 7000 **e** 1000

4 a 4 000 000 **b** 2 000 000 **c** 3 000 000 **d** 12 000 000 **e** 1 000 000

5 a 15 **b** 13 **c** 44 **d** 73 **e** 7 **f** 168

6 a 1.5 **b** 14.8 **c** 45.3 **d** 73.7 **e** 2.4 **f** 44

7 a 1100 students **b** 1060 students

8 a 300 **b** 290

9 a 9000 **b** 9000 **c** 8970

10

Attendance	To nearest 100	To nearest 1000
45 336	45 300	45 000
96 022	96 000	96 000
68 047	68 000	68 000
40 823	40 800	41 000
28 548	28 500	29 000
65 829	65 800	66 000
76 462	76 500	76 000
49 975	50 000	50 000

11 a 150 seconds **b** 100 seconds

12 a

Earth	Jupiter	Mars	Mercury	Neptune	Pluto	Saturn	Uranus	Venus
13 000	143 000	7000	5000	50 000	2000	121 000	51 000	12 000

The order of size
Pluto, Mercury, Mars, Venus, Earth, Neptune, Uranus, Saturn, Jupiter

b

Earth	Jupiter	Mars	Mercury	Neptune	Pluto	Saturn	Uranus	Venus
10 000	140 000	10 000	10 000	50 000	Zero	120 000	50 000	10 000

13 Rounding 45 to the nearest 100 gives 0 not 100, since the tens value is 4.

14 i. The lowest 835 000 The highest 844 999 **ii.** The lowest 839 500 The highest 840 499

15 455, 456, 457, 458, 459, 460

16 23 bags

17 a 288 **b** 205

9.2 Multiplying and dividing by powers of 10

1 a 270 **b** 6300 **c** 9700 **d** 420 **e** 38

2 a 3.7 **b** 0.53 **c** 0.088 **d** 0.52 **e** 0.0038

3 a 34.2 **b** 1970 **c** 13 400 **d** 3740

4 a 0.7 **b** 0.075 **c** 0.00 583 **d** 0.0374 **e** 0.485

5 a 30 **b** 100 **c** 0.3 **d** 100

6 a 45.1 **b** 6.2 **c** 3.2 **d** 5180 **e** 0.37

7 a 0.85 **b** 0.03 **c** 0.108 **d** 0.0079 **e** 0.0004

8 a 3 **b** 1000 **c** 30 **d** 30 **e** 0.03 **f** 0.003

9 a 0.03 **b** 100 **c** 300 **d** 10 **e** 30 **f** 0.03

10 a < **b** > **c** < **d** > **e** > **f** >

11 No, she isn't. She should move the digits three places to the left

12 Multiply by moving digits 6 places to the left
Divide by moving digits 6 places to the right

13 Multiply by moving the digits 12 places to the left
Divide by moving the digits 12 places to the right

14 1 000 000 000 000 000 000 000

15 **a** 1300 **b** 15 500 **c** 440 **d** 70

16 84.9

17 **a** 390 **b** 175 **c** 2350 **d** 70 **e** 25 **f** 208

18 **a** 0.375 **b** 0.075 **c** 4.55 **d** 5.25 **e** 0.615 **f** 2.008

9.3 Putting decimals in order

1 **a**

Thousands	Hundreds	Tens	Units	Tenths	Hundredths	thousandths
			4	5	7	
		4	5			
			4	0	5	7
			4	5		
			0	0	4	5
			0	5		
			4	0	5	

b 0.045, 0.5, 4.05, 4.057, 4.50, 4.57, 45

2 **a** 0.073, 0.7, 0.709, 0.73, 0.8
b 1.03, 1.203, 1.4, 1.404, 1.405
c 0.034, 0.34, 2.34, 3.4, 34

3 **a** £0.07, 56p, £0.60, £1.25, 130p
b $0.04, $0.35, $1, $1.04, $10

4 25 minutes, half an hour, 1 hour 10 minutes, 1.5 hours

5 **a** –0.82, –0.708, –0.7, 0.8, 0.82
b –5.44, –5.14, –5.12, 5.07, 5.11
c –1.73, –1.7, –1.65, 1.8, 1.82

6 1.6 cm, 3 cm, 13.4 cm, 14 cm, 170 cm

7 0.32 m, 2.69 m, 6 m, 27 m, 34 m

8 0.056 kg, 0.467 kg, 1 kg, 5 kg, 5.500 kg

9 98 cl, 800 cl, 876 cl, 1700 cl, 8300 cl

10 0.0670 t, 0.6509 t, 0.678 t, 6 t, 6.09 t, 6.6 t

11 **a** < **b** > **c** > **d** > **e** < **f** <

12 **a** < **b** > **c** < **d** > **e** > **f** >

13 **a** 3.14 is greater than 3.1 and less than 3.142 **b** 32p is greater than £0.07 and less than £0.56

14 Yes, since 0.508 < 0.510

15 Sam, Bran, Kay, Gill.

16 Ben Nevis, Snowdon, Scafell Pike, Clisham, Sawell, Yes Tor

17 0.2357

18 **a** Kate **b** Roy **c** Les

9.4 Estimates

1 **a** 3150 **b** 4100 **c** 7800

2 **a** 3250 **b** 3100 **c** 1790

3 **a** 2800 **b** 4600 **c** 32 000 **d** 34 000 **e** 180 625 **f** 60 000 **g** 82 000 **h** 10 200

4 **a** 5 **b** 20 **c** 20 **d** 20 **e** 30 **f** 50 **g** 100 **h** 100

5 **a** 4 **b** 5 **c** 15 **d** 4

6 **a** 10 **b** 23 **c** 162 **d** 32

7 **a** 16 **b** 90 **c** 10 **d** 4 **e** 2500 **f** 10 **g** 15 **h** 6

8 **a** 5 **b** 1 **c** 5 **d** 2

9 **a** 2 **b** 10 **c** 10 **d** 120

10 **a** 18.7 **b** 5.5 **c** –0.5

11 **a** Units should be 8 **b** Estimation is $50 \times 70 = 3500$ **c** Estimation is $\frac{35 + 65}{10} = \frac{100}{10} = 10$ **d** $\frac{36}{8} = 4$; $4 \times 10 = 40$ **e** Units are 4 – 7, so must end in a 7 = 10 so the calculation is wrong

12 Price of one can is 86p; it can be rounded up to £1, so 6 cans cost £6 at most

13 Price of one cake is 47p; can be rounded up to £0.50, so 8 cakes cost £0.50 × 8 = £4

14 The seller made mistake of 54p, and considered it £54

15 $1200

16 30 000

17 **a** Round up £3.65 to £4; £5.92 to £6; £7.99 to £8. So the total will be at most 4 + 6 + 8 = £18 **b** yes he can afford 45p, since remains at least 20 – 18 = £2

18 Approximate wages per year
Thomas $300 × 50 = 15 000
Dechia $6000 × 12 = 72 000
Joseph $500 × 50 = 25 000
Sheena $1000 × 12 = 12 000
Total $15 000 + $72 000 + $25 000 + $12 000 = $124 000
They will qualify to the bursary, since they bring home a total of less than $128 000

9.5 Adding and subtracting decimals

1 **a** 8.2 **b** 8.9 **c** 15.4 **d** 44.12

2 **a** 0.8 **b** 5.72 **c** 3.2 **d** 29.82

3 **a** 1.62 **b** 3.71 **c** 5.92 **d** 8.91

4 **a** 3.12 **b** 2.94 **c** 1.84

5 **a** 7.08 **b** 6.64 **c** 9.96

6 **a** 14.05 **b** 8.08 **c** 7.1 **d** 13.53

7 9.6 km

8 4.83 kg

9 c has the greatest value, **a** 22.38 **b** 22.16 **c** 242

10 1.58 m

11 1.765 kg

12 2.77 litres

13 3.44 m

14 He forget to put 0 in the first decimal place; the correct answer is 5.038

15 0.9 litre

16 **a** 20.1 km
b As every path is interconnected and to visit each path Sean has to visit D twice.
c 11.7 km

17 **a** 12 + 6 + 14 + 12 + 13 + 4 + 4 + 4 + 4 + 18 = £91
So the bill is about right. **b** Actual sum of the bill is £85.60 so the bill is actually not correct

18 **a** 10.65 kg **b** Horse, zebra, wolf and rabbit **c** 7.85 kg **d** 0.15 kg

9.6 Multiplying and dividing decimals

1 **a** 15.7 **b** 13.84 **c** 20.1 **d** 51.03

2 **a** True **b** False **c** True **d** True **e** False **f** True

3 **a** 2.13 **b** 6.12 **c** 6.47 **d** 1.48

4 **a** True **b** False **c** True **d** False **e** True **f** False

5 c

6 c

7 a

8 **a** 29.6 **b** 31.6 **c** 79.1 **d** 31.2

9 **a** 238.23 **b** 111.2 **c** 17.15

10 0.56 m

11 11.75 kg

12 £0.78

13 £4.35

14 £2.49

15 £8

10.1 Coordinates

1 A(2, 4), B(5, 2), C(–4, 3), D(–4, 0), E(–5, –4), F(0, –5), G(2, –3), H(5, –4)

2 **a**

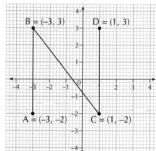

b The letter is "N"

Answers

3 a

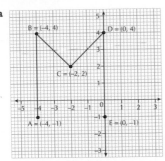

b The letter is "M"

4 Triangle

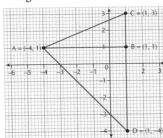

5 Kite

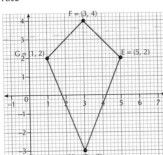

6 Parallelogram

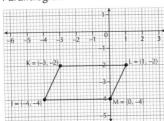

7 Square

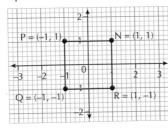

8 Trapezium

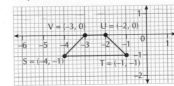

9 Rhombus

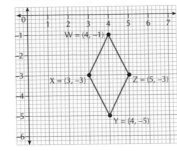

10 a and **b**

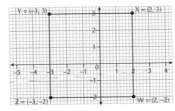

c Z(–3, –2) **d** Point of diagonals intersection (–0.5, 0.5)

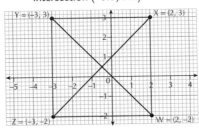

11 a and **b** for example

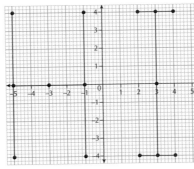

c (–5,4), (–5,0), (–5,–4), (–3,0),
(–1,4), (–1,0), (–1,–4)/ (2,4), (3,4),
(4,4), (3,0), (2,–4), (3,–4), (4,–4)
d Student's answer

12 a i. A(2, 8), B(1, 6) **ii.** D(5, 8), L(5, 6), C(7, 6) **iii.** E(5, 4) **iv.** J(0, 3), H(1, 1) **v.** F(7, 2), G(6, 1), K(7, 0) **vi.** I(2, 4)
b i. Hedge **ii.** Pond **iii.** Lawn **iv.** Vegetable plot **v.** Flower bed **vi.** Lawn **vii.** Patio

13 a (–2,–2) **b** (1, 1) and (2, 0) will both be winning next moves, so Tasha cannot block both moves
c (3, 2), (–2, 2), (–2, –3), (3, –3)

14 (–2, –3)

15 (5, 4)

10.2 Graphs from formulae

1

x	y	Coordinates
–2	–2	(–2,–2)
0	0	(0,0)
2	2	(2,2)

2

x	y	Coordinates
–2	1	(–2,1)
0	3	(0,3)
2	5	(2,5)

3

x	y	Coordinates
–2	3	(–2,3)
0	5	(0,5)
2	7	(2,7)

4

x	y	Coordinates
–2	–8	(–2,–8)
0	0	(0,0)
2	8	(2,8)

5

x	y	Coordinates
–2	–12	(–2,–12)
0	0	(0,0)
2	12	(2,12)

6

x	y	Coordinates
–2	–1	(–2,–1)
0	0	(0,0)
2	1	(2,1)

7 i. & ii.

a

x	y	Coordinates
–1	1	(–1, 1)
0	2	(0, 2)
1	3	(1, 3)
2	4	(2, 4)
3	5	(3, 5)
4	6	(4, 6)

b

x	y	Coordinates
−1	3	(−1, 3)
0	4	(0, 4)
1	5	(1, 5)
2	6	(2, 6)
3	7	(3, 7)
4	8	(4, 8)

c

x	y	Coordinates
−1	−4	(−1, −4)
0	−3	(0, −3)
1	−2	(1, −2)
2	−1	(2, −1)
3	0	(3, 0)
4	1	(4, 1)

iii.

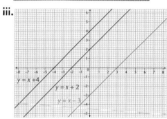

8 i. and ii.

a

x	y	Coordinates
−1	−2	(−1, −2)
0	0	(0, 0)
1	2	(1, 2)
2	4	(2, 4)
3	6	(3, 6)
4	8	(4, 8)

b

x	y	Coordinates
−1	−3	(−1,− 3)
0	0	(0, 0)
1	3	(1, 3)
2	6	(2, 6)
3	9	(3, 9)
4	12	(4, 12)

c

x	y	Coordinates
−1	−5	(−1, −5)
0	0	(0, 0)
1	5	(1, 5)
2	10	(2, 10)
3	15	(3, 15)
4	20	(4, 20)

iii.

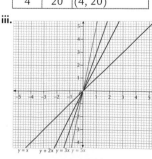

9 iii. a (−2,−8), (0,−2), (3 7)
b (−2,−2), (0, 2), (3, 8)
c (−2,−7), (0, 1), (3, 13)
iv. a, **b** and **c**

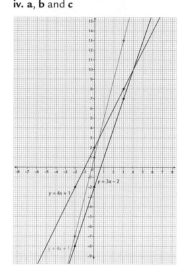

10 a

x	y = −2x + 3	Coordinates
−2	7	(−2, 7)
0	3	(0, 3)
2	−1	(2, −1)

b

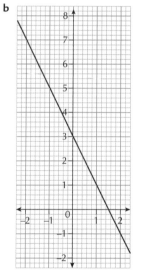

c i. (−2, 1), (0, −1), (2, −3)
ii. (−2, 7), (0, 1), (2, −5)

11 They are parallel
12 They all pass through (0, 0)
13 a

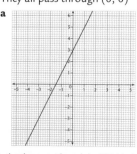

b

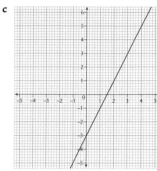

c

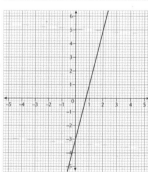

d

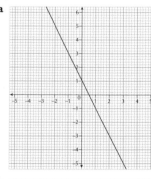

14 a

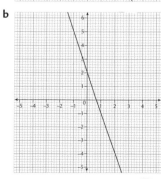

b

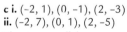

Answers

c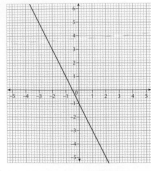

d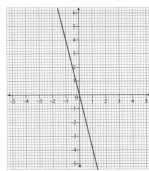

15 $y = x$ **16** $y = x + 1$
17 $y = 2x + 1$

10.3 Graphs of $x = a$, $y = b$, $y = x$ and $y = -x$

1 **a** $y = 3$ **b** $x = 3$ **c** $x = 4$ **d** $y = 0$
  e $y = -x$ **f** $y = -3$ **g** $x = 0$ **h** $y = 2$
  i $x = -1$ **j** $y = x$

2 $y = -1$ (orange), $y = -4$ (Red),
  $y = x$ (Brown), $x = -1$ (green)
  $x = 3$ (black), $x = -2$ (bright red)
  $y = -x$ (purple)

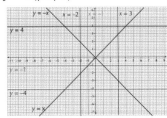

3 **a** I, D **b** J, I **c** C, H **d** E, D
  e E, H **f** B, J **g** K, G **h** C, F **i** D, A
  j B, G, H

4 $x = 2$ **5** $x = 4$
6 $y = x$ **7** $y = -x$
8 $x = 0$
9 **a i.** $x = -1$, $y = 3$

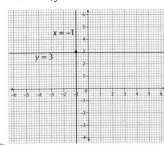

ii. $x = 4$, $y = -1$

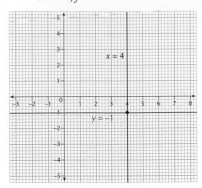

iii. $y = -5$, $x = -6$

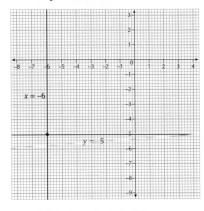

b i. (–9, 5) **ii.** (28, –15)
iii. (–48, –23)

10 **a**

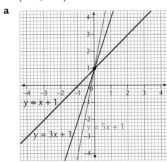

b They all pass through the point
(0, 1) **c** $y = -2x + 1$

11 **a** All parallel, because it is a
displacement of the original line
$y = 2x$ **b** $y = 2x - 5$

12 **a** $y = -3$ **b** $y = -5$ **c** $x = 7$ **d** $x = -2$
  e $y = x$ **f** $y = -6$ **g** $x = -3$ **h** $x = 7$
  i $y = -x$ **j** $y = -4$

13 20

10.4 Graphs of the form $x + y = a$

1

x	y	Total	Coordinates
1	4	5	(1, 4)
2	3	5	(2, 3)
–1	6	5	(–1, 6)

2

x	y	Total	Coordinates
1	6	7	(1, 6)
2	5	7	(2, 5)
–1	8	7	(–1, 8)

3

x	y	Total	Coordinates
1	–2	–1	(1, –2)
2	–3	–1	(2, –3)
–1	0	–1	(–1, 0)

4

x	y	Total	Coordinates
1	–1	0	(1, –1)
2	–2	0	(2, –2)
–1	1	0	(–1, 1)

5 (8, 0) and (0, 8)
6 (2, 0) and 0, 2)

7

x	y	Total	Coordinates
2	7	9	(2, 7)
0	9	9	(0, 9)
13	–4	9	(13, –4)

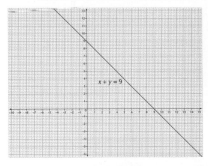

8 **a**

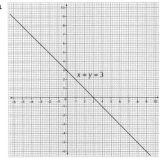

b

9 a

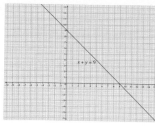

b

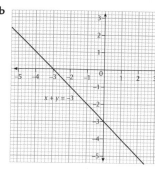

10 Because they are parallel

11 $x = y + 6$, because the other two equations are the same line.

12 **a** $x + y = 2$ **b** $x + y = 6$ **c** $x + y = 1$
d $x + y = -3$ **e** $x + y = -6$ **f** $x + y = 0$
g $x + y = 3$ **h** $x + y = -3$

13 a, d and e

14 18

15 **a** $(2, 2)$ **b** $(5.5, 5.5)$ **c** $(-4, -4)$

10.5 Conversion graphs

1 **a i.** 4.8 km **ii.** 7.2 km **iii.** 1.6 km
b i. 1.2 miles **ii.** 2.5 miles
iii. 3.7 miles

2 **a** after 3 hours it should be 36 miles **b** 2.5 hours

3 **a i.** £8 **ii.** £20 **b i.** $15 **ii.** $22

4 **a** 4.5 km **b** 1.1 minutes

5

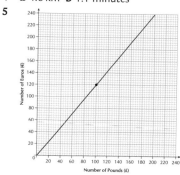

6

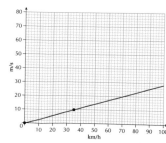

7

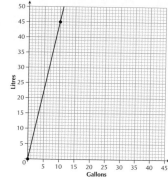

8 25 miles

9 £56

10 Distance in 1st minute = 0.9 km
Distance in 3rd minute = Distance after 3 minutes – distance after 2 minutes

11 £60

12 £6

13 400 euros

11.1 Fractions, decimals and percentages

1 **a** 50% **b** 25% **c** 75% **d** 30% **e** 35%

2 **a** 50% **b** 5% **c** 80% **d** 8%

3 **a** $\frac{1}{5}$ **b** $\frac{3}{10}$ **c** $\frac{9}{10}$ **d** $\frac{19}{20}$

4

Percentage	10%	70%	60%	36%	35%
Fraction	$\frac{1}{10}$	$\frac{7}{10}$	$\frac{15}{25}$	$\frac{9}{25}$	$\frac{7}{20}$
Decimal	0.1	0.7	0.6	0.36	0.35

5 $66\frac{2}{3}$%

6 $\frac{4}{5} = 80\%$, $\frac{7}{10} = 70\%$, $\frac{7}{20} = 35\%$, $\frac{37}{50} = 74\%$

7 **a** 120% **b** 190% **c** 153% **d** 174% **e** 274%

8 **a** 1.3 **b** 1.4 **c** 1.85 **d** 2.85 **e** 2.16

9 **a** $1\frac{1}{2}$ **b** $1\frac{1}{4}$ **c** $1\frac{3}{10}$ **d** $1\frac{3}{5}$ **e** $2\frac{3}{4}$

10 **a** $0.2 + 0.75 = 0.95$
b $\frac{1}{5} + \frac{3}{4} = \frac{19}{20}$

11 **a** 40% **b i.** Blue **ii.** Red **iii.** Green
By comparing the sectors areas after realizing that red and green make up half **c** $\frac{1}{10}$, $\frac{1}{5}$, $\frac{3}{10}$, $\frac{2}{5}$
d 80% **e** Green

12 He is wrong. 1 is equal to 100%, so anything more than 100% is more than 1.

13 **a** $\frac{17}{50}$ **b** 0.34 **c** 34%

14 **a** 0.185, $\frac{37}{200}$ **b** 0.205, $\frac{41}{200}$
c 0.208, $\frac{26}{125}$ **d** 0.218, $\frac{109}{500}$

15 **a** 12.5% **b** 37.5%, 62.5%, 87.5%

16 **a** 50%, 60%, 70%, 80% **b** $\frac{1}{10}$, $\frac{1}{5}$, $\frac{3}{10}$, $\frac{2}{5}$, $\frac{1}{2}$, $\frac{3}{5}$, $\frac{7}{10}$, $\frac{4}{5}$, $\frac{9}{10}$ **c** 50%

11.2 Fractions of a quantity

1 **a** £60 **b** £180 **c** £420 **d** £540

2 **a** £24 **b** £48 **c** £32 **d** £4

3 **a** 24 cm **b** 72 cm **c** 36 cm **d** 30 cm

4 **a** £0.60 **b** £0.30 **c** £0.15 **d** £0.75

5 **a** 4 m **b** 2 m **c** 1 m **d** 3 m

6 **a** 10 minutes **b** 40 hours
c 80 seconds

7 $\frac{3}{5}$ of £600 = £360; remains:
£600 – £360 = £240
$\frac{3}{4}$ of £240 = £180; remains:
£240 – £180 = £60

8 **a** 120° **b** 150°

9 No, $\frac{3}{4}$ of £2000 = £1500; remains:
£2000 – £1500 = £500
$\frac{7}{10}$ of £500 = £350; remains:
£500 – £350 = £150
Which is not enough for the speakers.

10 **a i.** 32 **ii.** 64 **iii.** 96 **iv.** 128
By comparing the area of sectors **b** Answers increase by 32 each time.

11 **a** 13 miles. **b** Yes, correct
$\frac{3}{4}$ of 26 = 19.5 which is less than 20 miles.

12 **a** £ 520 **b** £ 1040

11.3 Percentages of quantities

1 **a** £10 **b** £15 **c** £150 **d** £30

2 **a** £10 **b** £14 **c** £100 **d** £125

3 **a** £5 **b** £6 **c** £8 **d** £12

4 **a** 160 kg **b** 80 kg **c** 64 kg **d** 32 kg

5 **a** 320 cm **b** 64 cm **c** 128 cm
d 896 cm

6 **a** 1800 g, 1.8 kg **b** 9000 g,
9 kg **c** 1080 g, 1.08 kg **d** 5400 g,
5.4 kg

7 **a** £0.63 **b** £1.26 **c** £6.30
d £12.60

8 **a** £3 **b** £27 **c** £57 **d** £297

9 **a** £180 **b** £144 **c** 121,2 **d** £242.4

10 **a** 3.2 kg **b i.** 6.4 kg **ii.** 9.6 kg
iii. 1.6 kg **iv.** 28.8 kg

11 **a i.** £7 **ii.** £2.8 **b i.** £9.8 **ii.** £4.2
iii. £8.4 **iv.** £1.4 **v.** £3.5

12 **a** 45% **b** 32% **c** 40%

13 **a** £315 **b** £15.75 **c** £217.5

14 **a** £396 **b** £924

15 £30.38

16 120

17 **a** £10.5 **b** £382.5

Answers

11.4 Percentages with a calculator

1. **a** 327.6 cm **b** 31.2 cm **c** 226.2 cm **d** 733.2 cm
2. **a** 147.73 kg **b** 5.61 kg **c** 67.32 kg **d** 50.49 kg
3. **a** £270.405 **b** £405.6075 **c** £1400.097 **d** £456.684
4. **a** 38.598 cm **b** 12.866 cm **c** 131.0494 cm **d** 51.0964 cm
5. **a** 3.9 km, 3900 m **b** 12.9 km, 1290 m **c** 9.6 km, 9600 m **d** 6.75 km, 6750 m
6. Red party: 4048; Blue party: 6336: Yellow party: 3344
7. 332
8. 535 g
9. **a i.** 731 **ii.** 2322 **iii.** 1247 **b** Because the percentages add up to 100%
10. **a** 24.32 g **b i.** Since $16\% = \frac{1}{2}$ of 32%, then 16% of 76g $= \frac{1}{2}$ of 24.32 = 12.16 g **ii.** Since $38 = \frac{1}{2}$ of 76, then 32% of $38 = \frac{1}{2}$ of 24.32 = 12.16 m **iii.** Since 16% is half of 32%, then 16% of 38 is half of 12.16 = £6.08
11. **a** Failed **b** Failed **c** Failed **d** Passed
12. **a** 0.015 **b** £69.02
13. **a i.** 243.2 kg **ii.** 121.6 kg **iii.** 60.8 kg **iv.** 30.4 kg **b** 15.2
14. **a** Conservative: 10.7 million; Labour: 8.6 million; Liberal Democrat: 6.8 million **b** 3.6 million

12.1 Naming and drawing 3D shapes

1.

	Number of faces	Number of edges	Number of vertices
Cube	6	12	8
Cuboid	6	12	8
Square based pyramid	5	8	5
Tetrahedron	4	6	4
Triangular prism	5	9	6
Pentagonal prism	7	15	10
Hexagonal prism	8	18	12

2.

	Number of faces	Number of edges	Number of vertices
Square-based pyramid	5	8	5
Pentagonal-based pyramid	6	10	6
Hexagonal-based pyramid	7	12	7

3. Cube and cuboid
4. Tetrahedron or square-based pyramid
5. Cone
6.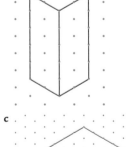
7. **a**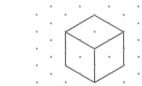
   b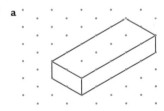
   c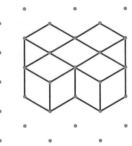
8. **a** Always true **b** Sometimes true **c** Never true **d** Always true
9. **a**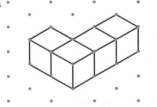
   b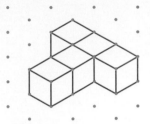
   c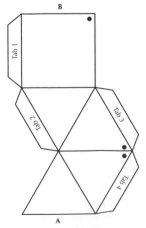
10. Because its sides are curved surfaces.
11. **a** 9 faces **b** 9 vertices **c** 16 edges
12. **a** 12 faces **b** 10 vertices **c** 20 edges
13. 10 cubes

12.2 Using nets to construct 3D shapes

1. Cube
2. Cube
3. Cube
4. Cuboid
5. Tetrahedron
6. Square- based pyramid
7. Triangular prism
8. Hexagonal prism
9. Triangles overlap, so one face will be missing
10. b, does not make a net of a cube
11. No, it is a net for a tetrahedron
12. **a**, **b** and **c**

13

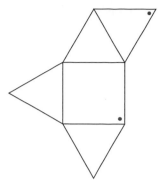

14

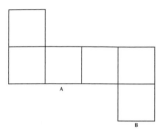

12.3 Volume of a cuboid

1   **a** 320 cm³ **b** 1680 cm³ **c** 16 m³
2   6000 m³
3   36 m³
4   9600 cm³
5   10 cm
6

	Length	Width	Height	Volume
A	6 cm	4 cm	1 cm	24 cm³
B	3.2 m	2.4 m	0.5 m	3.84 m³
C	8 cm	5 cm	3 cm	120 cm³
D	20 mm	16 mm	3 mm	960 mm³
E	40 m	5 m	2 m	400 m³

7   **a** 8 cm³ **b** 125 cm³ **c** 1728 cm³
8   6 cubes
9   **a** 16 m³ **b** 11 520 m³
10 Volume = 30 × 40 × 50 =
    60 000 cm³ = 60 litre
    So the amount of water
    = 60 ÷ 2 = 30 litre
11 Cube volume = $(5)^3$ = 125 cm³
    Cuboid volume =
    4 × 5 × 6 = 120 cm³
    So, the cube is bigger
12 96 packets
13 B
14 No, the volume of B is bigger
    than A
15 No, the height of A is bigger than B

12.4 Surface area of a cuboid

1   **a** 208 cm² **b** 164 cm² **c** 290 cm²
    d 40 cm²
2   214 cm²
3   318 cm²
4   468 cm²
5   1238 cm²

6   **a** 96 cm² **b** 216 cm² **c** 384 cm²
    d 486 cm²
7   **a** 6 cm² **b** 150 cm² **c** 2166 cm²
    d 864 cm²
8   62 cm²
9   60 cm²
10 24 cm²
11 Volume = 27 cm² → so edge length
    equals 3 cm
    Surface area = 3 × 3 × 6 = 54 cm²
12 Surface area = (2 × 30 × 40) +
    (2 × 30 × 100) + (2 × 40 × 100) =
    2400 + 6000 + 800 = 16 400 cm²
13 Surface area of the cuboid =
    (2 × 4 × 4) + (2 × 4 × 9) +
    (2 × 4 × 9) = 32 + 72 + 72 =
    176 cm²
    Surface area of the cube =
    5 × 5 × 6 = 150 cm²
    So the cuboid has a greater surface
    area, by 176–150 = 26 cm²
14 Surface area of the cuboid =
    (2 × 4 × 4) + (2 ×4 × 11.5) +
    (2 × 4 × 11.5) = 32 + 92 + 92 =
    216 cm²
    And the surface area of the cube
    = 6 × 6 × 6 = 216 cm²
15 Because there are two faces not
    included in the surface
16 44 m²
17 4416 m²
18 250 cm²

13.1 Probability words

1   A → b, B → c, C → a, D → f, E →
    d, F → e, G → g
2   **a** Triangle, 2 **b** Square, 2
    c Circle, 1 or 2 **d** Rectangle, 1
    e Triangle, 1
3   C, D and E, B, A
4   **a** Very likely **b** Unlikely **c** Very
    unlikely **d** Likely
5   **a** False **b** True **c** True **d** True
6   **a** A pupil with glasses **b** Boy
    c A pupil aged twelve **d** A pupil
    with fair hair
7   **a** Very unlikely **b** Likely
    c Impossible
8   **a** True **b** False **c** True **d** False
9   **a** Very unlikely **b** Evens **c** Unlikely
    d Very likely
10 E, C, B, A, D
11 **a** Bag B as evens chance from bag
    A whereas in bag B it's a very likely
    chance **b** Bag A as an unlikely
    chance from bag A but a very
    unlikely chance from bag B **c** Bag
    A as an unlikely chance of green
    from bag A but an impossible
    chance from bag B
12 It's fifty fifty as the chance
    does not depend on any of the
    previous flips

13 She is not correct as they all have
    the same chance
14 He is incorrect. Just because
    there are two options does
    not necessarily mean there is a
    50:50 chance
15 ABC, ABD, ABE, ACD, ACE, ADE,
    BED, BEC, DCE and DCB. Three of
    these combinations do not make a
    triangle, ABD, ABE and ACE, and
    the other seven combinations make
    a triangle. Hence it is more likely to
    choose the rods making a triangle
16 Find out the number of ways each
    event can happen – there are six
    ways he can roll a double. There
    are eight ways in which he can roll
    a total greater than eight. Hence
    there is more chance he loses the
    game than winning
17 Sunday as that day has three
    snows, more than any other day.
18 There are 8 prime numbers and 9
    even numbers, hence more chance
    of choosing an even number

13.2 Probability scales

1   **a i.** $\frac{1}{2}$ **ii.** $\frac{1}{6}$ **iii.** $\frac{1}{4}$ **iv.** $\frac{1}{12}$ **v.** 0

    b i. 0 **ii.** $\frac{1}{6}$ **iii.** $\frac{1}{2}$ **iv.** $\frac{1}{4}$ **v.** 1

2   **a** $\frac{1}{2}$ **b** $\frac{1}{4}$ **c** $\frac{1}{4}$
3   **a** 0.1 **b** 0.5 **c** 0.3 **d** 0.2 **e** 1 **f** 0.5
4   **a** 0.5 **b** 0.3 **c** 0.2 **d** 0 **e** 0.8
5   **a** $\frac{3}{8}$ **b** $\frac{1}{4}$ **c** $\frac{1}{4}$ **d** $\frac{1}{8}$
6   **a** 0.2 **b** 0.08 **c** 0.16 **d** 0.56
7   **a** $\frac{1}{4}$ **b** $\frac{1}{6}$ **c** $\frac{1}{4}$ **d** $\frac{1}{4}$ **e** $\frac{1}{12}$ **f** $\frac{11}{12}$
8   **a** $\frac{17}{30}$ **b** $\frac{5}{30}$ **c** $\frac{8}{30}$
9   **a** $\frac{12}{25}$ **b** $\frac{3}{25}$ **c** $\frac{8}{25}$ **d** $\frac{2}{25}$
10 **a** $\frac{3}{10}$ **b** $\frac{1}{5}$ **c** 0 **d** $\frac{1}{2}$
11 **a i.** $\frac{4}{52}$ **ii.** $\frac{13}{52}$ **iii.** $\frac{8}{52}$ **iv.** $\frac{1}{52}$

    v. $\frac{8}{52}$ **vi.** $\frac{26}{52}$

    b Since the cards have four Jacks
    and only one ace of spades, so
    the probability to choose a Jack
    is higher
12 **a i.** $\frac{1}{6}$ **ii.** $\frac{1}{2}$ **iii.** $\frac{1}{3}$ **iv.** $\frac{1}{2}$ **v.** $\frac{1}{6}$

    vi. $\frac{1}{3}$

    b Primes are 2, 3 and 5 so

    P(prime) = $\frac{1}{2}$ and as P(odd) = $\frac{1}{2}$

    Then P(prime) = P(odd)
13 **a** Nazir **b** Either add a 3 to
    Emma's spinner or remove a 3
    from Nazir's spinner

Answers

14 $\frac{6}{50} = 0.12$ if it had eight faces

then, $P(6) = \frac{1}{8} = 0.125$

As 0.12 is very close to 0.125 then Briony is correct

15 $\frac{4}{5}$

16 Tim has a $\frac{60}{300}$ chance of winning,

Kath has a $= \frac{35}{150} = \frac{70}{300}$ chance of winning

So Kath has the best chance of winning

17 **a** $\frac{1}{8}$ **b** $\frac{85}{100}$ **c** $\frac{1}{10}$

18 **a** $\frac{3}{10}$ **b** $\frac{24}{90}$

13.3 Experimental probability

1 **a** $\frac{26}{50}$ **b** $\frac{24}{50}$

2 **a** 0.16 **b** 0.19 **c** 0.16 **d** 0.17

3 **a** 0.66 **b** 0.34

4 **a** 0.26 **b** 0.23 **c** 0.24 **d** 0.27

5 **a** $\frac{11}{31}$ **b** $\frac{20}{31}$

6 **a** $\frac{1}{27}$ **b** $\frac{2}{27}$ **c** $\frac{24}{27}$

7 **a** 0.1 **b** 0.3 **c** 0.6

8 **a** $\frac{1}{34}$ **b** $\frac{4}{34}$ **c** $\frac{11}{34}$ **d** $\frac{18}{34}$

9 **a** Yes, the dice shows five over twice as likely as 1 **b** Roll the dice many more times **c** 0.2 **d** 0.25

10 Yes as you would expect a fair coin to land on heads about the same number of times as tails but the number of trials is too small

11 Yes, as the frequency of the red color is the highest

12 No as the frequency of the bus being on time is only 8, which is less than 12 the frequency of the bus being up to 5 minutes late, and so not generally on time

13 **a** $\frac{1}{36}$ **b** $\frac{6}{36}$ **c** $\frac{6}{36}$

14 **a** 0.48 **b** 0.52 **c** Not raining

15 Factor of 6

16 5

14.1 Introduction to ratio

1 The ratio of white beads to black beads is 5:1

2 **a** Bea has saved 4 times as much as Ade **b** 1:4 **c** 4:1

3 **a** 2:1 **b** 1:2 **c** 4:1 **d** 1:2

4 **a** 2:1 **b** 4:1 **c** 1:8

5 **a i.** 3:1 **ii.** 6:1 **iii.** 5:1 **iv.** 1:2 **b** 1.5

6 1:4

7 **a** 1:2 **b** 2:3

8 **a** 4:1 **b** 1:3 **c** 3:2

9 3:1

10 1:4

11 75% equal the ratio 75:100 and dividing both sides by 25 simplifies the ratio to 3:4

12 Yes correct, $20 \times 60 = 1600$ million

13 C is correct, there are 97 men for every 100 women so more women than men

14 **a** 5 times **b i.** 1:5 **ii.** 2:1 **iii.** 3:2 **c.** $1.4 \times 3 = 4.2$

15 **a i.** 3:1 **ii.** 1:3 **iii.** 2:1 **iv.** 8:1
b Pakistan 9p
Nigeria 8p
Germany 4p
UK 3p
Iraq 1.5p

16 **a** 12 black pieces **b** 1:2

14.2 Simplifying ratios

1 **a** $\frac{2}{3}$ **b** $\frac{1}{5}$ **c** $\frac{3}{2}$ **d** $\frac{3}{5}$ **e** $\frac{1}{30}$

2 **a** 1:4 **b** 2:3 **c** 2:1 **d** 8:1 **e** 2:3

3 **a** 2:5 **b** 5:2

4 2:7

5 **a** 5:2 **b** 2:3 **c** 3:5

6 2:3

7 4:3

8 49:38

9 **a** 2:1 **b** 4:3 **c** 3:5

10 **a** $\frac{3}{2}$ **b** $\frac{3}{4}$ **c** $\frac{5}{2}$ **d** $\frac{5}{6}$ **e** $\frac{4}{1}$

11 65:35 simplifies to 13:7

12 28 yellow and 21 white , so 28:21 simplifies to 4:3

13 1:3

14 **a** $\frac{1}{3}$ **b** 1:2

15 **a i.** 40:1 **ii.** 5:2 **iii.** 5:4
b Sulphur to oxygen

14.3 Ratios and sharing

1 **a** 1:2 gives 20kg and 40 kg **b** 3:1 gives £30 and £10 **c** 5:1 gives 150 litres and 30 litres **d** 5:4 gives 50 kg and 40 kg **e** 2:3 gives 12 cm and 18 cm **f** 3:4:1 gives 9:12:3 hours

2 **a** 60 cm **b** 60 litres **c** £12

3 **a** 5 kg **b** 8 miles **c** 24 g

4 **a** $\frac{5}{6}$ **b** 25 children

5 **a** $\frac{4}{5}$ **b** 500

6 **a** $\frac{2}{5}$ **b** £18

7 **a** $\frac{1}{9}$ **b** 1080 book

8 20 children

9 48

11 30 men

11 £54

12 44 item

13 **a** 1:400 000 **b** 24 km

14 **a** 1:200 000 **b** 3.5 cm

15 **a** Red 4, green 16 **b** 1:4 so ratio still remains the same **c** 1:5

14.4 Ratios in everyday life

1 140 girls, 210 boys

2 3:1

3 2:1

4 500

5 15

6 2

7 120

8 35

9 7.5m

10 Labour charge = £8000 – £3200 = £4800 The ratio 3200:4800 which simplifies to 2:3

11 **a** 2:3 **b** $\frac{60}{5} = 12$, so Pierre is $2 \times 12 = 24$ years

12 **a** 10:1 **b** 3:2 **c** 5:6

13 **a** 2:3 **b** 2:5 **c** 1:2 **d** 1:3

14 £5.75

15 1:6

16 **a** 4 **b** 26

15.1 Reflection symmetry

1

Isosceles triangle | Equilateral triangle | Square | Rectangle | Parallelogram | Kite

2

a | b | c | d | e | f

3

a | b | c | d | e | f | g

4

a | b | c | d | e | f | g | h | i

MATHSVIEW

5 **a** 1 **b** 2 **c** 8 **d** 1 **e** 5 **f** 4 **g** 6 **h** 3

6 **a** 1 **b** 0 **c** 2 **d** 0

7 **a**

b

c The sequence formed from shapes, where each shape consists of an integer associated with its reflection

8

a | b | c

9 a

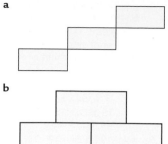

b

c

10 a

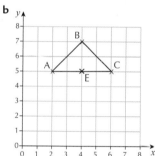

b

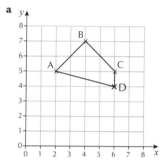

c

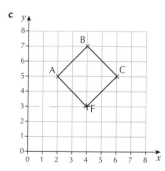

11 Examples are CHICK, CHOCK, DOCK, and HIKE

12 Examples are MOM, TAT, WOW, and OXO

15.2 Rotation symmetry

1 a 2 **b** 1 **c** 2 **d** 2 **e** 2
2 a 2 **b** 4 **c** 2 **d** 5 **e** 8
3 a 6 **b** 3 **c** 4 **d** 2 **e** 8
4 a 2 **b** 1 **c** 2 **d** 2 **e** 1 **f** 2

5

Shape	Number of lines of symmetry	Order of rotational symmetry
a Equilateral triangle	3	3
b Square	4	4
c Regular pentagon	5	5
d Regular hexagon	6	6
e Regular octagon	8	8

The number of lines of symmetry is the same as the order.

6 a Order 2 **b** order 1 **c** order 2 **d** order 1

7 a 4 **b** 6 **c** 12

8

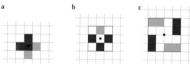

9

10 Ziz is correct if he uses capital letters, ZIZ

11 a

b

c

12

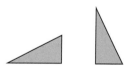

13 H, I, N, O, S, X, Z
14 Examples are 8, 88, 808, and 888

15.3 Properties of triangles and quadrilaterals

1 a Scalene **b** Obtuse – angled **c** Right – angled **d** Isosceles **e** Equilateral

2 You can see that AB = AC = BC, and that ∠ABC = ∠ACB = ∠BAC, so the triangle is equilateral

3 Has one obtuse angle and two acute angles, and have no equal sides

4 Has no equal sides and no equal angles

5 Has a right angle and two acute angles

6 a Square, rhombus **b** Rectangle, parallelogram, kite **c** Square, rectangle, parallelogram, rhombus **d** Trapezium

7 Has two pairs of parallel and equal sides

8 a Has 4 right angles and 4 equal sides **b** Has 4 equal sides

9 a Has 4 right angles two different pairs of equal sides **b** Has one pair of parallel sides

10 Has two different pairs of adjacent equal sides

11 a The square has 4 right angles whereas the rhombus has two different pairs of equal angles **b** The rhombus has 4 equal sides whereas the parallelogram has two different pairs of equal sides **c** The trapezium has one pair of parallel sides whereas the parallelogram has two pairs of parallel sides

12 i. a, b and d **ii.** a, b, c and f **iii.** a, e and f

13 Yes he is correct, no dots will create an equilateral triangle, but many isosceles triangles are available to draw

14 She is wrong as you can draw both

15 16, square, rectangle, parallelogram, trapezium, arrowhead

16

a 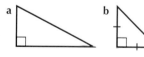 **b**

17

a 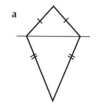 **b**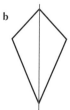

Answers

18

16.1 Finding unknown numbers

1 **a** 2 **b** 6 **c** 14 **d** 22
2 **a** 6 **b** 14 **c** 12 **d** 20
3 **a** 8 **b** 16 **c** 14 **d** 10
4 **a** 2.9 **b** 0.5 **c** 6.2 **d** 7.9
5 **a** –3 **b** –5 **c** –1 **d** –7
6 **a** 1 **b** 4 **c** 0 **d** 2
7 **a** 36 **b** 24 **c** 14 **d** 63
8 **a** 5 **b** 6 **c** 6 **d** 7
9 **a** 12 **b** 60 **c** 4 **d** 3
10 $a = 7$, $b = 13$, $c = 6$
11 $d = 6$, $e = 9$, $f = 3$
12 $a = 13$, $b = 6$, $c = 2$
13 **a** $5J + 3$ **b** $J = 2$ **c** $J = 4$ **d** $J = 7$
14 **a** $(k + 6) \times 2$ **b** $k = 3$ **c** $k = 6$
   d $k = 14$
15 **a** $a – 4$ **b** $a = 13$ **c** $a = 21$ **d** $a = 39$
16 **a** $2k + 3$ **b** $k = 6$ **c** $k = 4$ **d** $k = 25$

16.2 Solving equations

1 **a** 8 **b** 8 **c** 19 **d** 9
2 **a** 4 **b** 9 **c** 4 **d** 5
3 **a** 12 **b** 40 **c** 24 **d** 80
4 **a** 25 **b** 14 **c** 41 **d** 49
5 **a** 23 **b** 12 **c** 19 **d** 28
6 **a** 20 **b** 22 **c** 24 **d** 25
7 **a** –4 **b** –4 **c** –7 **d** –5
8 **a** 12 **b** 3 **c** 4 **d** 7
9 **a** 252 **b** 51 **c** 14 **d** 7
10 **a** $f + 25$ **b i.** $f + 25 = 37$ **ii.** $f = 12$
   c i. $f + 25 = 45$ **ii.** $f = 20$
11 **a** $8r$ **b i.** $8r = 88$ **ii.** $r = 11$ **c i.** $8r = $
   120 **ii.** $r = 15$
12 **a**

		$x + 48$		
	$x + 18$		30	
x		18		12

b i. $x + 48 = 54$ **ii.** $x = 6$ **c i.** $x + 48$
= 65 **ii.** $x = 17$ **d i.** $x + 48 = 90$ **ii.** x
= 42
13 **a** $y + 42$ **b i.** $y + 42 = 50$ **ii.** $y = 8$
   c i. $y + 42 = 100$ **ii.** $y = 58$ **d** 44
14 **a** $290 + x$ **b** $x = 70$
15

		100		
	$x + 40$		43	
$x + 15$		25		18
x	15	10	8	

The equation is $x + 83 = 100$,
$x = 17$
16 $4h = 28$
   $h = 28 \div 4$
   $h = 7\,\text{cm}$

17 $125 + z = 180$
   $z = 180 - 125$
   $z = 55°$

16.3 Solving more complex equations

1 **a** 4 **b** 4 **c** 8 **d** 2
2 **a** 9 **b** 5 **c** 15 **d** 3
3 **a** 4 **b** 3 **c** 6 **d** 12
4 **a** 5 **b** 9 **c** 13 **d** 14
5 **a** 4 **b** 3 **c** 8 **d** 5
6 **a** 18 **b** 15 **c** 20 **d** 9
7 **a** 5 **b** 6 **c** 36 **d** 6
8 **a** –3 **b** –7 **c** –12 **d** –6
9 **a** –1 **b** 1 **c** –1 **d** –5
10 **a** –2 **b** –8 **c** –3 **d** –10
11 **a** $4x + 80$ **b** Yes , $4x + 80 = 152$, so
   $4x = 72$, giving $x = 18$
12 **a** $8y + 30$ **b** $214 = 8y + 30$
   c $y = 23$, so the longest side is
   $23 + 15 = 38$
13 No, $2n + 15 = 99$, so $2n = 84$,
   $n = 42$
14 8
15 **i. a** $11 + 2x = 21$ **b** $14 + 2x = 26$
   c $12 + 2x = 26$ **d** $10 + 3x = 19$
   e $7x = 21$ **f** $3x + 7 = 40$
   ii. a $x = 5$ **b** $x = 6$ **c** $x = 7$ **d** $x = 3$
   e $x = 3$ **f** $x = 11$
16 **i. a** $7 + 3x = 31$ **b** $9 + 4x = 33$
   ii. a $x = 8$ **b** $x = 6$
17 **1.** $3x + 7 = 22$
   Subtract 7 $3x = 15$
   Divide by 3 $x = 5$
   Check $3 \times 5 + 7 = 15 + 7$
               $= 22$
   2. $\frac{x}{5} - 8 = 2$
   Add 8 $\frac{x}{5} = 10$
   Multiply by 5 $x = 50$
   Check $\frac{50}{5} - 8 = 10 - 8 = 2$
   3. $\frac{x}{2} - 5 = 6$
   Add 5 $\frac{x}{2} = 11$
   Multiply by 2 $x = 22$
   Check $\frac{22}{2} - 5 = 11 - 5 = 6$
   4. $3(x + 6) = 18$
   Divide by 3 $x + 6 = 6$
   Subtract 6 $x = 0$
   Check $3(0 + 6) = 3 \times 6 = 18$
18 **a** $x + 5$ **b** $5(x + 5) = 35$ **c** $x = 2$
   d 24 cm

16.4 Setting up and solving equations

1 **a** $2d + 5$ **b** $2d + 5 = 59$, $d = 27$
2 **a** $m – 12$ **b** $2m – 12 = 48$, $m = 30$
   c Mike has 30 pencils, Jon has
   18 pencils
3 **a** $t + 24$ **b** $2t + 24 = 122$ **c** $t = 49$
   d Josie is 73 years old

4 **a** $(14 + m) \times 6$
   b $(14 + m) \times 6 = 120$ **c** $m = 6$
5 **a** $2y + 28 = 1176$, where $y = $ the
   number of girls **b** The number of
   girls = 574, the number of boys
   = 602
6 **a** $6x – 11$ **b** $6x – 11 = 37$, $x = 8$
7 **a** $m – 9$ **b** $2m – 9$ **c** 24
8 **a** $a – 15$ **b** $2a – 15 = 73$, $a = 44$,
   Joy = 44, James = 29
9 **a** $w – 19$ **b** $2w – 19 = 71$, $w = 45$,
   male teachers = 26
10 **a** $8m + 27$ **b** $m = 7$
11 **a** $6a + 14$ **b** $6a + 14 = 29$ **c** $a = 2.5$,
   area of the left rectangle = $6a = $
   $6 \times 2.5 = 15\,\text{m}^2$, area of the right
   rectangle = $4 \times 3.5 = 14\,\text{m}^2$
12 **a** $2k – 10$ **b** $2k – 10 = 30$ **c** $k = 20$
   d 65 cm
13 **a** $4(x – 3) = 52$, $x = 16$
   b $4(y – 3) = 96$, $y = 27$
14 **a** £$(4b + 10.74)$
   b $4b + 10.74 = 19.34$ **c** $b = 2.15$
15 $2m + 12 = 56$, $m = 22$
16 $5c = £8.00 - £0.75$, $c = £1.45$

17.1 Interpreting pie charts

1 **a** 225 **b** 150 **c** 525
2 **a** 75 **b** 75 **c** 50 **d** 100
3 **a** 10 **b** 5 **c** 15
4 **a** 45 **b** 45 **c** 60 **d** 90 **e** 75 **f** 135
   g 90
5 **a** 6 **b** 6 **c** 4 **d** 8
6 **a** 60 **b** 20 **c** 100 **d** 30 **e** 30
7 **a** 45 **b** 72 **c** 45 **d** 54
8 **a** 15 **b** 50 **c** 45 **d** 40
9 **a** True $\frac{120}{360} = \frac{1}{3}$
   b True; both have equal angles.
   c False; 36 children voted for
   Sci Fi; 90° is the angle size, not
   the frequency
10 **a** Cricket **b** No, they are both
   the same
11 **a** Blue **b** No; $3 \times 60 = 180$; a
   sector of 180° would be a semicircle; the yellow sector is clearly
   more than half a circle
12 **a** Tech Net: 120 letters; Infoflow:
   44 letters
   b Tech Net did; Tech Net sent
   160 emails, while Infoflow sent
   154 emails
   c 1; Tech Net made 20 mobile
   calls, while Infoflow made 33
   mobile calls
   2; Tech Net made 180 Office
   telephone calls, while Infoflow
   made 33 Office telephone calls

13 The angle of Y7 sector is 120°, so they raised £563.30
The angle of Y8 sector is 90°, so they raised £422.50
The angle of Y9 sector is 60°, so they raised £281.70
The angle of Y10 sector is 45°, so they raised £211.25
The angle of Y11 sector is 45°, so they raised £211.25

14 Northern Ireland: 19.37 megalitres
Wales: 93.34 megalitres
Scotland: 102.14 megalitres
England: 419.14 megalitres

15 The angle of Collie sector is 30°, there are 6
The angle of Greyhound sector is 150°, five times the Collie so there are 30
The angle of Lurcher sector is 120°, four times the Collie so there are 24
The angle of Whippet sector is 60°, so there are 12
Total = 72 dogs

16 The angle of the German sector is 80°, so 1080 people attended all together

17.2 Drawing pie charts

1

Transport	Frequency	As 120 pupils are represented by 360°, each pupil will be represented by 360 ÷ 120 = 3°
Car	23	23 × 3 = 69°
Bus	17	17 × 3 = 51°
Train	25	25 × 3 = 75°
Bicycle	15	15 x 3 = 45°
Walk	40	40 × 3 = 120
Total	120	360°

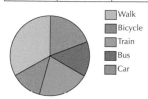
Walk, Bicycle, Train, Bus, Car

2

Bird	Crow	Thrush	Starling	Magpie	Other	Total
Frequency	19	12	8	2	19	60
Angle (°)	114	72	48	12	114	360

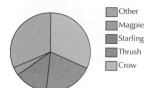
Other, Magpie, Starling, Thrush, Crow

3

Subject	Maths	English	Science	Languages	Other	Total
Frequency	12	7	8	4	5	36
Angle (°)	120	70	80	40	50	360

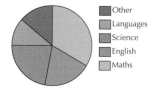
Other, Languages, Science, English, Maths

4

Food	Cereal	Toast	Fruit	Cooked	Other	None	total
Frequency	11	8	6	9	2	4	40
Angle (°)	99	72	54	81	18	36	360

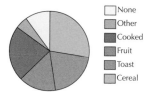
None, Other, Cooked, Fruit, Toast, Cereal

5

Goals	0	1	2	3	4	5 or more	Total
Frequency	3	4	7	5	4	1	24
Angle (°)	45	60	105	75	60	15	360

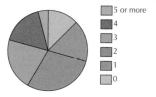
5 or more, 4, 3, 2, 1, 0

6

Colour	Red	Green	Blue	Yellow	Other	Total
Frequency	17	8	21	3	11	60
Angle (°)	102	48	126	18	66	360

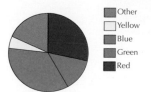
Other, Yellow, Blue, Green, Red

7 a

Size	8	10	12	14	16	18	Total
Frequency	3	7	10	12	6	2	40
Angle (°)	27	63	90	108	54	18	360

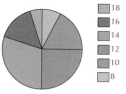
18, 16, 14, 12, 10, 8

b size 12 **c** $\frac{8}{40}=\frac{1}{5}$ **d** 30%

8 a

	Frequency	Angle (°)
Early	4	36
On time	18	162
Up to 5 minutes late	14	126
5 to 10 minutes late	3	27
Over 10 minutes late	1	9
Total	40	360

Over 10 minutes late, 5 to 10 minutes late, Up to 5 minutes late, On time, Early

b 18 **c** $\frac{4}{40}$ (10 %)

d Only one train was late for more than 10 minutes so the fraction is $\frac{1}{40}$ (or 2.5%) 2.5 % < 5 %, while 4 trains were late for more than 5 minutes $\frac{4}{40}$ (or 10%), so they achieved it.

9 Paul is correct as the angle for Amber will be 1°, which is too small to show on a pie chart.

10 Helen is correct about the 15° for travelling, but sleeping is 9 × 15 = 135° and not 130°.

11 a Angle = $\frac{12}{120} \times 360° = 36°$

b It is not possible to tell (it is < 10 %)

Answers

12 a

Eagle	1	20°
Birdie	4	80°
Par	11	220°
Bogey	2	40°
Double bogey	0	0°
Total	18	360°

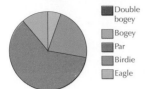

■ Double bogey
■ Bogey
■ Par
■ Birdie
■ Eagle

b Yes, she is under par by 4 **c** The tournament par = 72 × 4 = 288, the player's round score = 4 under par = 72 – 4 = 68, the total score = 68 × 4 = 272 of 288 (16 under par)

13 a The data for some type of flats would require a very small angle to be drawn.

b

Type of flat	Number of applicants
1 bedroom	646
2 bedrooms	1344
More than 3 bedrooms	88

c

Type of flat	Number of applicants	Angle (°)
1 bedroom	646	About 112
2 bedrooms	1344	About 233
More than 3 bedrooms	88	About 15

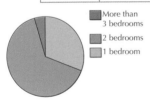

■ More than 3 bedrooms
■ 2 bedrooms
■ 1 bedroom

14

Donations	Legacies	Profit on sale of assets	Investment income	Trading income	Total
815 000	247 000	108 000	87 000	956 000	2 213 000

■ Trading income
■ Investment income
■ Profit on sale of assets
■ Legacies
■ Donations

15 1 with B, 2 with C, 3 with D, 4 with A

17.3 Grouped frequencies

1 a

Number of texts	Tally	Frequency
0–4	ℋℋ ℋℋ	10
5–9	ℋℋ ℋℋ ℋℋ //	17
10–14	ℋℋ ////	9
15–19	ℋℋ ℋℋ ℋℋ ℋℋ /	21
	Total: 57	57

b

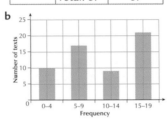

c 15–19

2 a

Number of minutes	Tally	Frequency
0–4	ℋℋ	5
5–9	ℋℋ ///	8
10–14	ℋℋ //	7
15–19	ℋℋ ℋℋ ℋℋ /	16
	Total: 36	36

b

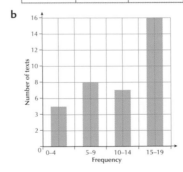

c 15–19

3 a

Number of times	Tally	Frequency
0–4	ℋℋ ℋℋ /	11
5–9	ℋℋ ℋℋ ℋℋ //	17
10–14	ℋℋ /	6
15–19	ℋℋ //	7
20–24	ℋℋ ///	8
25–29	ℋℋ //	7
	Total: 56	56

b

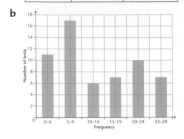

c 5–9

4 a

Minutes wait	Tally	Frequency
0–4	ℋℋ ℋℋ ///	13
5–9	ℋℋ ////	9
10–14	ℋℋ	5
15–19	////	4
20–24	///	3
25–29	//	2

b

0–4	5–9	10–14	15–19	20–24	25–29
13	9	5	4	3	2

c

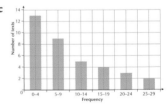

d 0–4

5 a

£0–£4	£4.01–£8	£8.01–£12	£12.01–£16	£16.01–£20	£20.01–£24
4	9	5	4	5	3

b

£0–£6	£6.01–£12	£12.01–£18	£18.01–£24
9	9	6	6

6 a i.

£0–£4	£4.01–£8	£8.01–£12	£12.01–£16	£16.01–£20
7	9	4	5	6

ii.

£0–£6	£6.01–£12	£12.01–£18	£18.01–£24
13	7	8	3

b i. £4.01–£8 **ii.** £0–£6

7 a Table (1) £4.01–£8; table (2) no mode **b** Class size of 4 **c** The class size shows a greater difference between frequencies

8 a Table 3, because the class size shows a greater difference between frequencies

b

Average age	Tally	Frequency
10–19	///	3
20–29	///// ///// //	12
30–39	///// ///// /	11
40–49	////	4
50–59	//	2

c

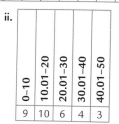

9 a i.

0–5	5.01–10	10.01–15	15.01–20	20.01–25	25.01–30	30.01–35	35.01–40	40.01–45
5	4	5	5	3	3	1	3	3

ii.

0–10	10.01–20	20.01–30	30.01–40	40.01–50
9	10	6	4	3

b i. No mode **ii.** 10.01 – 20
c Class size 10 as it shows the distribution

10 a

21–25	26–30	31–35	36–40	41–45	46–50
5	12	20	18	10	4

b 31 – 35 **c** 69 **d** 55
e Minimum range = 46 – 25 = 21
Maximum range = 50 – 21 = 29

11 a 239 minutes, 3 hours and 59 minutes **b** Spring **c** 1 in 31 **d** 3 days

12 a 41 – 60 **b** 23 **c** 21

13 a 31 – 40 **b** 55 **c** 29 **d** 4

17.4 Continuous data

1 a $10 < T \le 20$ **b** $0 < T \le 10$

2 a

Height, h (metres)	Frequency
$1.40 < h \le 1.50$	2
$1.50 < h \le 1.60$	4
$1.60 < h \le 1.70$	5
$1.70 < h \le 1.80$	7
$1.80 < h \le 1.90$	2

b $1.70 < h \le 1.80$

3 a $0 < T \le 10$ **b** $0 < T \le 10$

4 a $0 < M \le 5$ **b** $5 < M \le 10$

5 a

Length, L (metres)	Frequency
$2.40 < L \le 2.50$	2
$2.50 < L \le 2.60$	3
$2.60 < L \le 2.70$	5
$2.70 < L \le 2.80$	5
$2.80 < L \le 2.90$	5

b No mode

6 a

Temperature, T (°C)	Frequency
$15 < T \le 17$	6
$17 < T \le 19$	11
$19 < T \le 21$	7
$21 < T \le 23$	4
$23 < T \le 25$	3

b $17 < T \le 19$ **c** $19 < T \le 21$

7 a

Height, H (cm)	Frequency
$15 < H \le 20$	6
$20 < H \le 25$	11
$25 < H \le 30$	9
$30 < H \le 35$	1
$35 < H \le 40$	1

b $20 < H \le 25$ **c** 2

8 a

Height, M (grams)	Frequency
$75 < H \le 80$	9
$80 < H \le 85$	7
$85 < H \le 90$	8
$90 < H \le 95$	2
$95 < H \le 100$	2

b $75 < H \le 80$ **c** 4

9 a Smallest is 45.7, largest is 59.8, so divisions of 5 seems sensible

b

$45 < L \le 50$	8
$50 < L \le 55$	12
$55 < L \le 60$	8

10 a Smallest is 50.7, largest is 60.9, so 4 divisions of 3 seems sensible

b

$50 < T \le 53$	13
$53 < T \le 56$	4
$56 < T \le 59$	5
$59 < T \le 62$	6

11 a Smallest is 5.17, largest is 9.92, so 5 divisions of 2 seems sensible

b

$5 < M \le 6$	3
$6 < M \le 7$	3
$7 < M \le 8$	10
$8 < M \le 9$	6
$9 < M \le 10$	6

12 a No **b** Dr Speed's consultation times were from 2 up to 15 minutes; Dr Bell's consultation times were from 6 up to 14 minutes (he managed to follow the practice manager's lower limit); Dr Khan's consultation times were from 3 up to 10 minutes (he managed to follow the practice manager's upper limit).

13

Petrol (Litres)	Frequency
$25 < V \le 30$	2
$30 < V \le 35$	4
$35 < V \le 40$	7
$40 < V \le 45$	6
$45 < V \le 50$	6
$50 < V \le 55$	3
$55 < V \le 60$	2

Modal class $35 < V \le 40$

14

Weight (Kg)	Frequency
$6.0 < W \le 6.2$	8
$6.2 < W \le 6.4$	10
$6.4 < W \le 6.6$	8
$6.6 < W \le 6.8$	3
$6.8 < W \le 7.0$	2

Modal class $6.2 < W \le 6.4$

15

Height (m)	Frequency
$0.90 < H \le 1.00$	2
$1.00 < H \le 1.10$	4
$1.10 < H \le 1.20$	10
$1.20 < H \le 1.30$	11
$1.30 < H \le 1.40$	4

Modal class $1.20 < H \le 1.30$

Answers

16

Volume (L)	Frequency
$0.90 < V \leq 0.95$	6
$0.95 < V \leq 1.00$	6
$1.00 < V \leq 1.05$	10
$1.05 < H \leq 1.10$	4
$1.10 < H \leq 1.15$	5

Modal class $1.00 < V \leq 1.05$

18.1 Short and long multiplication

1 a

×	20	1
6	120	6	$120 + 6 = 126$

b

×	30	7
6	180	42	$180 + 42 = 222$

2 a 108 **b** 144 **c** 198 **d** 154

3 a 135 **b** 160 **c** 256 **d** 152

4 a 168 **b** 378 **c** 171 **d** 57

5 a

×	20	1	
10	200	10	210
6	120	6	126
			336

b

×	50	8	
10	500	80	580
6	300	48	348
			928

6 a 391 **b** 736 **c** 345 **d** 1288

7 a 405 **b** 480 **c** 576 **d** 342

8 a 588 **b** 798 **c** 361 **d** 247

9 a 768 **b** 1696 **c** 954 **d** 1431

10 a 5916 passengers **b** 612 miles

11 30

12 a If the number is (abc) the final answer will be (abcabc) **b** We always multiply by 1001

13 a i. 374 **ii.** 781 **iii.** 286 **iv.** 495
b $XY \times 11 = XZY$, where $Z = X + Y$
c 176 **d** 935

14 No he is not correct; the numbers are in different place values 465 ≠ 663

15

×	7	0.2	
10	70	2	72
2	14	0.4	14.4
0.7	4.9	0.14	5.04
			91.44

Still split up the numbers by place value, so put the 0.7 and 0.2 in their own row or column.

16 £1015

17 £8.67

18 a £188.6 **b** student's answer

18.2 Short and long division

1 a 14 **b** 21 **c** 28 **d** 52

2 a 18 **b** 28 **c** 31 **d** 38

3 a 12 **b** 17 **c** 15 **d** 24

4 a 4 **b** 8 **c** 40 **d** 8

5 a 15 r5 **b** 17 r8 **c** 30 r10 **d** 31 r3

6 a $15\frac{1}{2}$ **b** $17\frac{5}{8}$ **c** $31\frac{1}{4}$ **d** $62\frac{1}{2}$

7 a 14.25 **b** 18.25 **c** 36.5 **d** 54.75

8 a 42.75 **b** 43.25 **c** 136.8 **d** 57.67 (to 2 dp)

9 a 22 teams **b** 26 teams

10 13 boxes of 16 and one box of 24; or 9 boxes of 24 and one box of 16

11 a 58 coaches **b** €39,440 **c** €13.15 (to 2 dp)

12 a 16 **b** 10 more sheets

13 a 14 runners **b** 3 kilometres

14 88

15 a 17 trips **b** 5 trips

18.3 Calculations with measurements

1 a 6 cm **b** 60 cm **c** 600 cm **d** 60 000 cm

2 a 0.456 km **b** 4.562 km **c** 45.62 km **d** 0.4562 km

3 a 340 mm **b** 3400 mm **c** 3400 mm **d** 340 000 mm

4 a 1.259 kg **b** 0.1259 kg **c** 0.001259 kg **d** 125 900 kg

5 a 4320 g **b** 43 200 g **c** 4.32 g **d** 432 g

6 a 2.37 l **b** 0.237 l **c** 2.37 l **d** 0.0237 l

7 a 3 650 000 ml **b** 356 000 ml **c** 356 ml **d** 3560 ml

8 a 86 200 cl **b** 862 cl **c** 86.2 cl **d** 8.62 cl

9 a 1 hour 25 min **b** 3 hours 5 min **c** 1 hour 48 min **d** 3 hours 28 min

10 a 8.17 m **b** 298 g **c** 348.6 cl **d** 389.5 minutes or 6 hours 29.5 minutes

11 a metres **b** kilograms **c** kilometres **d** hours

12 Jug A = 1 litre, Jug B = 800 ml, Jug C = 500 ml
Fill up jug B with water, then transfer the 800 ml into jug A
Fill up jug B again, and use it to fill up jug A
That leaves 600 ml of water in jug B
Use jug B to fill up jug C
Now we have 100 ml left in jug B

13 No (total mass = 1 kg + 0.750 kg + 1.2 kg = 2.95 kg)

14 No, she only gets 7.5 hours, which is 450 minutes.

15 232.57 grams

16 UK

17 1 hour and 30 minutes

18.4 Multiplication with large and small numbers

1 a 1200 **b** 1200 **c** 12 000 **d** 120 000

2 a 2.8 **b** 28 **c** 2.8 **d** 28

3 a 12 **b** 120 **c** 1.2 **d** 12

4 a 0.18 **b** 0.018 **c** 0.48 **d** 0.0048

5 a 540 **b** 5400 **c** 540 **d** 54

6 a 2.1 **b** 2100 **c** 210 **d** 0.00021

7 a 0.012 **b** 12 **c** 12 000 **d** 120

8 a 63 **b** 6.3 **c** 63 000 **d** 0.063

9 a 52 900 **b** 5.29 **c** 0.0529

10 a 77 440 000 **b** 0.7744 **c** 77.44 **d** 0.007744

11 a 4.8 **b** 4.8 **c** 480 **d** 0.00048

12 a 0.16 m³ **b** 160 000 cm³

13 a 60 km **b** 180 km **c** 0.006 km

14 a 4800 cm² **b** 480 000 mm² **c** 0.4800 m² **d** 0.000 00048 km²

15 a 0.06 m³ **b** 0.000 000 000 06 km³

18.5 Division with large and small numbers

1 a 30 **b** 300 **c** 150 **d** 15

2 a 20 **b** 2 **c** 0.2 **d** 0.02

3 a 0.4 **b** 0.04 **c** 0.4 **d** 40

4 a 0.06 **b** 0.03 **c** 0.02 **d** 0.01

5 a 0.06 **b** 0.06 **c** 0.6 **d** 0.006

6 a 30 **b** 0.3 **c** 0.03 **d** 3
c (smallest), b, d, a

7 a 70 **b** 700 **c** 7 **d** 0.07
d (smallest), c, a, b

8 a 1.3 **b** 0.13 **c** 1.3 **d** 130

9 b, because b = 3, but a = c = d = 30

10 d, because d = 4, but a = b = c = 0.4

11 a Since area = $20 \times x = A$ cm², then $x = \frac{A}{20}$ **b** 0.02 **c** 50

12 400 000 hits

13 2333

14 a 0.000 000 012 g
b 0.000 000 000 000 000 000 019 g

15 40 075 km

19.1 Reflections

1

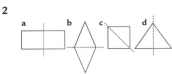

2

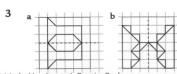

3

4

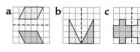

5

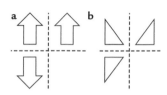

mirror line

mirror line

6

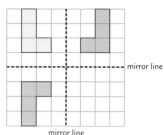

7 a

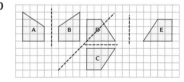

b Student's answer

8

9

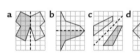

10

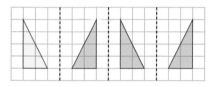

11 a and b

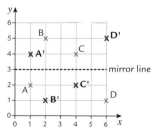

c A'(1,4) – B'(2,1) – C'(4,2) – D'(6,5) **d** (12,0)

12 a, b and d

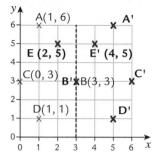

c A'(5,6) – B'(3,3) – C'(6,3) – D'(5,1)

13 a A(2,4); B(1,1); C(4,1)

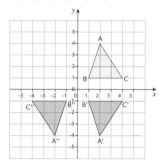

b A'(2,–4), B'(1,–1), C'(4,–1)
c A"(–2,–4), B"(–1,–1), C"(–4,–1)

14 a A'(1,–5); B'(4,–1); C'(1,–1)
b A"(–1, 5); B"(–4, 1); C"(–1, 1)
c

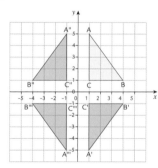

d A'''(–1,–5) – B'''(–4,–1) – C'''(–1,–1)

15 a

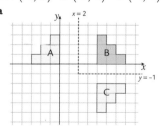

b $y = -1$, a horizontal line, 1 unit below the x-axis

16 No, we don't get back to the original shape

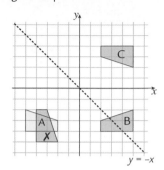

17

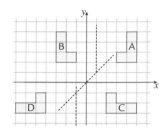

18

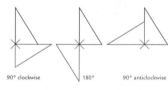

19.2 Rotations

1

90° clockwise   180°   90° anticlockwise

90° clockwise   90° anticlockwise   180° clockwise

2

a 180° clockwise   **b** 90° anticlockwise

c 180° anticlockwise   **d** 270° clockwise

3

a 90° clockwise   **b** 90° anticlockwise   **c** 90° anticlockwise

Answers

4 i. a

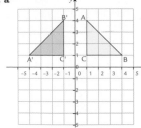

b A(1,4), B(4,1), C(1,1) **c** A'(-4,1), B'(-1,4), C'(-1,1)

ii. a

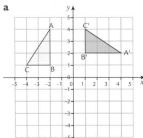

b A(-2,4), B(-2,1), C(-4,1) **c** A'(4,2), B'(1,2), C'(1,4)

iii. a

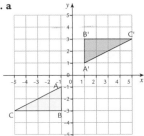

b A(-1,-1), B(-1,-3), C(-5,-3) **c** A(1,1), B'(1,3), C'(5,3)

iv. a

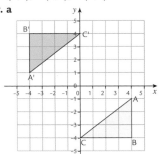

b A(4,-1), B(4,-4), C(0,-4) **c** A'(-4,1), B'(-4,4), C'(0,4)

5 i. a

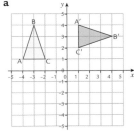

b A(-4,1), B(-3,4), C(-2,1)
c A'(1,4), B'(4,3), C'(1,2)

ii. a

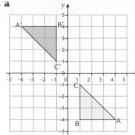

b A(4,-4), B(1,-4), C(1,-1)
c A'(-4,4), B'(-1,4), C'(-1,1)

iii. a

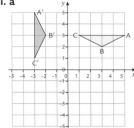

b A(5,3), B(3,2), C(1,3)
c A'(-3,5), B'(-2,3), C'(-3,1)

iv. a

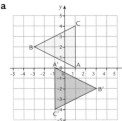

b A(1,0), B(-3,2), C(1,4)
c A'(-1,0), B'(3,-2), C'(-1,-4)

6 a

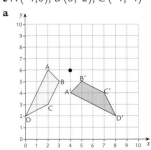

b A'(4,4), B'(5,5), C'(7,4), D'(8,2)

7 a and b

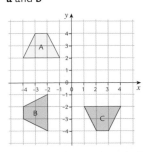

8 a Clockwise 90° about (-2,3)
b Anticlockwise 90° about (1,5)

9 a

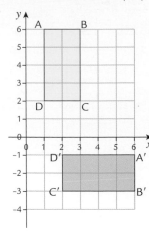

b A'(6,-1), B'(6,-3), C'(2,-3), D'(2,-1) **c** Anticlockwise 90° about (0,0)

10 a

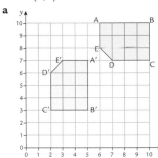

b A'(5,7), B'(5,3), C'(2,3), D'(2,6), E'(3,7) **c** Anticlockwise 90° about (4,9), or clockwise 270° about (4,9)

11 a and b

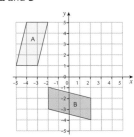

c clockwise 90° about (1,2), or anticlockwise 270° about (1,2)

12 90° clockwise and 270° anticlockwise; 90° anticlockwise and 270° clockwise; 180° clockwise and 180° anticlockwise; 90° clockwise and 450° clockwise Each pair of these rotations will move the shape into the same position

13 **a**, **b** and **c**

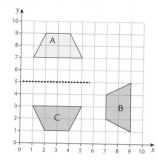

14 **a**, **b** and **c**

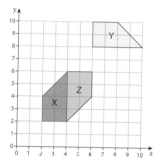

d Rotation 180° clockwise about (4,4)

15 **a**, **b** and **c**

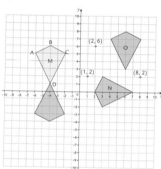

d Yes. Rotate 180° clockwise about (1,2)

16 **a**, **b** and **c**

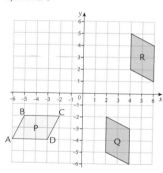

d No. The parallelogram has no symmetry. Look back at the shapes in Q15 (possible reflection), Q16 (no possible reflection)

19.3 Translations

1 **a** 6 units right **b** 3 units down and 3 units right **c** 6 units down **d** 5 units down and 7 units right **e** 6 units down and 6 units left **f** 2 units down and 4 units right **g** 1 unit up and 7 units right **h** 5 units up and 7 units left

2 **a** 6 units right **b** 5 units down and 1 unit left **c** 2 units down and 3 units left **d** 1 unit down and 6 units right **e** 2 units up and 3 units left **f** 3 units down and 4 units left **g** 5 units up and 7 units right

3 **a** P: A(1,4), B(4,2), C(1,2)
b and **d**

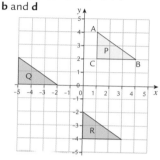

c Q: A'(−5,2), B'(−2,0), C'(−5,0)
e R: A"(0,−2), B"(3,−4), C"(0,−4)
f 6 units up and 1 unit right

4 **a**

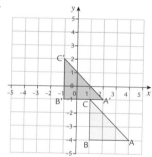

b 3 units up and 2 unit left **c** 3 units down and 2 unit right

5 **a** and **c**

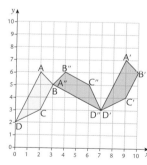

b A'(9,7), B'(10,6), C'(9,4), D'(7,3) **d** Rotation 90° clockwise about (3, 6)

6 **a** and **b**

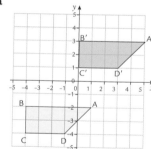

c 5 units up and 2 units right

7 **a** 3 units down and 2 units left **b** 6 units up and 4 units right **c** 4 units down and 2 units right

8 **a**, **b**, **c** and **d**

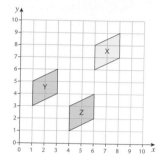

e Reflection in the line $x = 3$

9 **a** 8 units right and 9 units up **b** 8 units left and 9 units down

10 **a**

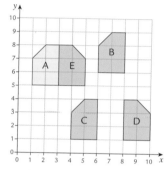

b A'(5,3), B'(0,3), C'(0,1), D'(3,1)
c Moved from one position to another

11 **a** and **b**

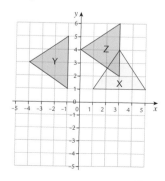

Answers

c Rotation 90° clockwise about (0,4), then translate 3 units right

12 2 units right and 3 units down

13

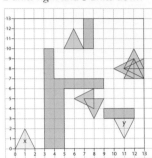

One possible solution:

1. Translate, up 10, right 5
2. Translate, down 3, right 5
3. Translate, up 1 right 0
4. Rotate, anticlockwise 90°, about (11,8)
5. Translate, right 1, down 0
6. Translate, down 3, right 4
7. Rotate, anticlockwise 90°, about (8,5)
8. Translate, down 2, right 3

14

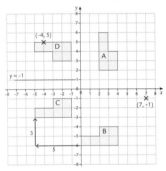

a Rotate anticlockwise 90° about (7,−1) **b** Translate 5 units left and 3 units up **c** Mirror line is at: $y = 1$ **d** Rotate clockwise 90° about (−4,5)

15
1. Rotate 90° clockwise about (3,5) and then translate 2 units right, 4 units down
2. Rotate 90° clockwise about (0,5) and then translate 5 units right, 1 unit down
3. Translate 5 units right, 1 unit down and then rotate clockwise 90° about (5,4)

16 One possible solution of 5 translations:
- 2 units up and 1 unit right, repeated three times
- 1 unit down and 2 units left
- 2 units up and 1 unit right

19.4 Tessellations

1

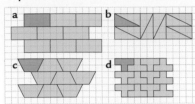

2

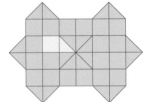

3 a i, iii, vi **b** ii triangle, iv parallelogram

4

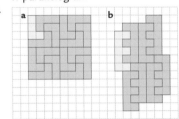

c Not possible **d** Not possible

5

6

7

8

9

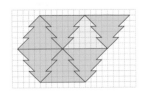

10

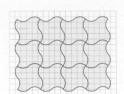

11

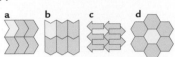

12 **i** c and d **ii** a **iii** b

13 a

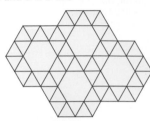

b The angles are 60°.

14

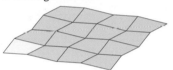

15 Triangle, parallelogram and stars could be used with the pentagon to tessellate, it depends on the way the shapes organised

1

2

3

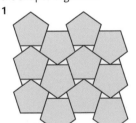

16 The angles will be 90°, 90°, 120°, 120°, and 120°. This will ensure that the vertices will meet accurately

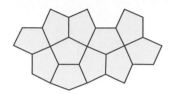

17

The shape between the circles is an equilateral triangle with concave sides

20.1 Powers and roots

1 a 324 b 5832 c 144 d 1728
e 169 f 2197

2 a and b

Cube	A	B	C
Area of each face cm²	1	4	9
Volume cm³	1	8	27

3 a 8 b 9 c 9 d 8

4 a 2.89 b 2.197 c –12.167 d 20.25
e 2.6 f 2.1

5 a 256 b 32 c 86.54 d 67.21
e 256 f 650.38

6 a 400 b 27 000 c 125 000
d 3 200 000 e 4900 f 8 000 000

7

Number	100	1000	10 000	100 000	1 000 000	10 000 000
Power of 10	10^2	10^3	10^4	10^5	10^6	10^7

8 a 4913 b 225 c 196 d 3375

9 a 4.913 b 0.0225 c 19 600
d 3.375

10 a 6 & –6 b 11 & –11 c 1.5 & –1.5
d 2.4 & –2.4 e 1.6 & –1.6
f 60 & –60

11 729

12 All answers = 1

13 a i. 1 ii. –1 iii. 1 iv. –1 v. 1
b i. –1 ii. 1 c If the power is an odd number then the answer will be negative, but if the power is an even number the answer will be positive

14 a 0.16, 0.064, 0.0256 b the sizes of the answers decrease, as we are multiplying decimals smaller than 1 and finding the fractional part of that less than 1 quantity.

15 Estimates between 7 and 7.5, true value is 7.1414

16 a 729 b $4096 = 64^2 = 16^3$

17 9 cm

18 23.04 m²

20.2 Powers of 10

1 a 530 b 79 c 2400 d 506.3 e 0.3

2 a 0.83 b 0.041 c 4.57 d 0.0604
e 347.81

3 a 6430 b 685 c 35 200 d 8074
e 2.1

4 a 0.941 b 0.00523 c 0.568
d 0.000715 e 45.892

5 a 31 b 678 c 0.034 d 8.23
e 5789

6 a 4250 b 567 c 0.023 d 0.00805
e 0.689

7 a i. 27 ii. 270 iii. 2700 b i. 0.5
ii. 5 iii. 50 c i. 380 ii. 3800
iii. 38 000 d i. 0.08 ii. 0.8 iii. 8.0

8 a i. 27 000 ii. 270 000 b i. 500
ii. 5000 c i. 380 000 ii. 3 800 000
d i. 80 ii. 800

9 a i. 0.073 ii. 0.0073 b i. 0.0004
ii. 0.00004 c i. 0.00028
ii. 0.000028 d i. 3.5842
ii. 0.35842

10 i.

	× 10	÷ 100	÷ 10	× 1000	× 100	÷ 1000
a 2000	20 000	20	200	2 000 000	200 000	2
b 7	70	0.07	0.7	7000	700	0.007
c 0.06	0.6	0.0006	0.006	60	6	0.00006

ii.

	÷ 100	× 10 000	÷ 10	× 1000	÷ 10	÷ 10 000
a 2000	20	20 000 000	200	2 000 000	200	0.2
b 7	0.07	70 000	0.7	7000	0.7	0.0007
c 0.06	0.0006	600	0.006	60	0.006	0.000006

11 143 000 000 m

12 273.3 m

13 a Remove the last three zeroes as 1 km = 1000 m b 149 600 000 km

14 a $0.678 \times 10^2 = 67800 \div 10^3$
b $253.8 \div 10^4 = 2.538 \div 10^2$
c $432008 \div 10^6 = 4.32008 \div 10$
d $3.268 \times 10^{23} = 326.8 \times 10^{21}$

15 a i. 300 000 hours ii. 30 000 hours
b 10 000 hours = 416.67 days

16 a 400 000 000 miles b 4 000 000 miles c 672 000 000 miles

17 a 0.000 000 000 000 000 000 000 911 grams
b 0.000 000 000 000 000 000 911 grams

18 a 10 000 000 000 000 000 000 000 00 000 b 600 000 000 000 000 000 000 000 000

20.3 Rounding large numbers

1 a 2600 b 900 c 83 700 d 4900

2 a 3260 b 1820 c 237 890
d 18 660

3 a 16 000 b 208 000 c 1000
d 12 000

4 a 3 550 000 b 9 720 000
c 3 050 000 d 15 700 000

5 a 7 000 000 b 2 000 000
c 1 000 000 d 10 000 000

6

Number	Nearest tens	Nearest hundreds	Nearest thousands	Nearest ten thousands
13 658	13 660	13 700	14 000	10 000
205 843	205 840	205 800	206 000	210 000
16 009 234	16 009 230	16 009 200	16 009 000	16 010 000
798	800	800	1000	0

7 The government, as the figure rounds to 2 million

8 It seems that Jenna received 7499 votes, Mickey received 7500 votes and the results were rounded to the nearest thousand

9 a Manchester and Stansted
b 160 000 000

10 For example if they have 24 500 members if we round as the president did to the nearest thousand he will be correct, but if we round to the nearest ten thousands also the opponent will be correct

Answers

11 Spain 41 millions
Germany 78 millions
Italy 58 millions
France 57 millions
Ireland 4 millions
Denmark 6 millions

12 **a** 10547 **b** 8450

13 **a** 66042996 **b** 66039000

14 **a** 52 and 48 **b** 582 and 584
c 4285 and 4258

20.4 Significant figures

1 **a** 4, 1.3 **b** 2, 320 **c** 3, 5.2 **d** 3, 0.51 **e** 1, 8.0 million

2 **a** 300 **b** 4000 **c** 60 **d** 0.9 **e** 0.09

3 **a** 5300 **b** 50 **c** 9.1 **d** 750 **e** 0.083

4 **a** 2150 **b** 9.56 **c** 0.266 **d** 268 **e** 1.88

5 **a** 3.2 **b** 8 **c** 0.064 **d** 0.008 **e** 83.7

6 **a** 0.1429 **b** 1.6 **c** 0.923 **d** 3.92 **e** 3.1

7 **a** 1 **b** 3 **c** 1 **d** 2

8 **a** 400 **b** 400 **c** 500 **d** 500

9 **a** 5 **b** 4 **c** 4 **d** 1

10 **a** 8 **b** 7 **c** 4 **d** 0.1 **e** 5 **f** 0.05

11 **a** £56540 **b** £38260 **c** £51290

12

Name	Mary	Chadrean	Sheila	Ismail	Sheena
Monthly salary (£)	1313	5429	4633	3039	3817

13 2.6

14 **a** 5, 6, 6, 6 **b** 12.03g

15 11 sq m

16 10000 bricks

17 1500 ml

18 1.91

20.5 Large numbers in standard form

1 **a** 10^2 **b** 10^6 **c** 10^4 **d** 10^1 **e** 10^{15}

2 **a** No, power of 10 part does not exist **b** No, part one should be not less than one **c** Standard form **d** No, the second part should be to the powers of 10 **e** No, part one should be less than 10

3 **a** 5.69×10^3 **b** 1.2×10^6 **c** 7.78×10^4 **d** 3.965×10^8 **e** 7.3×10^1

4 **a** 3.4×10^6 **b** 5.6×10^3 **c** 4.5×10^4 **d** 2.58×10^8

5 **a** 8×10^9 **b** 1.2×10^{10} **c** 1.5×10^{11} **d** 6.7×10^9

6 **a** 2300000 **b** 456 **c** 9000000 **d** 34780

7 **a** 2.5×10^5 **b** 1.764×10^7 **c** 1.369×10^5 **d** 8.1×10^7 **e** 4.225×10^5 **f** 9.0×10^8

8 **a** 7.3×10^7 **b** 2.56×10^4 **c** 2.59×10^5 **d** 1.873×10^{10}

9 **a** 2.01×10^8 **b** 4.7×10^7 **c** 2.976×10^7 **d** 6.8×10^6 **e** 3.3×10^6 **f** 7.98×10^5

10 **a** 5.2×10^8 **b** 6.0×10^9 **c** 7.0×10^{17} **d** 3.6×10^6

11 $7.45 \, km^3$

12 **a** 6.4×10^4 **b** 8.0×10^{18} **c** 4.2875×10^{13}

13 **a** 2.25×10^3 **b** 2.25×10^{13} **c** 1.5×10^3

14 San Marino 3.37×10^4
Russia 1.44×10^8
Germany 8.24×10^7
Ukraine 4.64×10^7
Greece 1.11×10^7
Order: Russia, Germany, Ukraine, Greece, San Marino

15 1.09×10^{30}

16 3.0×10^8

17 6.7 times (about 7 times)

18 1.496×10^8

21.1 Percentage increases and decreases

1 **a** 55 **b** 352 **c** 8.8 **d** 2750 **e** 38.5 **f** 135.3

2 **a** 60 **b** 33 **c** 225 **d** 900 **e** 57 **f** 17.25

3 **a** 112 **b** 2954 **c** 266 **d** 1092

4 **a** 169 **b** 26 **c** 2964 **d** 442

5 **a** 121.6 **b** 798.72 **c** 7436.8 **d** 7.68

6 **a** 104.88 **b** 538.89 **c** 3920.58 **d** 6.21

7 **a** £745.92 **b** 778.08km **c** 74.52 litres **d** 6.12 tonnes

8 **a** $207.92 **b** 86.68m **c** 50.12kg **d** 3.36g

9 **a** £188.23 **b** £80.49 **c** £244.30

10 **a** 775.2 tonnes **b** 583.68 tonnes **c** 485.64 tonnes

11 **a** £444 **b** £14356

12 522

13 The first time we have 10% of 200 = 20, so (200 + 20 = 220); but for the decrease we have 10% of 220 = 22, so (220 – 22 = 198). Since, we found the 10% decrease of a number other than the number we started with, we get a different answer.

14 **a** £38064 per year **b** £122

15 164.565m

16 **a** $\frac{2}{5}$ off **b** £260, £240, £300

17 **a** £150 **b** £5150 **c** £400

18 **a** £725, £520.38 **b** £179.38

21.2 Using a multiplier

1 **a** 1.2 **b** £103.2

2 **a** £38.40 **b** £73.20 **c** £220.80 **d** £11.28

3 **a i.** 1.3 **ii.** 1.36 **iii.** 1.43 **iv.** 1.06 **b i.** 93.6 **ii.** 97.92 **iii.** 102.96 **iv.** 76.32

4 **a** 48.3 **b** 56.7 **c** 77.7 **d** 81.9

5 **a** £16.47 **b** £53.31 **c** £173.24 **d** £469.82

6 **a** 49.98kg **b** 58.38kg **c** 66.78kg **d** 80.64kg

7 **a** 272.7cm **b** 353.7cm **c** 407.7cm **d** 461.7cm

8 **a** 0.85 **b i.** £53.55 **ii.** £44.63 **iii.** £222.70 **iv.** £50.99

9 **a** 0.9 **b** 0.7 **c** 0.63 **d** 0.57 **e** 0.25

10 **a** £69 **b** £23.4 **c** £17.7 **d** £45.48

11 £360 increased by 10% is 360 × 1.1 = £396
£396 decreased by 10% is 396 × 0.9 = £356.40, which is not equal to £360

12 **a** £76.8 **b** £92.16 **c** Different multipliers were used with different prices; the first time it was 80% and then it changed to 120%

13 **a** Because his new salary will be less than his salary before the pay cut **b** 35750 decreased by 2% is 35750 × 0.98 = 35035
35035 increased by 2% is 35035 × 1.02 = 35735.7, which is less than his last salary

14 **a** 244922 **b** The increase of 8% is of a smaller number than the 8% of the population in 10 years' time, so the town is losing more than it is gaining

15 Will, £238

16 **a** 328 **b** 279

17 **a** The 1264 MB file **b** The 1264 MB file

18 **a** 54720 **b** 59280

21.3 Calculating a percentage change

1 15%

2 25%

3 7.5%

4 26%

5 30%

6 4%

7 3%

8 2%

9 30% decrease

10 35%

11 **a** The multiplier is $\left(\frac{142025}{123500}\right)$ = 1.15 **b** 1.15 as a percentage is 115%
The increase is 115% – 100% = 15%

12 **a** The multiplier is 295/250 = 1.18 **b** The percentage is 18%

13 a The multiplier is 1260/1500 = 0.84 **b** The percentage is 16% decrease

14 Red party, 5% increase
White party, 55% decrease
Blue party, 6% increase

15 a 0.74 **b** 26%

16 a 62% **b** 34% **c** 17%

17 a 28% **b** 50% **c** 8% increase

18

Job	Cost	Income	Profit	Percentage Profit
4 Down Close	120	162	42	35%
High Birches	50	91	41	82%
27 Bowden Rd	25	42	17	68%
Church hall	20	34	14	70%

22.1 Graphs from linear equations

1 a

x	-2	-1	0	1	2	3
y = x + 3	1	2	3	4	5	6

b, c

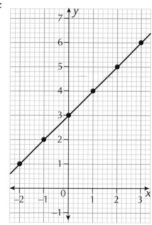

2 a

x	-2	-1	0	1	2	3
y = x - 2	-4	-3	-2	-1	0	1

b, c

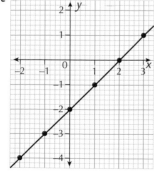

3 a

x	-2	-1	0	1	2	3
y = 3x	-6	-3	0	3	6	9

b, c

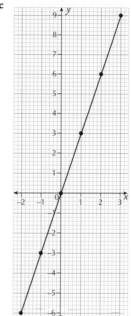

4 a

x	-2	-1	0	1	2	3
4x	-8	-4	0	4	8	12
y = 4x + 1	-7	-3	1	5	9	13

b, c

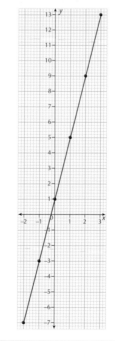

5 a

x	-2	-1	0	1	2	3
4x	-8	-4	0	4	8	12
y = 4x - 1	-9	-5	-1	3	7	11

b, c

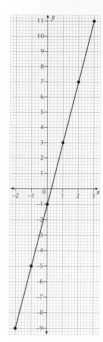

6 a, and b

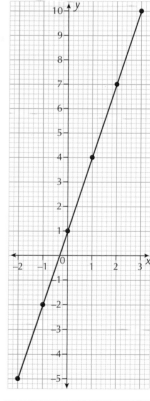

7 a

x	-4	-2	0	2	4	6
0.5x	-2	-1	0	1	2	3
y = 0.5x + 2	0	1	2	3	4	5

b, c

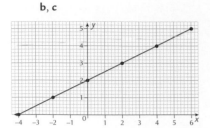

d We find that for each *x*-value, there's a definite *y*-value, the different lines passes through the coordinates formed by the *x*-value and its respective *y*-value. We also see that the coordinates fit their respective equations, viz, $y = x$, $y = 2x$ and $y = 4x$.

10 Coefficient of *x* is the same and they are parallel

b, c

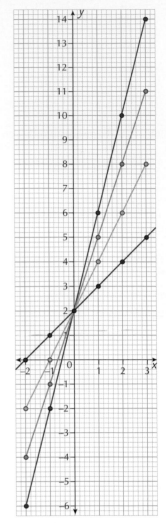

8 a

x	0	1	2	3	4	5	6	7	8	9	10
5	5	5	5	5	5	5	5	5	5	5	5
$y = 5 - x$	5	4	3	2	1	0	-1	-2	-3	-4	-5

b, c

11 a

x	-2	-1	0	1	2	3
$y = 2x$	-4	-2	0	2	4	6
$y = 2x + 2$	-2	0	2	4	6	8
$y = 2x - 2$	-6	-4	-2	0	2	4
$y = 2x - 4$	-8	-6	-4	-2	0	2

b, c

9 a

x	-2	-1	0	1	2	3
$y = x$	-2	-1	0	1	2	3
$y = 2x$	-4	-2	0	2	4	6
$y = 4x$	-8	-4	0	4	8	12

b, c

d Parallel line because of the same coefficient of *x*.
The point where the line cuts through the *y*-axis has the same value as the number added to the multiple of *x*, in each line

12 a

x	-2	-1	0	1	2	3
$y = x + 2$	0	1	2	3	4	5
$y = 2x + 2$	-2	0	2	4	6	8
$y = 3x + 2$	-4	-1	2	5	8	11
$y = 4x + 2$	-6	-2	2	6	10	14

d Every line cuts the same point at *y* axis at (0,2) **e** Coefficient of *x* are different about the lines

f

13 **a**
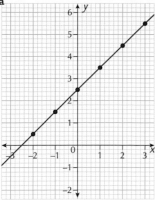

b
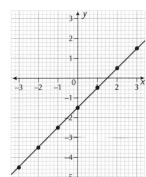

14 **a**
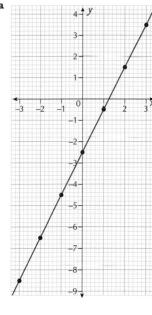

b
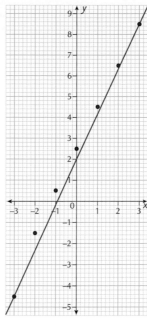

15 **a**

x	-2	-1	0	1	2	3
$y = 5x - 1$	-11	-6	-1	4	9	14
$y = 2x - 4$	-8	-6	-4	-2	0	2

b (-1, -6)

16
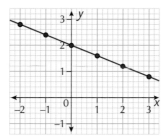

22.2 Gradient of a straight line

1 **a** 3 **b** 2 **c** 4 **d** 1
2 **i. a** 3 **b** 1 **c** 3 **d** 2 **ii. a** 3
   b 1 **c** 0 **d** 4
3 **a** $3x + 3$ **b** $x + 1$ **c** $3x$ **d** $2x + 4$
4 **a** 7 **b** 2 **c** 1
5 **a** $3x + 5$ **b** $2x + 7$ **c** $x + 4$ **d** $7x + 15$
6 **a** (0,1) **b** 2 **c** $y = 2x + 1$
7 **i. a** 2 **b** 3 **c** 4 **d** $\frac{2}{3}$ **ii. a** (0,3)
   b (0,1) **c** (0,2) **d** (0,3) **iii. a** $2x + 3$
   b $3x + 1$ **c** $4x + 2$ **d** $\frac{2}{3}x + 3$ (or)
   $0.6x + 3$
8 $y = 3x - 3$
9 $y = 2x + 1$
10 $y = 5$

11 **a** Yes, the straight line $y = 2x - 5$ is parallel to the one already drawn which is $2x + 2$. The coefficients of x are same, this makes the line parallel **b** No, $y = 2x - 5$ cut the y-axis at (0,-5) but the one already drawn cut the y-axis at (0,2)
12 **a** 3 **b** (0,-1) **c** $y = 3x - 1$
13 **a** 3 **b** (0,-5) **c** $y = 3x - 5$
14 Rhombus
15 **a** $y = 3x + 2$ **b** $y = 4x - 1$
   c $y = 2x + 1$
16 **a** $y = x + 6$ **b** $y = 2x - 3$ **c** $y = 4x + 1$
17 a and c lie on (M, N)

18 **a**

x	-2	0	4	5
$y = x + 2$		2	6	7
$y = 2x$	-4	0	8	10

b
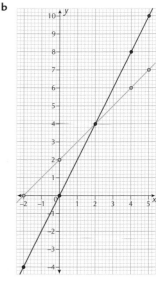

c (2,4)

22.3 Graphs from quadratic equation

1 **a**

x	-3	-2	-1	0	1	2	3
x^2	9	4	1	0	1	4	9
$y = x^2 + 1$	10	5	2	1	2	5	10

b, c
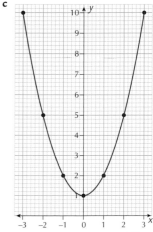

Answers

2 a

x	−3	−2	−1	0	1	2	3
x^2	9	4	1	0	1	4	9
$y = x^2 + 3$	12	7	4	3	4	7	12

b, c

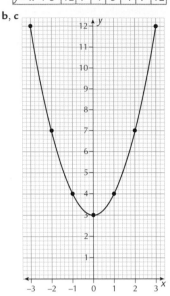

3 a

x	−3	−2	−1	0	1	2	3
$y = x^2 + 2$	11	6	3	2	3	6	11
$y = x^2 + 4$	13	8	5	4	5	8	13
$y = x^2 + 5$	14	9	6	5	6	9	14

b, c

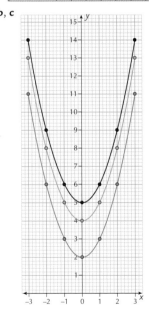

4 a

x	−3	−2	−1	0	1	2	3
x^2	9	4	1	0	1	4	9
$y = 2x^2$	18	8	2	0	2	8	18

b, c

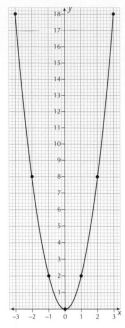

5 a

x	−3	−2	−1	0	1	2	3
x^2	9	4	1	0	1	4	9
$y = 2x^2 + 1$	19	9	3	1	3	9	19

b, c

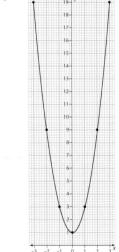

6 a

x	−3	−2	−1	0	1	2	3
x^2	9	4	1	0	1	4	9
$y = 0.5x^2$	4.5	2	0.5	0	0.5	2	4.5

b, c

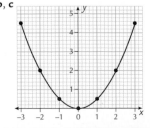

7 a

x	−3	−2	−1	0	1	2	3
x^2	9	4	1	0	1	4	9
$y = 0.5x^2 - 3$	1.5	−1	−2.5	−3	−2.5	−1	1.5

b, c

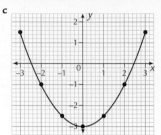

8 The lines are symmetric about the axis. The lines open upward. At $x = 0$, we have $y = x^2 + 2$, for which the intercept is shifted by 2 units, while for $y = x^2 + 4$, the intercept is shifted by 4 units.

9 All graphs have the same shape but are steeper for higher coefficients of x.

10 a

x	−3	−2	−1	0	1	2	3
x^2	9	4	1	0	1	4	9
$y = x^2 - 1$	8	3	0	−1	0	3	8

b When squaring an x, both negative x and positive x becomes x^2. The equation of $y = x^2 - 1$ is $y = (-2)^2 - 1 = 4 - 1 = 3$ and $y = (2)^2 - 1 = 4 - 1 = 3$

11 Emma is correct. When $x = -2$, $y = 8$
$8 = (-2)^2 + 4 = 4 + 4$

12

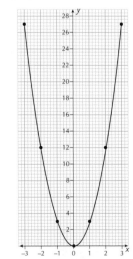

13 a

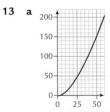

b 18.03 km **c** 44.16 km

14 a

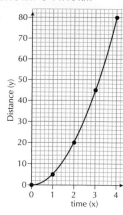

b $2 \times \sqrt{3}$ or 3.46 seconds

15 Line 1 $y = x^2 + 10$
    Line 2 $y = x^2 - 2$
    Line 3 $y = 9x^2$
    Line 4 $y = x^2 - 10$
    Line 5 $y = 3x^2 + 4$

23.1 Scatter graphs and correlation

1 **a** Strong negative correlation
    b Moderate positive correlation
    c None
    d Moderate negative correlation
    e Strong positive correlation
    f Weak negative correlation

2 **i.** e **ii.** a **iii.** e **iv.** c **v.** d **vi.** b **vii.** c

3 **i.** As the daily temperature increases the number of ice cream sold also increases
    ii. When the distance increases amount of fuel decreases
    iii. When the age of child increases height also increases
    iv. Heights of adults are not related to their age
    v. When the daily temperature increases the number of coats sold decreases
    vi. When the amount spent on revision increases, the exam score also increases
    vii. Price of perfume and amount of rainfall has no correlation

4 **a** Strong positive correlation
    b The sample is small and correlation does not imply causation

23.2 Creating scatter graphs

1 a

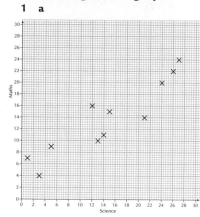

b Moderate positive correlation between Science and Maths scores

2

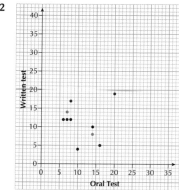

No relation between oral and written test

3 a

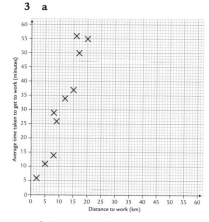

b Strong positive correlation between the time taken to get to work and the distance to work
c Answers may vary from student to student. Sample answer: It is because the route one person takes is longer than what the other person takes.

4

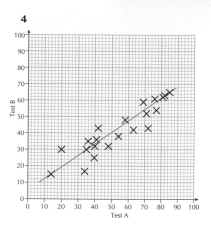

Expected scores: Test A = 8, Test B = 11
Test A = 64, Test B = 50
Test A = 51, Test B = 41
Test A = 33, Test B = 28

24.1 Congruent shapes

1 **a** Congruent **b** Not congruent
    c Congruent **d** Congruent
    e Not congruent

2 a and e are congruent b and j are congruent c and k are congruent d and f are congruent

3 F = I = L
    D = M = J
    C = K
    G = N

4 A = H   G = M
    B = D = N   O = K
    C = J - L   F = I

5 a and c are congruent

6 B = D

7 A and C are congruent – SAS
    B and E are congruent – ASA
    D and F are congruent – SSS

8 **a** Two triangles are congruent because all three sides have same length in both triangles by the condition SSS. 42 mm = 4.2 cm, $2\frac{1}{2}$ cm = 2.5 cm and $3\frac{1}{4}$ cm = 3.25 cm **b** Two triangles are congruent because all three sides have same length in both triangles by the condition SSS.
    $3\frac{3}{8}$ cm = 3.375 cm, $2\frac{5}{8}$ cm = 2.625 cm and $4\frac{1}{8}$ = 4.125 cm
    c Two triangles are not congruent because only two sides are same. It doesn't satisfy any condition.
    32 mm = 3.2 cm; $1\frac{1}{2}$ cm = 1.5 cm but $2\frac{1}{4}$ cm ≠ 2.4 cm

Answers

9   All three triangles are congruent. We know the sum of angles of a triangle is 180°. Therefore, the third angle in the first triangle is (180° – (100°+50°)) = 30°. The first and the second triangles are congruent by the condition ASA. The first and third triangles are congruent by the condition SAS.

10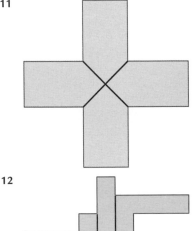

11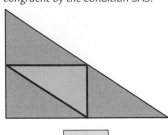

12

13 **a** Two triangles are congruent because two sides have same length and the angle between them is the same size in both triangles by the condition SAS **b** Two triangles are congruent because all three sides have same length in triangles that satisfy the condition SSS **c** Two triangles are congruent because two angles have same size and the side between them is the same length in both triangles that satisfy the condition ASA **d** Two triangles are congruent because two angles have same size and the side between them is the same length in both triangles that satisfy the condition ASA

14 **a**

P   Q

S   R

b PQRS is a parallelogram, we know that PQ and SR are parallel, PS and QR are parallel. SQ is common to both triangles hence the triangles SPQ and SRQ have same sides that satisfy the condition SSS and hence they are congruent

15 **a**

b

16 **a** 16 **b** 7 **c** 3

17 **a** A, C, D and H are must be congruent to it **b** B, F and G could be congruent, but more information is needed **c** E is definitely not congruent to it

24.2 Enlargements

1 **a**

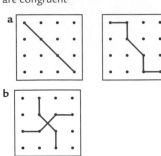

b

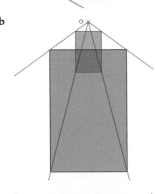

c

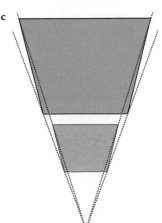

d

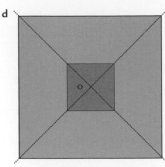

2 **a**

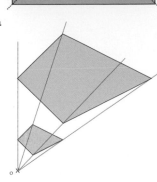

b

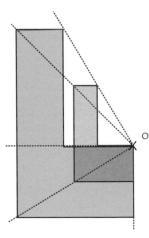

3 **a**

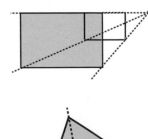

b

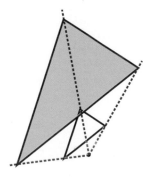

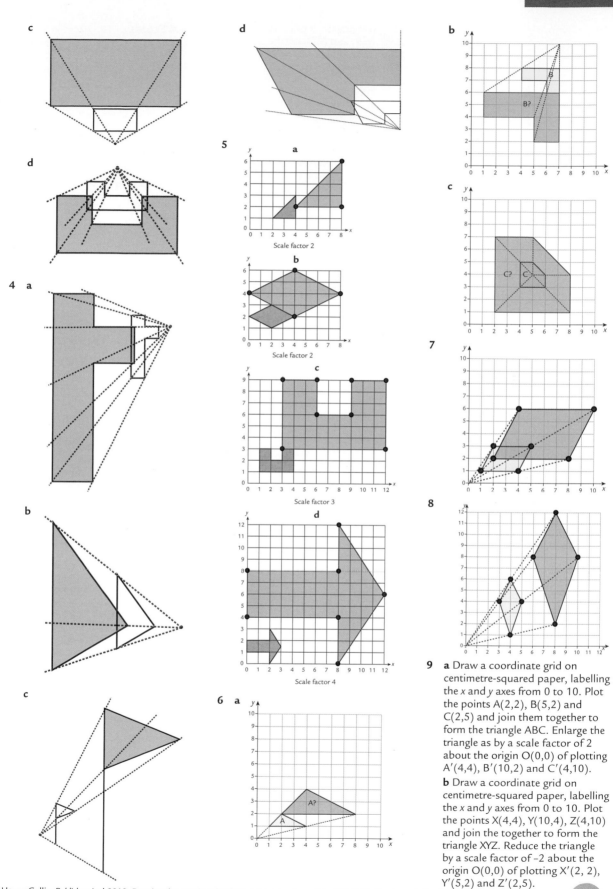

9 a Draw a coordinate grid on centimetre-squared paper, labelling the x and y axes from 0 to 10. Plot the points A(2,2), B(5,2) and C(2,5) and join them together to form the triangle ABC. Enlarge the triangle as by a scale factor of 2 about the origin O(0,0) of plotting A'(4,4), B'(10,2) and C'(4,10).

b Draw a coordinate grid on centimetre-squared paper, labelling the x and y axes from 0 to 10. Plot the points X(4,4), Y(10,4), Z(4,10) and join the together to form the triangle XYZ. Reduce the triangle by a scale factor of –2 about the origin O(0,0) of plotting X'(2, 2), Y'(5,2) and Z'(2,5).

Answers

10
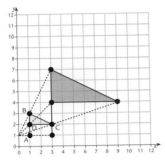

11
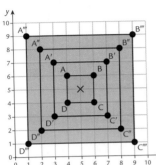

a (3,7) (7,7) (7,3) and (3,3) are the coordinates of A′B′C′ and D′, respectively **b** (2,8) (8,8) (8,2) and (2,2) are the coordinates of A′B′C′ and D′, respectively **c** (1,9) (9,9) (9,1) and (1,1) are the coordinates of A″B″C″ and D″, respectively **d** From point A to A′, A″, A‴, A⁗ the x coordinate decreases by 1 and y coordinate increases by 1.
From point B to B′, B″, B‴, B⁗ both coordinates are same and increases by 1.
From point C to C′, C″, C‴, C⁗ x coordinate increases by 1 and y coordinate decreases by 1.
From point B to D′, D″, D‴, D⁗ both coordinates are same and both decreases by 1.

12 By checking the length of the sides. Each side length of big shape is twice the length of smaller one.

13 a
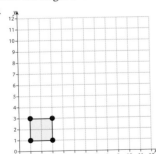

b 4 cm² **c** 16 cm² **d** 36 cm²
e 64 cm² **f** The area scale factor is the square of the length scale factor **g** Yes

14 a Pentagon **b** (0,0), (0,2), (3 4), (6,2) and (6,0) are the coordinates of A′, B′, C′, D′ and E′, respectively **c** (0, 0), (0, 3), (4.5, 6), (9, 3), and (9, 0) are the coordinates of A″, B″, C″, D″ and E″, respectively **d** (0, 0) does not move its position. Yes, this happens with any enlargement

e

Original point	Enlargement scale factor 2	Enlargement scale factor 3
A(0,0)	A′(0,0)	A′(0,0)
B(0,1)	B′(0,2)	B′(0,3)
C(1.5,2)	C′(3,4)	C′(4.5,4)
D(3,1)	D′(6,2)	D′(9,3)
E(3,0)	E′(6,0)	E′(9,0)

15 a AD = 10 cm **b** CB = 3.5 cm **c** 4:1 **d** 65 **e** ABC and BCD

16 a 2 **b** (9, 1)

17 (0, 8)

18 a (6,6), (9,15), (18,21) and (21,12) are the coordinates of enlarged shape **b** (2,1), (3,4), (4,5) and (6,3) are the coordinates of enlarged shape

24.3 Shape and ratio

1 a 2 : 5 **b** 1 : 10 **c** 4 : 5 **d** 1 : 5 **e** 1 : 4

2 a 9 : 2 **b** 4 : 9 **c** 3 : 4 **d** 9 : 20 **e** 13 : 16 **f** 3 : 20

3 1 : 3

4 4 : 9

5 1 : 10

6 A, C and F has sides in the ratio 1 : 3.

7 a 1 : 3 **b** 1 : 3 **c** 1 : 9

8 a 1 : 2 **b** 1 : 2 **c** 1 : 4

9 a 1 : 3, 1 : 4, 3 : 16 **b** 1 : 4 **c** 1 : 16

10 a 2 : 3 **b** 2 : 9 **c** $\frac{2}{9}$ **d** 9 : 20 **e** $\frac{2}{9} < \frac{9}{20}$

11 a 1200 m² **b i.** 30 000 m² **ii.** 3 Hectares **c** 1 : 5 **d** 1 : 25 **e** $\frac{1}{25}$

12 a 24 litres **b** 18 litres **c** 3 : 4

13 a 1 : 2 **b** 1 : 4 **c** 1 : 8 **d i.** 1 : 9 **ii.** 1 : 27

14 a 1 : 5 **b** 1 : 25 **c** 1 : 125 **d i.** 1 : 16 **ii.** 1 : 64

15 a 3 : 2 **b** 40 m² **c i.** 120 m² **ii.** 80 m² **iii.** £ 2240

16 a 23 : 13 **b** 30 m² **c** 90 m² **d** 3 : 1

17 a 2 : 3 **b** 4 : 5

18 a 2 : 3 **b** 4 : 9 **c** 8 : 27 **d** Area ratio is length ratio squared, volume ratio is length ratio cubed.

24.4 Scales

1 a 20 m **b** 55 m **c** 40 m **d** 90 m **e** 70 m

2 a 90 cm **b** 5.5 m **c** 9.45 m **d** 12 m

3 a 5.6 m **b** 3.5 m **c** 6.5 m

4 a length = 5.5 cm, height = 2.25 cm **b** length = 3 cm, width of blade = 0.24 cm **c** length = 3 cm, width = 1.5 cm

5 a 1020 m **b** 600 m **c** 612 000 m²

6 a scale 1 cm to 1.78 m **b** 6 m **c** 48

7

	Actual length	Length on scale drawing
a	4 m	16 cm
b	1.5 m	6 cm
c	50 cm	2 cm
d	3 m	12 cm
e	2.5 m	10 cm
f	1.2 m	4.8 cm

8

	Scale	Scaled length	Actual length
a	1 cm to 3 m	5 cm	15 m
b	1 cm to 2 m	12 cm	24 m
c	1 cm to 5 km	9.2 cm	46 km
d	1 cm to 7 miles	6 cm	42 miles
e	5 cm to 8 m	30 cm	48 m

9

	Scale	Scaled length	Actual length
a	1 cm to 6 m	3 cm	18 m
b	1 cm to 3 km	8 cm	24 km
c	1 cm to 5 km	5.2 cm	26 km
d	1 cm to 5 miles	8 cm	40 miles
e	1 cm to 6 miles	15 cm	90 miles

10 a i. On the scale drawing, the length of the kitchen is 3.5 cm, so the actual length of the room is 7 m.
On the scale drawing, the width of the kitchen is 2.5 cm, so the actual width of the room is 5 m.
On the scale drawing, the width of the window is 1.1 cm, so the actual width of the window is 2.2 m.
ii. On the scale drawing, the length of the bathroom is 2.3 cm, so the actual length of the room is 4.6 m.
On the scale drawing, the width of the bathroom is 1.2 cm, so the actual width of the room is 2.2 m.
On the scale drawing, the width of the window is 0.6 cm, so the actual width of the window is 1.2 m.

iii. On the scale drawing, the length of the bedroom 1 is 3.5 cm, so the actual length of the room is 7 m. On the scale drawing, the width of the bedroom 1 is 2.3 cm, so the actual width of the room is 4.6 m. On the scale drawing, the width of the window is 1.1 cm, so the actual width of the window is 2.2 cm.

iv. On the scale drawing, the length of the bedroom 2 is 2.9 cm, so the actual length of the room is 5.8 m. On the scale drawing, the width of the bedroom 2 is 2.3 cm, so the actual width of the room is 4.6 m. On the scale drawing, the width of the window is 1.1 cm, so the actual width of the window is 2.2 m.

b 58.28 m²

11 No, he's not correct. The scale is 12 cm to 54 miles, which is simplified to 2 cm to 9 miles

12 15 cm → 45 km
5 cm → 15 km
1 cm → 3 km

13 For example 1 cm = 10 yards

14 a i. and **ii.**

Section	Dimension in cm	Dimension in m	Scaled area	Real life area
Reception	2.5×2	5×4	5	20
Storeroom	8×2	16×4	16	64
Toilet	1.5×1	3×2	1.5	6
Office	$(3 \times 1.5) + (2 \times 4)$	$(6 \times 3) + (4 \times 8)$	12.5	50
Shop	$(8 \times 2) + (10.5 \times 2)$	$(16 \times 4) + (21 \times 4)$	37	148

b i. 1 : 2 **ii.** 1 : 4

15 Map ratio of 1 : 300 000 means
1 cm → 300 000 cm
300 000 cm = 3000 m = 3 km
So the distance on the map will be 12 ÷ 3 = 4 cm

16 Map ratio of 1 : 150 000 means
1 cm → 150 000 cm
150 000 cm = 1500 m = 1.5 km
So the distance on the map will be 45 ÷ 1.5 = 30 cm

17 Map ratio of 1 : 500 000 means
1 cm → 500 000 cm
500 000 cm = 5000 m = 5 km
So the distance on the map will be 60 ÷ 5 = 12 cm

18 a Scale: 1 cm to 250 m **b i.** 5 cm = 1250 m **ii.** 3.5 cm = 875 m **c** 8.5 cm = 2125 m = nearly 4.5 minutes. So they would have 30 and a half minutes to spare.

25.1 Algebraic notation

1 a $4k$ **b** $3t$ **c** $2x$ **d** $20y$ **e** ab **f** nm **g** $(\frac{1}{2})e$ **h** $(\frac{1}{4})r$

2 a $2ab$ **b** 40 **c** $10e$ **d** $80w$ **e** abc **f** $3n$ **g** $7.5x$ **h** $(\frac{3}{4})bh$

3 a a^2 **b** $4x^2$ **c** $9n^2$ **d** $1.4t^2$ **e** $3.2a$ **f** $(\frac{1}{5})d^2$ **g** $21x^2$ **h** $(\frac{2}{3})m^2$

4 a $2(n+1)$ **b** $4(t+12)$ **c** $8(3+k)$ **d** $0.5(t-6.4)$ **e** $4(k+2)$ **f** $2(12-y)$ **g** $8(x-2)$ **h** $6(40-y)$

5 a $2+3a$ **b** $9x$ **c** $7.7w$ **d** $ab-1.4$ **e** $5+4a$ **f** $8.5d$ **g** $12-7n$ **h** $13f+9$

6 a $6n$ **b** $20b$ **c** $2d$ **d** $0.5q$ **e** $10k$ **f** $24g$ **g** $12t$ **h** $1.0h$ or h

7 a $4n^2$ **b** $12d^2$ **c** $8p^2$ **d** $3.0a^2$

8 b $\frac{y}{4}$ **c** $\frac{t}{1.5}$ **d** $\frac{24}{n}$

9 h $\frac{m}{3}$ **c** $\frac{n}{4}$ **d** $\frac{ab}{2}$

10 b $\frac{3cd}{2}$ **c** $\frac{5ed}{4}$ **d** $\frac{12fg}{7}$

11 a Rosi has incorrectly calculated $5c \times 3c = 15c$ rather than multiplying $c \times c$ to get c^2. Correct answer is $5c \times 3c = 15c^2$ **b** This is correct **c** Rosi has incorrectly calculated $\frac{5}{g}$ rather than $\frac{g}{5}$. Correct answer is $\frac{g}{5}$ **d** Rosi has incorrectly calculated $2 + 4$ rather than 2×4. Correct answer is $2u \times 4v = 8uv$. **e** Correct **f** Rosi has incorrectly calculated $(9-5) \times y$ rather than $9 - (5 \times y)$. Correct answer is $9 - (5 \times y) = 9 - 5y$

12 a $3y + 2x$ **b** $5xy$ **c** $6xy$ **d** $x + 5y$

13 $3w \times 2h$ because the only term $3w \times 2h = 6wh$ others cannot simplify further.

14 $\frac{36xy}{2xy}$. All the other simplify to $18xy$

15 24 years old

16 23 and 17

17 48

18 $x = 8$, $8 - 7 = 1$

25.2 Like terms

1 a $11h$ **b** $5p$ **c** $6u$ **d** $2b$ **e** $5j$ **f** $5pr$ **g** $6k$ **h** $8y^2$ **i** $8h + 5g$ **j** $2g + 8m$ **k** $8f + 10d$ **l** $11x + 5y$ **m** $6 + 2r$ **n** $4 + 2s$ **o** $3c + 3$ **p** $14b + 7$

2 a $14w - 7$ **b** $6bf + 5g$ **c** $7d + 3d^2$ **d** $4t^2 + 5t$ **e** $2t - 3s$ **f** $2h - 2h^2$ **g** $4y - 9w$

3 a $16e + 6$ **b** $19u - 6$ **c** $4b + 3d$ **d** $7 + 7c$ **e** $6 - g$ **f** $2h + 2$

4 a and **d** will not simplify further

5 all

6 a $5a^2b$ **b** $2x^2y + 3xy^2$ **c** $5cd^2$ **d** $5e^2f$

7 a $5a^2b + ab$ **b** $2x^2y + 7xy^2$ **c** $10cd^2$ **d** 0

8 a $6x - 3y$ **b** $5x - y$ **c** $5x - y$ **d** $6x - 3y$

9 a $5x - y$ **b** $2x + 4y$ **c** $5x - y$ **d** $2x + 4y$

10 $3x + 6y - 3x + 6y$; all the other expressions result to 0, whereas this expression results to $12y$.

11 +, −, −

12 a $3b$, $4a$ **b** $8a$, $5b$ **c** $4a$, $5b$ **d** $4a$, $5b$

13 a y, $3x$ **b** $9y$, x **c** $8y$, x **d** $10y$, $2x$

14 b $2x + y + 3x - 2y = 5x - y$ **c** $2x + y + 2x + y + 4x - y = 8x + y$ **d** $2x + y + 3x - 2y + 4x - y = 9x - 2y$

25.3 Expanding brackets

1 a $5p + 10$ **b** $4m - 12$ **c** $2t + 2u$ **d** $4d + 8$ **e** $5b + 25$ **f** $6j - 24$ **g** $10 + 2f$ **h** $10 - 10n$

2 a $2a + 2b$ **b** $3q - 3t$ **c** $4t + 4m$ **d** $5x - 5y$ **e** $2f + 2g$ **f** $6 - 4f$ **g** $1.5h + 15$ **h** $14 - 4f$

3 a $5w + 2$ **b** $3d + 10$ **c** $9h + 15$ **d** $6x + 12$ **e** $2m - 13$ **f** $4 + 3q$

4 a $6a + 8$ **b** $8i + 22$ **c** $5p + 4$ **d** $8d - 7$ **e** $6e + 2$ **f** $8x + 2$ **g** $8m - 3$ **h** $9u - 22$

5 a $4a + 6$ **b** $6x - 15$ **c** $12t + 20$ **d** $50n - 30$ **e** $30 + 12a$ **f** $8 - 12y$ **g** $60 + 80r$ **h** $35 - 10m$

6 a $10a + 11$ **b** $15c + 20$ **c** $14e + 11$ **d** $13f + 31$

7 a $6g + 7$ **b** $6h + 6$ **c** 6 **d** $4k + 10$

8 a $5m - 18$ **b** $18n + 4$ **c** $5p - 27$ **d** $-2q + 14$

9 $5a + 30 + 3a + 6 = 4(2a + 9)$
$8a + 36 = 8a + 36$ are equivalent

10 $12x + 18 - 9x - 6 = 3(x + 4)$
$3x + 12 = 3x + 12$ are equivalent

11 c and e are equivalent

12 a $5x + 18$ **b** $-5x - 18$ **c** Both are same numbers but differs in part a and part b with positive and negative signs respectively

13 $a = 15$, $b = 21$

14 $a = 2$, $b = 2$

15 a Mistake 1: $(3x)$ as $5 + 3 = 8$ rather than $5 \times 3x = 15x$
Mistake 2: $-2(-3) = -6$ rather than 6 **b** $13x + 26$

25.4 Using algebraic expressions

1 $2x + 2$ cm

2 a $4x$ cm **b** x^2 cm²

3 a $4t + 2$ **b** $6w$ **c** $6m + 2$

4 a $t^2 + t$ **b** $2w^2$ **c** $2m^2 + m$

5 a $3t + 6$ **b** $4k + 8$ **c** $4c + 16$ **d** $3a + 24$ **e** $5x + 10$

6 a $9x\,\text{cm}^3$ b $10ab\,\text{cm}^3$ c $16t^2\,\text{cm}^3$

7 a i. $4x\,\text{cm}^2$ ii. $4x + 16\,\text{cm}^2$
iii. $2x + 2\,\text{cm}^2$ iv. $2x + 6\,\text{cm}^2$
b $12x + 24\,\text{cm}^2$

8 a $2x + 8\,\text{cm}$ b $2x + 16\,\text{cm}$
c $2x + 6\,\text{cm}$ d $2x + 10\,\text{cm}$
e $4x + 20\,\text{cm}$

9 a $5a + 10\,\text{cm}^2$ b $4a + 8\,\text{cm}^2$
c $9a\,\text{cm}^2$ d $18a + 18\,\text{cm}^2$

10 a $2a + 14\,\text{cm}$ b $2a + 12\,\text{cm}$
c $2a + 18\,\text{cm}$ d $4a + 22\,\text{cm}$

11
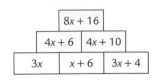

12
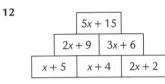

13
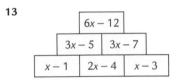

14 a $3t + 6$ b $4k + 8$ c $4c + 16$
d $3a + 24$ e $5x + 10$

15 a $(2 \times x) + (7 \times 3)$
b $(9 \times 3) + 2 \times (x - 3)$
c $2x + 21 = 27 + 2x - 6$
$2x + 21 = 2x + 21$

16
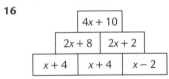

17
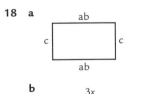

18 a
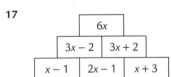

b

c

d
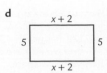

25.5 Using index notation

1 a 4^3 b 3^6 c 10^5

2 a $2 \times 2 \times 2 \times 2$ b $5 \times 5 \times 5$
c $6 \times 6 \times 6 \times 6 \times 6$
d $9 \times 9 \times 9 \times 9 \times 9 \times 9$

3 b 2^4 c 2^5 d 2^6

4 a a^4 b r^3 c b^5 d m^6 e $12a^2$ f $2p^2$
g $12g^3$ h $24k^4$

5 a $5f$ b w^4 c $7c$ d k^4 e $6D$

6 a a^2b b $6xy^2$ c u^2t^2 d $2cd^2$ e $2ab^2$
f $2xw^2$ g $4ut^2$ h $252e^2f$

7 a c^3 b $15c^5$ c c^6 d $63c^6$

8 a t^4 b $6t^4$ c t^4 d $30t^4$

9 a $20w^2 - 4w^3$ b $21x^3$ c $24y^3z - 40yz^2$

10 a $10a^2b^8$ b $64c^6$ c $16d^8e^4$

11 $6w = w + w + w + w + w + w$
Adding 6 times of w is $6w$
$w^6 = w \times w \times w \times w \times w \times w$;
multiplying 6 times of w is w^6

12 Height of the triangle is $\frac{b}{2}$ cm

13 $2^6 = 2 \times 2 \times 2 \times 2 \times 2 \times 2 = 64$
$4^3 = 4 \times 4 \times 4 = 64$

14 $2a^2$

15 a $4x^2\,\text{cm}^3$ b $y^3\,\text{cm}^3$ c $2t^3\,\text{cm}^3$
d $4k^3\,\text{cm}^3$ e $6n^3\,\text{cm}^3$

16 a 512 b 2048

17 $2^3 \times 2^4$; add the powers. So answer is $2^{3+4} = 2^7$

26.1 Adding and subtracting fractions

1 a $\frac{7}{8}$ b $\frac{11}{12}$ c $\frac{1}{12}$ d $\frac{1}{20}$

2 a $4\frac{1}{4}$ b $3\frac{1}{8}$ c $5\frac{7}{12}$ d $8\frac{11}{12}$

3 a $1\frac{3}{4}$ b $\frac{5}{8}$ c $2\frac{5}{8}$ d $\frac{1}{12}$

4 a $\frac{23}{24}$ b $\frac{5}{24}$

5 a $7\frac{7}{8}$ b $7\frac{7}{12}$ c $7\frac{2}{9}$ d $7\frac{7}{18}$

6 a $3\frac{2}{3}$ b $3\frac{3}{5}$ c $2\frac{5}{6}$

7 a $3\frac{1}{8}$ b 1 c $4\frac{3}{8}$

8 a $\frac{13}{40}$ b $\frac{3}{20}$ c $4\frac{3}{8}$

9 a $6\frac{3}{20}$ b $1\frac{7}{20}$ c $9\frac{9}{10}$ d $5\frac{1}{10}$

10 $12\frac{5}{12}$

11 $5\frac{7}{20}$

12 a $3\frac{5}{6}$ b $25\frac{2}{3}$

13 $2\frac{7}{8}$

14 a $2\frac{2}{3}$ and $2\frac{5}{12}$ b $2\frac{1}{4}$ and $2\frac{1}{6}$

15 a $3\frac{5}{6}$ b $7\frac{1}{18}$ c $8\frac{1}{18}$ d $1\frac{5}{6}$

16 $\frac{23}{24}$ m

17 $\frac{19}{24}$ litres

26.2 Multiplying fractions

1 a $\frac{3}{4}$ b $\frac{9}{4}$ c $\frac{12}{5}$ d $\frac{3}{2}$ e $\frac{3}{5}$ f $\frac{12}{5}$
g $\frac{8}{3}$ h $\frac{5}{3}$

2 a $\frac{1}{6}$ b $\frac{3}{10}$ c $\frac{1}{12}$ d $\frac{2}{9}$

3 a $\frac{1}{3}$ b $\frac{1}{2}$ c $\frac{4}{9}$ d $\frac{8}{15}$ e $\frac{1}{12}$ f $\frac{5}{9}$

4 a $\frac{1}{6}$ b $\frac{9}{20}$ c $\frac{1}{6}$ d $\frac{1}{3}$

5 a $\frac{1}{4}$ b $\frac{4}{9}$ c $\frac{9}{25}$ d $\frac{9}{16}$

6 a $\frac{1}{2}$ b $\frac{1}{12}$ c $\frac{7}{12}$ d $\frac{5}{8}$

7 a $\frac{1}{16}$ b $\frac{9}{16}$ c $\frac{4}{9}$ d $\frac{16}{25}$

8 a $\frac{9}{20}$ b $\frac{1}{4}$ c $\frac{1}{4}$

9 a $\frac{5}{12}$ b $\frac{5}{12}$ c $\frac{25}{64}$ d $\frac{4}{9}$

10 a 4 tins b 7 hours

11 a $\frac{2}{9}$ b $\frac{1}{18}$ c $\frac{2}{9}$

12 a $\frac{6}{7} \times \frac{5}{6} \times \frac{4}{5} \times \frac{3}{4} \times \frac{2}{3} \times \frac{1}{2} = \frac{1}{7}$
b $\frac{1}{10}$

13 Tom is wrong because half of the amount given to Jenny will be shared among *three* children while another half of the amount given to Mike will be shared only among *two* children. Jenny's children will get $\frac{1}{6}$ of the original amount and Mike's children will get $\frac{1}{4}$ of the original amount.

14 $a = 3$ and $b = 4$

15 a 8 ml of red, 6 ml of blue and 22 ml of yellow b 27 ml

16 a Jamie does $18\,\text{m}^2$ and Ollie does $12\,\text{m}^2$ b Jamie gets £54 and Ollie gets £36 c $\frac{5}{6}$ d Jamie plaster $\frac{18}{35}$

17 a 576 km b $\frac{1}{9}$ th part

26.3 Multiplying mixed numbers

1 a $\frac{3}{4}$ b $\frac{5}{8}$ c $\frac{5}{8}$ d $\frac{7}{8}$

2 a $4\frac{1}{6}$ b $4\frac{3}{8}$ c $5\frac{5}{6}$ d 12

3 a $4\frac{4}{9}$ b $6\frac{2}{9}$ c $4\frac{2}{3}$ d $14\frac{2}{3}$

4 a $\frac{9}{16}$ b $5\frac{1}{16}$ c $18\frac{1}{16}$ d $3\frac{1}{16}$

5 a 10 b 12 c $3\frac{3}{5}$ d $10\frac{16}{27}$

6 **a** $13\frac{1}{5}$ **b** $33\frac{1}{4}$ **c** $14\frac{2}{3}$ **d** $26\frac{3}{7}$

7 **a** $\frac{1}{5}$ **b** 4 **c** $6\frac{1}{4}$ **d** $\frac{8}{27}$

8 **a** Looking at the whole number parts 4 and 5, the answer should be at least $4 \times 5 = 20$. So Ana's answer is more likely to be correct.
b $22\frac{14}{28}$ or $22\frac{1}{2}$

9 **a i.** $2\frac{1}{4}$ **ii.** $1\frac{7}{9}$ **iii.** $1\frac{9}{16}$ **iv.** $1\frac{24}{25}$
b iv. Is closest to 2

10 **a i.** $1\frac{11}{25}$ **ii.** $1\frac{13}{36}$ **iii.** $1\frac{15}{49}$ **iv.** $1\frac{17}{64}$
b Closer to 2 **c** No, it will not be less than 1 because given every x values are greater than 1. So x^2 should be greater than 1.

11 Squaring a number is multiplying the number by itself. So both answers remains same.
Squared individual number and then multiplied altogether is as same as multiplied all numbers and then whole squared.

12 **a** 330 minutes **b** 170 minutes
c 168 minutes

13 **a i.** $\frac{3}{10}$ **ii.** $\frac{5}{10}$ **iii.** $\frac{7}{10}$ **iv.** $\frac{9}{10}$
b $\frac{1}{5} \times 5\frac{1}{2} = 1\frac{1}{10}$
$\frac{1}{5} \times 6\frac{1}{2} = 1\frac{3}{10}$

14 **a** $1\frac{2}{5}$ **b** $\frac{7}{40}$ **c** $\frac{21}{40}$

15 **a i.** 1 **ii.** 1 **iii.** 1 **iv.** 1
b $2\frac{1}{3} \times \frac{3}{7} = 1$ and $1\frac{1}{4} \times \frac{4}{5} = 1$

26.4 Dividing fractions

1 **a** 6 **b** 9 **c** $7\frac{1}{2}$ **d** 5

2 **a** 8 **b** $7\frac{1}{2}$ **c** 16 **d** $4\frac{4}{5}$

3 **a** 2 **b** $1\frac{1}{2}$ **c** $2\frac{1}{2}$ **d** 4

4 **a** $1\frac{1}{4}$ **b** $\frac{1}{3}$ **c** 4 **d** $\frac{1}{4}$

5 **a i.** $\frac{3}{4}$ **ii.** $1\frac{1}{3}$ **b i.** 2 **ii.** $\frac{1}{2}$

6 **a** 5 **b** 10 **c** $3\frac{1}{3}$ **d** 4

7 **a** $2\frac{1}{4}$ **b** $4\frac{1}{3}$ **c** $6\frac{3}{4}$ **d** $12\frac{1}{2}$

8 **a** 9 **b** $4\frac{1}{3}$ **c** $\frac{1}{3}$ **d** $\frac{7}{9}$

9 **a** $1\frac{3}{7}$ **b** $1\frac{2}{3}$ **c** $\frac{3}{4}$ **d** $2\frac{5}{8}$

10 **a** $2\frac{2}{5}$ **b** $\frac{4}{15}$

11 **a i.** 10 and $\frac{1}{10}$ **ii.** 12 and $\frac{1}{12}$
iii. 80 and $\frac{1}{80}$ **b** Reciprocal of
numbers **c** Bigger **d** Smaller

12 When we change the order of fraction, the answer is also reciprocated.

13 **a** $2\frac{1}{4}$ **b** $5\frac{1}{4}$

14 **a** 636mph **b** Hussain is right and 1017.6 km/h

15 $6\frac{5}{6}$ bottles

16 230 tiles

27.1 Parts of a circle

1 **a** 12 cm **b** 15 m

2 **i a** 1.2 cm **b** 1.6 cm **c** 2.4 cm
ii a 2.4 cm **b** 3.2 cm **c** 4.8 cm

3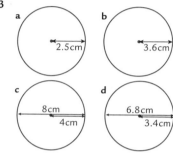

4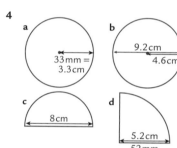

5 **a** Mark a point O on a sheet of paper, where a circle is to be drawn
Take a pair of compasses and measure the given radius using scale
Without disturbing the opening of the compasses, keep the needle at point O and draw a complete arc holding the compasses from its end

b Use the same procedure as a and draw a circle of given radius. Then draw the second circle keeping its centre straight to first circle, such that they touch each other
Repeat the same step for the third circle of given radius so that it touches the second circle

6 **a i.** Chord **ii.** Chord **iii.** Tangent
iv. Radius **b i.** Triangle **ii.** Triangle
iii. Parallelogram **iv.** Trapezium

7 Student's answer

8 Student's answer

9 **a** Draw a circle of radius 5 cm using compasses.
Mark the end points and centre as a, b, and o, respectively.
Draw another circle of diametre 2.5 (half of the large circle's radius) so that it must touch points a and o (i.e., a and o are the end points of the new circle).
b Draw a dotted line of 6 cm long, for radius. Construct a right angle at the left side and draw a 6 cm dotted line, perpendicular to the original line. Use compasses to draw the arc with a radius of 6 cm. Using the same centre, draw another arc of radius 3 cm and join the end points of the arcs by drawing lines.
c Draw a square of side 4 cm. Place the compasses on the right side top point and then draw the arc of radius 4 cm to the left side such that the arc touches the two ends of the square.

10 Match **a** C **b** I **c** J **d** H **e** G **f** B
g E **h** A **i** F **j** D

11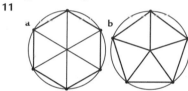

12 **a** Octagon **b** Use a protector to draw ten radii that form angles of 36° at the centre of the circle.
Join the ten points where the radii meet the circumference.

13 Use a protector to draw twenty radii that form angles of 18° at the centre of the circle.
Join the twenty points where the radii meet the circumference to get twenty-sided polygon.

14 No, because if we draw radii that form angles of 20° at the centre of the circle then we will get only 18 sided polygon and not 20 sided polygon.

15 **a–d** Student's answer
e Perpendicular bisector

16 **a–d** Student's answer **e** It must be a centre point of the circle

17 **a**
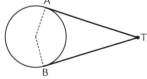
b Perpendiculars meet at the centre

18 Right angle triangles and similar triangles

Answers

27.2 Formula for the circumference of a circle

1. **a** 31.4 m **b** 25.12 m
2. **a** 22 cm **b** 3.5 cm or 34.5 mm
   c 6.6 cm or 65.9 mm **d** 15.1 m
   e 8.8 cm
3. **a** 188.4 cm **b** 108 cm
4. **a** 8.8 cm or 88 mm **b** 65.9 mm or 6.6 cm
5.

Radius	Diametre	Circumference
7 cm	14 cm	43.96 cm
2.5 cm	5 cm	15.7 cm
50 cm	100 cm	314 cm

6. 20.6
7. 14.3
8. 20.5
9. 94 cm
10. 263 cm
11. 942 million kilometres
12. 31.8 m
13. 377 m
14. 458.73 m
15. 353 cm
16. 71.429 mm
17. 47.1 cm
18. 16.82 m

27.3 Formula for area of a circle

1. **a** 78.5 m^2 **b** 50.24 m^2
2. **a** 201 cm^2 **b** 153.9 cm^2 **c** 63.6 cm^2
   d 10.2 cm^2
3. **a** 3.1 cm^2 **b** 153.9 mm^2 **c** 3.5 m^2
   d 38.5 cm^2 **e** 95 m^2
4. 346.185 cm^2
5. Circumference: 37.7 cm.
   Area: 113 cm^2
6. 19.6 cm^2
7. 190.8 cm^2
8. Area of semicircle $\frac{1}{2}\pi r^2 = \frac{1}{2}\pi 4^2$
   $= 8\pi$ cm^2
9. 50π cm^2
10. 36π cm^2
11. Yes, he is correct. Area of circle is πr^2 therefore $\pi 15^2 = 706.5$ cm approximately 707 cm.
12. Given diametre is 14 cm, therefore radius is 7 cm and so the area will be $\pi 7^2 = 49\pi = 153.86$ which is approximately 154 cm^2 so his assumption is correct
13. Area of the circle is πr^2 and it is not πd, therefore his result is wrong and the correct answer is 16π ˜ cm^2
14. Area of circle of radius 5 cm is 25π and area of circle of radius 10 cm is 100π cm^2.
    No, he is wrong, because two lots of 25π is 50π and it is not equal to 100π.
15. 8163 m^2

16. **a** 576π cm^2 **b** 48π cm^2 **c** 9.6π cm^2
17. 25.12 cm^2
18. $100 - 12.5\pi$ m^2

28.1 Using probability scales

1. **a** $\frac{1}{2}$ **b** $\frac{1}{10}$ **c** $\frac{7}{25}$ **d** $\frac{3}{10}$ **e** $\frac{4}{25}$
   f $\frac{7}{50}$
2. **a** $\frac{1}{4}$ **b** $\frac{1}{5}$ **c** $\frac{2}{5}$ **d** $\frac{1}{10}$ **e** $\frac{1}{20}$ **f** $\frac{9}{10}$
3.

| 0 | 0.1 | 0.2 | 0.3 | 0.4 | 0.5 | 0.6 | 0.7 | 0.8 | 0.9 | 1.0 |

A B C D
D' C' B' A'
0 0.4 0.7 0.8

4.

Outcome	Probability of outcome occurring (P)	Probability of event not occurring (1 – P)
A	$\frac{1}{4}$	$\frac{3}{4}$
B	$\frac{1}{3}$	$\frac{2}{3}$
C	$\frac{3}{4}$	$\frac{1}{4}$
D	$\frac{1}{10}$	$\frac{9}{10}$
E	$\frac{2}{15}$	$\frac{13}{15}$

Outcome	Probability of outcome occurring (P)	Probability of event not occurring (1 – P)
A	$\frac{2}{3}$	$\frac{1}{3}$
B	0.35	0.65
C	8%	92%
D	0.04	0.96
E	$\frac{5}{8}$	$\frac{3}{8}$
F	0.375	0.625

5. 0.4
6. **a** $\frac{1}{4}$ **b** $\frac{3}{5}$ **c** $\frac{3}{4}$ **d** 0 **e** $\frac{3}{4}$ **f** $\frac{2}{5}$ **g** $\frac{1}{4}$
   h 1
7. 0.991
8. **a** $\frac{1}{10}$ **b** $\frac{4}{5}$ **c** $\frac{17}{50}$ **d** $\frac{16}{25}$
9. **a** $\frac{1}{4}$ **b** $\frac{5}{6}$ **c** $\frac{7}{12}$ **d** $\frac{11}{12}$ **e** 0 **f** $\frac{1}{6}$
10. The probability that a ticket is a winning ticket is $\frac{1}{100}$
    The probability that a ticket is a losing ticket is $\frac{99}{100}$
    So, the number of losing tickets is $\frac{99}{100} \times 2400 = 2376$
11. **a** P(even) $= \frac{4}{9} >$ P(>6) $= \frac{3}{9}$
    b P(prime) $= \frac{4}{9} <$ P(odd) $= \frac{5}{9}$
    c P(multiple of 5) $= \frac{1}{9} <$ P(multiple of 4) $= \frac{2}{9}$
    d P(triangular number) $= \frac{3}{9} =$ P(square number) $= \frac{3}{9}$

12. No, since the total amount of discs in a bag is not given
13. As the number of blue and black pens is not given, the probability of blue pens cannot be calculated. However, since there are only two types of pens and the number of blue pens is higher than the number of black pens, the probability of selecting a blue pen is more than 50% or $\frac{1}{2}$.
14. **a** $\frac{1}{3}$ **b** 3
15. **a** $\frac{1}{5}$ **b** $\frac{2}{5}$ **c** $\frac{1}{4}$ **d** $\frac{1}{2}$
16. Bel assumed that there is no other probability than losing or winning the game, and hence, the total probability of winning or losing the game is 1.

28.2 Mutually exclusive outcomes

1. **a** Red shirts are not blue shirts, so they are mutually exclusive
   b Striped shirts are shirts containing blue, so they are not mutually exclusive
2. A and C are mutually exclusive
3. A and B are mutually exclusive
4. **a** No **b** Yes **c** No
5. **a**
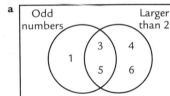
   b No **c** $\frac{1}{3}$
6. **a**
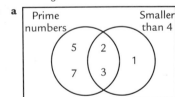
   b No **c** $\frac{1}{5}$
7. **a** Yes **b** No **c** No **d** No **e** No **f** Yes
8. **a** Yes **b** No **c** Yes **d** No **e** No **f** No
9. **a**
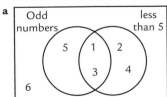
   b Not mutually exclusive as 1 and 3 are in both sets
10. 35 is a multiple of 5 and multiples of 7, so they are not mutually exclusive

11 a

Spinner 1	Spinner 2	Total score
+2	0	2
+2	−1	1
+2	+1	3
−3	0	−3
−3	−1	−4
−3	+1	−2
+4	0	4
+4	−1	3
+4	+1	5

b A and C are mutually exclusive so there are no outcomes occurring together

12 a B and C are mutually exclusive so there are no outcomes occurring together **b** Because there are some numbers in the intersection, this shows that the outcomes are not mutually exclusive

13 An even number, a number greater than 7, a number less than 4

14 4 pairs of these outcomes are mutually exclusive

15 5 pairs of these outcomes are mutually exclusiv.

28.3 Using sample spaces to calculate probabilities

1 a

	1	2	3	4	5	6
Head	H,1	H,2	H,3	H,4	H,5	H,6
Tail	T,1	T,2	T,3	T,4	T,5	T,6

b i. $\frac{1}{12}$ ii. $\frac{1}{4}$ iii. $\frac{1}{3}$

2

DIE 1	DIE 2	SUM
1	1	2
1	2	3
2	1	3
1	3	4
2	2	4
3	1	4
1	4	5
2	3	5
3	2	5
4	1	5
2	4	6
3	3	6
4	2	6
3	4	7
4	3	7
4	4	8

3 a

	1p	2p	5p	10p	20p	50p	£1
Kim	Kim, 1p	Kim, 2p	Kim, 5p	Kim, 10p	Kim, 20p	Kim, 50p	Kim, £1
Franz	Franz, 1p	Franz, 2p	Franz, 5p	Franz, 10p	Franz, 20p	Franz, 50p	Franz, £1

b i. $\frac{1}{14}$ ii. $\frac{3}{14}$ iii. $\frac{1}{7}$ iv. $\frac{13}{14}$ v. $\frac{5}{7}$

4 a

Clyde	Delroy
plain	plain
plain	cheese
plain	beans
cheese	plain
cheese	cheese
cheese	beans
beans	plain
beans	cheese
beans	beans

b i. $\frac{1}{3}$ ii. $\frac{1}{3}$ iii. $\frac{1}{9}$ iv. $\frac{1}{9}$ v. $\frac{1}{9}$ vi. $\frac{1}{3}$ vii. $\frac{4}{9}$ viii. $\frac{2}{3}$

5 a

	1	2	3	4	5	6
1	2	3	4	5	6	7
2	3	4	5	6	7	8
3	4	5	6	7	8	9
4	5	6	7	8	9	10
5	6	7	8	9	10	11
6	7	8	9	10	11	12

b i. $\frac{1}{12}$ ii. $\frac{1}{9}$ iii. 0 iv. $\frac{1}{36}$ v. $\frac{5}{12}$ vi. $\frac{7}{12}$ vii. $\frac{1}{6}$ viii. $\frac{1}{2}$ ix. $\frac{5}{18}$ x. $\frac{13}{18}$

6 a

	1	2	3	4	5	6
1	1	2	3	4	5	6
2	2	4	6	8	10	12
3	3	6	9	12	15	18
4	4	8	12	16	20	24
5	5	10	15	20	25	30
6	6	12	18	24	30	36

b 6, 12 **c** $\frac{5}{18}$

7 a

	1	2	3	4	5	6
1	2	3	4	5	6	7
2	3	4	5	6	7	8
3	4	5	6	7	8	9
4	5	6	7	8	9	10
5	6	7	8	9	10	11

b 30 **c** i. $\frac{2}{15}$ ii. 0 iii. $\frac{1}{2}$ iv. $\frac{13}{15}$ v. $\frac{1}{2}$ vi. $\frac{1}{3}$

8 Listing all the outcomes: HHH, HHT, HTH, THH, TTT, TTH, THT, HTT
Probability of three heads is $\frac{1}{8}$

9 a Listing all the outcomes:
Apples, Bananas
Apples, Pears
Bananas, Pears
b There are more pears.

10 3157, 3517, 3197, 3917, 3597, 3957

11 a

	0	1	2	3
0	0	1	2	3
1	1	2	3	4
2	2	3	4	5
3	3	4	5	6

b $\frac{3}{16}$ **c** There are 16 scores, out of which 3 scores are less than 2

12 $\frac{1}{10}$

13 An even score

29.1 Equations with and without brackets

1 a 33 **b** 48 **c** 252 **d** 4

2 a $x = 13$ **b** $y = 9$ **c** $k = 2\frac{3}{5}$ **d** $n = 4\frac{5}{8}$

3 a $x = 11$ **b** $x = 3$ **c** $y = 18$ **d** $a = 44$

4 a $x = 48$ **b** $x = 36$ **c** $x = 45$ **d** $x = 140$

5 a $y = 24$ **b** $t = 18$ **c** $t = 27$ **d** $x = 22$

6 a $3x - 15 = 11$ **b** $x - 5 = \frac{11}{3}$
$3x = 15 + 11; x = \frac{11}{3} + 5$
$3x = 26; x = \frac{11 + 15}{3}$
$x = 8\frac{2}{3}; x = \frac{26}{3}$ or $8\frac{2}{3}$

7 a $x = 6\frac{3}{4}$ **b** $x = 5\frac{1}{4}$ **c** $y = 6\frac{3}{4}$ **d** $r = 6\frac{2}{5}$

8 a $x = 3.6$ **b** $w = 2.8$ **c** $t = 16.1$ **d** $t = 62.5$

9 a $a = 14$ **b** $b = 4$ **c** $c = 35$ **d** $d = 39$

10 a $p = 4.1$ **b** $q = 4.9$ **c** $w = 2.7$

11 a second line error, should be
$9x - 2 = 25$
$9x = 25 + 2 = 27$
$x = 27/9, x = 3$
b third line error, should be
$4y = 19 - 3$
$4y = 16$
$y = 4$
c third line error, should be
$4x - 6 = 14$
$4x = 14 + 6$
$4x = 20, x = 5$

12 a second line error, should be
$4x = 9 - 3; 4x = 6; x = \frac{3}{2}$
b second line error, should be
$2x + 3 = \frac{24}{4}; 2x = 6 - 3; x = \frac{3}{2}$

13 Ann is not correct. $2(b + 2) = 20$
$b + 2 = 10$
$b = 10 - 2; b = 8$

14 Peter is correct

15 **a** $3x + 12\frac{1}{2} = 37\frac{1}{2}$ **b** $x = 8\frac{1}{3}$

16 **a** (1,7), (2,6), (3,5), (5,3), (6,2), (7,1) **b** 22, 34, 62, 26
c $a = 149$ **d** Since $p = 2a+b$, 35 is an odd number which is not a multiple of 2.

17 **a** $2(4y - 3) + 4y = 24$ **b** $y = 2\frac{1}{4}$

18 **a** $x + (3x - 5) + (2x + 1) = 11$
b $x = 2\frac{1}{2}$ cm

29.2 Equations with the variable on both sides

1 **a** $x = 15$ **b** $x = 15$ **c** $t = 12$ **d** $x = 11$

2 **a** $x = 9$ **b** $y = 6$ **c** $t = 4\frac{1}{2}$ **d** $k = 18$

3 **a** $y = 20$ **b** $z = 7$ **c** $x = 11$ **d** $x = 5\frac{1}{2}$

4 **a** $m = 8$ **b** $n = 21$ **c** $p = 6$ **d** $x = 8$

5 **a** $x = 18$ **b** $t = 10\frac{1}{2}$ **c** $x = 3\frac{1}{3}$
d $n = 34$ **e** $x = 7$ **f** $d = 13$ **g** $x = 5$
h $x = 20$

6 **a** $x = 12$ **b** $x = 8$ **c** $x = 6$ **d** $x = 4$

7 **a** $x = 15$ **b** $x = 9$ **c** $x = 4$ **d** $x = 12$
e $x = 4$ **f** $x = 24$ **g** $x = 2$ **h** $x = 5$

8 **a** $x = 4\frac{1}{2}$ **b** $x = 2\frac{1}{4}$ **c** $x = 6\frac{1}{2}$
d $x = 3\frac{1}{4}$

9 **a** $y = -5$ **b** $k = -2$ **c** $r = -3$ **d** $n = -3$

10 **a** $x = -11$ **b** $x = 2$ **c** $x = \frac{2}{3}$ **d** $x = \frac{3}{7}$

11 Joel is not correct. $4x - 3 = x + 9$
$4x - x = 9 + 3$
$3x = 12$
$x = 4$

12 **a** $4x + 2x = 9$
$6x = 9$
$x = \frac{9}{6}$ or $\frac{3}{2}$
b $3x - 9 = 4x$
$3x - 4x = 9$
$x = -9$

13 Kathy is not correct
$b + 8 = 3b$
$8 = 3b - b$
$8 = 2b; b = 4$

14 Jack is not correct
$8x - 5 = 12x$
$- 5 = 12x - 8x$
$x = -1\frac{1}{4}$

15 **a** $2x + 35 = 3x + 12$ **b** $x = 23$ **c** $2 \times 23 + 35 = 81$ and $3 \times 23 + 12 = 81$

16 **a** $8 - x$ **b** $10 - 5x$ **c** $8 - x = 10 - 5x$ and a stamp cost £0.50
d £$7\frac{1}{2}$ or 7.5

17 **a** 114 cm² **b** 225 cm²

29.3 More complex equations

1 **a** $x = 9$ **b** $x = 8$ **c** $x = 18$ **d** $x = 2$

2 **a** $y = 24$ **b** $a = 9$ **c** $t = 8$ **d** $x = 21$
e $n - 22$ **f** $x = 8$

3 **a** $x = 3\frac{3}{5}$ **b** $x = 2$ **c** $x = \frac{5}{2}$ **d** $x = 5$

4 **a** $x = 2$ **b** $g = 4$ **c** $w = -9$

5 **a** $x = 24$ **b** $a = 21$ **c** $t = 14$ **d** $p = 5$
e $x = 0$ **f** $x = 20\frac{1}{2}$

6 **a** $a = 3$ **b** $b = 8$ **c** $c = 4$ **d** $d = 6$ **e** $e = 2$ **f** $f = 7$

7 **a** $x = 2$ **b** $x = 11$ **c** $x = 3$ **d** $x = 6$ **e** $x = -4$ **f** $x = -3$

8 **a** $w = 19$ **b** $x = 13$ **c** $y = 11$

9 **a** $x = 11$ **b** $y = 12$ **c** $t = 4$ **d** $x = 23$
e $n = 32$ **f** $x = 28$

10 **a** $x = 2$ **b** $y = -1$ **c** $t = 1$

11 **a**

x	5	6	7	8	9
$2(x + 1)$	12	14	16	18	20
$3(x - 2)$	9	12	15	18	21

b 8 **c** 8

12 **a** Perimetre of hexagon is $6(y + 6)$ and pentagon is $5(y + 8)$ **b** $y = 4$
c 60

13 a and d have the same answer
$w = 6$
b and f have the same answer
$w = 15$
c and e have the same answer
$w = 2.5$

14 **a** $5(x - 4) = 3(x + 2)$ **b** $x = 13$
c $5(13 - 4) = 45$ and $3(13 + 2) = 45$

15 **a** $3(t + 6)$ **b** $4t$ **c** $3(t + 6) = 4t$
d $t = 18$ **e** Sides of triangle are 24, 24, 24 and sides of squares are 18, 18, 18

16 **a** $2(x + 5)$ **b** $4(x - 2)$ **c** $2(x + 5) = 4(x - 2)$ **d** $x = 9$ **e** $2(9 + 5) = 28$ and $4(9 - 2) = 28$

17 **a** $a + 6$ **b** $\frac{a + 6}{4} = 12$ **c** $a = 42$

29.4 Rearranging formulae

1 **a** $t = s - 25$ **b** $t = w + 6.5$ **c** $t = \frac{a}{4}$
d $t = \frac{m}{21}$

2 **a** $n = \frac{T - m}{b}$ **b** $n = q - t + 12$
c $n = \frac{y - a}{3}$ **d** $n = \frac{y - 3}{3}$

3 **a** $m = r + 3$ **b** $m = \frac{r}{4} + 3$
c $m = \frac{r + 3}{4}$ **d** $m = 4(r + 3)$

4 **a** $b = a - c$ **b** $b = a + c$ **c** $b = \frac{a}{c}$
d $b = ca$

5 **a** $x = y - 9$ **b** $x = \frac{y}{4}$ **c** $x = \frac{y + 1}{5}$
d $x = \frac{y}{6} - 5$ **e** $x = 3y - 1$ **f** $x = \frac{8 - y}{3}$

6 **a** $p = \frac{A}{9}$ **b** $x = y - 5$

7 **a** $x = \frac{y - 12}{5}$ **b** $x = \frac{y + 2}{31}$ **c** $x = 50 - y$
d $x = \frac{y - 45}{25}$ **e** $x = 20 - 3y$
f $x = \frac{18 - y}{4}$

8 **a** $k - 3b + 1 = a$ **b** $b = \frac{k - a + 1}{3}$

9 **a** $a = \frac{p}{2} - b$ **b** $b = \frac{p}{2} - a$ **c** $a = \frac{A}{b}$
d $b = \frac{A}{a}$

10 **a** $p = \frac{t - 20q}{10}$ **b** $q = \frac{t - 10p}{20}$

11 **a** $m = 8$ **b** $m = 131$
c $m = \frac{x + y}{2}$
$2m = x + y$
$x = 2m - y$
d $x = 34$ **e** $y = 2m - x$ **f** $y = 10.7$

12 **a** $D = 5$ **b** $M = DV$ **c** $M = 48$
d $V = \frac{M}{D}$ **e** $V = 8$

13 **a** $\frac{A}{h} = a + b$
If $A = 2a$, then $2\frac{a}{h} - b = a$
b $b = a\frac{(2 - h)}{h}$

14 Second line error, should be $\frac{w}{3 - x}$
$= 5y$
$\frac{w - 3x}{3} = 5y$
$\frac{w - 3x}{15} = y$

15 **a** $P = x + x + y$ **b** $P = 38$ **c** $y = P - 2x$
d $x = \frac{p - y}{2}$

16 **a** $a = (4 \times x) + (6 \times 5)$
$a = 4x + 30$
b i. $a = 58$ **ii.** $a = 110$
c $x = \frac{a - 30}{4}$
d i. $x = 5$ **ii.** $x = 10$

17 **a** $P = u + u + u + u + u + 9 + 9$
b $u = \frac{P - 18}{5}$ **c** $u = 31$

30.1 Direct proportion

1

Time taken (minutes)	5	10	20	30	45
Distance (km)	13	26	52	78	117

2 **a** 168 pence **b** 252 pence **c** 42 pence **d** 28 pence

3 **a** 54 g **b** 108 g **c** 216 g **d** 13.5 mg

4 **a** 6 kg **b** 24 small loaves

5

Miles	5	40	100	15	25	125
Kilometres	8	64	160	24	40	200

6 **a** 22 **b** 33 **c** 110 **d** 5.5

7 **a** 320 drips **b** 150 mins or 2.5 hours

8 **a i.** £1.92 **ii.** £3.20 **iii.** 16 pence
b i. 200 g **ii.** 1 kg **iii.** 50 g

9

Bar	2.1	0.7	1.4	8.4	12.6
Psi	30	10	20	120	180

10 **a** 32.8 kg **b** 30 m

11

Post height (cm)	128	134	175	209
Shadow length (cm)	64	67	87.5	104.5

12

Pounds (£)	42	66	210	228	300
US dollars (US$)	63	99	315	342	450

13 a

Kilocalories (kcal)	38	190
Kilojoules (kJ)	160	800

b 800kJ:160kJ = 5:1,
190kcal:38kcal = 5:1

14 a £18 **b** 10 light bulbs

15 No; in the last two columns of the table the values of the degree Celsius doubles, whereas the degree Fahrenheit does not. This means that the temperature in degree Celsius is not proportional to temperature in degree Fahrenheit.

16 a i. 5 litres **ii.** 30 litres **b** 315 km

17 a

Pounds (£)	50	150
New Zealand (NZ$)	96	288

b 1:3 **c** 1:3

30.2 Graphs and direct proportion

1 a

Side (x cm)	2	5	7	8	10
Perimetre (y cm)	8	20	28	32	40

b

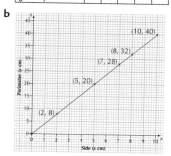

2 a

Drink (x ml)	200	100	500	1000
Sugar (y g)	20	10	50	100

b x = 200 and y = 20, then
$20 = m \times 200$
$m = 20 \div 200$
$m = 0.1$
The formula is $y = 0.1x$

c

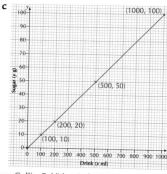

3 a

Time taken (minutes)	5	10	15	20	25
Distance (km)	7	14	21	28	35

b 1.4 **c** $D = 1.4T$

4 a

Side (x cm)	2.5	4	6	9	15
Perimetre (y cm)	10	16	24	36	60

b $y = 4x$ (or) $x = \frac{y}{4}$

5 a

Mass (x kg)	0.5	1	1.5	2	3
Price (y pence)	24	48	72	96	144

b 48 **c** $y = 48x$ **d** 360 pence

e

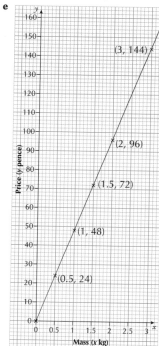

6 a

Pounds (£x)	25	50	75	100
Hong Kong dollars (HK$$y$)	300	600	900	1200

b $y = 12x$ **c** HK$15 240

7 a 36 km/h **b** $\frac{18}{5}$ **c** $y = \frac{18}{5}(x)$

8 a

Side (x mm)	5	10	15	20	25
Diagonal (y mm)	7	14	21	28	35

b $y = 1.4x$ **c i.** 16.8 mm **ii.** 26.6 mm **iii.** 43.4 mm

9 a $y = 2x$

b

AB (x cm)	4.1	5.2	12.9	9.4	3.2
AC (y cm)	8.2	10.4	25.8	18.8	6.4

c

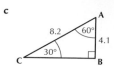

10 a $y = 2.5x$
b

Inches (x)	1	2	3	4	5	6
Centimetres (y)	2.5	5	7.5	10	12.5	15

c

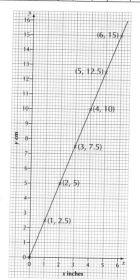

11 a No; in the first two columns of the table the values of radius doubles, whereas the area does not. This means that the area is not proportional to the radius.

b

Radius (cm)	3	6	9	12
Radius (cm²)	9	36	81	144

c 9:28.3 (or 1:3.14) and 81:254.5 (or 1:3.14) **d** The area of a circle is proportional to Radius² **e** 452

12 a $270 **b** £20 **c** $d = 1.8p$

13 a $y = \frac{x}{12}$ **b** 5 litres

c Yes. It is a 300 mile journey. So, $\frac{300}{12}$ = 25 litres of diesel is required to complete the journey. So he will have to refuel 5 more litres. **d** £35

14 a Red line graph. It shows that y is directly proportional to x, because directly proportional line should be a straight line and should pass through the origin. **b** $y = mx$

30.3 Inverse proportion

1 a

a metres	40	50	30
b metres	60	48	80

b $ab = 2400$ m²

2 a 36 **b** $x = 9$ **c** $y = 36$ **d** (2,18) and (6,6)

3 40 minutes

4 a 24 hours **b** 48 hours **c** 16 hours

5 **a** 6 hours **b** 4 hours

c

Speed (*x* km/h)	100	150	120	200	300
Time (*y* hours)	6	4	5	3	2

d If the speed is multiplied by any number, the time must be divided by the same number
For example, $100 \times 1.5 = 150$ and $6 \div 1.5 = 4$
$150 \times 2 = 300$ and $4 \div 2 = 2$
If we multiply each pair of *xy* values, the value will be same and equal to 600
Therefore, *x* and *y* are inversely proportional **e** $xy = 600$

f

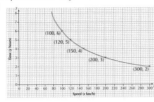

6 **a** 100 books **b** 200 books

c

Cost of a book (£*x*)	2	2.50	5	10	20	25
Number bought (*y*)	500	400	200	100	50	40

d If the cost is multiplied by any number, the number bought must be divided by the same number,
e.g. $2 \times 1.25 = 2.50$ and $500 \div 1.25 = 400$
$2 \times 5 = 10$ and $500 \div 5 = 100$
If we multiply each pair of *xy* values, the value will be same and equal to 1000. Therefore, *x* and *y* are inversely proportional. **e** $xy = 1000$

f

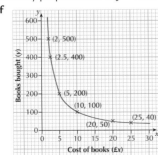

7 **a** 5 hours **b** 1000 km/h
c $xy = 4000$

8 **a** 24 paces **b**

Length of pace (*p* metres)	0.5	0.6	1	1.2
Number of paces (*n*)	24	20	12	10

c If the length of pace is multiplied by any number, the number of

paces must be divided by the same number, e.g.
$0.6 \times 2 = 1.2$ and $20 \div 2 = 10$
$0.5 \times 2 = 1$ and $24 \div 2 = 12$
If we multiply each pair of *xy* values, the value will be same and equal to 12. Therefore, *p* and *n* are inversely proportional. **d** $pn = 12$

9 **a** $bh = 200$ **b**

Base (*b* cm)	20	16	12.5	10	8
Height (*h* cm)	10	12.5	16	20	25

c

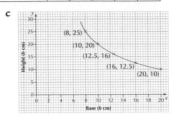

d 13.33 **e** $bh = 200$
$h = \dfrac{200}{15}$
$h = 13.33$

10 60 toys

11 **a** 36 km/h **b** 11 **c** 6.6 m/s

12 **a** £1500 **b** £1000 **c**

Number of families (*n*)	10	20	30	40	50	60
Cost for each family (£*c*)	3000	1500	1000	750	600	500

d Yes, if the number of families is multiplied by any number, the cost for each family must be divided by the same number, e.g.
$10 \times 2 = 20$ and $3000 \div 2 = 1500$
$20 \times 1.5 = 30$ and
$1500 \div 1.5 = 1000$
If we multiply each pair of *xy* values, the value will be same and equal to £30 000. Therefore, number of families and cost for each family are inversely proportional.

e

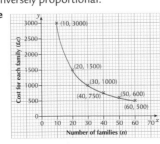

f 38 families **g** $nc = 30000$

13 **a** $xy = 25.6$ **b**

x	1	2	3	4	5	6	7	8
y	25.6	12.8	8.533	6.4	5.12	4.267	3.657	3.2

c

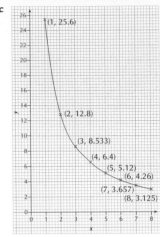

d $y = 7.69$

30.4 Comparing direct and inverse proportion

1 **a** i **b** iv

2 **a** direct **b** direct **c** inverse **d** direct **e** inverse **f** direct

3 **a** yes **b** $d = 80t$ **c** 680 metres

4 **a** 4 hours **b** 2.5 hours **c** If the time is multiplied by any number, the speed must be divided by the same number, e.g. $5 \times 1.6 = 8$ and $4 \div 1.6 = 2.5$
If we multiply each pair of *xy* values, the value will be same and equal to 20. Therefore, time and speed are inversely proportional. **d** $tw = 20$

5 **a** $y = 0.25x$ **b** $xy = 400$

6 **a** (5,200), (10,400) **b** $y = 40x$

7 **a** (1,12), (2,6), and (3,4) **b** $xy = 12$

8 **a** $y = 13x$ **b** $y = 3.5x$ **c** $y = 4x$

9 **a** $xy = 60$ **b** $xy = 75$ **c** $xy = 96$

10 a

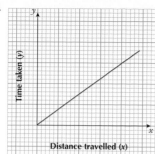

Time taken (y) vs Distance travelled (x)

b

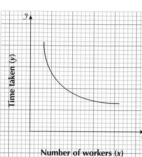

Time taken (y) vs Number of workers (x)

c

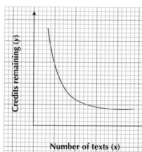

Credits remaining (y) vs Number of texts (x)

d

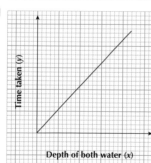

Time taken (y) vs Depth of both water (x)

11 a Direct proportional $r = 13.5f$
b Neither

12 Because a direct proportional line must pass through the origin.

13 11 hours

14 $\dfrac{xy}{(x+y)}$ minutes

31.1 Step graphs

1 a £1.6 **b** £2.2 **c** £4.2 **d** £4.8
2 a up to 5 km £3 **b i.** £3 **ii.** £4
iii. £5 **iv.** £6 **v.** £6.5

3 a up to 3 weeks **b i.** £0.80
ii. £2.40 **iii.** £8 **c i.** 5 weeks **ii.** 7
weeks **iii.** 8 weeks
4 a £4 **b i.** £4 **ii.** £20 **iii.** £6 and
£11 **c** Up to 6 miles
5 a

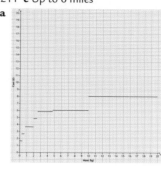

b

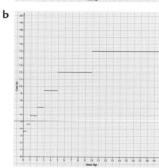

6 a

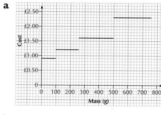

b

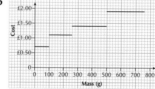

7 a £3.5 **b** Cheapest ticket = £3
Twice the ticket = 2 × £3 = £6. Fare
of 10 km = £5.50. So he can travel
using this ticket.
8 a £8 **b** Price of 6.4 mile = £14.
Double the price of 3.2 mile = £12.
Hence the price of 6.4 mile is more
than double the price of 3.2 mile.
9 a Short stay car park **b** Long stay
car park cost for 1½ and 3 hrs is 4.
So it is the cheapest cost compared
to short stay car park.
10 a Matches graph 3 **b** Matches
graph 2 **c** Matches graph 1
d Matches graph 4

11 a Horizontal axis-Amount spent;
vertical axis-Postage and packaging

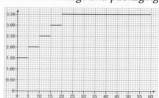

b Cost of DVD = £9, so cost of
packaging 3 DVDs separately =
2 + 2 + 2 = £6. Cost of 3 DVDs =
£27. Cost of packaging together =
£3.50. So £2.50 can be saved.

12 a 60p
b

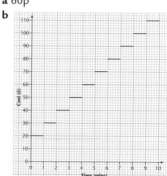

13 a £10 **b** £13 **c** 16 hrs 36 mins

31.2 Distance–time graphs

1 a,b,c 30 miles in the vertical Y-axis

d 120 miles **e** 10.45am

2 a,b

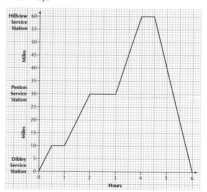

c i. 20 miles **ii.** 45 miles
iii. 40 miles

Answers

3

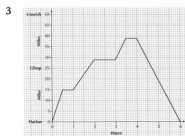

4

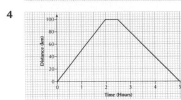

5

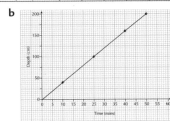

6 a

Time (minutes)	0	10	25	40	50
Depth (cm)	0	40	100	160	200

b

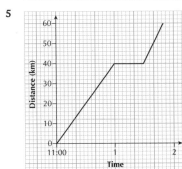

7 **a** 2 times **b** 200 km

8 **a** 400 m **b** 7.5 minutes

9 **a** 750 m **b** 2 minutes **c** 2000 m

10 **a** 09:05 **b** 800 m

11 80 miles/hour

12 **a** 50 km/hr **b** 40 km/hr

13 40 km/hr

14 First part of the walk is faster. Because he takes only 5 minutes to reach the shop but he takes 12.5 mins to reach home from shop.

15 24 minutes

16 Without the stop, the average speed will be 4.16 m/s. With stop average, speed will be 3.3 m/s. So, 0.86 m/s more than the average speed with the stop.

17 a

Time (minutes)	0	50	100	150	200	250	300
Water left (litres)	9000	7400	5800	4200	2600	1000	0

b

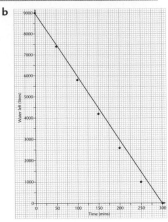

c 281 minutes 15 seconds

31.3 More time graphs

1 **a** 20 minutes **b** 50 °C **c** about 7 minutes

2 **a** 10 minutes **b** 3.5 miles

3 **a** 9 am **b** 1 hour **c** 12:45 pm **d** $1\frac{1}{4}$ hrs **e** 40 km

4 **a** Coach A **b** Coach B **c** 3 hours

5 **a** 300 m **b** 75 m **c** 41 secs **d** D:40 secs E: 60 secs

6 **a** 8000 **b** 17 000 **c** 23 000

7 **a** 2 mins **b** after 4 mins **c** 3mins

8 **a** 800 m **b** 9:03 am **c** 2 mins **d** 9:07 am

9 **a** Graph 2 **b** Graph 3 **c** Graph 1

10 **a** Graph 2 **b** Graph 1 **c** Graph 3 **d** Graph 4

11 Paul's graph is a straight line. This means that he ran at the same speed throughout the race. Paul won the race, finishing about 20 secs before Ron. The shape of Ron's graph indicates that he started quickly and then slowed down. He was in the lead for the first 850 metres, before Paul overtook him. Jeff started slowly, but then picked up speed to overtake Paul, staying ahead of him for a minute before running out of steam and slowing down to come in last, about 30 secs behind Ron.

12

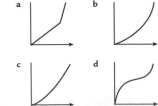

13

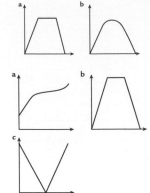

14

31.4 Graphs showing growth

1 **a** 3000 **b** 11 000 **c** 36 000 **d** 173 sec **e** 215 secs

2 **a** £10 000 **b** 10 years **c** 24 years **d** £37 000 **e** £75 000

3 **a** 60 km/hr **b** 22 km/hr **c** 14 secs

4 **a** 6 cm **b** 92 cm **c** 9 years old **d** 28 years old

5 **a** 800 **b** 1201 **c** 23¼ years **d** 1800

6 **a** 475 **b** 5200 **c** 175 secs

7 **a** £14 135 **b** £57 089 **c** 46 years

8 **a** 2.5 **b** 17 **c** 67 **d** 286 **e** 69 days

9 **a** 85 °C **b** 46 °C **c** 23.5 °C **d** 20 minutes **e** 5 minutes

10 **a** 1.5 **b** 1.5
c $24\,000 \times 1.5 = 36\,000$
$36\,000 \times 1.5 = 54\,000$
$54\,000 \times 1.5 = 81\,000$
$81\,000 \times 1.5 = 121\,500$
$121\,500 \times 1.5 = 182\,250$

11 **a** 25 years **b** After 10 years

12 1320

13 21°C

14 a

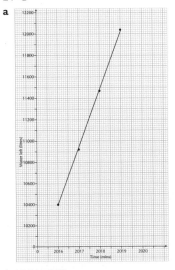

b 12 641.265

15 **a i.** $y = 3$ **ii.** $y = 24$ **b** $x = 5$

16 **a** $a = 4$ **b** $b = 5$ **c** $c = 100$

32.1 Grouped frequency tables

1 a $10 < T \le 20$ b $0 < T \le 10$

2 a $4.5 \le L < 5$ b 13 c 8 d 25
   e Cannot be determined exactly; could be 9

3 a

Height, h (metres)	Frequency
$1.40 < h \le 1.50$	2
$1.50 < h \le 1.60$	4
$1.60 < h \le 1.70$	5
$1.70 < h \le 1.80$	7
$1.80 < h \le 1.90$	2

   b $1.70 < h \le 1.80$

4 a

Mass, M (kilograms)	Frequency
$0 < M \le 1$	4
$1 < M \le 2$	2
$2 < M \le 3$	1
$3 < M \le 4$	3
$4 < M \le 5$	2
$5 < M \le 6$	2

   b $0 < M \le 1$ c $\frac{3}{14}$

5 a

Temperature, T (°C)	Frequency
$8 < T \le 10$	3
$10 < T \le 12$	5
$12 < T \le 14$	3
$14 < T \le 16$	3
$16 < T \le 18$	2

   b $10 < T \le 12$ c 9

6 a

Volume of liquid (V cl)	Tally	Number of coconuts
$10 \le V < 10.5$	||	2
$10.5 \le V < 11$	Ｗ	5
$11 \le V < 11.5$	|||	3
$11.5 \le V < 12$	||	2
$12 \le V < 12.5$	||||	4
$12.5 \le V < 13$	||||	4

   b 2.8 c $10.5 \le V < 11$ d $\frac{7}{20}$ e $\frac{2}{5}$

7 a

Results (R marks)	Tally	Number of students
$30 \le R < 35$	||||	4
$35 \le R < 40$	Ｗ |||	8
$40 \le R < 45$	Ｗ	5
$45 \le R < 50$	Ｗ Ｗ |	11
$50 \le R < 55$	||	2

   b 19 c $45 \le R < 50$ d $\frac{3}{5}$ e $\frac{13}{30}$

8 a

Errors, E	Frequency
$8 \le E < 12$	9
$12 \le E < 16$	4
$16 \le E < 20$	4
$20 \le E < 24$	3

   b $8 \le E < 12$ c $\frac{13}{20}$ d $\frac{3}{20}$

9 a

Magazines, M	Frequency
$42 \le M < 47$	3
$47 \le M < 52$	4
$52 \le M < 57$	8
$57 \le M < 62$	9
$62 \le M < 67$	5
$67 \le M < 72$	1

   b $57 \le M < 62$ c 17%

10 a

Height, H (metres)	Frequency
$152 \le H < 161$	2
$161 \le H < 170$	7
$170 \le H < 179$	6
$179 \le H < 188$	2
$188 \le H < 197$	3

   b $161 \le H < 170$ c $\frac{1}{4}$

11 a

Scores, S	Frequency
$4 \le S < 9$	5
$9 \le S < 14$	3
$14 \le S < 19$	7
$19 \le S < 24$	5

   b 20 c 2 d $14 \le S < 19$ e No; all mode scores (8 and 21) lie outside the modal class, $14 \le S < 19$. From the table we can find that the highest frequency lies corresponding to the interval $14 \le S < 19$ but the mode values (8 and 21) are repeated maximum number of times, which lie outside the modal class. f $\frac{2}{5}$

12 a False; there were 14 in the group b False; there is a modal class not a single mode c False; probability of the child eating more than 8 pieces fruits $\frac{3}{14}$ but the probability of the child eating less than 6 pieces is $\frac{9}{14}$. Therefore, they are not same.

13 a Last length in the table is not specified in class interval. All intervals are given in inclusive type (that is the upper bound of one class and the next class are not same).
   b

Length, L (mm)	Frequency
$40 \le L < 45$	5
$45 \le L < 50$	2
$50 \le L < 55$	4
$55 \le L < 60$	3
$60 \le L < 65$	6

   c $60 \le L < 65$

14 a 11–15 b 18 c Yes, it is possible d 6% e 14 f 20

15 a

Money (£)	Tally	Frequency
$0 < M \le 20$	|	1
$20 < M \le 40$	||||	4
$40 < M \le 60$	Ｗ ||	7
$60 < M \le 80$	Ｗ |	6
$80 < M \le 100$	||	2

   b 100, actual range 70
   c Actual mode is 70. It does not lie within the modal class which is $40 < M \le 60$.
   d $\frac{3}{5}$

16

Litres, L	Frequency
$27.5 \le L < 32.5$	4
$32.5 \le L < 37.5$	6
$37.5 \le L < 42.5$	5
$42.5 \le L < 47.5$	7
$47.5 \le L < 52.5$	5
$52.5 \le L < 57.5$	3

   Modal class $42.5 \le L < 47.5$

17 a No b Dr Speed didn't follow any of the manager's advice because his minimum range of consultation is 2 minutes and the maximum range is 15 minutes.
   Dr Bell partially followed the manager's advice because his minimum range of consultation is more than 5 minutes but his maximum range of consultation exceeds 10 minutes which violates the manager's advice.
   Dr Khan partially followed the manager's advice because his maximum range of consultation is less than 10 minutes as per manager's advice but his minimum range of consultation less than 5 minutes which violates the manager's advice.

Answers

32.2 Drawing frequency diagrams

1 a
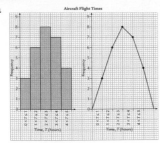
Aircraft Flight Times

b
Temperatures of European capital cities

c
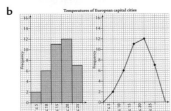
Lengths of metal rods

d
Masses of animals on a farm

2
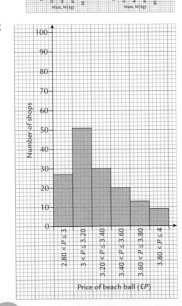

3 a
Amount of coffee (V ml)

b

Amount of coffee (V ml)	Number of cups	Percentage
$485 \leq V < 490$	3	6%
$490 \leq V < 495$	7	14%
$495 \leq V < 500$	13	26%
$500 \leq V < 505$	16	32%
$505 \leq V < 510$	11	22%

4 a City B **b** City B **c** 10 months **d** 5

5 a
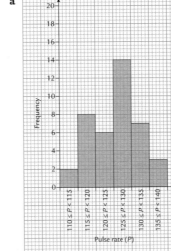
Pulse rate (P)

b $125 \leq P < 130$

6 a Orchestra **b** Orchestra

7

Time (minutes)	Number of students
$0 \leq T < 10$	4
$10 \leq T < 20$	20
$20 \leq T < 30$	15
$30 \leq T < 40$	7
$40 \leq T < 50$	12
$50 \leq T < 60$	5

8 a
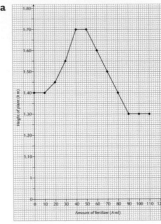

b 25 ml or 70 ml **c** 1.61 m **d** 40 ml of fertilizer is advisable because it gives maximum height to the plant. 50 ml gives the same height, but 40 ml is sufficient and cost-effective to the farmer.
e Levels greater than 50 ml give decreased heights.

9 a
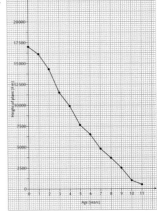

b £3200 **c** 2 years 8 months
d Because the price is depreciating every year, so after 13 years the car will lose its worth

10 a

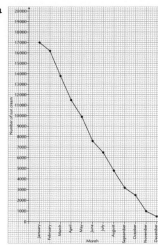

b December, because in the month of December it will be winter
c Since December to February is summer in Australia, during these months the sale will be high and during June to August the sales will get reduced (winter) and during September to November the sales will gradually increase

11

Money (£)	Frequency		
	Class A	Class B	Class C
$0 < x \le 5$	4	3	4
$5 < x \le 10$	9	9	9
$10 < x \le 15$	7	8	9
$15 < x \le 20$	9	9	9
$20 < x \le 25$	0	3	1

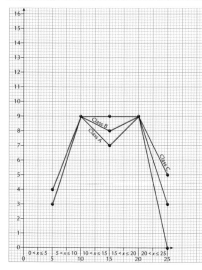

12 a

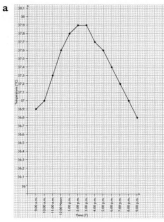

b 11.40 a.m., 6.00 p.m. **c** Line
d Fever lasted for 6 hours 20 minutes. 6 hours is a quarter of a day. Hence, the fever laster for more than a quarter of a day.
e After 3 pm

13 a

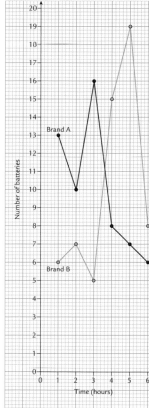

b Brand B as more batteries last 4 hours or more
c Brand A as more batteries fail in 3 hours or less
d Brand B as battery life is generally higher

32.3 Comparing data

1 **a** Range: 17 mins **b** Mean: 16.4
2 **a** Range: 10, 11, 12.5, 13
 b In Northumberland and Leicestershire, the month of April is spring, but for Oxfordshire and Surrey it is partly winter and partly spring; so Oxfordshire and Surrey will have a colder climate than Northumberland and Leicestershire
3 Matt scores constantly well, when Jon's scores vary a lot
4 **a** Yes **b** No **c** Yes
5 **a** QuickDrive: 3, $3\frac{2}{3}$, 7

 GroundWorks: 2&3, $3\frac{1}{3}$, 3

 b The mean of Quickdrive is greater than that of Groundworks. It shows that Quickdrive takes more time to complete the work when compared to Ground Works.
 c Range of QuickDrive is greater than that of GroundWorks and it shows that GroundWorks is more consistent in completing the work in **a** shorter span of time.
6 **a** Philippa is a good runner compared to Mark because she has completed many races in lesser time when compared to that of Mark and hence she has a mean lesser than that of Mark **b** The range of Philippa is higher than that of Mark's which indicates that Mark is more consistent in the races.
7 **a** Margaret Wix school has scored higher when compared to Beaumont school since the median is greater for Margaret Wix school.
 b Margaret Wix school's range is also lower than Beaumont school and hence we could say that they were more consistent.
8 **a**

Group	Mode	Range	Mean
8A	153	11	157
9A	163	11	166

 b The difference between the average of 8A and 9A is 9, average of the group 9A is higher than the group 8A, hence, 9A children are considerably taller than 8A children. **c** There is no change in range, it is consistent
9 **a** Range (minutes): 13. On Tuesdays and Fridays, the time taken to put away the books is higher compared to the other days. **b** Mean (hours): 20.4

Answers

10 **a** Range: 642, 609, 590, 592
b When comparing the range of four weeks, week 1 is higher than rest of all weeks while week 3 has lowest range value when compared with others

11 **a** Mean: 25 and range: 8 **b** Mean: 7.5 and range: 0.8 **c** Mean: 550 and range: 64 **d** Yes. Arrange all the numbers in ascending order without duplicates. Since they all have odd number of numbers, we need to take the middle-most value to find the mean.

12 **a** Mean: 6, 4.2, 8.8; Range: 2, 3, 1 **b** Ever last and electro because they last for long time since their means are higher and more reliable since their ranges are the least.

13 **a** Mean 350; Range 130 **b** 300 **c** The mean will also increase by £30 since each value increases by £30 hence the total increases by (4×30).
There is no change in range because each value is equally raised by 30 so the difference between the highest and lowest value will remain the same.
d Both the mean and the range increase by 10%. New mean = 385 which is 350 increased by 10% and the new range = 143 which is 130 increased by 10%.

14 The mean of Larmindor's average weekly rainfall is greater than that of Tutu Islands, indicating that Larmidor receives more rainfall than Tutu Island resort. The higher range of Tutu Island indicates that is doesn't have a very predictable weather. It rains suddenly on some days and shines brightly on the other days. Whereas Larmindor has a predicable pleasant weather.

15

	Mean	Range
Cardiff	13.6	14
Edinburgh	11.8	12
London	13.8	16

Edinburgh is generally the coldest, while London being slightly warmer than Cardiff but with a more variable temperature as London has both the least as well as the highest temperatures.

16 **a** 10, 14 **b** Because the mean requires two numbers equidistant from 12 **c** There are three possible solutions: 6, 12, 12; 7, 10, 13; 8, 8, 14 **d** Because to go further above or below these solutions will go outside of the range **e** The totals all make 30, which is three times the mean

17 **a** Range 8; Mean 6 **b** 7, 9, 11, 13, 15 range mean is $11 = 6 + 5$ (when 5 is added to each number then the mean also increases by 5) **c** 4, 8, 12, 16, 20 range is 16 and mean is when the numbers are doubled; the range is also doubled. **d** Range $8 + 2x$; Mean $6 + x$

18 7, 7, 9, 15

32.4 Which average to use

1 **i. a** 5.8 **b** 2 **c** 8 **d** 10 **e** 2 **f** 8.7
ii. a Yes **b** Yes **c** Yes **d** No, because the model is an extreme value and the mean is just 6.8. **e** No because the numbers below the median are close to the median but the values above it are far away. **f** No, 30 is an extreme value

2 **a**

Time (T) seconds	Tally	Frequency
$10 < T \le 12$	|||	3
$12 < T \le 14$	||||	4
$14 < T \le 16$	⊞|	6
$16 < T \le 18$	||	2

b The value which is repeated the most number of times is the mode. But in the given data, no value is repeated twice. Hence, mode is not suitable whereas a modal class becomes suitable.

3 **i. a** 9 **b** 9 **c** 12
ii. a Yes because there is no extreme value **b** No because the values are not spread through the range **c** Yes, because there are no extreme values.

4 **i. a** 14 **b** 64.6 **c** 43
ii. a Yes, because there are no extreme values. **b** No, because there is an extreme value 1. **c** No because there is an extreme value 89

5 **i. a** 11 **b** 4 and 78 **c** 25 **ii. a** Yes because the repeating value is not an extreme value **b** No, since the modes are 4 and 78 which are both extreme values **c** Yes because the repeated value is not an extreme value.

6 **i. a** Median 22, Mean 22.8 **b** Median 4, Mean 5 **c** Median 22, Mean 17.8 **ii. a** Median, because more number of the values given are closer to the median than the mean value. **b** Median, because more number of the values given are closer to the median than the mean value. **c** Mean

7 **a** Mode 12, Median 16, Range 29, Mean 17.5 **b** Median

8 **a** Range 33, Mean 21.2, Median 15 **b** 46 is an extreme value in the given data. So, neither mean nor range is sensible to use for this data, however the median ignores extreme values so is valuable for this data.

9 **a** In none of the days do the same number of people fail, therefore mode is unsuitable **b** Median 29, Mean 26.4 **c** Because median value will keep differing depending on the position of the data in the given data range and it will be the middle value always which will be unpredictable. But mean will give a value considering the whole data range into consideration.

10 **a**

Donation, M ($)	Tally	Frequency
$0 < M \le 1$	⊞ ||	7
$1 < M \le 2$	||	2
$2 < M \le 3$	⊞	5
$3 < M \le 4$	⊞	5
$4 < M \le 5$	|||	3
$5 < M \le 6$	⊞	5

b $0 < M \le 1$ **c** 2.90 **d** The modal amount is better than the median in this case because most of the data given fall under the modal class.

11 **a** Mode **b** Mean **c i.** Mode, because mode will indicate that the workers are getting paid less, i.e. 250. **ii.** Mean, it will indicate that the combined pay of managers and workers are less to the owners. **iii.** Median, will be chosen by the owners because they want to show that all are getting paid above average, i.e. 250.

12 **a** Because mean value is higher than Jimmy's mean value **b** Jimmy will argue that he has scored highest range when compared to Joe's range. Jimmy's range = 94, Joe's range = 40. **c** Joe, because mean and median are higher than Jimmy's values and range is lesser than Jimmy's proving that Joe is a consistent player when compared to Jimmy.

13 **a** Mike: Mean 64.7% Range 58% Bev: Mean 78.5% Range 3% **b** Bev is better person to help

14 **a** Shop A: Mean 4212.5, Range: 4400 Shop B: Mean 5187.5, Range: 6950 **b** On comparing the means, we see that Shop B spends more than Shop A and it is more consistent **c** Shop B

15 **a** Periwinkle **b** Periwinkle has one extreme size in their list of sizes which will affect the mean value. Jenny's indication of sizes is good and it won't affect the mean value. **c** Yes. Periwinkle can use median which is 16 so that it won't be affected by the extreme size of 30 in its list of sizes. Jenny's can use mean or median. Both will be 16 in this case for Jenny's. Hence, median would suit both the shops.

33.1 Simple interest

1 £5850
2 £928.80
3 **a** £16 **b** £144
4 **a** £95 **b** £1140
5 £40.88
6 £713.92
7 £10 278.40
8 £3489.20
9 £4 3120
10 1.8%
11 1.4%
12 1.8%
13 Total interest Cameron has paid is £453.60; total interest Mary has paid is £907.20; Mary pays more interest by £453.60
14 **a** £153.60 **b** 19.2%
15 **a** £1176 **b** Ruby must pay £336 in total more than Shannon
16 Total interest Rebecca has paid is £158.76; Total interest Ntuse has paid is £171.36; total interest Kezia has paid is £165.90; Ntuse pays more interest than others

33.2 Percentage increases and decreases

1 **a** £500 **b** £300 **c** £344
2 **a** £136 **b** £76.50 **c** £63.75 **d** £40.80
3 **a** £354.69 **b** £236.02 **c** £518.30
4 **a** £602.55 **b** £719.55 **c** £571.55
5 **a** £6240 **b** £9880 **c** £10 920 **d** £12 220
6 13.2 seconds

7

Age	Mass (kg)		
8 weeks	2	5	6
10 weeks	2.5	6.25	7.5

8 **a** £13 536 **b** £9212 **c** £22 278
9 £24.48 million
10 938 billion cubic km.
11 **a** 1.07 is larger than 1, so it represents an increase; the percentage increase was 7% **b** 0.7 is less than 1, so it represents a reduction; the percentage reduction was 30%

---

c 0.13 is less than 1, so it represents a reduction; the percentage reduction was 87% **d** 2.8 is larger than 1, so it represents an increase; the percentage increase was 180%
12 Peter is wrong
The width 20 is reduced to 16 which is 20% and not 25%. The length 16 is reduced to 12 which is 25%. So his statement is partially correct and partially wrong.
13 **a** £478.40 **b** £377.30 **c** £467.23 **d** £235.80
14

Place	Percentage change	Population in 1980	Population in 2000
Smallville	30% decrease	1857	1300
Lansbury	8% increase	46 000	49 680
Gravelton	2% decrease	20 204	19 800
Smithchurch	48% increase	97 297	144 000
Deanton	19% increase	28 000	33 320
Tanwich	5% decrease	680	646

15 **a** 1.24 **b** 24%
16 **a** 0.85 **b** 15%
17 5%
18 **a** 8.93% **b** 15.38% **c** 37.9% **d** 135.14%

33.3 Calculating the original value

1 £64
2 £225
3 **a** £20 **b** £60
4 **a** 1.22 **b** 1.48 **c** 1.06 **d** 1.085 **e** 2.2 **f** 4.3
5 **a** 0.78 **b** 0.52 **c** 0.94 **d** 0.915
6 **a** £250 **b** £450 **c** £180 **d** £720
7 **a** £84 **b** £324 **c** £225 **d** £675
8 £48
9 £18200
10 **a** £56.40 **b** £456.80 **c** £879
11 **a** £54.30 **b** £143.80 **c** £281.53
12 **a** £285 **b** £83.45 **c** £1860
13 £650
14 £55
15 £21 150
16 **a** 280 **b** 290 **c** 280 **d** 300 **e** 300 **f** 290; the matching pairs are: AC, BF, DE

---

17 74 kg

33.4 Using percentages

1 **a** 39.78% **b** 31 people
2 **a** 1.35% **b** 18.83%
3 **a** 75% **b** 52.94% **c** 60%
4 39.13%
5 **a** 80 elephants **b** 80% **c** 125%
6 **a** 44.44% **b** 80% **c** 125%
7 **a** 28.26% **b** 38.1% **c** 76.92%
8 **a** 60% **b** 240 students **c** 66.67%
9 **a** 1742 visitors **b** 56.52% **c** 1218 visitors
10 **a** 25% **b** 55% **c** 60%
11 **a** 8.65% **b** £407.15 **c** £84
12

Item	Price before VAT (£)	Price including VAT (£)
Fit gas cooker	85.00	97.75
Fix broken window	44.00	50.60
Install computer	29.80	34.27
Service boiler	69.40	79.81
Replace radiator	142.80	164.22

13 810 pupils
14 **a** 480 cushions **b** 520 yellow cushions **c** 35.84%
15 **a** 62.5% **b** 60% **c** 435 adult tickets **d** £8439
16 **a** 40% **b** 10 000 km **c** 4000 km
17 **a** 24% **b** £51.30 **c** £22.50
18 **a** 300 people **b** 1:4

34.1 Angles in polygon

1 **a i.** 360° **ii.** 360° **b i.** 102° **ii.** 55°
2 **a** 80° **b** 110° **c** 52°
3 **a** 141° **b** 72°
4 **a**

Hexagon   Pentagon

b

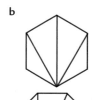

Sum of the Interior angles of hexagon is 720°

Sum of the Interior angles of octagon is 1080°

5 **a** 540° **b** 172°
6 116°
7 **a** 100° **b** 150° **c** 285°

Answers

8 **a** 900° **b** 40°

9 **a** $a = 50°, b = 130°, c = 50°, d = 130°$
   b $e = 60°, f = 120°, g = 120°, h = 60°$

10 **a** and **d** – could not be the sum of the interior angles of a polygon because $1180 \div 180 = 6.55$ and $6488 \div 180 = 36.04$; these angles could not be split into triangles using diagonal from the same vertex

11 **a** Amil divides each polygon into several triangles by connecting a point inside the polygon to each of its vertex but a polygon should be split into triangles from one of its vertices; the sum of the interior angles for each triangle is 180°; there is an excess of two triangles, so he subtracted 360°

b

	Hexagon	**Octagon**
Number of triangles	6	8
Sum of the interior angles	$6 \times 180° - 360° = 720°$	$8 \times 180° - 360° = 1080°$

12 Jessie's mistakes: second line is not correct; $13 - 2$; it should be $13 + 2$
    Ben's mistakes: subtracted by 2 and then he multiplied by 180; he should have divided by 180 and then added 2; the answer should be $2340 \div 180 = 13$, $13 + 2 = 15$ sides

13 20°

14 17°

15 13 sides

16 310°

34.2 Constructions

1 **a**

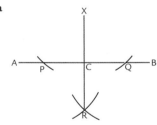

Set the compasses to any suitable radius; draw arcs from X to intersect AB at P and Q; with the compasses still set at the same radius, draw arcs centred on P and Q to intersect at R below AB; join XR; XR is perpendicular to AB and intersects AB at a point C; CX is the shortest distance from X to the line AB

b

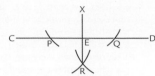

Set the compasses to any suitable radius; draw arcs from X to intersect CD at P and Q; with the compasses still set at the same radius, draw arcs centred on P and Q to intersect at R below CD; join XR; XR is perpendicular to CD and intersects CD at a point E; EX is the shortest distance from X to the line CD

2 **a**

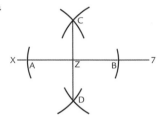

Set the compasses to a radius that is less than half the length of XY; with the Centre at Z, draw arcs on either side of Z to intersect XY at A and B; set the compasses to a radius that is greater than half the length of XY and with centre at A and then B, draw arcs above and below XY to intersect at C and D; Join CD; CD is the perpendicular from the point Z

b

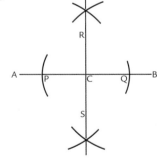

Set the compasses to a radius that is less than half the length of AB; with the Centre at C, draw arcs on either side of C to intersect AB at P and Q; wst the compasses to a radius that is greater than half the length of AB and, with centre at P and then Q, draw arcs above and below PQ to intersect at R and S; join RS; RS is the perpendicular from the point C

3

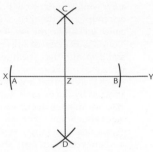

Draw a horizontal line $XY = 10\,\text{cm}$ long. Mark a point $Z = 4\,\text{cm}$ from one end; set the compasses to a radius that is 4 cm length of XY; with the Centre at Z, draw arcs on either side of Z to intersect XY at A and B; set the compasses to a radius that is 6 cm length of XY and, with centre at A and the B, draw arcs above and below XY to intersect at C and D; join CD; CD is the perpendicular from the point Z

4

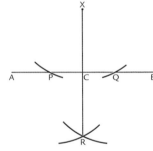

Draw a horizontal line $AB = 10\,\text{cm}$ long. Mark a point about $X = 5\,\text{cm}$ from the line; set the compasses to radius 5 cm; draw arcs from X to intersect AB at P and Q; with the compasses still set at the same radius, draw arcs centred on P and Q to intersect at R below AB; join XR; XR is perpendicular to AB and intersects AB at a point C; CX is the shortest distance from X to the line AB

5

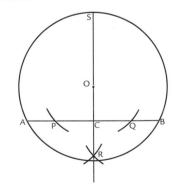

Set the compasses to draw a circle of radius 6 cm and label the centre

O; draw a line AB of any length across the circle; set the compasses to any suitable radius; draw arcs from O to intersect AB at P and Q; with the compasses still set at the same radius, draw arcs centred on P and Q to intersect at R below AB; Join OR; OR is perpendicular to AB and intersects AB at a point C; extend the perpendicular from OS to make a diametre of the circle 12 cm

6

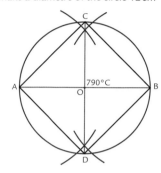

7 **a** Construct a triangle ABC given that AB = 5 cm, BC = 5 cm and AC = 5 cm; draw a line segment BC = 5 cm; with B as centre, draw an arc of radius 5 cm above the line BC; with C as centre, draw an arc of 5 cm to intersect the arc at A; join AB and AC; now ABC is the required triangle **b** Construct a triangle ABC given that AB = 6 cm, BC = 4 cm and AC = 6 cm; draw a line segment BC = 4cm; with B as centre, draw an arc of radius 6 cm above the line BC; with C as centre, draw an arc of 6 cm to intersect the arc at A; join AB and AC; now ABC is the required triangle **c** Construct a triangle ABC given that AB = 4 cm, BC = 8 cm and AC = 6 cm; draw a line segment BC = 8 cm'; with B as centre, draw an arc of radius 4 cm above the line BC; with C as centre, draw an arc of 6 cm to intersect the arc at A; join AB and AC; now ABC is the required triangle

8 **a** Construct a triangle ABC given that AB = 7 cm, BC = 6 cm and AC = 8 cm; draw a line segment BC = 6 cm; with B as centre, draw an arc of radius 7 cm above the line BC; with C as centre, draw an arc of radius 8 cm to intersect the arc at A; join AB and AC; now ABC is the required triangle **b** Construct a triangle ABC given that AB = 5 cm, BC = 9 cm and AC = 6 cm; draw a line segment BC = 9 cm; with B as centre, draw an arc of radius 5 cm above the line BC; with C as centre, draw an arc of radius 6 cm to intersect the arc at A; join AB and AC; now ABC is the required triangle

9 **a** Draw a line segment AC = 6cm; make angle CAX = 90° at the left-hand end; with C as centre, draw an arc of radius 7 cm long, BC = 7 cm; join A and B; thus, ABC is the required right angled triangle **b** Draw a line segment AC = 7.5cm; make angle ACX = 90° at the right-hand end; with A as centre, draw an arc of radius 8.8 cm long, BA = 8.8cm; join A and B; thus, ABC is the required right-angled triangle **c** Draw a line segment AC = 6 cm; make angle CAX = 90° below the line; with C as centre, draw an arc of radius 10 cm long, BC = 10 cm; join A and B; thus, ABC is the required right-angled triangle

10 **a** Draw a line segment AC = 12 cm; make angle ACX = 90° at the right-hand end; with A as centre, draw an arc of radius 13 cm long, BA = 13 cm; join A and B; thus, ABC is the required right-angled triangle **b** Draw a line segment AC = 6 cm; make angle ACX=90° at the right-hand end; with A as centre, draw an arc of radius 7.5 cm long, BA = 7.5 cm; join A and B; thus, ABC is the required right-angled triangle

11 Measurements are sufficient to draw a triangle; you must know at least two side length and one angle or two angles and one side length or three side length; so, a and c cannot be drawn accurately

12 **a**

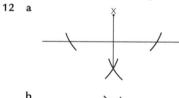

b

13

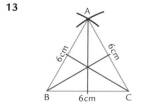

Draw a line segment BC = 6 cm; with B as centre, draw an arc of radius 6 cm above the line BC; with C as centre, draw an arc of

radius 6 cm to intersect the arc at A; join AB and AC; now ABC is the required equilateral triangle; the centre of the triangle is the point where the perpendicular from each vertex to the opposite side meet which is the centroid; the centre of the triangle is the point where the perpendicular bisectors of the three sides meet which is the circumcentre

14 Helen is correct; set the compasses of same radius for all the arc that she draws; then only she can construct a perpendicular from a point to the line.

15

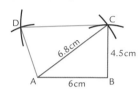

Draw AB = 6 cm; with A as centre and radius 6.8 cm, draw an arc; with B as centre and radius 4.5 cm draw another arc, cutting the previous arc at c; join BC and AC; with A as centre and radius 4.5 cm, draw an arc; with C as centre and radius 6cm draw another arc, cutting the previously drawn arc at D; join DA and DC

16 7.2 cm
17 7.2 cm
18 Yes

34.3 Angles in regular polygons

1

Regular Polygon	Number of sides	Sum of exterior angles	Size of each exterior angle	Size of each interior angle
Hexagon	6	360°	60°	120°
Octagon	8	360°	45°	135°
Nonagon	9	360°	40°	140°
Decagon	10	360°	36°	144°

2 **a** $x = 45°, y = 135°$
　　b $z = 60°, w = 120°$
3 **a** 360° **b** Square **c** 90°
4 **a** 7 **b** 1260° **c** 140°
5 **a** 10 triangles **b** 1800° **c** 150°
6 360°
7 **a** 30° **b** 150° **c** 1800°
8 **a** 36 sides **b** 360 **c** 3600 sides **d** 360 000 sides
9 **a** 90 sides **b** 40 sides **c** 13 sides
10 **a** 177° **b** 21240° **c** 120 sides

Answers

11 **a** False; a regular quadrilateral is called a square **b** False, 360 divided by the number of sides of the triangle which gives the size of exterior angle of triangle is 120° **c** True; e.g. the sum of the interior angles of a regular pentagon with 5 sides is 540° = 108° × 5 **d** False; the size of the exterior angle of a regular decagon is 36°. 36 ≠ 17.

12 25° is not a factor of 360°; so Joanne is wrong; a polygon splits into at least one triangle which has the size of the interior angle 60°; so Ben is wrong

13 **a** 60° **b** 90° **c** 108° **d** 72°

14 **a** Triangle ADE that has two sides of equal length **b** Exterior angles of pentagon is 360 ÷ 5 = 72°; 180° – exterior angle = interior angle: 180° – 72° = 108°; interior angle splits into 3 parts: 108° ÷ 3 =36°

15 Angle of AFB = 22.5°

16 **a** a = 120°, b = 30°, c = 30° **b** Interior angles of hexagon is 120°; at A, 120° – b = 120° – 30° = 90°; at E, 120° – c = 120° – 30° = 90°; same applies to B and D; the size of each interior angles of is 90°; sum of the interior angles is 360°; therefore, ABDE is a rectangle

17 a = 135°, b = 22.5°, c = 90°, d = 45°, e = 45°; since triangle FDH is an equilateral triangle in which all three sides are equal; so c = d = e = 60°; interior angles of hexagon are 135°, 135° – 60° = 75°; 75° ÷ 2 = 37.5°; therefore, b = 37.5°; since triangle HGF is an isosceles triangle in which two sides are equal same length and same angle, 180° – 75° = 108°; therefore, a = 108°

18 a = 132°

34.4 Regular polygons and tessellations

1

2 a

b Three pentagons have angle measures that sums to 324° (108° + 108° + 108°), leaving a gap of 36°.; four pentagons surrounding a point on the plane would have angle measuring 432°, which would cause an overlap of 72°; therefore, regular pentagon does not tessellate

3 a

b Two octagons have angle measures that sums to 270° (135°+135°), leaving a gap of 90°; three octagons surrounding a point on the plane would have angle measuring 405°, which would cause an overlap of 45°; therefore, regular octagon does not tessellate

4

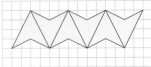

5

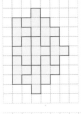

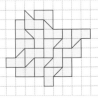

6

7

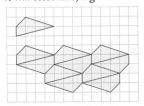

8

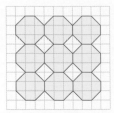

Make a tessellation using octagons and squares as well as in a different way using octagons and triangles

9 Peter is correct; he needs to place two yellow angles and two black angles together at a point; then he can make tessellation; two yellow angles = 2 × 138° =276°; two black angle = 2 × 42° = 84°; 276° + 84° = 360°

10 a i.

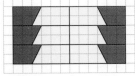

ii.

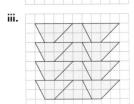

iii.

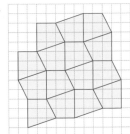

iv.

b It will tessellate; e.g.

c Yes; a quadrilateral will always tessellate because the sum of angles in any quadrilateral is 360°

11 The octagon on the left will tessellate, but regular octagons cannot fit together in a pattern without gaps

12 a

Regular polygon	Size of each interior angle	Does this polygon tessellate?
Equilateral triangle	60°	Yes
Square	90°	Yes
Regular pentagon	108°	No
Regular hexagon	120°	Yes
Regular octagon	135°	No

b Equilateral triangles, squares and regular hexagons are the only regular polygons that will tessellate; each have interior angles whose measures are factors of 360°; therefore, there are only three regular tessellations

13 Lauren makes triangle, quadrilateral and hexagon

14 Holly tessellates 4 small shapes to make a rectangle; George cannot tessellate the small shape to make a bigger similar shape; he can fit only 3 shapes and leave a gap

15 a 4 tiles **b** 9 tiles

16 Ben has budget to buy 10 tiles; an area that is 120 cm wide and 60 cm tall he can fit 6 tiles of 40 cm wide and 30 cm tall; for the remaining area of 120 cm wide and 40 cm tall, he can rotate and fit 4 tiles that is 30 cm wide and 40 cm tall

35.1 Multiplying out brackets

1 a $6a$ **b** $20b$ **c** $36c$ **d** $50d$
2 a $2a + 10$ **b** $4t + 28$ **c** $5x - x^2$ **d** $y^2 + 4y$
3 a $8n + 4$ **b** $6t + 21$ **c** $15k - 6$ **d** $20x - 50$ **e** $3c + 13.5$ **f** $4z + 6$ **g** $30 - 15y$ **h** $3 - x$
4 a $-3a - 6$ **b** $-3m - 15$ **c** $8f - f^2$ **d** $-4x + x^2$
5 a $4t + 11$ **b** $8g + 3$ **c** $5t - 13$ **d** $7m - 8$ **e** $n - 12$ **f** $5k - 6$ **g** $2x + 7$ **h** $3a + 3$
6 b is the odd one
7 a $4x + 8$ **b** $2x^2 + 2x$ **c** $7y + 1$
8 a $2x + 20$ **b** $12y + 2$ **c** $4t - 12$ **d** $15 - 5x$ **e** $7g - 6$ **f** $5p - 8$
9 a $6x + 8$ **b** $8i + 1$ **c** $22n - 12$ **d** $2r + 14$ **e** $11f + 26$ **f** $23u + 6$
10 a Always true since it is commutative property under addition **b** Never true since $3(p + 2) = 3p + 6$ not $3p + 2$ **c** Sometimes true; when $a = 10$ then $10 - a = a - 10$; other cases it won't be true; **d** Sometimes, when $m = 0$ or 2
11 a $3(2y + 3) = 6y + 9$ here $6y$ not $5y$ **b** $x(x + 3) = x^2 + 3x$ not $3x^2$ **c** $16x^2 - 6x$ not $10x + 6x$

12 Perimetre of rectangle = $2(l + b) = 2(2x + 5 + 5x) = 2(7x + 5) = 14x + 10$; Sam missed the multiple of 2 in his calculation; Eamon calculated area of rectangle
13 $2x(6x + 5)$; $x(12x + 10)$; $10x(\frac{6}{5}x + 1)$; $12x(x + \frac{5}{6})$
14 a $12s + 20$ cm **b** 104 cm
15 12 cm, 13 cm, 13 cm
16 a i. $a = x - 2y$ **ii.** $b = 2x - 2$ **b i.** $2xy$ **ii.** $2a$ **iii.** $ab + 2xy$
17 $8a$, $7a - 3$, $4a - 2$

35.2 Factorising algebraic expressions

1 a 5 **b** 4 **c** 6 **d** 4
2 a $2x + 6 = 2(x + 3)$ **b** $3d - 12 = 3(d - 4)$ **c** $4y + 8 = 4(y + 2)$ **d** $5e - 20 = 5(e - 4)$
3 a 4 **b** $2a$ **c** $2y$ **d** $3xy$
4 a $5(m + 4)$ **b** $3(x - 7)$ **c** $6(n - 8)$ **d** $4(2 + c)$
5 a $2(2x + 3)$ **b** $3(3x - 2)$ **c** $4(3y + 4)$ **d** $5(2y - 3)$ **e** $6(3 - 2x)$ **f** $6(5 - 3t)$ **g** $5(5 + 4t)$ **h** $9(5m + 3)$
6 a $3x$ **b** 5 **c** $5c$ **d** $4xy$
7 a $5(3x + 2)$ **b** $4(y - 5)$ **c** $6(3a + 4b)$ **d** $4(3p - 5q)$
8 a $2(a + 2b)$ **b** $2(3a + b)$ **c** $5d(c + 4)$ **d** $6cd$ **e** $6(4x + 5y)$ **f** $2(3x - 4y)$ **g** $-9(3a + 2b)$ **h** $-8(3f + 2g)$
9 a $4(2m - 3n)$ **b** $m(2m - 3n)$ **c** $m(10m - 3n)$ **d** $-3(2m - 3n)$; c is the odd one out
10 a $5a + 15 = 5(a + 3)$ **b** $10x - 10 = 10(x - 1)$ **c** $4x - 8 = 4(x - 2)$
11 $3x + 5$
12 Both are correct; sum of number = $n + n + 6 + n + 12 + n + 18 = 4n + 36$; since $4(n + 9) = 4n + 36$, $2(2n + 18) = 4n + 36$
13 a $2x + xy = x(2 + y)$ **b** $10a + 12b = 2(5a + 6b)$ **c** $5a - 10b = 5(a - 2b)$
14 a 10, 13, 22 **b** −10, −7, 2 **c** $3a + 15 = 3(a + 5)$ **d** 150
15 a $3x + 12 = 3(x + 4)$ **b** $6n + 3 = 3(2n + 1)$ **c** $6x + 4 = 2(3x + 2)$
16 a $4t + 4 = 4(t + 1)$ **b** $6r + 7$ **c** $5p + 30 = 5(p + 6)$
17 $1 + 6b$; y, $(25 \quad x)$
18 a i. $10x + a$ **ii.** $100x + a$ **b** $(2x \times 10) + x$ **c** 42, 21, 84 **d** $10 \times 2x + x = 20x + x = 21x$; divisible by 21

35.3 Equations with brackets

1 a i. $x = 6$ **ii.** $x = 3$ **iii.** $y = 5\frac{1}{2}$ **iv.** $y = 7$ **b ii.** $x + 6 = \frac{18}{2}$; $x + 6 = 9$; $x = 3$ **iv.** $y - 2 = \frac{20}{4}$; $y = 2 + \frac{20}{4}$; $y = 7$
2 a $2(y - 8) = 20$; $y - 8 = \frac{20}{2}$; $y - 8 = 10$; $y = 18$
b $2(y - 8) = 10$; $y - 8 = 5$; $y = 8 + 5$; $y = 13$ **c** $40 = 5f - 85$; $40 + 85 = 5f$; $125 = 5f$; $f = \frac{125}{5} = 25$ **d** $\frac{20}{4} = w - 9$; $5 = w - 9$; $5 + 9 = w$; $14 = w$
3 a $x = 3\frac{2}{3}$ **b** $y = 1\frac{3}{4}$ **c** $t = 7\frac{1}{6}$ **d** $c = \frac{1}{8}$
4 a $5x + 3$ **b** $x = 7$
5 a $8x - 14$ **b** $x = 5\frac{1}{2}$
6 a $13x - 19$ **b** $x = 3\frac{10}{13}$
7 a $x = 3\frac{1}{2}$ **b** $x = 7$ **c** $t = 9$
8 a $x = 9$ **b** $x = 7$ **c** $x = 13$
9 a $v = 4$ **b** $d = -11$ **c** $k = \frac{1}{2}$
10 a $52 - 4m - 8 = 3m - 3$; $44 + 3 = 3m + 4m$; $47 = 7m$; $\frac{47}{7} = m$; $m = 6\frac{5}{7}$ **b** $7d + 14 = 21d + 4d + 2$; $7d - 4d = 23 - 14$; $3d = 9$; $d = 3$ **c** $9 + 6b = 33 - 8 + 4b$; $6b - 4b = 25 - 9$; $2b = 16$; $b = 8$
11 a $3(x + 5) = 12$; $x + 5 = 4$ not 9; $x = -1$ not 4 **b** $5y - 11 = 2y + 25$; $3y = 36$; $y = 12$ **c** $4(z + 3) = 8z + 10$; $z + 3 = 2z + \frac{10}{4}$; $z = \frac{2}{4}$ or $\frac{1}{2}$
12 No he is not correct $-4x = 2x + 10$ Subtract 14 on both sides; $-6x = 10$; subtract $2x$ on both sides; $x = -\frac{10}{6}$ or $-\frac{5}{3}$
13 a Divide both sides by 5 **b** $x - 3 - 3x + 6 = 2$ Expand the brackets; $-2x + 3 = 2$; $x = \frac{1}{2}$ **c** $6(x + 1) + 12 (x - 2) = 0$; $x + 1 + 2(x - 2) = 0$; $x + 1 + 2x - 4 = 0$; $3x - 3 = 0$; $x = 1$
14 a $5x - 15 - 4x + 8 = x - 7$ **b** $5(x - 3) - 4(x - 2) = 10$; $x - 7 = 10$; $x = 17$ **c** $x - 7 = -7$; $x = 0$
15 a $P = 12h + 6$ Since regular hexagon has six equal sides; side = $\frac{12h + 6}{6} = 6\frac{2h + 6}{6} = 2h + 1$ **b** $P = 48$ cm, $S = 8$ cm **c** $h = 2$ cm
16 $x = \frac{43}{7}$ cm
17 $5(2n - 11) = 35$; $n = 9$
18 a $x + x + 1 + x + 2 = 54$; $x = 17$; 17, 18, 19 **b** $2x + 1 + 2x + 3 + 2x + 5 = 63$; $x = 9$; 19, 21, 23

35.4 Equations with fractions

1 a $x = 30$ **b** $y = 21$ **c** $t = 16$ **d** $n = 30$ **e** $m = 30$
2 a $x = 20$ **b** $x = 10$ **c** $x = \frac{20}{3}$ or $6\frac{2}{3}$ **d** $x = 5$
3 a $x = 4$ **b** $x = 2$ **c** $y = 2\frac{1}{2}$ **d** $t = 6$
4 a $a = 8$ **b** $b = 16$ **c** $c = 3\frac{1}{5}$ **d** $k = 14$ **e** $t = 15$

Answers

5   **a** $x = 18$ **b** $x = 10$ **c** $x = 11$ **d** $x = 33$

6   **a** $d = 10$ **b** $u = -3\frac{3}{4}$ **c** $m = -15$

    d $7\frac{1}{2}$

7   **a** $x = 0$ **b** $y = 7$ **c** $k = 11$ **d** $d = 6$

8   **a** $x = 19$ **b** $t = 7\frac{1}{3}$ **c** $x = 11\frac{1}{3}$

    d $c = 13\frac{3}{5}$

9   **a** First method: divide both sides by 4; $\frac{x+1}{3} = 4$; $x + 1 = 12$; $x = 11$; second method: multiply by 3 on both sides; $4(x + 1) = 48$; $4x + 4 = 48$; $x = 11$ **b** First method: divide both sides by 3; $\frac{t-3}{4} = 3$; $t - 3 = 12$; $t = 15$; second method: multiply by 4 on both sides; $3(t - 3) = 36$; $3t - 9 = 36$; $t = 15$ **c** First method: divide both sides by 6; $\frac{2+x}{5} = \frac{4}{6}$; $2 + x = \frac{20}{6}$; $x = \frac{8}{6}$ or 1 $\frac{2}{6}$; second method: multiply by 5 on both sides; $6(2 + x) = 20$; $12 + 6x = 20$; $x = \frac{8}{6}$ or $1\frac{2}{6}$

    d First method: divide both sides by 5; $\frac{c-1}{9} = \frac{10}{15}$; $(c - 1) = \frac{10}{15}9$; $c - 1 = 6$; $c = 7$; second method: multiply by 9 on both sides; $15c - 15 = 90$; $15c = 105$; $c = \frac{105}{15}$; $c = 7$

10   **a** Area of rectangle $= l \times b$; $l = \frac{3n+4}{5}$ cm   **b** $13 \times 5 = 3n + 4$; $n = 20\frac{1}{3}$

11   **a** $A = l \times b$; $l = \frac{A}{b} = \frac{2n+3}{6}$ cm

    b $9 = \frac{2n+3}{6}$; $n = 25\frac{1}{2}$

12   **a** $A = \frac{1}{2} \times b \times h$; $\frac{34}{n+5}$ cm $= h$

    b $n + 5 = 3$; $n = -2$

13   **a** $\frac{x+x+4+x-3+2x}{4} = 11$; $x = 8\frac{3}{5}$

    b $\frac{43}{5}, \frac{63}{5}, \frac{28}{5}, \frac{86}{5}$

14 $\frac{2}{3}(2x + 1) = 6$; $x = 4$

15 $3\frac{x-50}{2} = 45$; $x = 80$

16 $\frac{3}{5}(x - 2) = 9$; $x = 17$

36.1 Metric units for area and volume

1   **a** $40\,000 \text{ cm}^2$ **b** $70\,000 \text{ cm}^2$
    c $200\,000 \text{ cm}^2$ **d** $35\,000 \text{ cm}^2$
    e 8000 cm^2 **f** 5.4 cm^2 **g** 0.6 cm^2

2   **a** 200 mm^2 **b** 500 mm^2 **c** 850 mm^2
    d 3600 mm^2 **e** 40 mm^2

3   **a** 2 m^2 **b** 8.5 m^2 **c** 27 m^2 **d** 1.86 m^2
    e 0.348 m^2

4   **a** 3000 mm^3 **b** $10\,000 \text{ mm}^3$
    c 6800 mm^3 **d** 300 mm^3
    e 480 mm^3

5   **a** 5 m^3 **b** 7.5 m^3 **c** 12 m^3

6   **a** 8 litres   **b** 17 litres   **c** 0.5 litres
    d 3000 litres   **e** 7200 litres

7   **a** 8.5 cl   **b** 120 cl   **c** 84 ml

8   **a** $2\,000\,000 \text{ cm}^3$ **b** $500\,000 \text{ cm}^3$
    c 0.078 cm^3 **d** 9.3 cm^3

9   **a** 8.4 litres   **b** 0.065 litres   **c** 4.8 cm^3
    d 0.2 litres   **e** 9000 cm^3 **f** 3750 ml

10   **a** 1 centimetre = 10 millimetre, by using areas of squares, $1 \text{ cm} \times 1 \text{ cm} = 10 \text{ mm} \times 10 \text{ mm}$, $1 \text{ cm}^2 = 100 \text{ mm}^2$ **b** By using volumes of cubes, $1 \text{ cm} \times 1 \text{ cm} \times 1 \text{ cm} = 10 \text{ mm} \times 10 \text{ mm} \times 10 \text{ mm}$, $1 \text{ cm}^3 = 1000 \text{ mm}^3$

11   **a** In order to convert the smaller units to larger units, divide the value by the conversion factor 10 000: $5000 \text{ cm}^2 = 5000 \div 10\,000 = 0.5 \text{ m}^2$ **b** In order to convert smaller units to larger units, divide the value by the conversion factor 100: $3500 \text{ mm}^2 = 3500 \div 100 = 35 \text{ cm}^2$ **c** In order to convert larger units to smaller units, multiply the value by the conversion factor 1 000 000: $7.5 \text{ m}^2 = 7.5 \times 1\,000\,000 = 7\,500\,000 \text{ mm}^2$

12   2000 lead

13   6 days

14   160 square paving slabs

15   6p

16   **a** 2300 cm^3 **b** 1.5 litres

17   **a i.** 3600 m^2 **ii.** 0.36 hectares
    b £360

36.2 Volume of a prism

1   **a** 24 cm^3 **b** 864 cm^3 **c** 72 m^3

2   **a** 24 m^3 **b** $30\,000 \text{ cm}^3$ **c** 4500 cm^3

3   **a** 24 000 litres   **b** 30 litres
    c 4.5 litres

4   Volume = 3375 cm^3, capacity = 3.375 litres

5   20 cm

6   8 cm

7   **a** 8000 mm^3 **b** 8 cm^3

8 3150 cm^3

9 465 cm^3

10   **a** 61.5 m^2 **b** 922.5 m^3
    c 922 500 litres

11   Yes. $\frac{1}{2} \times 12 \times (15 + 25) \times 40 = 9600$

12   **a** $262\,500 \text{ cm}^3$ **b** 0.63 tonnes

13   **a** 1:2:3   **b** Volume of mini box = 288 cm^3; Volume of medium box = 2304 cm^3; volume of giant box = 7776 cm^3 **c** 1:8:27   **d** The ratio of the volumes are cube of the ratio of the heights   **e i.** 8   **ii.** 27   **f** Since the dimensions of medium box is double the dimensions of mini box and the dimensions of giant box is thrice the dimensions of mini box   **g** 1:6:15   **h** Prices does not depend on volumes, so the ratio of prices is different from the ratio of volumes

14   **a** Volume of a container = 2000 cm^3 but volume of water is 2100 cm^3.; since the volume of container is less than the volume of water, Ben can't be right   **b** Volume of water is 1100 cm^3; height of water $= \frac{1100}{10} \times 10 = 11$cm; so Jordan is correct

15   216 litres

16   **a** 300 cm^3 **b** 48   **c** 2 layers, 24 cm
    d 28800 cm^3 **e** £110.4

17   52 cm

18   **a** 7.5 cm   **b** 10 cm

36.3 Surface area of a prism

1   **a** 684 cm^2 **b** 146.8 m^2 **c** 736 cm^2

2   **a** 312 cm^2 **b** 810 mm^2

3   **a** 0.0312 m^2 **b** $0.000\,81 \text{ m}^2$

4 5400 cm^2

5   **a** 275 cm^2 **b** 3100 cm^2

6   **a** 540 cm^2 **b** $43\,200 \text{ cm}^3$
    c 43.2 litres   **d** 1.036 m^2

7 1728 cm^2

8 13.44 m^2

9   **a** 655.4 cm^2 **b** 692.5 cm^3

10 940 cm^2

11   **a** 80 cm^2 **b** 736 cm^2 **c** 960 cm^3

12   **a** 312 cm^2, 1248 cm^2, 2808 cm^2 **b** 1:4:9   **c** Since the dimensions are in the ratio 1:2:3, we must square this ratio to get the ratio of areas   **d** Since the ratios of their lengths of the corresponding sides are equal, these two shapes are similar

13   **a** No; if we cut the prism in half, its surface area will not be halved because the surface area of the original component and the part of the prism is nearly equal   **b** True

14

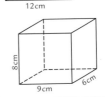

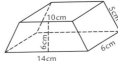

15   4 m

16 $l = 25$ cm

17   **a** 11.28 m^2 **b** 11.46 m^2 **c** 2.64 m^2

18 **a** Expression on the left is not correct; if we find the area of all sides, we get $6x + 6x + 6 \times 11 + x \times 11 + 10 \times 11 = 23x + 176$ **b** 5.9 cm

36.4 Volume of a cylinder

1 **a** 45 cm³ **b** 6 m³ **c** 4.32 cm³
2 **a** 502.4 cm³ **b** 117.8 m³ **c** 1695.6 cm³ **d** 415.4 m³
3 **a** 1130.4 cm³ **b** 226.1 cm³ **c** 75.4 m³ **d** 37.7 m³ **e** 9.8 cm³
4 1061.32 mm³
5 **a** 1.57 m³ **b** 1178 litres
6 35.3 litres
7 **a** 1st cylinder: 381 510 cm³; 2nd cylinder: 254 340 cm³; 3rd cylinder: 127 170 cm³ **b** 0.76 m³
8 **a** 111 600 cm³ **b** 0.1116 m³
9 **a** 640π cm³ **b** 160π cm³ **c** 640π m³ **d** 2560π m³
10 **a** 3768 cm³ **b** 282 600 cm³ **c** 785 m³ **d** 8.0384 m³
11 The tin with the greatest volume is 'c' - cylindrical tin (2356 cm³)
12 **a** 4200 litres **b** 2800 litres
13 **a** 12 cm **b** 1.4 m **c** 40 cm
14 **a** Always true; we know that $V = \pi r^2 h$. If we substitute $h = 2h$, then we get $V = 2(\pi r^2 h)$ which is double the original volume **b** Never true; we know that $V = \pi r^2 h$. If we substitute $r = 2r$, then we get $V = 4\pi r^2 h$, which is 4 times of original volume **c** Never true; we know that $V = \pi r^2 h$; double the diametre gives $4r$; if we substitute $r = 2r$, then we get $V = 4\pi r^2 h$
15 **a i.** 20 cm **ii.** 1256 cm² **iii.** 125 600 cm³ **iv.** 126 litres **b** 33 000 litres
16 **a** 8607 mm³ **b** 28.3%, 1/3 **c** 180.7 g
17 **a** 18 086 ml **b** 15 826 ml **c** 87.5% **d** 351 cups
18 **a** 10 000 cm³ **b** 250 cm² **c** 17.8 cm

36.5 Surface area of a cylinder

1 **a** 351.7 cm² **b** 133.5 m² **c** 791.3 cm² **d** 308.6 m²
2 **a** 387.9 m² **b** 534.1 cm² **c** 565.2 cm² **d** 120.9 m² **e** 108.3 m²
3 **a** 282.6 cm² **b** 13.6 m² **c** 267.0 cm²
4 330.6 cm²
5 13.2 m²
6 **a** 132π m² **b** 80π m² **c** 132π m²
7 A has the larger surface area: 351.9 cm²
8 **a** 56π cm² **b** 40π cm² **c** 56π cm²
9 **a** 125π cm² **b** 75π m² **c** 7500π cm² **d** 312 500π cm²
10 107.8 cm² nearest to 108 cm²

11 No, I do not agree with Tom; area of the circle = $2\pi r^2 = 2 \times \pi \times 20^2 = 800\pi$ cm²; curved surface area = $2\pi rh = 2 \times \pi \times 20 \times 10 = 400\pi$ cm²; total surface area = $2\pi rh + 2\pi r^2 = 800\pi$ cm² + 400π cm² = 1200π cm²
12 75.4 m²
13 **a** B has the largest surface area; cylinder A: total surface area = $20R\pi + 2R^2\pi$; cylinder B: total surface area = $20R\pi + 8R^2\pi$; **b** Substitute $R = 5$ cm; cylinder A: total surface area = $150\pi = 471.2$ cm²; cylinder B: total surface area = $300\pi = 942.5$ cm²; so still correct
14 **a** 12.06 m² **b** 19.6%
15 **a** 653.12 cm² **b** 78.1 cm² **c** 760 cm²
16 415 cm²
17 **a** 3.18 cm or 6.37 cm **b** 831.7 cm² or 927.5 cm²

37.1 Speed

1 16 km/h
2 **a** 1.5 km/minute **b** 90 km/h
3 **a** 6.4 m/second **b** 1.5 km/minute **c** 800 km/h **d** 375 m/minute
4 **a** 18 km **b** 72 km **c** 27 km **d** 45 km
5 **a** 4.5 km **b** 1.5 km **c** 0.75 km **d** 0.05 km
6 **a** 360 m **b** 20 km **c** 3150 km **d** 0.5 m
7 **a** 5 seconds **b** 10 seconds **c** 6.25 seconds **d** 25 seconds
8 **a** $\frac{1}{3}$ hour **b** $5\frac{1}{2}$ hours **c** 40 seconds **d** 1250 seconds
9 **a**

Time (t hours)	0.5	1	1.5	2	2.5
Distance (d km)	3	6	9	12	15

b $1\frac{1}{3}$ hours
10 **a** 2975 km **b** 5.9 hours
11 **a** 32 km **b** Because the relation between distance and speed represented in the graph as a straight line **c** 1.6 km/minute **d** 31.25 minutes
12 0.75 mm/h
13 **a i.** 5.6 m/s **ii.** 20 km/h **b** 26.7 m/s, 96 km/h **c** 50, 1200, 2.5, 6, 30
14 Graph b, because it has a part in which Michael travels 4 km in 0 minutes, which is impossible
15 **a** 24 km/h **b** 2 minutes 18 seconds **c** Sarah, Gemma, Lucy
16 5.9 seconds
17 **a** 263.8 cm **b** 5763 m
18 56 km/h

37.2 More about proportion

1 2.8 litres/second
2 0.9 litres/h
3 **a** 360 litres **b** 10800 litres **c** 2 minutes 48 seconds
4 **a** 48 litres **b** 5 seconds
5 **a** 4 g **b** 6.25 cm³
6 286 g
7 **a** 21.7 cm³ **b** 27.6 g **c** 230 g
8 **a** 9 g/cm³ **b** 13.3 cm³ **c** 1125 g
9 **a** 16000 kg **b** 8000 kg/m³ **c** Volume = $\frac{\text{mass}}{\text{density}} = \frac{400}{8000} = 0.05$ m³ **d** $\frac{8000 \text{ kg}}{1000000 \text{ cm}^3} = \frac{400 \text{ kg}}{50000 \text{ cm}^3}$, since $400 \times 1000000 = 400000000 = 8000 \times 50000$
10 **a i.** 9420 cm³ **ii.** 9.42 litres **b** 38 seconds
11 **a** $m = \frac{3}{10}V$, or $V = \frac{10}{3}m$ **b** 0.3 g/cm³ **c** a, because the point (12, 40) makes the formula m = $\frac{3}{10}V$ correct, while the others don't
12 **a** 135 litres **b** 10.8p
13 **a** 120 litres/minute **b** 44 litres **c** 40 seconds **d** Car A
14 **a i.** 67.55 g **ii.** 3.5 g **b** 12953 cm³
15 9.25 g/cm³

37.3 Unit costs

1 **a i.** £0.25 **ii.** 0.25p **b i.** 4 g **ii.** 400 g
2 **a i.** 34.8p **ii.** 0.348p **b i.** 2.87 g **ii.** 287 g
3 **a** £1.16 **b** £0.97 **c** The pack of four has a better value, as it has the lower cost
4 **a** 49.3p **b** The cost of 100 g of tomato puree in a jar is 54.5p, so the tubes has a better value as it has the lower cost
5 £2.40
6 **a** 52 kB **b i.** 2320 kB **ii.** 21 653 kB
7 A 100 g of the 225 g cans costs 36.9p, and 100 g of the 435 g cans costs 33.3p, so the 435 g can has a better value, as it has a lower cost
8 A 100 g of the 600 g box costs 44.2p, and 100 g of the 850 g box costs 46.5p, so the 600 g box has a better value, as it has a lower cost
9 1 ml of the 125 ml tube costs 2.4p, and 1 ml of the 75 ml tube costs 2.5p, so the 125 ml tube has a better value, as it has a lower cost
10 The large pot has a larger quantity and a lower cost, so it has a better value
11 **a** 20p **b** 50% **c** 38.9%
12 £8.63

Answers

13 **a** Rapids, because it doesn't charge any additional fees besides the cost of the hire period **b** £2.50 **c** $H = 2.5p$ **d i.** £11.25 **ii.** 7 hours 12 minutes

38.1 Graphs from equations of the form $ay \pm bx = c$

1 **a**
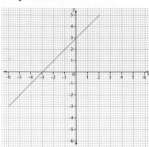

b
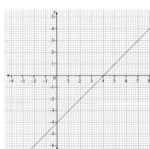

c
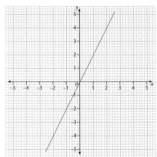

d
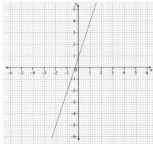

2 **a**
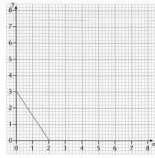

b
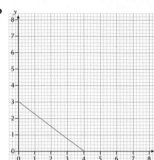

c
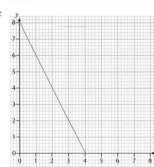

d
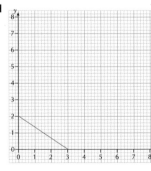

e
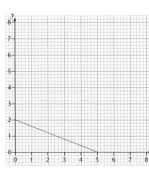

3 **a**
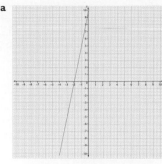

b
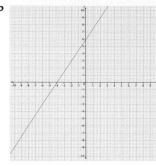

c
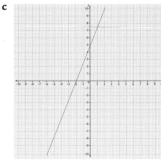

d
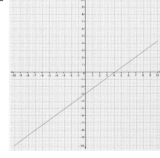

e
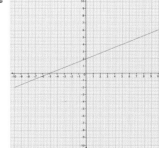

4 **a** (0, 9), (6, 0) **b** (0, 3), (−15, 0)
   c (0, −7), (4, 0)

5 **a**

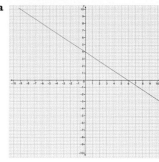

b

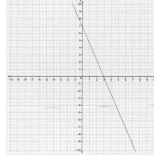

c

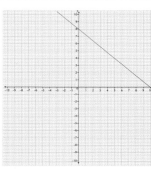

d

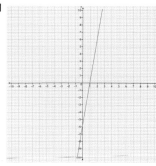

e

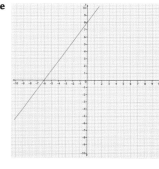

6 **a**

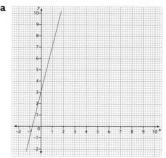

   b i. $x = 1$ **ii.** $x = \frac{3}{2}$ **iii.** $x = \frac{1}{2}$

7 **a**

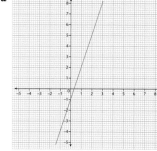

   b i. $y = -4$ **ii.** $y = -1$ **iii.** $y = 8$ **c i.** $x = 0$ **ii.** $x = \frac{2}{3}$ **iii.** $x = 2$

8 **a**

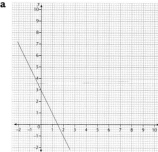

   b i. $\left(\frac{3}{2}, 0\right)$ **ii.** (0, 3) **c** $x = -\frac{3}{2}$

9 **a**

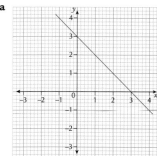

   b $p = -0.5$

10 **a**

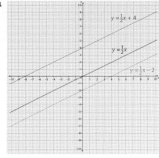

   b All the lines are parallel; $y = \frac{1}{2}x - 2$ is a translation of $y = \frac{1}{2}x$ with 2 units down, and $y = \frac{1}{2}x + 4$ is a translation of $y = \frac{1}{2}x$ with 4 units up **c** Translate $y = \frac{1}{2}x$, 6 units up

11 **a**

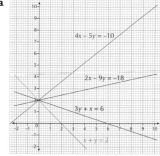

   b All the lines intersect at (0, 2)

12 **a**

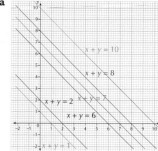

   b The line $x + y = a$, intersects with the axes at (0, a) and (a, 0) **c** Plot the points (−5.3, 0) and (0, −5.3), then draw a line that passes through them

13 **a**

x	1	2	3	4	5	6
y	6	3	2	$\frac{3}{2}$	$\frac{6}{5}$	1

   b If $x = 0$, then $y = \frac{6}{0}$, which is an undefined value

Answers

c

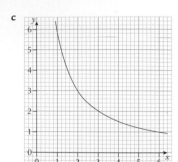

14 a

b £ 8 per hour **c** 12.5 hours
d $T = 100$ means she works 100 hours in the month, which is not realistic since she is working only on Saturdays

15 a 30 metres per second
b

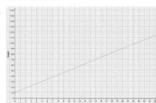

c 7.5 seconds

16 a

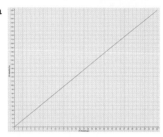

b 20 days **c** Because he will need to use a grid from 0 to more than 750 for the number of days, which is not realistic

17 a, b Student's own answers

38.2 Graphs from quadratic equations

1 a

x	−3	−2	−1	0	1	2
x^2	9	4	1	0	1	4
2x	−6	−4	−2	0	2	4
$y = x^2 + 2x$	3	0	−1	0	3	8

b, c

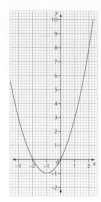

2 a

x	− 5	− 4	− 3	− 2	− 1	0	1
x^2	25	16	9	4	1	0	1
4x	− 20	− 16	− 12	− 8	− 4	0	4
$y = x^2 + 4x$	5	0	− 3	− 4	− 3	0	5

b, c

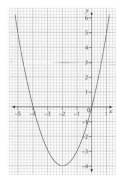

3 a

x	−4	−3	−2	−1	0	1
x^2	16	9	4	1	0	1
3x	−12	−9	−6	−3	0	3
2	2	2	2	2	2	2
$y = x^2 + 3x + 2$	6	2	0	0	2	5

b, c

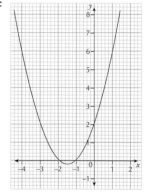

4 a

x	−4	−3	−2	−1	0	1	2
x^2	16	9	4	1	0	1	4
2x	−8	−6	−4	−2	0	2	4
−3	−3	−3	−3	−3	−3	−3	−3
$y = x^2 + 2x −3$	5	0	−3	−4	−3	0	5

b, c

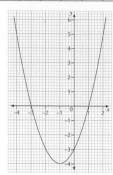

5

x	−1	0	1	2	3
x^2	1	0	1	4	9
x	−1	0	1	2	3
$y = x^2 + x$	0	0	2	6	12

6 a

x	−7	−6	−5	−4	−3	−2	−1	0	1
x^2	49	36	25	16	9	4	1	0	1
6x	−42	−36	−30	−24	−18	−12	−6	0	6
$y = x^2 + 6x$	7	0	−5	−8	−9	−8	−5	0	7

b, c

7 a

x	−2	−1	0	1	2
x^2	4	1	0	1	4
$2x^2$	8	2	0	2	8
$y = 2x^2 + 1$	9	3	1	3	9

b, c

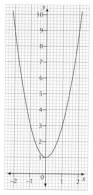

8 a

x	y
–2	–6
–1	–4
0	0
1	6
2	14

b

x	y
–2	–1
–1	–1
0	1
1	5
2	11

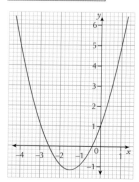

c

x	y
–2	–7
–1	–6
0	–3
1	2
2	5

9 a

x	–2	–1	0	1	2
x^2	4	1	0	1	4
$3x^2$	12	3	0	3	12
$y = 3x^2 - 5$	7	–2	–5	–2	7

b, c

10 a

x	–4	–3	–2	–1	0	1	2	3	4
$y = x^2 - 10$	6	–1	–6	–9	–10	–9	–6	–1	6
$y = 2x^2 - 10$	22	8	–2	–8	–10	–8	–2	8	22
$y = 3x^2 - 10$	38	17	2	–7	–10	–7	2	17	38
$y = 4x^2 - 10$	54	26	6	–6	–10	–6	6	26	54

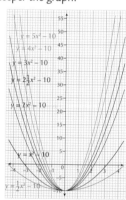

b All pass through (0, 10); the greater the coefficient of x^2, the steeper the graph.

c

11 a

x	–2	–1	0	1	2
$y = 3x^2 - 2$	10	1	–2	1	10
$y = 3x^2$	12	3	0	3	12
$y = 3x^2 + 1$	13	4	1	4	13
$y = 3x^2 + 3$	15	6	3	6	15

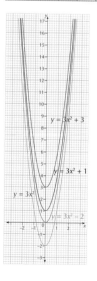

Answers

b The graphs are the same shape, however they have been translated up and down in the y-axis; $y = 3x^2$ translated 1 unit up is $y = 3x^2 + 1$; $y = 3x^2$ translated 2 units down is $y = 3x^2 - 2$

c

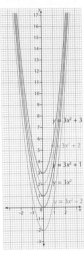

12 a

x	-5	-4	-3	-2	-1	0	1	2	3
8	8	8	8	8	8	8	8	8	8
$-2x$	10	8	6	4	2	0	-2	-4	-6
$-x^2$	-25	-16	-9	-4	-1	0	-1	-4	-9
y	-7	0	5	8	9	8	5	0	-7

b, c

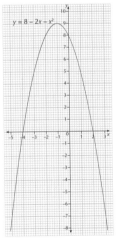

d The graph is open downward and intersects with the x-axis at $x = -4, 2$

13 a

x	-5	-4	-3	-2	-1	0	1	2	3
$x + 1$	-4	-3	-2	-1	0	1	2	3	4
$(x + 1)^2$	16	9	4	1	0	1	4	9	16

b

x	-5	-4	-3	-2	-1	0	1	2	3
$y = x^2$	25	16	9	4	1	0	1	4	9
$y = (x-1)^2$	36	25	16	9	4	1	0	1	4
$y = (x-2)^2$	49	36	25	16	9	4	1	0	1

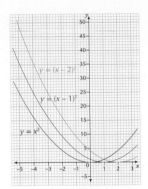

c The graphs are the same shape, however translated to the right in the x-axis; $y = x^2$ translated 1 unit right is $y = (x - 1)^2$; $y = x^2$ translated 2 units right is $y = (x - 2)^2$

d

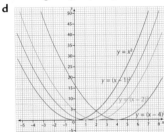

14 a

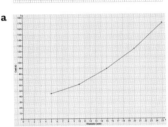

b i. £72 **ii.** £160 **c** 16.6 mm

15 a

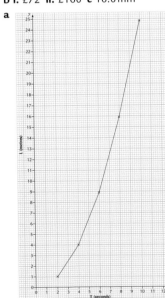

b 14 m **c** 4.4 seconds

16 a

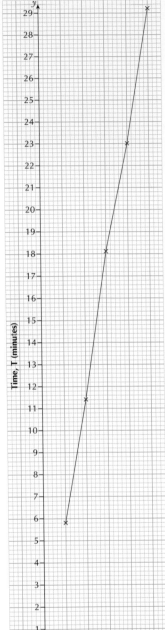

b 14.6 minutes **c** 3.4 km
d Because the speed is changing and not fixed through the run, so we need more information to estimate the time she will take to run 10 km

17 a

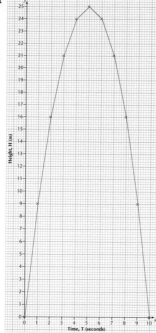

b 10 seconds **c** 25 m
d 4.4 seconds

38.3 Solving quadratic equations by drawing graphs

1 a

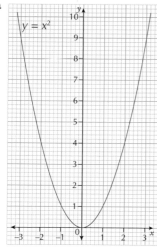

b $y = 4.4$ **c i.** ±1.7 **ii.** ±2.4 **iii.** ±2.7

2 a

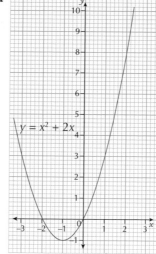

$y = x^2 + 2x$

b $y = 1.9$ **c i.** 0.7, –2.7 **ii.** 0.4, –2.4
iii. 0, –2

3 a

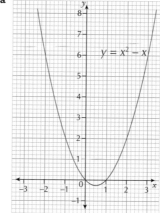

$y = x^2 - x$

b $y = 1.7$ **c i.** 2.3, –1.3 **ii.** 1.8, –0.8
ii. 1.3, –0.3

4 a

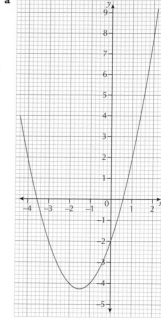

b $y = 5.4$ **c i.** 0.6, –3.6 **ii.** 0.3, –3.3
iii. 1.2, –4.2

5 a

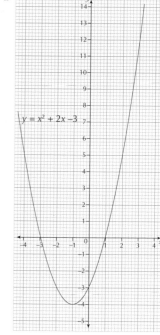

$y = x^2 + 2x - 3$

b $y = -3.5$ **c i.** 1.2, –3.2 **ii.** 1, –3
iii. 0.7, –2.7

Answers

6

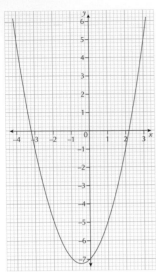

$x = 2.2$ or -3.2

7

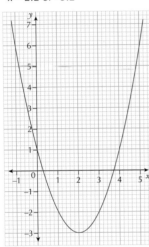

$x = -0.3$ or 3.7

8 a

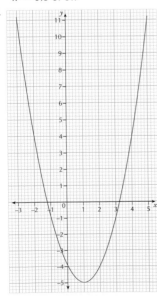

b i. 3.2, –1.2 **ii.** 2.4, –0.4
iii. 4.3, –2.3

9 a

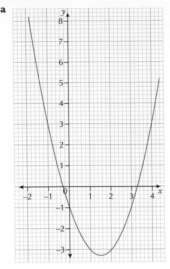

$x = -0.3$ or 3.3

b

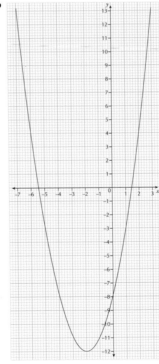

$x = -5.2$ or 1.2

c

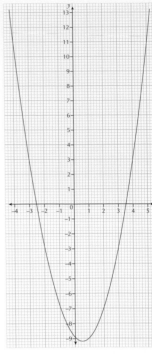

$x = -2.5$ or 3.5

10

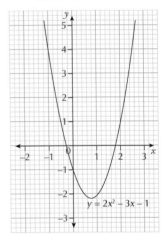

$y = 2x^2 - 3x - 1$

$x = -0.6$ or 1.6

11 Because the values of $y < -5$ are
below the graph of the equation,
hence any horizontal line we draw
from these values will not cut
the graph

12 a Because if we draw a horizontal
line from $y = -\frac{13}{4}$, it will cut the
graph in one place
b Each value of $y < -\frac{13}{4}$ is below
the graph of the equation, so when
drawing a horizontal line from
these values it will not cut
the graph

13 a False **b** False **c** True

14 a

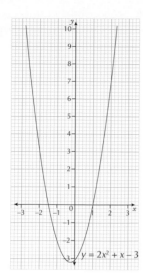

b $y = 1$ **c i.** 1, –1.5 **ii.** 1.6, –2.1
iii. 0.8, –1.3

d i.

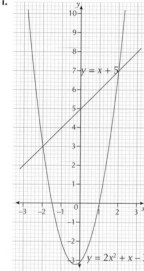

ii.

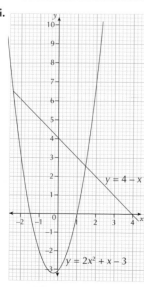

15 a Area = width × length =
$x(2x – 2) = 2x^2 – 2x$

b

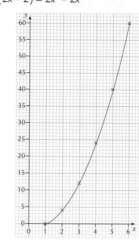

c Width = $x = 4.5$ cm,
length = $2x – 2 = 7$ cm

16 a Area = $\frac{1}{2}$ × height × base =
$\frac{1}{2} × 2x × (3x + 5.5) = 3x^2 + 5.5$

b

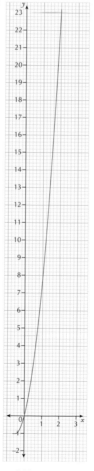

c $x = 1.5$
d Height = $2x = 2 × 1.5 = 3$ cm;
base = $3x + 5.5 = 3 × 1.5 + 5.5 =$
10 cm

17 a Area = $(x+1)(x+1) = x^2 + 2x + 1$,
but area = 30, so $x^2 + 2x + 1 = 30$,
which gives $x^2 + 2x – 29 = 0$

b

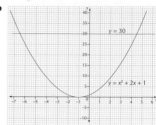

$x = 4.5, –6.5$ **c** The side length =
$x + 1 = 4.5 + 1 = 5.5$ cm

39.1 Calculating the length of the hypotenuse

1 a 8.5 cm **b** 9.1 mm **c** 2.1 m
   d 130 mm
2 a 5.8 cm **b** 9.2 cm **c** 12 **d** 13.4
   e 10.9
3

a	b	c	a^2	b^2	$a^2 + b^2$	c^2
2	3	3.6	4	9	13	13
10	9	13.5	100	81	181	181
8	15	17	64	225	289	289
12	18	21.6	144	324	468	468

4 b, d and e
5 10.8 cm
6 7.1
7 a $QR = 4$ **b** $PR = 5$
8 a $AC = 420$, $AB = 360$, $BC = 100$
   b $AC = 300$, $AB = 410$, $BC = 450$
   c $AC = 540$, $AB = 500$, $BC = 280$
9 5 cm
10 15.6 cm
11 a Cannot exist **b** Cannot exist
   c Can exist **d** Can exist
12 5, 12, 13
   7, 24, 25
   9, 40, 41
   11, 60, 61
13 $xz = 12.2$
14 a 4.5 **b** 3.6 **c** 4.2 **d** 9.8
15 a 75 **b** $\frac{20}{200}$
16 a 37 **b** 20 **c** £630
17 a 44 **b** About $\frac{1}{2}$ minute
18 47.2 km

39.2 Calculating the length of a shorter side

1 a 6.2 m **b** 10.2 cm **c** 6.6 cm
   d 8.7 cm **e** 4.5 m
2 a 19.7 cm **b** 9.4 mm **c** 223.6 mm
   d 2.4 m
3 11.2 cm

Answers

4 **a** $4^2 + 9.6^2 = 10.4^2$
$16 + 92.6 = 108.16$
So $108.16 = 108.16$
b $3^2 + 2.25^2 = 3.75^2$
$9 + 5.0625 = 14.0625$
So $14.0625 = 14.0625$
c $11.7^2 + 15.6^2 = 19.5^2$
$136.89 + 243.36 = 380.25$
So $380.25 = 380.25$

5 $11.6\,m$

6 If $10^2 + 16^2 = 20^2$
$100 + 256 = 400$
So $356 \neq 400$

7 $4.6\,m$

8 $x = 5.66\,cm$

9 $x = 9.90\,cm$

10 **a** $x = 7.07\,cm$ **b** $p = 24.14\,cm$

11 $3.8\,m$

12 Yes, because the sides of the triangle consists of the mast and the cable and the ground form a right triangle, since $5.25^2 + 7^2 = 76.56 = 8.75^2$

13 $x = 8.48$, so the perimetre $= 28.96$

14 $a = 4$ so the area $= 4 \times 8 = 32$

15 $DB = 5.7$ so $DC = 2.7$

16 **a** $8.66\,cm$ **b** $43.3\,cm^2$

17 $34.8\,cm^2$

18 **a** 54 **b** 4 sections

39.3 Using Pythagoras' theorem to solve problems

1 $d = 238.5$

2 **a** $x = 14.8, y = 8.6$
b $x = 132.3, y = 269.3$

3 $x = 96.6\,cm$

4 **a** $x = 424.3\,mm$ **b** 0.8 **c** 89.4
d $x = 50, y = 64$

5 107.6

6 10.8

7 105.5

8 $8.66\,cm$

9 **a** $CB = 6.9, PQ = 25.2$
b In the triangle ABC, the angle $B = 90 - 17 = 73°$
And in the triangle PQR, the angle $R = 90 - 73 = 17°$
This makes the two triangles have the same size of angles, which makes them similar

10 **a** The angle P is common, $S = Q = 90°$, $T = R$ (corresponding angles) So, the two triangles are similar since they have three equal angles
b $0.7\,m$ **c** $4.6\,m$

11 **a** 36.1 **b** 22.36

12 $2.6\,m$

13 **a** $h = 7.4$ **b** 22.2

14 4.5

15 $EF = 16.97$

40.1 Negative powers of 10

1 **a** 634 **b** 0.0634 **c** 6340
d 0.00634

2 **a** 8700 **b** 87 000 **c** 873 100 **d** 87.1
e 87310

3 **a** 637 000 **b** 63 700 **c** 63 700 000
d 6 307 000 **e** 6307

4 **a** 1.24 **b** 0.0124 **c** 12.45
d 0.000 124 **e** 1245

5 **a** 0.00785 **b** 0.000 785 **c** 0.0785
d 7.850 **e** 0.78501

6 **a** 0.083 **b** 0.0083 **c** 0.83 **d** 8.3
e 0.000 083

7 **a** 0.0127 **b** 0.000 127 **c** 0.12701
d 0.00127 **e** 1.2071

8 **a** 376 **b** 0.0023 **c** 0.003 09 **d** 235
e 0.000 001

9 **a i.** 0.1 **ii.** 0.01 **iii.** 0.001
iv. 0.0001 **b i.** 0.92 **ii.** 0.00071
iii. 0.42 **iv.** 0.0098 **v.** 0.00214

10 **a i.** 600 000 **ii.** 6 000 000 000
iii. 60 **iv.** 0.6 **b i.** 5000
ii. 50 000 000 **iii.** 0.5 **iv.** 0.005
c i. 23 000 000 **ii.** 230 000 000 000
iii. 2300 **iv.** 23 **d i.** 600 **ii.** 6
000 000 **iii.** 0.06 **iv.** 0.0006
e i. 25 200 **ii.** 252 000 000 **iii.** 2.52
iv. 0.0252

11 **a** 2 **b** –2 **c** 479.8
d 0.000 000 047 98

12 All the calculations move the digits two places to the left

13 73820×10^{-3}
$73820 \div 10^3$
0.078320×10^3
$0.078320 \div 10^{-3}$

14 **a** 4 milligrams **b** 8 megawatts
c 7.5 nanometres

15 $0.07\,g$

16 $960\,m$

17 3.7

18 140

40.2 Standard form

1 **a** 8.5×10^{-1} **b** 1.27×10^{-2}
c 4.32×10^{-1} **d** 512×10^{-3}

2 **a** 0.00641 **b** 0.000903 **c** 0.080
d 0.00071

3 **a** 0.003142 **b** 0.000501
c 0.000 000 985 2 **d** 3.8×10^{-3}
e 7.09×10^{-4}

4 **a** 6.4×10^{-3} **b** 2.25×10^{-6}
c 1.44×10^{-6} **d** 1.6×10^{-9}

5 **a** 9×10^2 **b** 1.6×10^5 **c** 9×10^6
d 1.6×10^3

6 **a** 1.3×10^{-2} **b** 7.1×10^{-1}
c 8.9×10^{-2}

7 **a** 6.2×10^{-7} **b** 1.1×10^{-2}
c 4.8×10^5

8 **a** 2.4×10^{-2} **b** 6.2×10^4
c 2.9×10^{-2}

9 **a** 9.9×10^{10} **b** 6.7×10^4 **c** 2.8×10^9

10 **a** 7.8×10^7 **b** 7.1×10^{-1} **c** 3×10^{-2}

11 **a** 4×10^4 **b** In standard form **c** In standard form **d** 1×10^9 **e** 8×10^4

12 387×10^{-6}, 0.00387, 38.7×10^{-3}, 0.0387×10^6

13 Because the powers of ten in the two numbers are not equal

14 $10^{10} = 10\,000\,000\,000$; when multiplying by 10^{10} the number becomes larger, so we move the digits ten places to the left
$10^{-10} = \dfrac{1}{10\,000\,000\,000}$;
when multiplying by 10^{-10} we divide the number by 10^{10}, so the number becomes smaller and we move the digits ten places to the right

15 **a** 9.2×10^{-28} **b** 7.36×10^{-27}
c 2.76×10^{-21}

16 **a** $2 \times 10^3\,mm = 2000\,mm = 2\,m$
b $2 \times 10^1\,mm = 20\,mm$

17 **a** $1 \times 10^{-4}\,m = 0.0001\,m = 0.1\,mm$
b $1 \times 10^{-1}\,m = 0.1\,m = 100\,mm$

18 **a** $5 \times 10^{-2}\,cm$ **b** $5\,cm$ **c** 2.56×10^7

40.3 Rounding appropriately

1 **a** 15 **b** 6 **c** 36 **d** 5

2 **a** 130 **b** 30 **c** 4000 **d** 4

3 **a** £60 **b** 1000 **c** 0.6 **d** 4

4 **a** 0.001 **b** 200 **c** 50 kg **d** 600

5 **a** 9 **b** 200 km **c** £30 **d** 0.002

6 **a** 60 mph **b** 24° **c** 50 kg

7 **a** 2.5 hours **b** 10 seconds
c 80 seconds

8 **a** 1.8 m **b** 700 000 000 bytes
c 60 000 000

9 **a** 20 **b** 5 **c** 0.03

10 **a** 0.34 **b** 0.02 **c** 200

11 $1.62\,km^2$

12 Billy rounded the number to one significant figure, but Isaac rounded the same number to two significant figures

13 3.8×10^4, 37842, 3.78×10^4
7.2×10^{-3}, 0.007234, 7.32×10^{-3}
3.8×10^{-4}, 0.000 378 4, 3.78×10^{-4}
7.2×10^3, 7234, 7.23×10^3

14 **a** 18.51 **b** Grace to one significant figure; Victoria to two significant figures; Zeenat to three significant figures

15 £29.98

16 505

17 **a** £596.73 **b** £0.02 **c** £92.70
d £221.90

18 **a** 25 **b** 516.5 kg **c** 24877 g **d** 11 g

40.4 Mental calculations

1 **a** 1000 **b** 2000 **c** 2160 **d** 4320
e 62.5 **f** 31.25 **g** 135

2 **a** 600 **b** 180 **c** 140 **d** 920 **e** 24
f 7.2 **g** 5.6

3 **a** 360 **b** 740 **c** 1080 **d** 4320
e 0.9 **f** 1.85 **g** 2.7

4 **a** 10 000 **b** 1800 **c** 2700 **d** 210
e 24 **f** 18 **g** 48

5 **a** 420 **b** 1230 **c** 1560 **d** 540

6 **a** 920 **b** 57.5 **c** 1150 **d** 46 **e** 540
f 450 **g** 0.72

7 **a** 5 **b** 12.8 **c** 13 **d** 0.52 **e** 8
f 1.28 **g** 24

8 **a** £59.76 **b** £1.245 **c** £74.70
d £0.996

9 **a** 35 (use 700 ÷ 20) **b** 2250 (use
90 × 24) **c** 28 (use 7 ÷ 0.25) **d** 14
(use 210 ÷ 15)

10 **a** 28800 **b** 430 **c** 400
d 303030300

11 **a** £60 **b** 525 kg **c** 43

12 £2.16

13 **a** £45 **b** 429 pupils

14 £12000

15 £56

16 0.75 m²

17 65 200 000 copies

18 146 cm

40.5 Solving problems

1 Adult's ticket: £4.50; child's ticket:
£3.50

2 **a** 2.7183 **b** $\frac{590}{217}$

3 3.7×10^5 mm

4 **a** 7 weeks **b** Wacko Magic 14
programmes and Country Facts
7 programmes

5 10.5 hours

6 £154.68

7 **a** 26, 42, 68 **b** 0.2, 0.3, 0.5, 0.8
1.3, 2.1, 3.4, 5.5, 8.9 **c** 0.25, 0.5,
0.75

8 The packs of 35 give the best
value, since:
0.74 ÷ 20 = £0.037
1.19 ÷ 35 = £0.034
1.26 ÷ 36 = £0.035

9 The box of 200, since:
1.72 ÷ 200 = 0.0086
2.61 ÷ 300 = 0.0087
4.45 ÷ 500 = 0.0089

10 £1.75

11 **a** €296 **b** 84 bottles **c i.** €2
ii. £1.35

12 The 12-packet box, since:
3 ÷ 12 = 0.250
5 ÷18 = 0.278
8 ÷ 30 = 0.267

13 The large bag, since:
0.28 ÷ 30 = 0.0093
0.90 ÷ 100 = 0.0090
2.30 ÷ 250 = 0.0092

14 Leisureways

15 33 jars

16 **a** £105 **b** 7 days

41.1 More about brackets

1 **a** $4x + 4$ **b** $3y - 21$ **c** $16 - 8x$
d $20 - 4t$

2 **a** $4x + 20$ **b** $x^2 + 5x$ **c** $7y + 70$
d $10y - y^2$

3 **a** $2x + 2y$ **b** $x^2 + 2x$ **c** $x^2 + xy$
d $m^2 - 7m$

4 **a** $4x + 2$ **b** $12a - 4b$ **c** $3x^2 - 3xz$
d $3x^2 + 2xy$

5 **a** $x^2 + 7x$ **b** $x^2 - 4x$ **c** $x^2 + 2x$ **d** $x^2 + x$

6 **a** $6x^2$ **b** $10t^2$ **c** $36x^2$ **d** $4x^2$ **e** $9x^2$

7 **a** $2t^2 + 10t$ **b** $4t^2 + 10t$ **c** $6t^2 + 10t$
d $10tx + 6t$

8 **a** $8x^2 - 2x$ **b** $6x^2 - 9x$ **c** $6x^2 + 2xy$
d $36t^2 - 6t$

9 **a** $g^2 + 4g$ **b** $g^2 - g$ **c** $3g^2 - 9g$
d $4g - g^2$

10 **a** $9x^2 + 15x$ **b** $8x^3 + 36x$ **c** $55x^2 - 5x^3$
d $60x^3 - 42x^2$

11 **a i.** $5 + s$ **ii.** $6(5 + s)$ **iii.** $n(5 + s)$
b $30 + 6s$, $5n + ns$

12 **a** $7x^2 + 21x$ **b** $3x^2 + 5xy$ **c** $2t^2 - 18t$

13 **a i.** $b + 10$ **ii.** $3(b + 10)$ **iii.** $b(b + 10)$
b ii. $3b + 30$ **iii.** $b^2 + 10b$

14 **a** $2 - a$
b Area1 $= a^2 + (2 - a)(a + b)$
Area2 $= 2(a + b) - ab$ **c** Expand the
brackets for both expressions
Area1 $= a^2 + (2 - a)(a + b)$
$= a^2 + 2a + 2b - a^2 - ab$
$= 2a + 2b - ab$
Area2 $= 2(a + b) - ab$
$= 2a + 2b - ab$

15 **a** x^2 **b** $7x$ **c** $x^2 + 7x$

16 **a** $x(x + 3) = x^2 + 3x$ **b** $y(y - 4) =$
$y^2 - 4y$ **c** $t(t + 8) = t^2 + 8t$

17 $r + 8$

18 $18x^2 + xy - x$

41.2 Factorising expressions containing powers

1 **a** $4(x + 2)$ **b** $3(4y - 5)$ **c** $7(2 - x)$
d $8(4y + 5)$

2 **a** $x(x - 3)$ **b** $t(t + 5)$ **c** $y(y - 4)$
d $n(6 + n)$

3 **a** $x(x + 6)$ **b** $n(2 - n)$ **c** $n(20 + n)$
d $x(3 - x)$

4 **a** $x(x + k)$ **b** $x(2c + x)$ **c** $x(3x + 1)$
d $n(4n - 1)$

5 **a** $6(x + 2)$ **b** $4(3x - 2y)$ **c** $3(3t^2 - 2)$
d $3(2a + 3c)$

6 **a** $6x(x + 2)$ **b** $3y(3 + 2y)$
c $8x(3x + 2)$

7 **a** $4x(x + 1)$ **b** $6y(y - 1)$ **c** $2t(t + 5)$
d $2(3x^2 - 1)$

8 **a** $2x(x + 2y)$ **b** $3a(2a - 3b)$
c $4q(3p - 4q)$

9 **a** $12x(x + y)$ **b** $4a(a - 2b)$ **c** $5x(y + z)$
d $2b(a + 6)$

10 **a** $a(a^2 + 2)$ **b** $x^2(1 - 2x)$ **c** $3n(2 + n^2)$
d $2x^2(2x - 1)$

11 He did not take the r outside the
brackets correct answer is
$2r(4n + 2)$ and he can factorize
further to $4r(2n + 1)$

12 **a** Not possible **b** $x(x - 6)$
c $2(2x - 5)$ **d** Not possible

13 **a** 105 **b** $u(4u + 1)$ **c** 105

14 **a** $a(5 + a)$ **b** Not possible
c $c(4c - 5)$ **d** Not possible
e $g(8 - g)$ **f** $7(h - 2)$ **g** Not possible

15 **a** $2x(x + 4)$ **b** $3x(x + 3)$ **c** $12x(x + 3)$
d $4x^2(2x - 1)$

16 **a** $5x(y + 2)$ **b** $3xy(2x + 1)$
c $a(4b - 3)$ **d** $5ab(2a + 3)$

17 We can't find any common factor
for 8 and 9

18 $14y - 6$

41.3 Expanding the product of two brackets

1 **a** $ab + a + b + 1$ **b** $cd + 3c + d + 3$
c $pq + 2p + 4q + 8$
d $st + 7s + 5t + 35$

2 **a** $ac + ad + bc + bd$
b $ac - ad + bc - bd$
c $ac + ad - bc - bd$
d $ac - ad - bc + bd$

3 **a** $x^2 + 2x$ **b** $a^2 - 3a$ **c** $p^2 + 5p$
d $y^2 - 5y$

4 **a** $x^2 + 4x + 3$ **b** $x^2 + 11x + 18$
c $x^2 + 9x + 20$ **d** $x^2 + x - 72$

5 **a** $a^2 + 3a + 2$ **b** $n^2 + 7n + 12$
c $x^2 - 7x + 12$ **d** $z^2 - 14z + 40$

6 **a** $x^2 + 5x + 6$ **b** $t^2 + 9t + 20$ **c** $15 -$
$2n - n^2$ **d** $y^2 - 10y + 16$

7 **a** $(a + 5)(a - 2) = a^2 + 3a - 10$
b $(b + 7)(b + 8) = b^2 + 15b + 56$
c $(c - 3)(c - 7) = c^2 - 10c + 21$
d $(4 + d)(9 + d) = d^2 + 13d + 36$
e $(e + 11)(e + 5) = e^2 + 16e + 55$
f $(f - 9)(f + 1) = f^2 - 8f - 9$

8 **a** $x^2 + 4x + 4$ **b** $x^2 + 2x + 1$
c $x^2 + 8x + 16$ **d** $x^2 + 14x + 49$

9 **a** $x^2 + 10x + 25$ **b** $x^2 - 16x + 64$
c $x^2 + 200x + 10000$
d $x^2 - 400x + 40000$

10 **a** 156 **b** $x^2 + 5x + 6$ **c** 156 **d** If we
substitute $x = 10$ in the expression
before expanding we will have
(13×12)

11 **a i.** $c^2 - 25$ **ii.** $f^2 - 16$ **iii.** $t^2 - 81$
b When simplify and combine like
terms, the sum of third term
equals zero

12 $(x + 3)^2 = (x + 3)(x + 3)$
$= x^2 + 3x + 3x + 9$
Simplify and combine like terms
$= x^2 + 6x + 9$

Answers

13 **a** Area = length × height
Where length = $p + q$ and
height = $r + s$
So, Area = $(p + q)(r + s)$ **b** The area
of the rectangle is divided into 4
small rectangles (sections)
Rectangle 1 (area = pr)
Rectangle 2 (area = ps)
Rectangle 3 (area = qr)
Rectangle 4 (area = qs)

14 **a** The entire length is w and the
length of the white part is x, so
the length of the blue part is $w - x$, and its height is $y + z$ **b** Area
of blue rectangle = area of the
whole rectangle – area of the
white rectangle
= $w \times (y + z) - x(y + z)$
= $wy + wz - xy - xz$

15 **a** $x^2 + 9x + 18$ **b** $x^2 - 3x - 10$

16 The area of the coloured rectangle
is given by $(x + 6)(x - 2)$. Working
out the area of each section in the
diagram and adding them gives
$x^2 + 6x - 2x - 12$, then collecting like
terms gives $x^2 + 4x - 12$. So, $(x + 6)$
$(x - 2) = x^2 + 4x - 12$

17 **a** $(5y - 1)(y + 2) = 5y^2 + 10y - y - 2 = 5y^2 + 9y - 2$ **b** $(y - 3)^2 = y^2 - 6y + 9$

18 $(x + 2)^2 + (x + 1)^2 = (x^2 + 4x + 4) + (x^2 + 2x + 1) = 2x^2 + 6x + 5$

41.4 Expanding expressions with more than two brackets

1 **a** $x^3 + 3x^2 - 4$ **b** $x^3 + 3x^2 - 6x - 8$

2 **a** $x^2 + 5x + 6$ **b** $x^3 + 6x^2 + 11x + 6$

3 **a** $x^2 + 3x - 4$ **b** $x^3 + 5x^2 + 2x - 8$

4 **a** $x^2 - 4x + 3$ **b** $x^3 - 6x^2 + 11x - 6$

5 **a** $x^3 + 5x^2 - x - 5$ **b** $x^3 + 3x^2 - 4x - 12$
c $x^3 - 4x^2 - x + 4$

6 **a** $x^3 + x^2 - x - 1$
b $x^3 - 3x^2 + 4$
c $x^3 + x^2 - 8x - 12$

7 **a** $x^3 + 5x^2 + 8x + 4$
b $x^4 + 6x^3 + 13x^2 + 12x + 4$

8 $x^3 + 4x^2 + 4x$

9 **a** $(x^2 - 9)(x + 1) = x^3 + x^2 - 9x - 9$
b $(x^2 - 2x - 3)(x - 3) = x^3 + x^2 - 9x - 9$

10 **a** $x^2 + 2x + 1$ **b** $x^3 + 3x^2 + 3x + 1$
c $x^3 + 6x^2 + 12x + 8$
d $x^3 - 3x^2 + 3x - 1$
e $x^3 - 6x^2 + 12x - 8$

11 **a** $(x + 2)^2(x - 2)$ **b** $x^3 + 2x^2 - 4x - 8$

12 Volume = $(2x + 1)^3$

13 $c = 4$

14 $4x^3 + 29x^2 + 54x + 37$

15 $7x^3 - 24x^2 + 6x + 37$

16 $(x - 5)$

42.1 Finding trigonometric ratios of angles

1

	opposite	adjacent	hypotenuse
a	a_1	a_2	a_3
b	b_3	b_1	b_2
c	c_1	c_3	c_2
d	d_3	d_2	d_1
e	e_2	e_3	e_1

2

a	$\cos x$		$\frac{3}{5}$	0.6
	$\tan x$		$\frac{4}{3}$	1.33
b	$\sin x$	$\frac{24}{30}$	$\frac{4}{5}$	0.8
	$\cos x$	$\frac{18}{30}$	$\frac{3}{5}$	0.6
	$\tan x$	$\frac{24}{18}$	$\frac{4}{3}$	1.33
c	$\sin x$	$\frac{36}{39}$	$\frac{12}{13}$	0.92
	$\cos x$	$\frac{15}{39}$	$\frac{5}{13}$	0.38
	$\tan x$	$\frac{36}{15}$	$\frac{12}{5}$	2.4

3 **i**

	opp	adj	hyp
a	8	6	10
b	15	20	25
c	24	10	26
d	7	24	25

ii.

	$\sin x$	$\cos x$	$\tan x$
a	$\frac{8}{10}$	$\frac{6}{10}$	$\frac{8}{6}$
b	$\frac{15}{25}$	$\frac{20}{25}$	$\frac{15}{20}$
c	$\frac{24}{26}$	$\frac{10}{26}$	$\frac{24}{10}$
d	$\frac{7}{25}$	$\frac{24}{25}$	$\frac{7}{24}$

4

	i	ii
a	$\tan x$	$\frac{5}{9}$
b	$\sin x$	$\frac{5}{6}$
c	$\cos x$	$\frac{5}{8}$
d	$\sin x$	$\frac{7}{11}$

5 **a** $\frac{12}{5}$ **b** $\frac{12}{13}$ **c** $\frac{5}{13}$ **d** $\frac{5}{12}$ **e** $\frac{5}{13}$
f $\frac{12}{13}$

6 **a** 0.8 **b** 0.6 **c** 0.6 **d** 0.8 **e** 1.33
f 0.75

7 **a** 0.5 **b** 0.866 **c** 0.577

8 Both equal 0.707
$\sin 45° = \cos 45°$

9 There is no tangent (not defined)

10 **a** 200 **b** 0 **c** 0 **d** 500 **e** 350 **f** 0

11

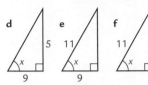

12 **a** $\sin 0° = 0$, $\sin 25° = 0.423$, $\sin 45° = 0.707$, $\sin 70° = 0.940$, $\sin 90° = 1$ **b** $\cos 90° = 0$, $\cos 65° = 0.423$, $\cos 45° = 0.707$, $\cos 20° = 0.940$, $\cos 0° = 1$ **c** The sine and cosine of two different angles are equal if the sum of the angles equals 90°
$\sin x = \cos(90 - x)$, and $\cos x = \sin(90 - x)$

13 **a** 0, 0.268, 0.577, 1, 1.73, 3.732 **b** 11.430, 57.29, 114.589, 286.478, 572.957, not defined
c The tangent increases rapidly, as we get closer to 90°

14 **a i.** $\sin x = \cos y$ **ii.** $\sin y = \cos x$
b $\tan x$ is the reciprocal of $\tan y$

15

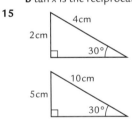

16

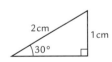

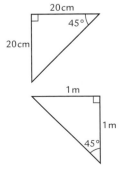

17

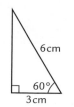

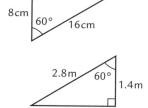

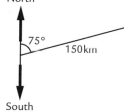

18 $A = 5.2 \div 2 = 2.6\,m$

To find the height of the triangle
(h)

$\tan 75° = \dfrac{h}{A} = \dfrac{h}{2.6}\ 3.7$

$= \dfrac{h}{2.6}$, so $h = 3.7 \times 2.6 = 9.6\,m$

Area of the triangle =
$\dfrac{1}{2} \times$ height $\times$ base

$= \dfrac{1}{2} \times 9.6 \times 5.2 = 24.96\,m^2$

42.2 Using trigonometric ratios to find the sizes of angles

1 **a** 55.9° **b** 7.3° **c** 39.2° **d** 30.8°
e 23.3° **f** 30.3° **g** 42.1° **h** 81.9°

2 **a** 32.7° **b** 42.1° **c** 38.9° **d** 57.3°
e 72.3° **f** 40.2° **g** 78.7° **h** 36.9°

3 **a** 64.7° **b** 45.2° **c** 82.6° **d** 64.5°
e 82.1° **f** 84.0° **g** 35.8° **h** 25.2°

4 **a** 53.1° **b** 61.0° **c** 6.4° **d** 21.3°
e 8.2°

5 **a** 40.6° **b** 60.9° **c** 48.8° **d** 9.5°
e 12.1°

6 **a** 55.2° **b** 62.2° **c** 83.6° **d** 72.1°
e 40.1°

7 **a** 56.4° **b** 62.7° **c** 46.7° **d** 36.9°

8 **a** 32.2° **b** 33.6° **c** 34.6° **d** 51.7°

9 **a** 60.3° **b** 31.0° **c** 32.9° **d** 49.4°

10 **a** 59.0° **b** 59.0° **c** 31.0° **d** 43.3°
e 44.9° **f** 38.7° **g** 28.6° **h** 55.2°

11 3 m

12 23.3 m²

13 **a** 12.4 m **b** 9.1 m

14 8.5°

15 53.1°

16 33.6°

17 21.4°

42.3 Using trigonometric ratios to find lengths

1 **a** 9.5 cm **b** 3.8 cm **c** 5.3 cm
d 6.3 cm

2 **a** 2.9 cm **b** 5.1 cm **c** 8.0 cm
d 6.5 cm

3 **a** 4.2 cm **b** 6.4 cm **c** 8.9 cm
d 8.9 cm

4 **a** 0.276 **b** 0.232 **c** 0.276 **d** 0.999
e 0.017 **f** 81.847

5 **a** 9.21 **b** 1.88 **c** 3.21 **d** 0.104
e 1.49 **f** 1.91×10^3

6 **a** 10.8 cm **b** 5.18 cm **c** 6.64 cm
d 18.0 cm

7 **a** 10.4 mm **b** 1.43 m **c** 2.36×10^3

8 1.10 m

9 38.9 mm

10 2.42 m

11 0.7 m

12 Since $\sin x = \cos (90 - x)$.
Therefore, $\sin 45° = \cos 45°$

13 3.8 m

14 10.8 m

15 6.2 cm²

16 4 m

17 **a**

North

75°

150 km

South

b 145 km

18 **a** 150 km **b** 260 km